Green Chemistry

Vinod K. Tiwari · Abhijeet Kumar ·
Sanchayita Rajkhowa · Garima Tripathi ·
Anil Kumar Singh

Green Chemistry

Introduction, Application and Scope

Vinod K. Tiwari
Department of Chemistry
Banaras Hindu University
Varanasi, Uttar Pradesh, India

Abhijeet Kumar
Department of Chemistry
Mahatma Gandhi Central University
Motihari, Bihar, India

Sanchayita Rajkhowa
Department of Chemistry
The Assam Royal Global University
Guwahati, Assam, India

Garima Tripathi
Department of Chemistry
T. N. B. College
Bhagalpur, Bihar, India

Anil Kumar Singh
Department of Chemistry
Mahatma Gandhi Central University
Motihari, Bihar, India

ISBN 978-981-19-2733-1   ISBN 978-981-19-2734-8   (eBook)
https://doi.org/10.1007/978-981-19-2734-8

© The Editor(s) (if applicable) and The Author(s), under exclusive license to Springer Nature
Singapore Pte Ltd. 2022

This work is subject to copyright. All rights are solely and exclusively licensed by the Publisher, whether
the whole or part of the material is concerned, specifically the rights of translation, reprinting, reuse
of illustrations, recitation, broadcasting, reproduction on microfilms or in any other physical way, and
transmission or information storage and retrieval, electronic adaptation, computer software, or by similar
or dissimilar methodology now known or hereafter developed.

The use of general descriptive names, registered names, trademarks, service marks, etc. in this publication
does not imply, even in the absence of a specific statement, that such names are exempt from the relevant
protective laws and regulations and therefore free for general use.

The publisher, the authors, and the editors are safe to assume that the advice and information in this book
are believed to be true and accurate at the date of publication. Neither the publisher nor the authors or
the editors give a warranty, expressed or implied, with respect to the material contained herein or for any
errors or omissions that may have been made. The publisher remains neutral with regard to jurisdictional
claims in published maps and institutional affiliations.

This Springer imprint is published by the registered company Springer Nature Singapore Pte Ltd.
The registered company address is: 152 Beach Road, #21-01/04 Gateway East, Singapore 189721,
Singapore

Foreword

The design of chemical reaction that reduces or eliminates the generation of hazardous substances demands the adoption of innovative and revolutionary green chemistry approach. The practice of green chemistry methodology in organic synthesis has certainly led to significant environmental benefits and has strengthened economic innovations. This book on *Green Chemistry—Introduction, Application, and Scope* provides a very interesting overview of the fundamental principles of green chemistry with their notable perspectives. This book would be useful to UG/PG students, academicians, researchers as well as chemists working in chemical industries because of the absence of any worthwhile book on this subject.

This book by Vinod K. Tiwari and coworkers begins with the introduction of green chemistry and its twelve principles coined by Paul Anastas and John Warner and moves forward with the exhaustive discussion on the important metrics of the green chemistry such as *E-factor, atom economy* to evaluate the greenness of a chemical process. Other important discussions in this chapter include principles of *inherently safer design* (ISD) and barriers in the direction of implementation of different green chemistry principles.

Various environmentally benign alternatives such as microwave, ultrasound, visible light, ball milling, *mechanocatalysis*, and electro-organic approaches to conventional C–C and C–X (X = heteroatom) bond-forming organic transformations have been elaborated in Chap. 2. Following this, the chapter deals with the introduction of greener solvents, i.e., water, ionic liquid, supercritical CO_2, switchable solvent, fluorous biphasic system, polyethylene glycol as well as using solvents derived from biobased feedstock such as vegetable oil. The *growing impact of ionic liquids in synthesis* mainly in heterocyclic synthesis and application in carbohydrate chemistry is elaborated in Chaps. 4 and 5, respectively.

A book on green chemistry today would not have been justified if this does not include catalytic reaction. The author has rightly included the application of various catalytic transformations such as oxidation, reduction, carbon–carbon, and carbon–heteroatom bond-forming reactions in this book. Emerging concept of organocatalysis, types of catalysts, and application in asymmetric synthesis is appropriately highlighted in Chap. 7.

v

Furthermore, the significant importance of enzyme-mediated transformation and notable applications of green chemistry in some real-world cases, e.g., the synthesis of pharmaceuticals, API, polymers, etc., have been discussed in the last two chapters. Also, the impact of employing green method and its comparison with the conventional approaches are highlighted in this book.

All the chapters provide relevant references to original papers, reviews, and some specialized books on the topic. The present book on green chemistry presents all essential and important concepts with clarity and provides up-to-date information on almost all aspects of green chemistry for UG/PG and doctoral students of chemistry. It becomes apparent that this book has basic objective of bringing the subject clarity, without compromising on most essential concepts, but avoiding less-frequently encountered components. This book should be successful in developing "fondness" for green chemistry among scientific community. This book should be useful to the chemistry academic faculty and industrial chemists equally. It is a much-welcome addition to repertoire of practicing researchers and academics. Academic and research background of the authors justify their book on the green chemistry.

Prof. Ganesh Pandey, FNA, FNASc, FASc
Distinguished Professor
Department of Chemistry
Institute of Science
Banaras Hindu University
Varanasi, India

Preface

In the present scenario, the development of a greener and environment-friendly approach to minimize and/or avoid the generation of waste material during the synthesis of a vast range of compounds with diverse applications has been considered to be an important area of investigation. Especially after realizing the short- and long-term adverse effect of conventional synthetic approaches on *safety, health, and environment (SHE)*, the attention of research community took a paradigm shift in achieving the synthesis from conventional to efficient synthetic methods with utmost care of lives and environment, the essentials of sustainable development. Therefore, we are presenting this book with the title *Green Chemistry—Introduction, Application, and Scope* for a better understanding of green chemistry fundamentals with notable perspectives, particularly about the various emerging synthetic approaches under environmentally benign conditions that would prove to be an important and useful piece of study material to the readers.

This book has been divided into nine chapters that cover the topics mentioned in the *green chemistry* course of most of the academic institutions across the world. Thus, it would offer the students of UG/PG as well as researchers from both academic institutions and industry a complete understanding of the different emerging topics using the latest notable examples. Chapter 1 titled *"Green Chemistry: Introduction to the Basic Principles"* provides the fundamental information about the green chemistry and its twelve principles jointly coined by Paul Anastas and John Warner. These principles act as a guiding tool for the development of an environmentally benign process. This chapter covers the exhaustive discussion about some of the important metrics like *E-factor, atom economy*, etc., to evaluate the greenness of a chemical process. Further, different ways to prevent and minimize the generation of waste material in a chemical process have also been discussed with the help of suitable examples. Other important topics such as the principle of *inherently safer design* (ISD) and barriers in the direction of implementation of different green chemistry principles have also been covered. In Chap. 2 titled *"Energy-Efficient Process in Organic Synthesis,"* various environmentally benign alternatives to the conventional heating methods adopted to achieve different types of organic transformations, for example, C–C bond-forming reactions, heterocyclic synthesis, etc., have

vii

been discussed. We have briefly covered the applications of microwave, ultrasound, visible light, and electro-organic method in achieving different types of organic transformations. Also, the application of ball milling method which is also emerging as an energy-efficient way in organic synthesis, in particular, for the *direct and indirect mechanocatalysis*, has been highlighted with the latest examples. In Chap. 3 titled *"Green Solvents: Application in Organic Synthesis,"* discussion is about the application of different types of greener alternatives to conventional organic solvents in organic synthesis have been included. Organic reactions explored using water, ionic liquid, supercritical CO_2, switchable solvent, fluorous biphasic system, polyethylene glycol as well as using solvents derived from biobased feedstock such as vegetable oil have been discussed elaborately. In Chap. 4 titled *"Growing Impact of Ionic Liquids in Heterocyclic Chemistry,"* the importance of ionic liquids as another greener alternative to conventional solvents and their application as a solvent in organic synthesis particularly for the synthesis of heterocyclic rings have been covered thoroughly. Chapter 5 titled *"Growing Impact of Ionic Liquids in Carbohydrate Chemistry,"* highlights the applications of ionic liquids in carbohydrate chemistry in great detail. Chapter 6 titled *"Catalysis: Application and Scope in Organic Synthesis"* covers the vast discussion on different types of catalysis and their applications in various types of organic transformations including oxidation, reduction, carbon–carbon, and carbon–heteroatom bond formations. *Organocatalysis* is now emerged as a valuable area of research and beyond doubt a truly greener alternative to the transition metal-based catalysis. Considering its immense importance in asymmetric organic synthesis, this year's Nobel Prize in Chemistry has been awarded to Prof. Benjamin List and Prof. David W. C. MacMillan, and thus, a separate Chap. 7 titled *"Organocatalysis: A Versatile Tool for Asymmetric Green Organic Syntheses"* has been presented. This chapter is dedicated to organocatalysis with brief discussions related to different types of organocatalysis and their applications in various asymmetric transformations. Furthermore, the significant importance of enzyme-mediated transformation is considered and included in Chap. 8 titled *"Enzymes in Organic Synthesis."* At the last, Chap. 9 titled *"Application of Green Chemistry: Examples of Real-World Cases"* focuses on the notable application of environmentally benign approaches in some real-world cases such as for the synthesis of pharmaceuticals, APIs, polymers, etc. In addition, the impact of employing green method and their comparison with the conventional approaches have also been highlighted in this book. Overall, this book presented a detailed scope of green chemistry to a wide range of readers and investigators.

We would like to express our heartfelt thanks and deep sense of appreciation to Prof. Ganesh Pandey, Distinguished Professor at the Department of Chemistry, Institute of Science, Banaras Hindu University, for his consistent support and moreover for writing the venerated foreword to the Springer book. The authors thank Prof. Richard R. Schmidt (Universitat Konstanz), Prof. Xi Chen (University of California-Davis), Prof. R. P. Tripathi (CSIR-Central Drug Research Institute, Lucknow), Prof. Pradeep K. Tripathi (CSIR-National Chemical Laboratory, Pune), and Prof. D. Basavaiah (University of Hyderabad) for their useful suggestions during the writing of this book. We also thank both Dr. Anand K. Agrahari (UC-Davis and Mr. Mangal Singh

Yadav (BHU, Varanasi) for their great care to the reference section. The authors are highly thankful to the *"Springer Nature"* who implicit the importance of the present book and considered the publishing with great interest. Last but not least, we express our special thanks to the entire editorial team associated with Springer Nature, in particular Ms. Priya Vyas, Associate Editor—Applied Sciences and Engineering, Ms. Sharmila Mary Panner Selvam, Project Coordinator, and Mr. Madanagopal Deenadayalan, Project Manager, who extended their support in publishing this theme with the high standard of publication maintained in bringing out this book on *Green Chemistry—Introduction, Application, and Scope* with Springer Nature.

Varanasi, India	Vinod K. Tiwari
Motihari, India	Abhijeet Kumar
Guwahati, India	Sanchayita Rajkhowa
Bhagalpur, India	Garima Tripathi
Motihari, India	Anil Kumar Singh

Contents

1 Green Chemistry: Introduction to the Basic Principles 1
   1 Introduction 1
   2 Principles of Green Chemistry 1
   3 Principle of Inherently Safer Design (ISD) 30
   4 Various Barriers in the Implementation of Green Chemistry 31
   References 34

2 Energy-Efficient Process in Organic Synthesis 37
   1 Introduction 37
   2 Microwave-Assisted Organic Synthesis 38
     2.1 Microwave-Assisted Coupling Reactions 39
     2.2 Microwave-Assisted Heterocyclic Synthesis 41
   3 Ball Milling Method for Organic Synthesis 48
     3.1 Application of Ball Mill Method in Synthesis
        of Heterocycles 49
     3.2 Asymmetric Synthesis Under Ball Mill Method Using
        Organocatalysts 51
     3.3 Coupling Reaction Under Ball Mill Condition 53
   4 Ultrasonic Methods in Organic Synthesis 55
   5 Photo-Induced Organic Transformations 58
     5.1 Photo-Catalyst Catalyzed C–C and C–X Bond Forming
        Reaction 59
     5.2 Photo-Induced Peptide Coupling Reaction 64
     5.3 Photo-Induced Decarboxylative Coupling 65
   6 Electrochemical Approach for Organic Synthesis 69
   7 Conclusions 72
   References 72

3 Green Solvents: Application in Organic Synthesis 79
   1 Introduction 79
   2 Types of Green Solvents 80
   3 Solvent Selection Guides 80

4	Green Solvents and Their Application		81
	4.1	Water as Solvent	81
	4.2	Ionic Liquids	86
	4.3	Switchable Solvents	87
	4.4	Supercritical CO_2 as Solvent	89
	4.5	Solvents from the Renewable Bio-Based Feedstock	92
	4.6	Water Extract of Agrowaste Ash (AWEs)	94
	4.7	Glycerol as Green Solvent	94
	4.8	Fluorous Biphasic Solvents	101
	4.9	Polyethylene Glycol [PEG] as a Solvent	104
5	Conclusions		107
References			107

4 Growing Impact of Ionic Liquids in Heterocyclic Chemistry 113

1	Introduction		113
2	Structure and Types of RTILs		114
3	Properties of ILs		121
	3.1	Melting Point (m. p.)	122
	3.2	Thermal Stability/Decomposition	124
	3.3	Viscosity	124
	3.4	Conductivity	125
	3.5	Density and Polarity	126
	3.6	Toxicity	126
4	Chemical Synthesis of Some ILs		127
5	Impact of RTILs in the Synthesis of Biologically Relevant Skeletons		130
	5.1	Impact of RTILs in the Synthesis of Biologically Relevant Heterocyclic Skeletons	131
	5.2	RTIL-Mediated Synthesis of Five-Membered Heterocycles	132
	5.3	Ionic Liquid-Mediated Synthesis of Six-Membered Heterocycles	144
6	Conclusions and Future Outlook		164
References			165

5 Growing Impact of Ionic Liquids in Carbohydrate Chemistry 177

1	Introduction		177
2	Representative Example for the Synthesis of Carbohydrate-Based Chiral ILs		178
3	Application of RTILs in Carbohydrate Chemistry		179
	3.1	Dissolution and Gelation of Carbohydrates in ILs	180
	3.2	IL-Mediated Some Common Reactions in Carbohydrate Chemistry	183
	3.3	Ionic Liquids in Enzyme-Induced Carbohydrate Modifications	188
	3.4	ILs in Glycosidic Bond Formation Methodology	191

Contents xiii

4	Conclusions and Future Outlook	198
References		200

6 Catalysis: Application and Scope in Organic Synthesis 207
- 1 Introduction 207
  - 1.1 Type of Catalysts 208
- 2 Catalytic Oxidation Process 209
  - 2.1 Oxidation of Alkenes 211
  - 2.2 Oxidation of Alkanes 220
  - 2.3 Oxidation of Aromatic Hydrocarbon 222
- 3 Catalytic Reduction 223
  - 3.1 Reaction Conditions in Heterogenous Catalysis 224
  - 3.2 Hydrogenation of Alkenes 224
  - 3.3 Hydrogenation of Alkynes 228
  - 3.4 Hydrogenation of Aldehydes and Ketones 230
  - 3.5 Catalytic Reductive Amination 231
- 4 Catalytic C–C Bond Formation 234
- 5 Catalytic C–N Bond Formation: Click Chemistry 240
- 6 Catalysis by Acidic Clays and Zeolites 241
  - 6.1 Acidic Clays 242
  - 6.2 Zeolites 245
- 7 Organocatalysis: General Consideration 253
- References 254

**7 Organocatalysis: A Versatile Tool for Asymmetric Green
Organic Syntheses** 261
- 1 Introduction 261
- 2 Classification of Organocatalysis 263
  - 2.1 Lewis Base Catalysis 263
  - 2.2 Lewis Acid Catalysis 264
  - 2.3 Brønsted Base Catalysis 266
  - 2.4 Brønsted Acid Catalysis 267
- 3 Iminium Catalysis 268
  - 3.1 Application of Iminium Catalysts in Organic Synthesis ... 270
- 4 Enamine Catalysis 285
  - 4.1 Asymmetric Aldol Reactions 286
  - 4.2 Asymmetric Michael Reaction 299
- 5 Carbohydrate-Based Asymmetric Organocatalysis 302
  - 5.1 Enantioselective Epoxidation 302
  - 5.2 Sugar-Based Prolinamide Catalysts 305
  - 5.3 Carbohydrate Based Pyrrolidine Catalysts 305
- 6 Conclusion 306
- References 308

8	**Enzymes in Organic Synthesis**	317
	1 Introduction	317
	2 Applications of Enzymes in Synthesis	318
	3 Conclusion and Future Perspectives	344
	References	345

9 Application of Green Chemistry: Examples of Real-World Cases ... 353

1 Introduction ... 353

2 Selected Examples of Real-World Applications of Green
Chemistry ... 355

   2.1 Greener Synthetic Pathway for the Synthesis of Ibuprofen 355

   2.2 Application of Surfactants for Liquid Carbon Dioxide 356

   2.3 Development of Environmentally Benign Marine
Antifoulant ... 358

   2.4 Use of Genetically Engineered Microbes
as Environmentally Benign Catalyst ... 360

   2.5 Polylactic Acids as Green Alternate of Plastics ... 363

   2.6 Rightfit™ Pigments: A Green Replacement of Toxic
Organic and Inorganic Pigments ... 366

   2.7 Healthier Fats and Oils by Enzymatic Interesterification 368

   2.8 Green Approach Toward the Synthesis of Sertraline
Hydrochloride (Zoloft) ... 371

3 Conclusion ... 371

References ... 373

About the Authors

Prof. Vinod K. Tiwari is currently as Professor of Organic Chemistry at the Department of Chemistry, Banaras Hindu University, India. He earned his M.Sc. degree from BHU (in 1998) and doctoral degree from CSIR-Central Drug Research Institute, Lucknow (Mentor: Dr. R. P. Tripathi), in 2004. He has postdoctoral experience at University of Florida, USA (Mentor: Prof. Alan R. Katritzky), University of California-Davis, USA (Mentor: Prof. Xi Chen), and Universitat Konstanz, Germany (Mentor: Prof. Richard R. Schmidt). He was offered the post of Lecturer at Bundelkhand University (2004–2005) before being appointed in BHU. With more than 21 years of research and 18 years of teaching experience, he has supervised 13 Ph.D. theses, 21 M.S. dissertations and contributed significantly to 156 peer-reviewed publications (Citations: 6400, h-index: 39, i_{10} : 101) in addition to several patents and invited book chapters of high repute. He completed 7 major research projects and delivered 201 invited lectures in India and abroad. His teaching interest covers almost all branch of organic chemistry, however current research focus on Synthetic Carbohydrate Chemistry and novel synthetic methodology. He holds Hony. Secretary, ACCT(I) and has wide editorial experience. His research has received many prestigious awards, notably Young Scientist Award (Dr. D. S. Bhakuni Award in 2004, ICRABTS-2005, UP-CST in 2010, 1st H. C. Srivastava Award-2012 and CRSI-2012); Most Cited Paper Award (2006, 2013); Dr. Arvind Kumar Memorial Award, ICC (2010); Prof. R. C. Shah Memorial Award, ISCA (2011); Dr. Ghanshyam Srivastava Memorial award, ICS (2012), Prof. A. S. R. Anjaneyulu 60th Birthday Commemoration Award, ICS (2013); 1st Prof. N. Roy Award for Excellence on Synthetic Carbohydrate Chemistry-2015; Dr. S. S. Despande National Award for High Impact Contribution in Chemistry-2016; ACCT(I)-Excellence in Carbohydrate Research Award-2019; BHU Most Productive Researcher Award-2019, listed in top 2% scientist, and CRSI Bronze Medal for the year 2021.

Dr. Abhijeet Kumar is currently working as Assistant Professor of Organic Chemistry at Mahatma Gandhi Central University (MGCU), Bihar, India. He has received his B.Sc. and M.Sc. degrees in Chemistry from Banaras Hindu University in 2006 and 2008, respectively. In 2016, he completed his doctoral research from

Indian Institute of Technology (IIT), Kanpur, with Prof. M. L. N. Rao, where he received research training in the development of metal-catalyzed cross-coupling reactions using organobismuth reagents as green organometallic reagents. He received 'Visiting Faculty Fellowship' to carry out research work at JNCASR, Bangalore. He has been involved in teaching the green chemistry to the UG/PG courses at MGCU since last four years. He has published several peer-reviewed publications and also contributed few book chapters of high repute.

Dr. Sanchayita Rajkhowa is currently working as Assistant Professor of Chemistry at the Department of Chemistry, The Assam Royal Global University, Guwahati, Assam, India. She has graduated from Miranda House, DU (in 2006); M.Sc. from University of Delhi (in 2008) and earned her Ph.D. degree from NEHU-Shillong (in 2017). She has research experience at CSIR-NEIST, Jorhat (2010–11) and taught chemistry to M.Sc. students at Gauhati University, Guwahati (2018–2019) followed by teaching at Jorhat Institute of Science and Technology, Jorhat (2020–21). She has delivered over 15 invited lectures in various institutions and was awarded 'Young Scientist Award' by CONIAPS XXIV on Innovations in Physical Sciences for the year 2019. She has published several research publications in peer-reviewed journals of high repute. In addition to wide editorial experience, she has also authored several book chapters on chemistry, environment and ecology with international publishers and her latest edited book is on *Environmental Sustainability and Industries. Technologies for Waste Treatment* with Elsevier.

Dr. Garima Tripathi is currently Assistant Professor of Chemistry and also the coordinator of Department of Biotechnology, T. N. B. College, TM Bhagalpur University, Bihar. She has earned her M.Sc. degree in Chemistry from Banaras Hindu University in 2008 and Ph.D. from Indian Institute of Technology (IIT), Kanpur with Prof. R. Gurunath, where she considered research in the area of peptide chemistry. She received DST-Women Scientist-B award from DBT, India, and worked as Women Scientist-B at the Department of Chemistry, IIT-Kanpur. Recently, she has received start-up grant from university grant commission (UGC), India to establish her research at T. N. B. College, Bhagalpur. She has awarded 'Visiting Faculty Fellowship' to carry out research work at JNCASR, Bangalore. She has significantly contributed 12 research publications and some book chapters of high repute.

Dr. Anil Kumar Singh is currently working as Assistant Professor in the Department of Chemistry, Mahatma Gandhi Central University, Bihar. He has completed his B.Sc. and M.Sc. degrees in Chemistry from Banaras Hindu University in 2009 and 2011, respectively. He earned his Ph.D. degree in 2015 from University of Delhi, New Delhi, India, under the supervision of Dr. Brajendra Kumar Singh. He has published several research articles of national and international repute. He has been awarded with prestigious Erasmus Mundus-Action 2 fellowship during his Ph.D. and worked at Ghent University, Belgium (Mentor: Prof. Johan van der Eycken).

Abbreviation

[admIm]Br	1-Allyl-2,3-dimethylimidazolium bromide
[amIm]Cl	1-Allyl-3-methylimidazolium chloride
[bdmIm]Cl	1-Butyl-2,3-dimethylimidazolium chloride
[bmIm][BF$_4$]	1-Butyl-3-methylimidazolium tetrafluoroborate
[bmim]MeSO$_4$	1-Butyl-3-methylimidazolium methanesulfate
[bmim]PO$_4$	1-Butyl-3-methylimidazolium phosphate
[bmpy]Cl	3-Methyl-N-butylpyridinium chloride
[C1mIm][Cl]	1,3-Dimethylimidazolium chloride
[C1OCH$_2$mIm] N(CN)$_2$]	1-Methoxymethyl-3-methylimidazolium dicyanamide
[C4-2,3-m2Im][Cl]	1-(1-Butyl)-2,3-dimethylimidazolium chloride
[C4mIm][BF$_4$]	1-(1-Butyl)-3-methylimidazolium tetrafluoroborate
[C4Py][PF$_6$]	N-(1-Butyl)pyridinium hexafluorophosphate
[emIm][ba]	1-Ethyl-3-methylimidazolium benzoate
[emim]BF$_4$	1-Ethyl-3-methylimidazolium tetrafluoroborate
[emIm]Cl	1-Ethyl-3-methylimidazolium chloride
[hmIm][ba]	1-Hexyl-3-methylimidazolium benzoate
[mIm]	1-Methylimidazolium
[mmIm]	1-Methyl-3-methylimidazolium
[moemim][OMs]	1-Methoxyethyl-3-methylimidazolium methanesulfonate
[Ms$_2$N]$^-$	Bis(methanesulfonyl) amide
[N(TFMS)$_2$]$^-$	Bis(trifluoromethanesulfonyl)imide
[NMM]$^+$[HSO$_4$]$^-$	N-Methylmorpholinium hydrogen sulfate
[NTf$_2$]$^-$	Bis(trifluoromethylsulfonyl)amide
[OTf]$^-$	Trifluoromethylsulfonate
[R(Rf)taz][Y]	1,4-Alkyl(polyfluoroalkyl)-1,2,4-triazolium ionic liquids
[taz][X]	1-Alkyl-4-polyfluoroalkyl-1,2,4-triazolium halide
3-DDM	Three-dimensional dealuminated mordenite
6-APA	6-aminopenicillanic acid
Ac	Acetyl

ACE	Angiotensin-converting enzyme
ADM	Archer Daniels Midland
ADME	Absorption, Distribution, Metabolism, and Excretion
API	Active pharmaceutical ingredient
BHT	Butylated hydroxytoluene
BINAP	2,2′-bis(diphenylphosphino)-1,1′-binaphthyl
Bn	Benzyl
BtRC	Benzotriazole ring cleavage
Bz	Benzoyl
CAL-B	*Candida antarctica* lipase-B
cEF	Complete E-factor
CFCs	Chlorofluorocarbons
COD	1,5-Cyclooctadiene
CuAAC	Cu(I)-catalyzed azide–alkyne cycloaddition
DABCO	Diazabicyclo[2.2.2]octane
DAHP	3-Deoxy-D-arabino-heptulosonic acid-7-phosphate
dba	dibenzylideneacetone
Dca/[N(CN)$_2$]$^-$	Dicyanamide anion
DCM	Dichloromethane
DCOI	4,5-dichloro-2-*n*-octyl-4-isothiazolin-3-one
DDT	Dichlorodiphenyltrichloroethane
de	Diastereomeric excess
DFT	Density functional theory
D-Gal	D-galactose
D-GalNAc	*N*-acetyl-D-glactosamine
D-Glu	D-Glucose
DHQ	3-Dehydroquinic acid
DHS	Dehydroshikimic acid
DIB	Diacetoxyiodobenzene
Diglyme	Diethylene glycol dimethyl ether
DIOP	2,3-O-Isopropylidene-2,3-dihydroxy-1,4-bis(diphenylphosphino)butane
DIPAMP	(2-methoxyphenyl)-[2-[(2-methoxyphenyl)-phenylphosphanyl]ethyl]-phenylphosphane
DIPEA	*N,N*-Diisopropylethylamine
DMA	*N,N*-dimethylacetamide
D-Man	D-Mannose
DMAP	4-Dimethylaminopyridine
DMF	Dimethyl formamide
DMM	Dimethoxymethane
DMSO	Dimethylsulfoxide
DPAT	Diphenylammonium triflate
dr	Diastereomeric ratio
DuPHOS	1,2-bis(phospholano)benzene
E4P	Erythrose 4-phosphate

EBHP	Ethylbenzene hydroperoxide
ee	Enantiomeric excess
EPA	US Environmental Protection Agency
FC	Friedel–Crafts
FDA	US Food and Drug Administration
GC	Gas chromatography
GCI	Green Chemistry Institute
GC-MS	Gas chromatographic-mass spectrometry
HCFCs	Hydrochlorofluorocarbon
HCs	Hydrocarbons
HMG	Hydroxymethylfurfural
H-MOR	H-Mordenite
HOMO	Highest occupied molecular orbital
HPB	Canadian health protection board
HPLC	High-performance liquid chromatography
ILs	Ionic liquids
IR	Infrared
ISD	Inherently safer design
ISP	Inherently safer process
LDL	Low-density lipid
LMCT	Ligand to metal charge transfer
LPG	Liquified petroleum gas
LUMO	Lowest unoccupied molecular orbital
MAEC	Maximum allowable environmental concentration
MCF	Mesocellular foam
*m*CPBA	*m*-Chloroperbenzoic acid
MIC	Methyl isocyanate
MLCT	Metal to ligand charge transfer
MSDS	Material safety data sheet
MUFA	Mono-unsaturated fatty acid
NaAsc	Sodium ascorbate
NBD	Norborna-2,5-dien
NIS	*N*-Iodo succinimide
NMO	4-methylmorpholine-N-oxide
NMR	Nuclear magnetic resonance
NSAID	Nonsteroidal antiinflammatory drug
NsNIPh	[(nosylimino)iodo]benzene
P_c	Critical pressure
PCA	Protocatechuic acid
PEG	Polyethylene glycol
PET	Poly(ethyleneterephthalate)
PLA	Polylactic acid
PMB	*p*-Methoxybenzyl
PMHS	polymethylhydrosiloxane
p-TSA	*Para*-Toluenesulfonic acid

PUFA	Polyunsaturated fatty acid
RERCs	Perchloroethylenes
RTILs	Room-temperature ionic liquids
RuAAC	Ruthenium-catalyzed azide–alkyne cycloaddition
SEC	Size exclusion chromatography
sEF	Simple E-factor
SFA	Saturated fatty acid
SHE	Safety, health, and environment
sn	Stereospecific numbering
SOMO	Singly occupied molecular orbital
SPS	Switchable polarity solvents
TAG	Triacylglycerol
TBAHS	Tetrabutylammonium hydrogensulfate
TBDMSCl	*Tert*-butyldimethylsilyl chloride
TBHP	*Tert*-butylhydroperoxide
TBTO	Tributyltin oxide
T_c	Critical temperature
TCA	Trichloroacetimidate
tda	Thiodiacetate
Tf	Trifluoroacetate
TFA	Trifluoro acetic acid
TFMS	Trifluoromethane sulfonate
TFMSA	Trifluoromethane sulfonate acid anhydride
TFSI	*Bis*(trifluoromethanesulfonyl)imide
Tg	Phase transition temperature
THF	Tetrahydrofuran
TNT	Trinitro toluene
TRIP	3,3′-bis(2,4,6-triisopropylphenyl)-1,1′-binaphthyl-2,2′-diylhydrogenphosphates
Ts	Tosyl
TS	Transition state
TSAC	2,2,2-Trifluoro-N(trifluoromethylsulfonyl) acetamide
US	Ultrasound
VOCs	Volatile organic compounds
ZSM-5	Zeolite Socony Mobil–5

Chapter 1
Green Chemistry: Introduction to the Basic Principles

1 Introduction

The chemists have always been passionate about the synthesis of chemicals of diverse varieties of structurally simple to the complex compounds. These scientific efforts led to the development of vast varieties of compounds having immense importance in the areas of medicinal and material sciences. Several drugs, natural products, polymers, and materials were discovered that made our life very comfortable. In last few decades, the synthetic approaches toward the development of different varieties of compounds of medicinal and material interest have taken a paradigm shift. In contrast to the conventional approaches in chemical synthesis were achieving the best product yield and construction of diverse range of a target compound was the prime focus without paying much attention toward the impact of the synthetic methodology adopted for it, on the safety, health, and environment (SHE), the development of environmentally benign and green synthetic reaction protocols and technology is the principal goal in the present scenario. In that direction, the development of the concept of green chemistry and its twelve basic principles (Fig. 1) formulated by Prof. Paul T. Anatsas and John C. Warner works as a guiding tool for the synthetic chemist [1]. These basic twelve principles have been discussed in details in the following sections.

2 Principles of Green Chemistry

Principle 1: "*It is better to prevent waste than to treat or clean up waste after it has been created.*" [2]

It is well known proverb that "*prevention is better than cure.*" Similar to that, it is advisable to prevent the generation of different varieties of waste materials that could be non-hazardous to highly hazardous in nature and gets generated in

---

© The Author(s), under exclusive license to Springer Nature Singapore Pte Ltd. 2022
V. K. Tiwari et al., *Green Chemistry*,
https://doi.org/10.1007/978-981-19-2734-8_1

Fig. 1 Different principles of green chemistry

chemical reactions and process either in the form of by-product, reagents, and auxiliary components like solvents, etc. It may include waste material which could be flammable such as paint, coating, organic solvents, and compressed gases like LPG, acetylene, hydrogen, etc. along with solid chemicals like lithium aluminum hydride, sodium, etc. or explosive like TNT or corrosive like acids, bases, etc. Some of them are highly toxic such as pesticides, insecticides, and various heavy metals As, Cr, Pb, Hg, Cd, etc. containing chemicals. In general, it is advisable to go through the information available in the material safety data sheet (MSDS) to get idea about the different properties including physical, chemical, toxicity, etc. of a chemical and waste material also if it is fully characterized.

Sometime different chemicals which are available in the form useful drugs or chemicals when gets expired are also considered as waste material. But every type of waste material imparts economic, social, and environmental impact of varied degree and these require proper disposal as per the strict guidelines formulated by various agencies. Importantly, these waste materials would have huge impact on safety and health of the chemists and workers involved in the complete process. Apart from that, waste generated would also have impact on environment, ecosystem and it will ultimately affect the life on the earth.

Cost of the target compound will greatly be affected by the cost involved in treatment of the waste material generated during its formation. Therefore, this principle emphasizes on the complete prevention of the waste material generation or if it is not possible then minimize the waste formation by developing and adopting an environmentally benign process. In order to compare the greenness of a method or process in terms of the generation of waste material, different metrics have been devised. For

2 Principles of Green Chemistry 3

example, Roger. A. Sheldon in 1980 introduced the concept of "*E-factor* or *Environmental factor*" which is the ratio of mass of waste (Kgs) to mass of product (Kg) in order to assess the impact of various different manufacturing processes on environment [3]. Ideally, the E-factor value should be zero which is practically difficult to achieve in most of the cases. The value of E-factor varies in different types of industry of industry (Table 1). As shown in Table 1, fine chemicals and pharmaceutical industry occupies the top slots in the production of waste material primarily due to the use of organic solvents as well as use of stoichiometric amount of reagents. Although for the calculation of E-factor, it is important to identify the waste material which will not necessarily be the by-product, regent only. In fact, the waste could be defined as *everything but the desired product* [4]. But generally, solvents in particular water and some other solvents which are a part of reaction mixture either in the form of solvent or by-product in a chemical reaction get ignored while calculating E-factor. Therefore, two boundaries have been suggested to quantify the E-factor. Simple E-factor (sEF) simply takes the mass of raw material, reagents and products into account and do not consider any solvent as waste material to calculate E-factor. On the other hand, in case of complete E-factor (cEF), all the solvents including water, assuming these have not been recycled gets considered as waste material. Therefore, the true value of environmental factor lies somewhere between these two values of sEF and cEF.

$$\text{Simple E factor (sEF)} = \Sigma \text{ m (raw materials)} + \Sigma \text{ m (reagents)}$$
$$- \text{m(product)}/\text{m(product)}$$

$$\text{Complete E factor (cEF)} = \Sigma \text{ m (raw materials)} + \Sigma \text{ m (reagents)}$$
$$+ \Sigma \text{ m(solvents)} + \text{m(water)} - \text{m(product)}/\text{m(product)}$$

Therefore, the value of E-factor varies and depends on various other factors such as the consideration of initial point in the synthetic pathway. In general, E-factor is calculated taking only those processes into account which are being carried out at the manufacturing site. But it is important to keep in mind that the synthesis of the target compound could be started by procuring an intermediate which itself requires multi-step synthesis. So the value of E-factor would vary according to our consideration of initial starting point which could be a commodity type and commercially available raw materials or it could be an intermediate which require synthesis involving various

S. No	Industry segment	Tons per annum	E-factor (kg waste per kg product)
1	Oil refining	10^6–10^8	<0.1
2	Bulk chemicals	10^4–10^6	<1–5
3	Fine chemicals	10^2–10^4	5–50
4	Pharmaceuticals	10–10^3	25– >100

Table 1 E-factors in the chemical industry [3]

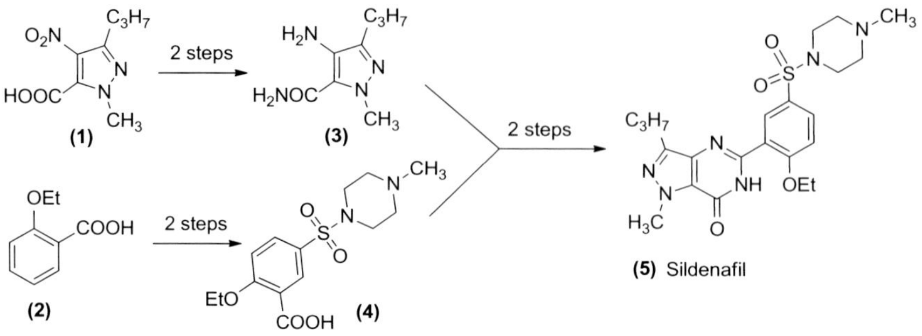

Scheme 1 Commercial method for the preparation of Sildenafil by Pfizer

steps. For example, the synthesis of sildenafil citrate which is popularly known with brand name Viagra is manufactured by Pfizer starting using pyrazole derivative (**1**) and 2-ethoxybenzoic acid (**2**) as starting materials to prepare another pyrazole (**3**) and piperazine derivatives (**4**) separately. These two combine together to produce the target compound sildenafil (**5**) (Scheme 1) [5].

In this case, the E-factor value reported by Pfizer was 6 kg/kg which was very low compared to the earlier reported processes. This improved process for the synthesis of sildenafil citrate developed by Pfizer also received the UK Award for Green Chemical Technology in 2003. But the synthesis of the starting material pyrazole derivative 1-methyl-4-nitro-3-propyl-1*H*-pyrazole-5-carboxylic acid (**1**) used in Scheme 1, itself requires five steps starting with commodity like starting material diethyl oxalate and 2-pentanone (Scheme 2). Therefore on taking this fact into consideration, the value of overall E-factor increases to 13.8 kg/kg. But the value itself does not reflect about the overall greenness and its environmental impact as the nature of waste material is also an important parameter to consider while comparing the greenness of different processes.

For example, environmental impact of 1 kg of NaCl would entirely be different from the same amount of organic solvent such as DCM, benzene, or heavy metal (Cr, Hg)-based waste material. Therefore to compare the greenness, another term "*Environmental quotient (EQ)*" was introduced by Sheldon. It is the product of E-factor and an unfriendliness quotient, Q which is an arbitrarily assigned number [4]. For example, the value of Q could be assigned 1 for NaCl, whereas it could be 100–1000 for salts of toxic heavy metals. Although Q is an arbitrary number and so its magnitude is debatable. But it does reflect about the environmental impact of a

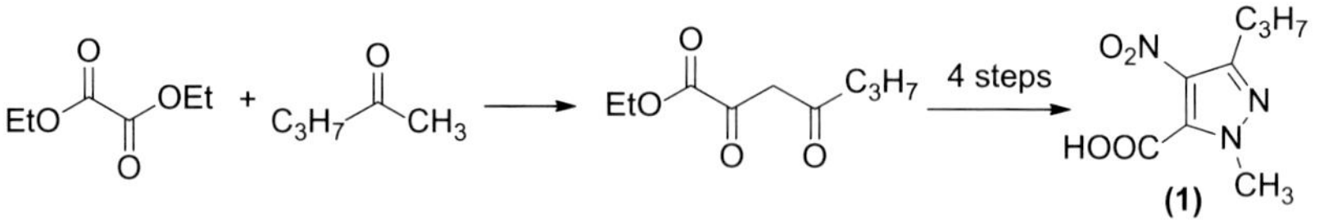

Scheme 2 Synthesis of pyrazole derivative **1**

2 Principles of Green Chemistry

chemical method. Various factors such as whether the chemical under consideration is being disposed or recycled also affect the value of EQ. Waste generation could be avoided or minimized by employing various innovative approaches, some of which have been discussed below:

(a)   **Step or Pot-economic synthesis**

This approach involves the synthesis of a compound by clubbing various synthetic steps in the same pot in case of a multistep synthesis as long as it is possible. It avoids the use of auxiliaries substances such as solvents, reagents required for performing the reaction, purification and isolation of intermediates in each individual steps. Apart from that, one-pot synthesis also saves time and energy which are also very important factors that contribute to the cost of the compound. Most importantly, this approach allows the minimization of waste generation which otherwise would have been generated in each individual steps. For example, the syntheses of *sertraline hydrochloride* (6) (Fig. 2) which is an antidepressant and is being marketed under the brand name *"Zoloft"* by Pfizer was performed employing green approach which not only minimized the waste generation but also at the same time reduced the cost, energy consumption, etc.

The aryl-substituted tetralone (9) is an important intermediate in the preparation of sertraline. Earlier synthetic route involves the four steps which begins with the formation of benzophenone derivative (6) using benzoyl chloride, o-dichlorobenzene as starting materials (Scheme 3).

It further reacts with diethyl succinate to provide another an acid intermediate (7), which after hydrogenation led to the formation of another intermediate (8). Finally, this intermediate cyclized in presence of thionyl chloride and furnish the desired aryl-substituted tetralone (9) (Method A, Scheme 3). Later, Finorga developed a new process; the synthesis of this tetralone derivative was achieved in one step involving Friedel–Crafts reaction between α-naphthol and o-dichlorobenzene (Method b, Scheme 3). This slight modification drastically minimized the waste generation. The tetralone derivative (9) was further used for the synthesis of sertraline (12) involving three different steps (Method A, Scheme 4) that involved the formation of different intermediates (10–11) which was isolated after completion of the reaction and further subjected to another step. Overall, it requires isolation at each step, use of different solvents such as toluene, hexane, THF, and ethanol. In

Fig. 2 structure of
Sertraline Hydrochloride (6)

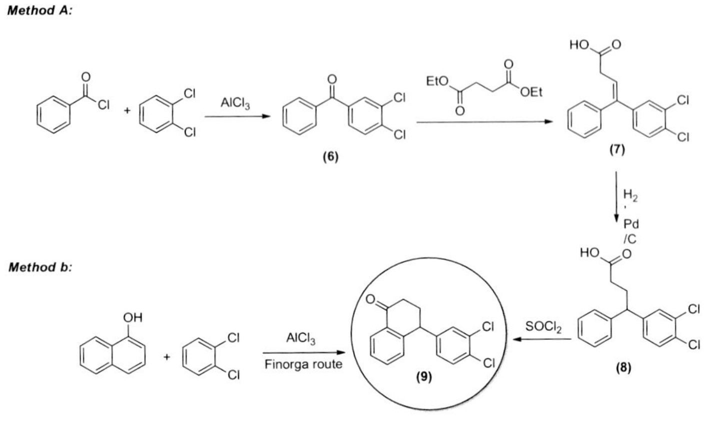

6   1 Green Chemistry: Introduction to the Basic Principles

Method A:

Scheme 3 Synthesis of tetralone derivative (9) using two different methods

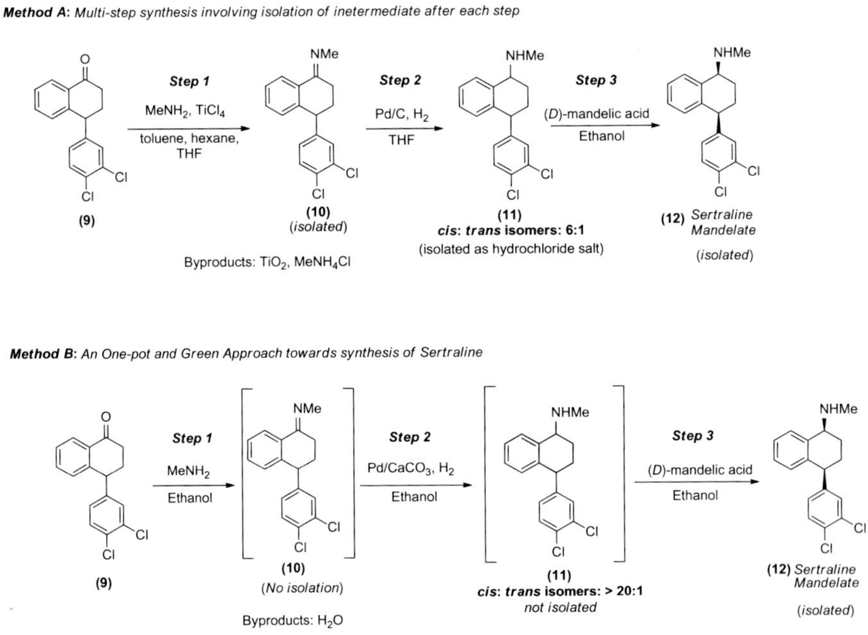

Scheme 4 Comparison between two different synthetic routes to sertraline

addition to that, generation of TiO2 as waste material is another drawback associated with this method. Therefore, Pfizer modified the synthetic route and developed a greener synthetic route for the formation of sertraline where the synthesis of sertraline (12) was achieved in one-pot operation using ethanol as single solvent (Method

2 Principles of Green Chemistry

B, Scheme 4). Importantly, no intermediate was isolated along with that there was no generation of waste material like TiO2.

Overall, this improved route established by Pfizer increased the product yield, energy efficiency, and time. Apart from that, this optimized synthetic routes decreased waste toxic waste generation either in the form of solvents or by-products. More importantly one-pot synthesis saved the time and energy and solvents in the isolation and purification of intermediates. Considering the immense importance of this method, it was awarded US EPA Presidential Green Chemistry Challenge award in year 2002.

(b) Development of catalytic methods

In order to prevent or minimize the generation of the chemical waste, the reaction protocol developed using a catalyst has been found to be superior compared to the conventional approach using stoichiometric amount of reagents. For example, the oxidation of secondary alcohol 1-phenylethanol could be achieved employing two different methods (Scheme 5). In conventional approach, the CrO_3 is used as an oxidizing agent in presence of H2SO4 as an acid source. The reaction produces $Cr_2(SO_4)_3$ as a by-product along with water (Method 1, Scheme 5). While in another approach, oxidation was achieved under catalytic condition using molecular oxygen (Method 2, Scheme 5). In this case, the only by-product is water and the E-factor value drastically reduced to 0.1. Importantly, it also avoided the use of Cr-based reagent which was used in stoichiometric amount and also resulted into large amount of waste material.

Similarly, small modification in the process could drastically reduce the waste production. For example, different synthetic routes could be adopted for the preparation of propylene oxide (Scheme 6). For example, in case of chlorohydrine route, the

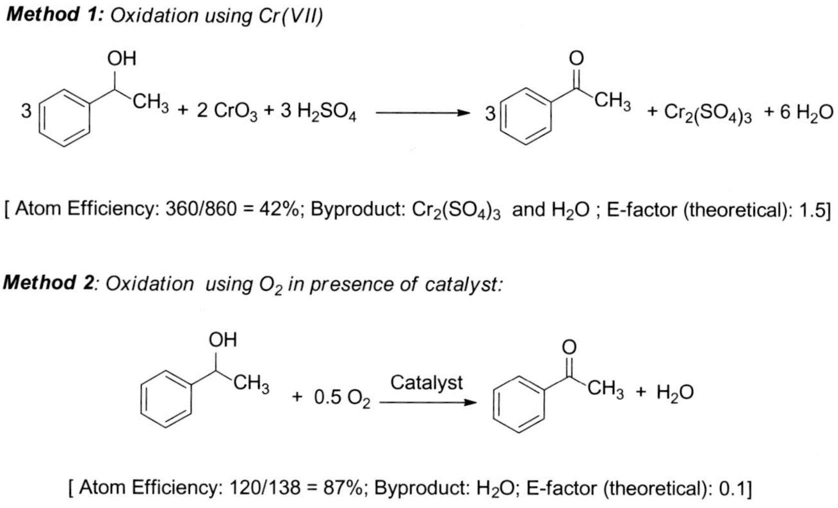

Method 1: Oxidation using Cr(VII)

[Atom Efficiency: 360/860 = 42%; Byproduct: $Cr_2(SO_4)_3$ and H_2O ; E-factor (theoretical): 1.5]

Method 2: Oxidation using O_2 in presence of catalyst:

[Atom Efficiency: 120/138 = 87%; Byproduct: H_2O; E-factor (theoretical): 0.1]

Scheme 5 Oxidation of acetophenone to benzophenone

Scheme 6 Preparation of propylene oxide using different methods

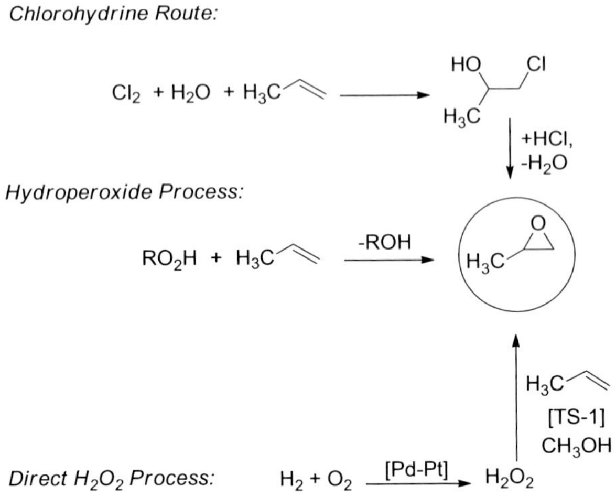

Chlorohydrine Route:

$Cl_2 + H_2O + H_3C$ ⟶ (HO, Cl, H_3C)

+HCl, -H_2O

Hydroperoxide Process:

$RO_2H + H_3C$ —-ROH→ (H_3C, O epoxide)

H_3C
[TS-1]
CH_3OH

Direct H_2O_2 Process: $H_2 + O_2$ —[Pd-Pt]→ H_2O_2

chlorohydrine produced through the reaction between alkene, chlorine, and water undergoes epoxide formation in presence of acid. In addition to that in another method, peroxy acid was used as source of oxygen for the expoxidation of alkene. Instead of these two steps process and using peroxy acid, epoxidation of alkene was achieved through the in situ generation of hydrogen peroxide using metal-catalyzed reaction condition (Scheme 6).

Synthesis of *caprolactum* which is a precursor for the synthesis of nylon-6 using Sumitomo process is another striking example where modification in the conventional process led to the generation of water as the only by-product compared to the conventional approach where ammonium sulfate was being produced in large amount. In conventional method, the transformation of cyclohexanone to its oxime derivative is achieved using hydroxylamine sulfate. The oxime further undergoes Beckmann rearrangement in presence of H_2SO_4 or oleum which is used in stoichiometric amount (Method 1, Scheme 7). In this process, for the production of each kg of caprolactum, 4.5 kg of $(NH_4)_2SO_4$ gets generated as a by-product. Compared to this conventional method, only water gets produced as the only by-product using Sumitomo method with E-factor value significantly reduces to 0.32 (Method 2, Scheme 7).

Similarly, the chemical transformation achieved using biocatalyst is also an attractive way for the reduction of chemical waste generation. For example, the synthesis of 6-aminopenicillanic acid (6-APA) (**15**) using conventional approach requires two different steps through the formation of an intermediate (**14**) using chlorinated reagents, solvents (Scheme 8). At the same time, it also needs to maintain reaction temperature at −40 °C. Compared to this approach, the improved synthetic route using *penicillin acylase* (Pen-acylase) as a biocatalyst, the enzymatic cleavage of penicillin G (**13**) to 6-APA (**15**) could be achieved in a single step at 37 °C [6, 7]. So for the production of 1 kg of 6-APA under conventional method, Me_3SiCl (0.6 kg), PCl_5 (1.2 kg), $PhNMe_2$ (1.6 kg), NH_3 (0.2 kg), *n*-BuOH (8.4 l Kg), and CH_2Cl_2 (8.4 l kg) were required, whereas for enzymatic cleavage, NH_3 (0.09 kg)

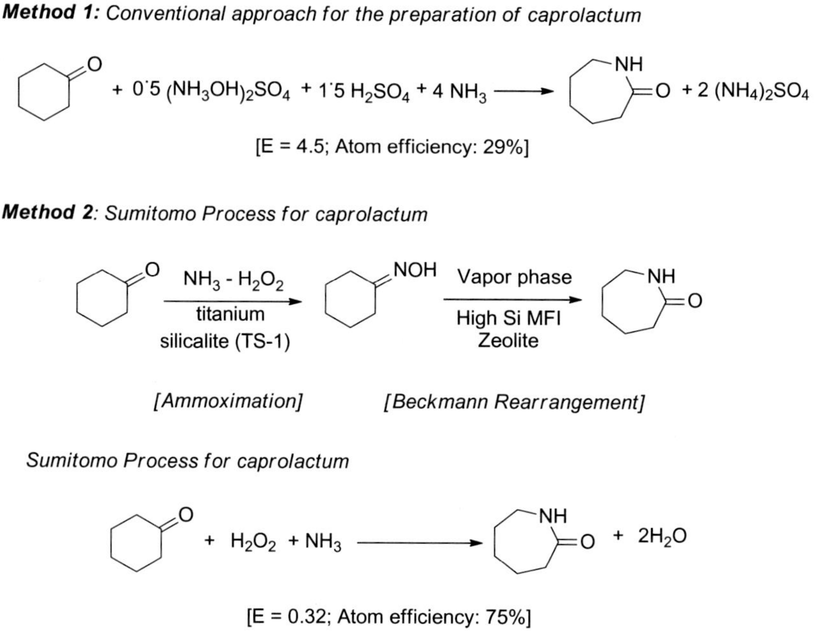

2 Principles of Green Chemistry

Method 1: *Conventional approach for the preparation of caprolactum*

$$\text{cyclohexanone} + 0\dot{.}5\ (NH_3OH)_2SO_4 + 1\dot{.}5\ H_2SO_4 + 4\ NH_3 \longrightarrow \text{caprolactam} + 2\ (NH_4)_2SO_4$$

[E = 4.5; Atom efficiency: 29%]

Method 2: *Sumitomo Process for caprolactum*

$$\text{cyclohexanone} \xrightarrow[\substack{\text{titanium}\\ \text{silicalite (TS-1)}}]{NH_3 - H_2O_2} \text{cyclohexanone oxime} \xrightarrow[\substack{\text{High Si MFI}\\ \text{Zeolite}}]{\text{Vapor phase}} \text{caprolactam}$$

[*Ammoximation*]   [*Beckmann Rearrangement*]

Sumitomo Process for caprolactum

$$\text{cyclohexanone} + H_2O_2 + NH_3 \longrightarrow \text{caprolactam} + 2H_2O$$

[E = 0.32; Atom efficiency: 75%]

Scheme 7 Comparison between Sumitomo and conventional methods for caprolactum synthesis

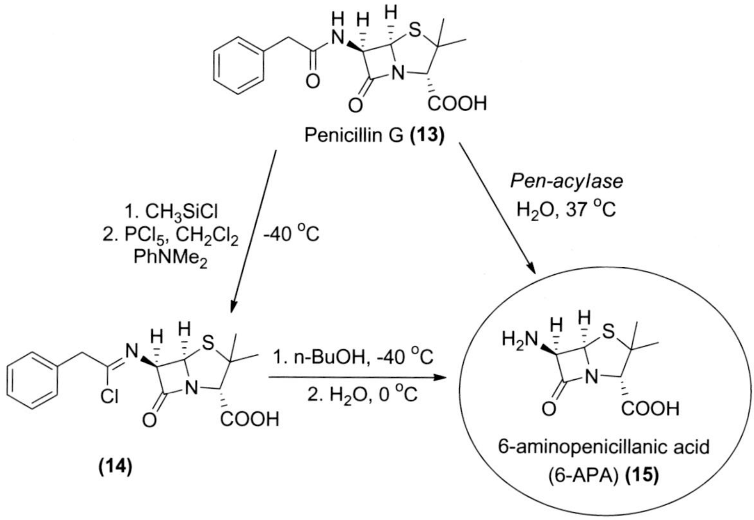

Penicillin G **(13)**

1. CH₃SiCl
2. PCl₅, CH₂Cl₂ / -40 °C
   PhNMe₂

Pen-acylase
H₂O, 37 °C

(14)

1. n-BuOH, -40 °C
2. H₂O, 0 °C

6-aminopenicillanic acid
(6-APA) **(15)**

Scheme 8 Deacylation of Penicillin G under chemical and enzymatic method was used as the only reagent to maintain the pH, and the conversion was achieved in single step (Scheme 8). Therefore, this environmentally benign biocatalytic method allowed the drastic reduction in production of toxic chemicals waste. Several such transformations are known in the literature and will also be discussed in the separate chapter.

(c) **Development of solvent-free method**

Among various types of wastes, solvents are a major contributor due to their multiple role in a chemical process. Most of the chemical reaction require organic solvents as a reaction medium due to its role in efficient mass transfer, control of localized temperature, reaction kinetics, and also for the stabilization of certain reaction intermediates. In addition to that, purification process either through column chromatography, crystallization, etc. also requires ample solvents. Therefore, in order to minimize the waste generation in the form of solvent, various synthetic methods have been explored under the solvent-free condition. For example, syntheses of diverse range of medicinally important heterocycles have been reported under solvent-free reaction condition. For example, the synthesis of functionalized pyrroles could easily be achieved under solvent-free reaction condition (Scheme 9) [8]. Similarly, the synthesis of benzofuran (**16**) could also be achieved through Rap–Stoermer reaction under solvent-free reaction condition (Scheme 10) [9].

Principle 2: *"Synthetic methods should be designed to maximize the incorporation of all materials used in the process into the final product."* [2]

The minimization of the waste could also be achieved via the efficient use of the raw materials used in a process. For the synthesis of any compound especially if the synthesis requires more than one step as generally happens in case of structurally complex natural products and drugs, different synthetic routes could be designed. And every route will demand a particular set of reactants, reagents, solvents, catalyst, etc. Apart from these, every synthetic scheme would also require different set of chemical transformations. Some of which may be common. Depending on the

Scheme 9 Preparation of 2,5-disubstituted pyrrole under solvent-free condition

(R^1: H, Cl, Br etc.; R^2: H, I, Br; R^3: H, Me, Br, OMe)   (Yield: Up to 98%)

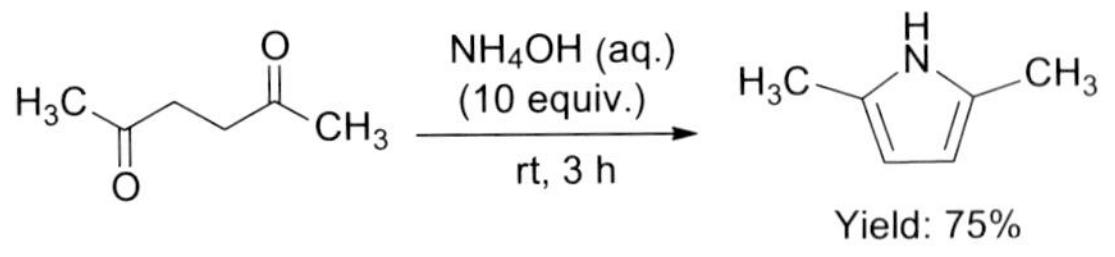

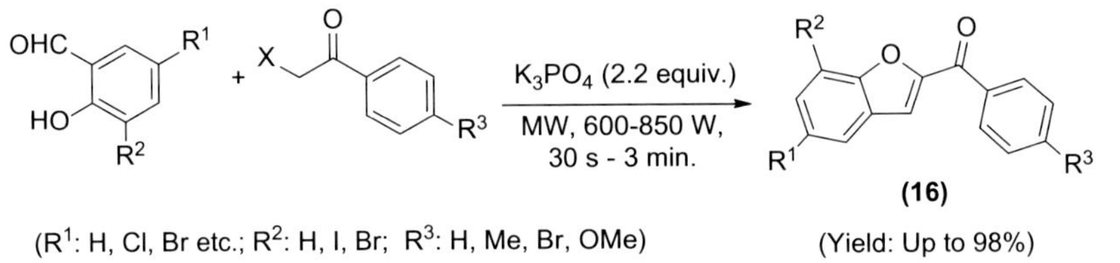

Scheme 10 Solvent free synthesis of benzofuran

number of synthetic steps, types of reactions and reagents involved, product yield, amount and types of waste generated would also differ. In general, the primary goal of a chemist was to achieve the synthesis and improving the desired product yield and selectivity of the product. But after realizing the environmental impact and deteriorating effect of some of the organic solvents, toxic reagents, by-products, etc. along with the fast depletion in the non-renewable resources especially petroleum, the approach is slowly moving toward the minimization of waste generation and maximum incorporation of the raw material into the product. In order to evaluate and compare the efficiency and greenness of a chemical process, various matrices have been proposed. Along with E-factor, *"atom economy"* which sometimes also referred to as *"atom efficiency"* was another important tool that was first introduced by B. M. Trost [10]. This value of % atom economy helps in evaluating the greenness of a process.

Atom economy could be represented as follows (Fig. 3)

Therefore, with the help of the E-factors value and % atom economy, the greenness of a process could be compared. Every metrics has some sort of limitation and the value does not fully reflect the greenness of a process. Unlike E-factor value that takes all the by-product, reagents, solvents, etc. into account, the value of atom economy only provides information about the conversion of starting material used into the product and do not account for the other substances which are not a part of the stoichiometric equation. With the help of atom economy value, the theoretical E-factor could be calculated. For example, an atom economy of 80% indicates the conversion of rest of the 20% of the starting material into by-product so the E-factor would be 20/80, i.e., 0.25. But practically, the E-factor value would be much higher than the theoretical one if reaction involves the use of excess amount of reagents, solvents, salt formation, etc. Therefore, in order to enhance the atom economy and reducing E-factor of an overall process, the synthetic routes could be designed in such a way that would include the synthetic steps having high atom economy. Different types of reactions which have high value of atom economy, i.e., maximum incorporation of atoms of starting material into product, have been developed under catalytic protocol that led to the higher value of atom economy (Fig. 7) which has been discussed below with few examples (Fig. 4).

(a) **Addition Reaction**: The atom economy in case of the addition reaction of alkene and alkyne is 100% as it simply involves the addition of the entire reagent. For example, in case of the addition reaction of HBr to 2-methylprop-1-ene, all the atoms of the starting materials get incorporated into the product 2-bromo-2-methylpropane without losing anything in the form of by-product (Scheme 11).

(b) **Pericyclic Reactions**: The atom economy in case of pericyclic reaction such as cyclcoaddition, electrocyclci reaction, and sigmatropic rearrangement is

Fig. 3 Calculation of % atom economy

$$\text{Atom Economy} = \frac{\text{Molecular weight of desired product}}{\text{Molecular weights of all reactants}} \times 100\%$$

Fig. 4 Types of reactions with high atom economy

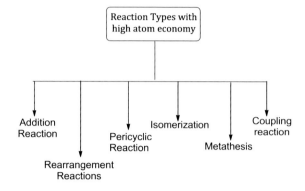

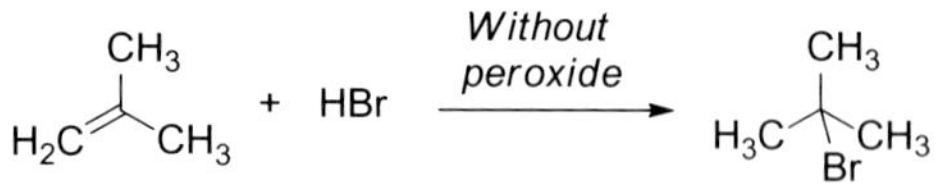

Scheme 11 Preparation of t-butyl bromide involving addition reaction

100%. For example, in case of Diels–Alder reaction which is basically a [4 + 2] cycloaddition reaction between and dienes and a dienophile, the atom economy is 100% as all the participating reactants converts into the product, for example, the formation of adduct through the [4 + 2] cycloaddition reaction between 1,3-butadiene and maleic anhydride (Scheme 12) [11].

Various other reactions including catalytic hydrogenation, oxidation, carbonylation, and hydroformylation with high atom economy values have been shown in Scheme 13.

For example in case of reactions such as Claisen rearrangement, catalytic hydrogenation, carbonylation, and hydroformylation, the atom economy is 100%. Similarly by including catalytic reaction protocols in the synthetic pathway, the waste production could be reduced drastically in the production of API. For example, the synthesis of ibuprofen which is an antiinflammatory drug is a classic example of waste reduction employing greener approach. For example, the synthetic route patented by Boots company, UK, in 1960 for the synthesis of Ibuprofen involves six steps starting from 2-methylpropylbenzene (Scheme 14) [12]. Reaction involves the formation of different intermediates (**17–21**, Scheme 14). This method not only generates lots of acidic wastes such as acetic acid, HCl, and aluminum-based waste material as well as ammonia, but also the atom economy was only 40%, i.e., rest of the 60% materials were not utilized and came out as waste material.

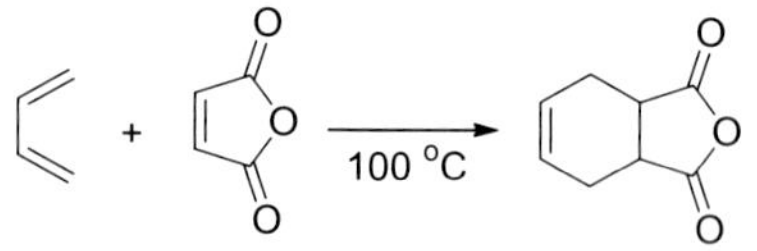

Scheme 12 [4 + 2] Cycloaddition reaction to form adduct

2 Principles of Green Chemistry

Scheme 13 Examples of Different catalytic processes with high atom economy

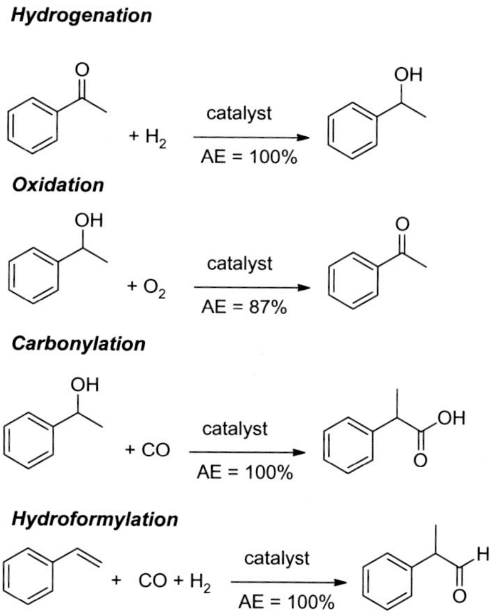

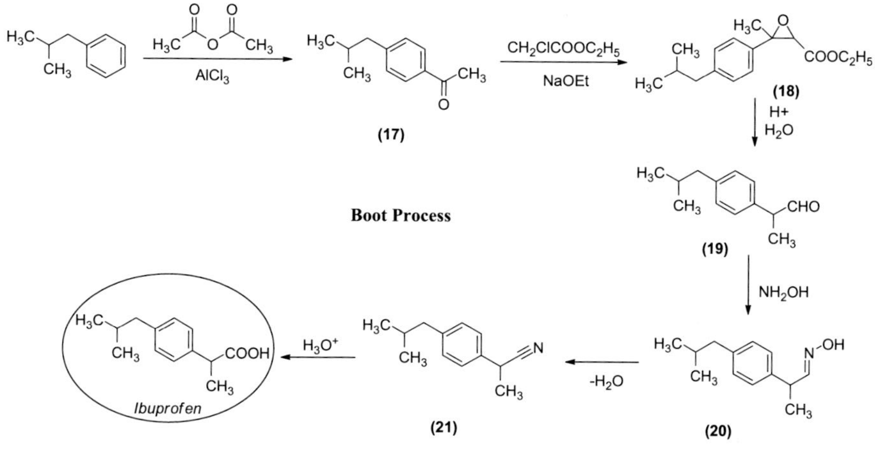

Hydrogenation

Oxidation

Carbonylation

Hydroformylation

Boot Process

Scheme 14 Boot process (Brown synthesis) for the synthesis of Ibuprofen

Later, BHC Company which is now known as BASF developed a green process involving only three steps including acylation and metal-catalyzed hydrogenation and carbonylation steps through the formation of alcohol (**22**) (Scheme 15). Out of which the catalytic hydrogenation and carbonylation steps were highly efficient with 100% atom economy (Scheme 15) [12]. This route gratifyingly enhanced the overall atom economy value from 40 to 77%. In addition to the reduction in the number of synthetic steps, this greener approach was found to be superior in the firsts step, i.e., in the acylation of isobutylbenzene which require use of stoichiometric amount of

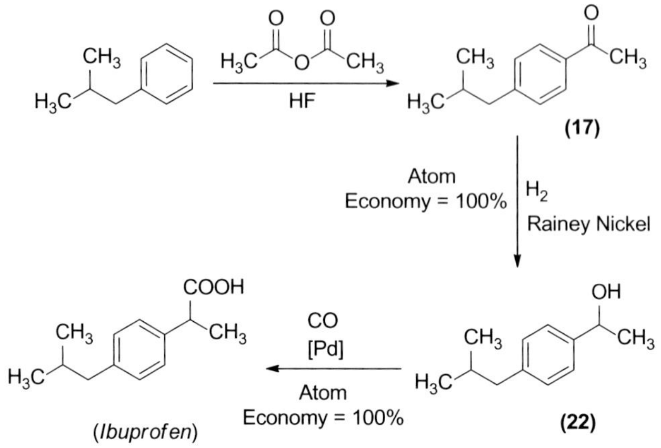

Scheme 15 Green synthetic route (BHC process) to the synthesis of Ibuprofen

aluminum chloride in the Boots process. This led to the generation of huge amount of waste material in the form of aluminum trichloride hydrate. In comparison to that, catalytic amount of hydrogen fluoride was for this step, and interestingly, HF was also recovered and employed repeatedly. Similarly, the Ni and Pd catalysts used in step 2 and 3 in BHC process were also reused.

Principle 3: "*Wherever practicable, synthetic methods should be designed to use and generate substances that possess little or no toxicity to human health and the environment.*" [2]

This principle emphasizes on very important aspect of the chemical synthesis which is about its impact on the human health as well as environment. So according to this principle, only the efficiency of synthetic methodology which mainly includes atom efficiency or E-factor should not only be the criteria to judge environmental acceptability and greenness of a method. Toxicity associated with the reactants, reagents, and intermediates as well as with the products is also another important aspect which requires serious consideration while designing a synthetic method. The chemicals could be hazardous to human leading to various types of toxic effect like carcinogenicity, neurotoxicity, hepatotoxicity, teratogenicity, mutagenicity, etc. Similarly, it could also be hazardous to the environment leading to toxic effect to the plants, aquatic animals, birds, amphibians, and also to various other mammals. In addition to that, chemicals which are explosive, inflammable, corrosive, oxidizer, and reducer in nature also create physical hazard that may lead to the some of the serious accidents. Similarly, some of the chemicals like chlorinated hydrocarbons, greenhouse gases, persistent chemicals, etc. are posing global hazard and frequent use of leading to the phenomenon like ozone depletion, global warming, acid rain, etc. Therefore, by reducing the use of hazardous chemicals, the risk which is a function of hazard and exposure, associated with a method, can also be minimized or completely avoided. Various strategies could be adopted to understand the probable hazardous

2 Principles of Green Chemistry

effect of a chemical used in different roles like starting material, reagents, products, by-products, etc.

Life cycle analysis or life cycle assessment (LCA) is a systematic study to assess the environmental impact of a product or processes at the various stages of its life [13]. It is basically considered as analysis from cradle to grave, i.e., assessment of environmental impact of the entire process starting from the extraction of raw material, production of product, and its distribution to the user along with the assessment of the material which will be disposed of to the environment as waste material. It would help in analyzing and identifying the probable hazardous substance that could be used or generated during the entire journey from raw material to waste material. Another important strategy is the *replacement of more hazardous chemical with environmentally benign alternative* while developing a method. For example, phosgene is a very reactive and useful C1 source for the carbonylation process used in the manufacture of drugs, polymers, dyes, insecticide, pesticides, etc. But due to its highly toxic nature and gaseous phase, it requires more precaution and is generally prepared and used immediately at the manufacturing site itself. Therefore to minimize the risk in using phosgene, various other greener alternative like organic carbonates such as dimethyl carbonates (DMC), ethylene carbonate (EC), dipehnylcarbonate (DPC), etc. have been also been developed. In particular, DPC has emerged as an attractive and environmentally benign alternative of highly toxic phosgene gas which is used in the production of polycarbonate, polyurethanes, etc. which are widely used for the manufacture of automobiles and various electronic appliances [14]. Similarly, dimethyl carbonate (DMC) is being used as an eco-friendly alternative to other popular methylating agents such as dimethyl sulfate and methyl iodide which are considered to be toxic to human [15].

Similarly, to achieve the synthesis of biologically active heterocyclic compounds such as benzothaizole and benzoxazole, benzotriazole ring cleavage (BtRC) method has emerged as a very attractive method. Generally, it requires butyl tin hydride (Bu$_3$SnH) along with AIBN in toluene solvent (Method A, Scheme 16).

Tiwari et al. developed a greener protocol using polymethylhydroxysilane (PMHS) which is biodegradable, non-toxic, and cost-effective reagent to achieve the similar transformation (Method B, Scheme 16). For example, the reaction of 6-O-(1,2:3,4-di-O-isopropylidine-α-D-galactopyranose)-1H-benzo[d][1,2,3]triazole-1-carbothioate (**23**) with polymethylhydrosiloxane (PMHS) in presence of 5 mol% AIBN provided 78% yield of respective 6-O-(benzothiazol-2'-yl)-1,2;3,4-di-O-isopropylidene-α-D-galactopyranose (**24**) (Scheme 16) [16]. Compared to the transformation using toxic tin based reagent, this transformation achieved using PMHS offers a greener alternative to it.

The conventional C–C bond forming reaction via metal-catalyzed coupling reaction using different organometallic reagents and organic halides is being replaced with direct C-H bond functionalization reaction to avoid the pre-functionalization in the form of halo and pseudoderivatives as coupling partner with organometallic reagents which are sometimes toxic like Sn and also not safe to handle. In order to minimize the metal-based waste generation, several transition metal-free reaction protocols have been developed. For example, the arylation of pyrazine which is a

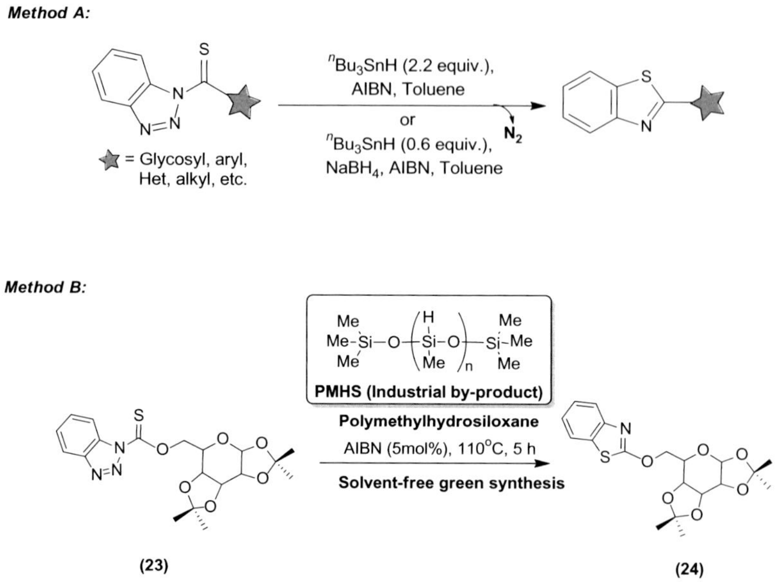

Method A:

= Glycosyl, aryl, Het, alkyl, etc.

$^nBu_3SnH$ (2.2 equiv.), AIBN, Toluene

or

$^nBu_3SnH$ (0.6 equiv.), $NaBH_4$, AIBN, Toluene

N_2

Method B:

Me
Me–Si–O–(Si–O)ₙ–Si–Me
Me   Me   Me
     H

PMHS (Industrial by-product)

Polymethylhydrosiloxane

AIBN (5mol%), 110°C, 5 h

Solvent-free green synthesis

(23)

(24)

Scheme 16 Polymethylhydrosiloxane-mediated Tiwari protocol for the benzotriazole ring cleavage (BtRC)

medicinally useful scaffold could be achieved through Pd-catalyzed Suzuki coupling protocol using chloropyrazine and arylboronic acid (Scheme 17).

In another greener approach, the pyrazine itself reacted with iodo phenol to produce the desired coupled product through C-H functionalization under potassium *tert*-butoxide (KO^{t}Bu)-mediated reaction condition without using any other transition metal-based catalyst. Interestingly, reaction worked under microwave-assisted

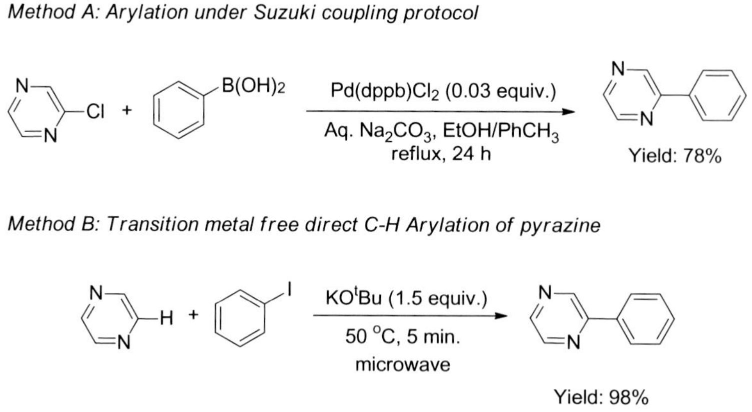

Method A: Arylation under Suzuki coupling protocol

$Pd(dppb)Cl_2$ (0.03 equiv.)

Aq. Na_2CO_3, EtOH/PhCH$_3$ reflux, 24 h

Yield: 78%

Method B: Transition metal free direct C-H Arylation of pyrazine

KO^{t}Bu (1.5 equiv.)

50 °C, 5 min. microwave

Yield: 98%

Scheme 17 KO^{t}Bu-mediated arylation of pyrazine

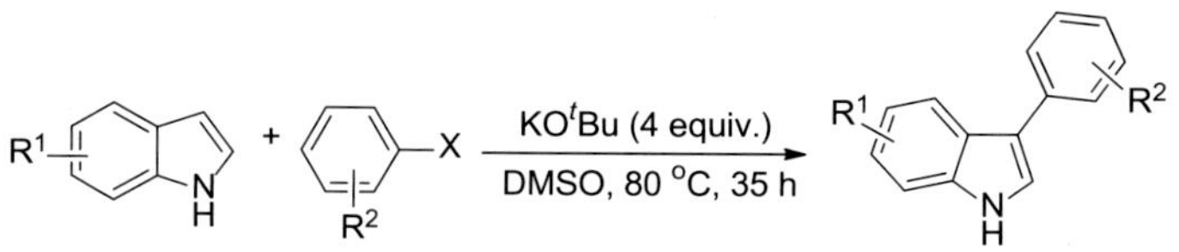

2 Principles of Green Chemistry   17

Scheme 18 KO^{t}Bu-mediated arylation of indole

reaction condition and completes within 5 min. compared to the Suzuki protocol that needs 24 h to complete. [17, 18]. Although in this case, the excess amount of pyrazine (40 equiv.) was used that is certainly a major drawback to improve. Similarly in another interesting example, the arylation of Indole at C3 position was achieved under transition metal-free reaction protocol using aryl halide, KO^{t}Bu, and DMSO as aryl source, base, and solvents, respectively (Scheme 18) [19]. In most of the reaction attempted, the formations of N-arylated products were also observed up to 21% along with C3-arylated indole as major products. Degassing of solvent was essential for achieving the best conversion to the product, i.e., C3-arylated indole (**25**) as low (34%) to no conversion of the C3-arylated indole was obtained when solvent was used without degassing or with saturated with oxygen. Among different halides such as iodo, bromo, chloro, fluoroarenes and pseudohalides such as triflate derivatives, iodoarenes were found as the most efficient aryl donor under given reaction condition.

Likewise, the arylation of oxindole was also achieved under metal-free and in absence of any organometallic reagent through the direct coupling between aryl halide and oxindole (**26**) itself. The reaction afforded good-to-excellent yield of the desired arylated product (**27**) under the reaction condition consisting of CsOH as base and acetonitrile as a solvent in presence of white LED light of 23 W. The developed methods were found to work efficiently with the aryl halides in particularly with aryl bromide and iodides having electronically diverse substituents (Scheme 19) [20].

Plethora of such methods involving direct C-H activation and functionalization are being developed. Compared to the conventional approach where the use of organometallic reagents and metal catalysts are essential, the above-mentioned methods developed under transition metal and organometallic reagent-free reaction

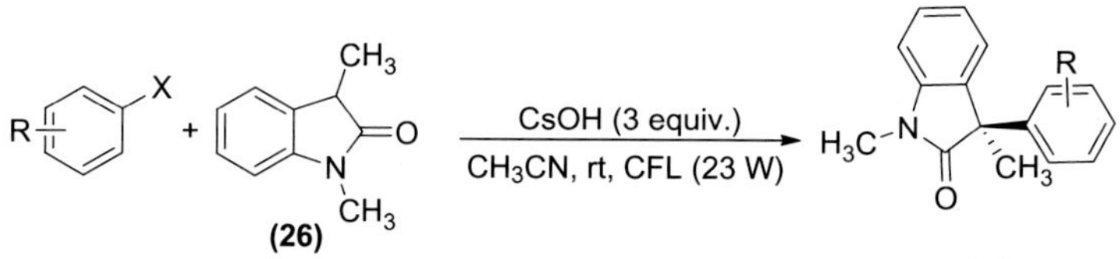

Scheme 19 Photo-catalyzed transition metal free arylation of oxindoles

condition provide environmentally benign ways to form the C–C bonds. Another important strategy to minimize the health and environmental impact of a toxic and reactive chemicals if it is essential to use in a synthetic process is its preparation at the time of its requirement without storing it for long time, i.e., *on-demand* preparation of such chemicals could prevent their potential hazardous impact. For example, the leakage of highly hazardous methyl isocyanate (MIC) gas that was stored in huge amount for the synthesis of *carbaryl*, a popular pesticide manufactured by UCIL, was considered as one of the prime reason for the Bhopal gas disaster. It led to one of the deadliest accidents in the world history in the night on December 2, 1984. More than 3000 people lost their lives. Therefore, identification of a toxic or potential a potential hazardous substance is essential so that greener alternative could be developed. Although for several compounds, the properties are well documented but the major challenge is to identify the toxic compounds or intermediates. Therefore, while designing a synthetic scheme sometimes, it becomes difficult to predict the probable hazard that may arise due to the use of a particular chemical or due to the generation of an intermediate. Few assessment tools such as Environmental Assessment Tool for Organic Synthesis (EATOS) are also available which considers not only the amount of reactant or waste material but also considers about the toxicity associated with them in a quantitative way. Likewise, several other tools such as USEtox have also been developed to analyze the human toxicity and ecotoxicity.

Principle 4: *"Chemical products should be designed to affect their desired function while minimizing their toxicity."* [2]

This principle emphasizes on the fact that while designing a new chemical which could be used for either material or pharmaceuticals purposes should be designed in such a way that the required efficacy could be maintained at the same time the toxicity could also be minimized. With the advancement in the science and technology, various computational methods have been developed which help in quantifying the different physicochemical, toxicity properties of a compound. Therefore in contrast to the conventional approach where the synthesis was followed by in vitro and in vivo studies to determine the biological effect of a newly developed compound, the present approach is to investigate the different properties through in silico studies. The computational approach allows prediction of various physicochemical properties including evaluation of Absorption, Distribution, Metabolism, and Excretion (ADME) of a compound with the help of various freely available software. For example, Swiss ADME is a freely available on-line tool for the evaluation of different aspects including pharmacokinetics and bioavailability, drug-likeness, etc. of a compound [21]. Likewise, molecular docking studies along with molecular dynamics simulation studies give idea about the 3D interaction of a compound with the target site (Fig. 5).

In case of several drugs especially those which are chiral, one enantiomeric form could be toxic or not effective whereas other one could be effective. These could be thoroughly studied before bringing them in use to avoid their toxic effect on human health or environment. For example, thalidomide which is a drug to treat morning sickness in pregnant women was launched in the market as racemic mixture. The use of this drug caused deformities in newly born baby [22]. Later, it was found that

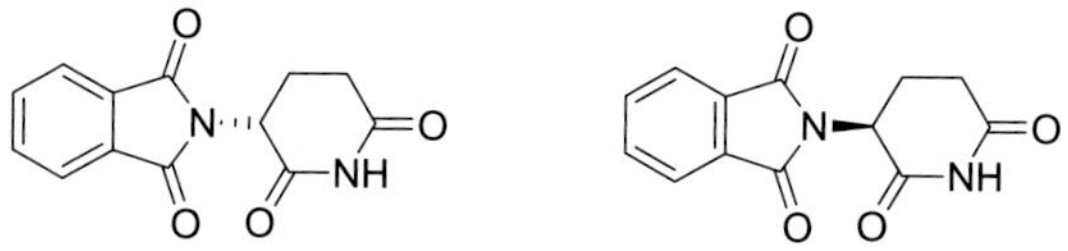

2 Principles of Green Chemistry

Fig. 5 Two enantiomers of thalidomide and their biological effect

(R)-thalidomide

Sedative

(S)-thalidomide

Teratogenic

(S)-entiomer is toxic and has teratogenic effect, whereas (R)- enantiomer has desired sedative effect. Initially due to the lack of toxicity profile of both these isomers, these drugs which were prescribed to the pregnant women led to the birth of baby suffered with various defects including phocomelia. Worldwide more than 10,000 cases were reported with similar symptoms. Later in 1960, it was removed from the market. To avoid such fatal issues at the later stage, it is important to investigate different aspects and properties of a compound and analyze the data before moving for the synthesis.

Principle 5: *"The use of auxiliary substances (solvents, separation agents, etc.) should be made unnecessary whenever possible and when used, innocuous."* [2]

In general, in any reaction apart from main reactants and reagents, other substances are also required either for providing suitable reaction medium or for the separation of the reaction mixture and isolation of desired products. These are considered as auxiliary substances. It mainly includes solvents along with any other reagents like acid, base, etc. which are required during the separation and purification of desired product from the reaction mixture. Here, it is important to highlight and discuss the role of various solvents in any reaction and their impact on the environment as well as on human life, and if these affect our environment and human life, then do we have any other alternative of the solvents mainly organic solvents which we commonly use in a conventional synthesis/classical chemical approach? Would it be possible to carry out a reaction without solvent? Or do we have any other mechanism to minimize the use of organic solvents if it is unavoidable and indispensable to use? In order to find the answer to these questions, it is important to understand the role of solvent in various stages during the synthesis of a compound. Just consider a reaction which involves solid reactants and reagents. Even after mixing these all together in a reaction vessel, since it is not a homogenous mixture, mass transfer would not be easy which is required for reaction to occur. And if it occurs, then the reaction rate would be slow; overall, it will affect all the factors like yield, rate of reaction, reaction time, etc. Therefore, the use of solvent becomes indispensable in some cases to control all these aspects. Most of the reaction especially organic transformations generally requires organic solvents such as chloro hydrocarbons (dichloromethane, chloroform, dichloroethane, etc.) and benzene or its derivatives like toluene, xylene, etc. Similarly, dimethylformamide (DMF), dimethylacetamide (DMA), *n*-methylpyrilidone (NMP), dimethylsulfoxide (DMSO), etc. are frequently used solvents in case of organometallic reactions like cross-coupling, C-H activation, etc. In fact, some reactions like metal-catalyzed cross-coupling reaction furnishes excellent yield in one particular solvent, while it does not work with other. And the use of solvent is not limited as reaction medium only but these are required for the separation and purification also either during column chromatography or crystallization. Despite of such excellent role of these solvents, their impact on environment and human health became matter of serious concern from past decades. For example, solvent is one of the major contributors of waste generated in fine chemicals and pharmaceutical industry leading to high E-factor values. In addition to that, since some of the organic solvents like chloro hydrocarbons like DCM, chloroform, benzene, etc. are volatile and carcinogenic, exposure to these for longer time will have serious effect on health and environment. The other factor which is important to consider is their source which is mostly non-renewable petroleum. Apart from the health and environmental issues, costs of these solvents are also important to discuss which will affect the overall cost of any compound. For mg scale reaction, the impact may not be very noticeable but if we imagine their impact on large scale, or at industrial scale, picture would be clearer. Therefore, it is important to avoid or minimize the use of these auxiliary substances, and if it is indispensable, then explore other alternative of these solvent which would have very low or no toxicity and at the same time could also be cost effective and environmentally benign. In accordance to this principle, different types of organic transformations have been developed under solvent-free reaction protocol.

Synthesis of heterocyclic compounds under solvent-free condition

Heterocyclic rings are core scaffold in diverse varieties of biologically active compound and drugs. In particular, N-heterocycles are core scaffold in more than 85% of the drugs. In general, organic solvents are used for the synthesis of such compounds. But several solvent-free methods have also been developed which would prevent the addition of waste material in the form of solvents. For example, imidazo [1,2-a]pyridines (**29**) has been prepared under solvent-free reaction protocol via multicomponent reaction among 2-aminopyrimidine (**28**), aldehyde, and isonitrile. Reaction completed within 5 min. in presence of Yb(OTf)$_3$ as a Lewis acid catalyst under the microwave-assisted condition and furnished yield upto 99% (Scheme 20) [23].

Similarly formation of pyrazoles ring (**31**) has also been achieved under solvent-free condition, through the cyclocondensation reaction between suitable hydrazine derivatives and 3-formylchromone (**30**) (Scheme 21).

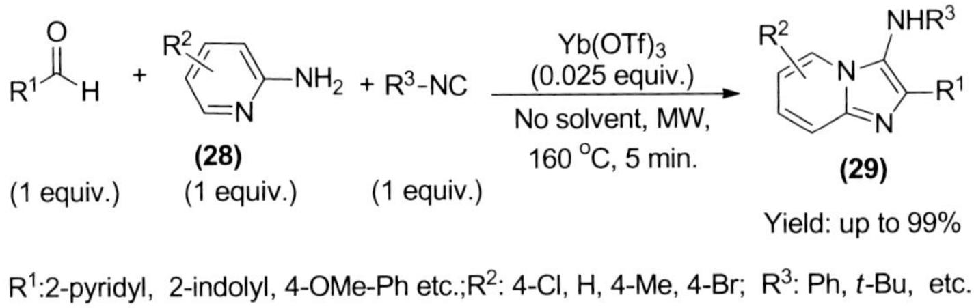

R^1:2-pyridyl, 2-indolyl, 4-OMe-Ph etc.;R^2: 4-Cl, H, 4-Me, 4-Br; R^3: Ph, t-Bu, etc.

Scheme 20 Yb(OTf)$_3$ catalyzed solvent-free synthesis of imidazo[1,2-a]pyridines

2 Principles of Green Chemistry   21

Scheme 21 Synthesis of substituted pyrazoles under solvent-free reaction condition

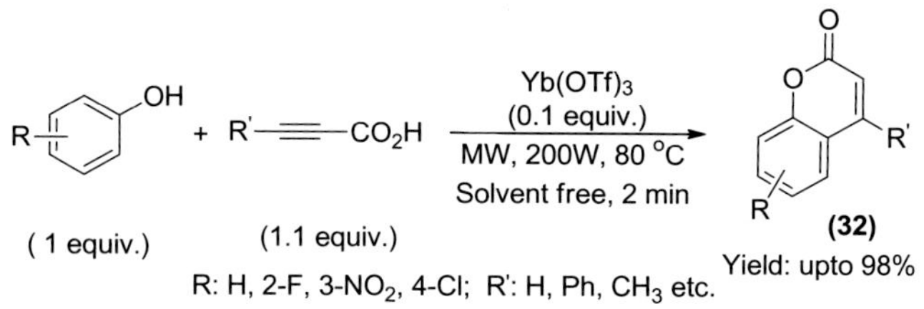

Scheme 22 Preparation of coumarin ring using Yb(OTf)$_3$ as a catalyst

Likewise, oxygen heterocycles such as coumarin ring have also been created under solvent-free reaction condition using Yb (OTf)$_3$ as a catalyst. Reaction between propiolic acid and phenol afforded coumarin derivatives (**32**) within 2 min. reaction time (Scheme 22) [24].

Greener alternative to Organic solvent

Considering the toxicity, volatility, renewability, and overall impact on the environment and life, some of the solvents which are being explored as alternative to the frequently used organic solvents and attracting the attention of scientific community are as follows:

Water: Approximately, earth's 70% part is covered with water. And, it is renewable also. Since water provides reaction medium to most of the reactions occurring in our body so unquestionably, it is important component of our life and therefore toxicity factor could also be ignored as long as it is pure and uncontaminated. Overall, it could be considered as least expensive, non-toxic, and eco-friendly solvent. Therefore, vast verities of reactions have been successfully explored in aqueous reaction medium. There are certain limitations where presence of water is strictly avoided. For example, in the reaction where some of the highly moister sensitive compounds like organolithium (RLi), Grignard's reagent (RMgX), lithium aluminum hydride (LiAlH$_4$), titanium tetrachloride (TiCl$_4$), aluminum chloride (AlCl$_3$), BF$_3$etherate, etc. are involved, water would not be a suitable solvent. In fact in some cases, it would have disastrous effect also. Some of the reports are available in the literature where water has been used as solvent in presence of such highly moisture sensitive reagents also [25]. In recent years, the concept of reaction "*on water*" started attracting the attention of scientific community which allows the use of water in reaction involving

Scheme 23 Formation of tetrahydrofurans through nucleophilic addition of organometallic reagents

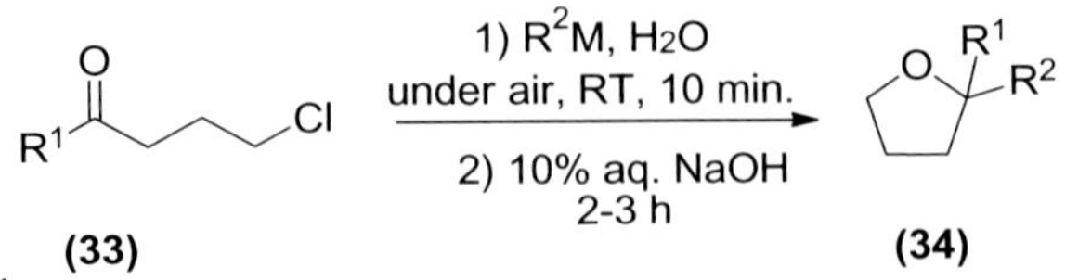

$$R^1 \overset{O}{\underset{(33)}{\diagdown}} Cl \quad \xrightarrow[\substack{2)\ 10\%\ aq.\ NaOH \\ 2\text{-}3\ h}]{\substack{1)\ R^2M,\ H_2O \\ under\ air,\ RT,\ 10\ min.}} \quad \overset{O \diagup R^1}{\underset{(34)}{\diagup R^2}}$$

highly water sensitive compounds such as organ lithium and Grignard reagent. For example, the formation of tetrahydrofuran ring (**34**) was achieved through the reaction between ketone (**33**) with either organolithium or Grignard reagents on water medium (Scheme 23) [26].

Similarly isolation of the water-soluble product from water could also be a cumbersome task. Similarly, recovery of water to obtain pure water would be either time consuming if natural evaporation method is opted or would require ultra-filtration method. Although it is important to highlight that any solvent cannot be considered as universal solvent for all different types of reactions. It means different reactions may or may not work in a particular solvent and it also depends on various other reaction parameters also. But a reaction can be attempted in above-mentioned solvents in order to minimize or avoid use of them.

Supercritical CO_2: Apart from water, another solvent which is present in atmosphere in ample amount is carbon dioxide (CO_2). Although it generally exists in gaseous form but as we know that gases can be liquefied at certain temperature and pressure. Likewise, liquefied CO_2 and supercritical CO_2 are also being used for the organic reactions [27]. Unlike any organic solvent, it is non-inflammable and considered to be non-toxic. Other properties such as low viscosity and mixing ability with gases allow it to be used as better alternative than some other solvents like water and organic solvent.

Likewise, various other greener alternatives like polyethylene glycol, ionic liquids, fluorous biphasic solvent, etc. to the conventional organic solvents are available in the literature. Importance and application of these green solvents have been discussed elaborately in the subsequent chapter.

Principle 6: *"Energy requirements of chemical processes should be recognized for their environmental and economic impacts and should be minimized. If possible, synthetic methods should be conducted at ambient temperature and pressure."* [2]

This principle underlines the overall economic and environmental impact of using different energy sources used for performing chemical reactions. The requirement of energy begins from the synthesis step to workup followed by purification and isolation of the desired product. In general in organic synthesis, after completion of reaction and work up, isolation of compound needs column chromatography or distillation. These both the techniques of chemical separation are energy intensive which consumes huge amount of energy. Apart from that if reaction works at high temperature, then there is need of thermal energy which generally obtained through conventional heating process. And energy requirement for each step does not only affect the overall economy or cost but also affect the environment. In addition to the conventional heating method which leads to loss of large amount of heat

2 Principles of Green Chemistry

energy, various other alternatives which are considered to be more efficient have been explored (Fig. 6). The importance and the various aspects along with different types of reactions performed using these methods have been discussed in detail in Chap. 3.

Principle 7: "A raw material or feedstock should be renewable rather than depleting whenever technically and economically practicable." [2]

This principle advocates the use of renewable sources over non-renewable petroleum feedstock or natural gases for manufacturing various products and as energy sources. Extensive use of these non-renewable sources may take our future generation to more critical and adverse condition. To procure more safe and sustainable energy sources, we should rely on natural renewable sources for material manufacturing and fulfill our energy demands [28].

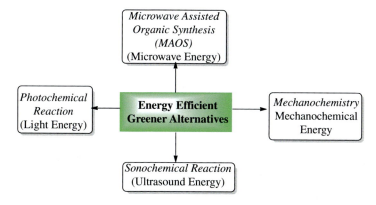

Fig. 6 Various Energy efficient approaches adopted in chemical synthesis

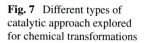

Fig. 7 Different types of catalytic approach explored for chemical transformations

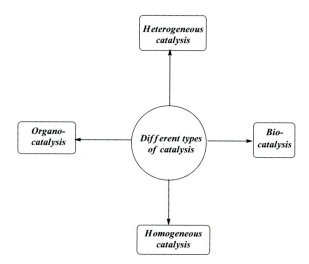

Some of the best renewable sources are the products which can be derived from plants like cellulose, starch, suberin, lignin, polyhydroxyalkanoates, chitin, lactic acid, glycerol, and oils. Chitin is also abundantly found in some arthropods (crustaceans) as main constituent of their exoskeleton. The use of these natural renewable livestock may lead to a sustainable energy source and other products. Lignin-based products, for instance, have a huge market comprising food additives, cement, dispersants, binders, batteries, and raw material for chemicals like humic acid and DMSO. Lignin can provide high strength in concrete and weather resistance. Chitin is a lead by-product in sea food industries can be reused by transforming it to chitosan by deacetylation. Chitosan found various applications in industries, water purification, and biomedical applications like use in high blood pressure, high cholesterol, obesity, wound healing, etc. A wise use of these by-products of seafood industries may fulfill our current need of energy and material manufacturing.

Principle 8: "Unnecessary derivatization (blocking group, protection/deprotection, temporary modification of physical/chemical processes) should be avoided whenever possible." [2]

In general, the synthesis of natural products and structurally complex drugs requires multiple steps for their synthesis. In particular, the challenges may arise in dealing with the more than one functional groups which having similar reactivity and affection toward a particular reagents.

In that case to achieve different types of selectivity such as chemoselectivity, regioselectility, stereoselectivity protection-deprotection, or blocking strategy is used. For example, in the reaction of Grignard reagent or organolithium reagent with 4-formylbenzonitrile,there are two functional groups aldehyde and CN and both could react with Grignard reagent so in order to make this reagent to react only with nitrile (CN), masking/protection of aldehyde group is required which involves two additional steps, i.e., protection and deprotection (Scheme 24). Similarly for the synthesis of 4-bromoaniline, two additional steps are required (Scheme 25) which could be avoided by adopting different synthetic methods where these additional steps could be avoided or minimized.

Similarly in order to achieve reaction at less reactive center in presence of more reactive one, generally reactive functional group is protected, and later at the end of the reaction, it is deprotected, for example, for the formation of alcohol (**38**) which is could not be achieved directly through the reaction of ethyl acetoacetate with phenyl magnesium bromide as it leads to the formation of hydroxy ester (**35**).

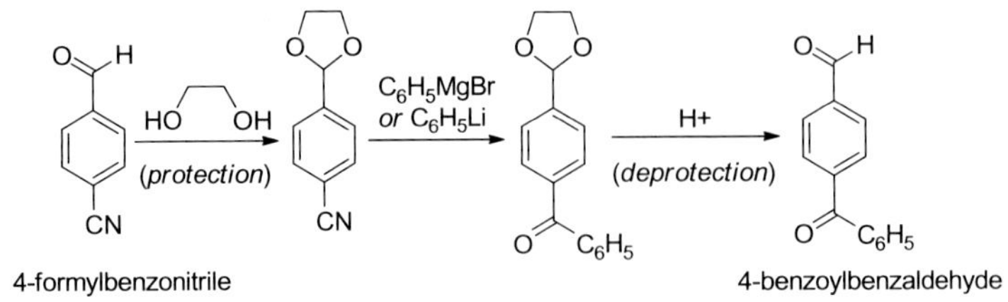

Scheme 24 Preparation of 4-benzoylbenzaldehyde

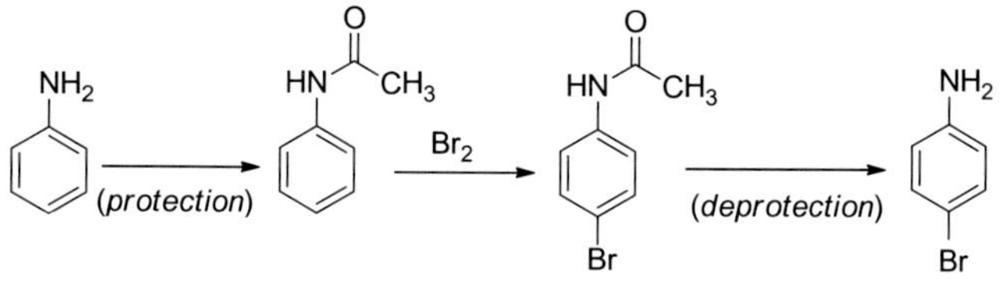

2 Principles of Green Chemistry 25

Scheme 25 Preparation of 4-bromoaniline

Therefore, it requires protection of ketone group with ethylene glycol to form ketal (**36**) (Scheme 26). This ketal (**36**) now reacts with the phenyl magnesium bromide to produce alcohol (**37**) which on deprotection of ketal ring delivers desired alcohol (**38**). Overall; two additional steps of protection and deprotection are involved.

It is important to highlight that this overall process leads to the generation of lots of waste materials in the form of by-product, solvents, reagents etc. as additional auxillary substances will be required for protection and deprotection step. In addition to that, additional consumption of energy, time, etc. will be required in the overall process. Several such protection and de-protection steps could be involved in case of the synthesis of a complex compound. In order to achieve such direct transformation, different catalytic methods have been developed which avoids these additional steps. In an excellent example of the synthesis of a complex natural product without involving protecting group, Baran et al. demonstrated the synthesis of marine products ambiguine H, hapalindole U, welwitindolinone A, and fischerindole I without using any protecting groups in the synthetic pathway involving seven to ten steps [29].

Principle 9: "*Catalytic reagents (as selective as possible) are superior to stoichiometric reagents.*" [2]

This principle highlights the advantages of catalytic methods over the stoichiometric method due to very explicable reason. As under catalytic methods unlike stoichiometric method, it requires substoichiometric amount of reagents which in general performs better job under mild condition or reduced reaction time with high

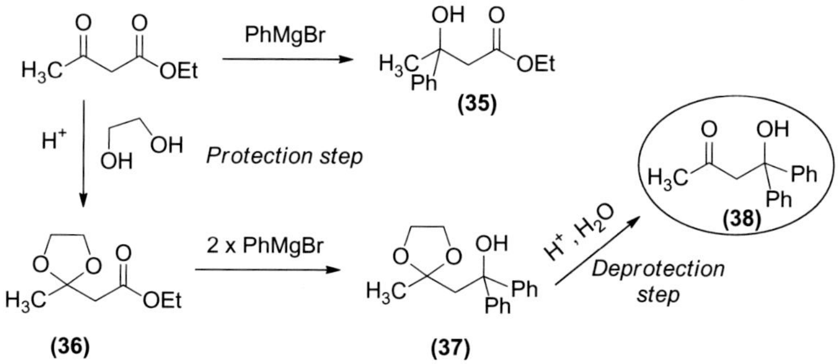

Scheme 26 Preparation of hydroxyl ketone involving protection and deprotection steps

Scheme 27 Preparation of sulfuric acid using two different catalytic methods

$$2SO_{2\,(g)} + O_{2\,(g)} \xrightleftharpoons{catalyst} 2SO_{3\,(g)}$$

$$SO_{3\,(g)} + H_2O_{(l)} \longrightarrow H_2SO_{4\,(l)}$$

selectivity. It overall reduces the amount of waste, energy requirement as well as enhances the atom economy.

Catalysis term was initially introduced by Berzelius in 1836. In general, compared to the non-catalyzed reaction, catalyst accelerates the rate of reaction as well as also enhances the selectivity. Various different types of catalysis have been developed. In general, catalyst could be divided into two categories including homogeneous and heterogeneous catalyst. In case of homogeneous catalysis, the catalysts are in the same phase with respect to the reaction medium. For example, sulfuric acid could be produced through the two different methods. In lead chamber process which is an example of homogeneous catalysis, both the reacting species $SO_{2\,(g)}$, $O_{2\,(g)}$ and catalyst NO (g) are in the same phase. In case of contact process which is an example of heterogeneous catalysis, the catalyst vanadium pentoxide (V_2O_5) is a solid (Scheme 27) [30].

Homogeneous catalysts may include different types of acids, metal complexes, enzymes, etc. Similarly, various metals complexes, metal-encapsulated covalent organic frameworks (COFs), and metal–organic frameworks are few examples of heterogeneous catalyst. Another catalytic approach using organic compounds like proline, quinidine, etc. as catalyst has attracted the attention which are generally classified as *organocatalysis* [31]. In year 2021, German scientist Benjamin List and Scotland-born scientist David W.C. MacMillan shared the Nobel Prize in Chemistry for their immense contribution in the area of asymmetric organocatalysis. Similarly, various organic transformations have been achieved using enzymes received from various biological sources as catalyst; it is termed as *Biocatalysis* [32]. Different types of catalysis and reactions explored using them have been discussed elaborately in another chapter.

Principle 10: *Chemical products should be so designed that at the end of their function, they do not persist in the environment and break down into innocuous degradation products.* [2]

This principle highlights very important aspects about the fate of a chemical after its use. It is very obvious that there is a purpose behind any chemical synthesis. This principle highlights the importance that during the designing which is very early stage of a chemical synthesis, one should not only focus on the application part but at the same time should also think about the behavior of these chemicals once these will reach to the environment after their use, i.e., complete life cycle assessment (LCA) is required. Because the impact of these chemicals or their by-products is very important since it directly or indirectly affect environment and ultimately life existing on the earth. For example, most commonly used artificial polymers are plastics, and we are aware about their frequent use. These polymers include

2 Principles of Green Chemistry

polyethylene either high-density polyethylene (HDPE) and low-density polyethylene (LDPE). Likewise polyvinyl chloride (PVC), polystyrene (PS) are some of the important polymers which gets synthesized on very large scale and are important part of our daily life. Although as stated earlier, these are now important and in some case unavoidable part of our life but after their disposal to the environment, most of them is non-biodegradable or non-degradable or their degradation rate is extremely slows. These chemicals are termed as *"persistent chemicals."* Because of that, these get accumulated under the soil or create land, water pollution which affects human as well as animal's life. Apart from that, these also affect productivity of crops, and in these days, these have become a matter of serious concern. For example, diclophenac which is a non-steroidal antiinflammatory drug (NSAID) used as antiinflammatory, pain killer, etc. was found to be the main cause behind the drastic reduction in the population of oriental white-backed vulture (OWBV) which was a very commonly occurred raptors. These vultures mainly died of renal failure mainly due to accumulation of diclofenac residues in the kidney [33] (Fig. 8).

It raised the serious concern and encouraged scientific community to design biodegradable materials using naturally occurring biopolymers like starch, cellulose, etc. As these biodegradable polymers would easily get decomposed or degraded by the action of microbes which includes bacteria, fungi, etc. through their enzymatic action, even simple hydrolysis process in the environment would convert these polymers into their monomers. In addition to the use of amylose and cellulose as biodegradable polymers, various other have also been developed) [34] (Fig. 9).

Fig. 8 Structure of diclofenac

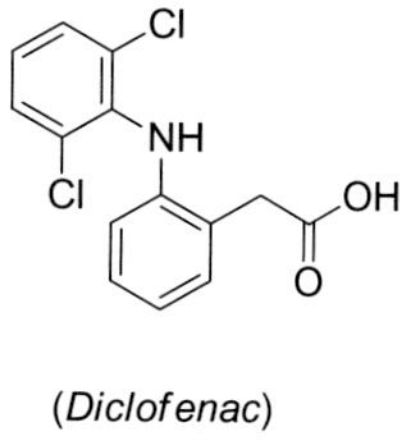

(Diclofenac)

Fig. 9 Examples of biodegradable polymers

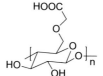

Polyhydroxybutyrate

Polylactic acid

Polycaprolactone

Carboxymethylcellulose (CMC)

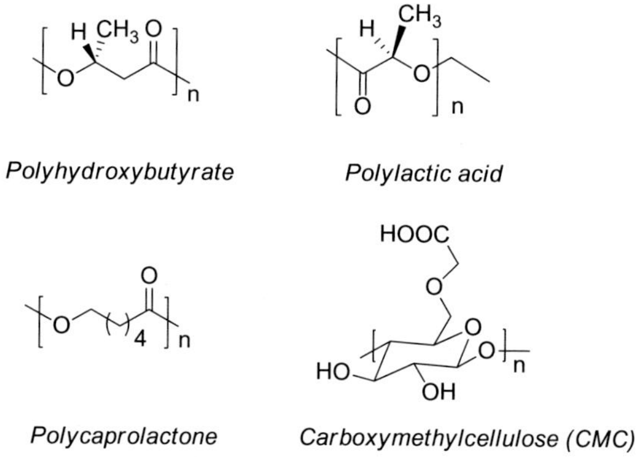

Principle 11: *"Analytical methodologies need to be further developed to allow for real-time, in-process monitoring and control prior to the formation of hazardous substances."* [2]

This principle mainly underlines the importance of real-time in-process monitoring of a reaction. According to this principle, the formation of hazardous substances could be controlled or their formation could be predicted, and accordingly, precautionary steps could be taken to avoid any untoward incidents by proper monitoring of the reaction using available analytical tools. It highlights the use of advanced analytical tools like infrared (IR), UV–VIS, gas chromatography (GC), gas chromatographic-mass spectrometry (GC–MS), high performance liquid chromatography (HPLC), nuclear magnetic resonance (NMR), etc. for better understanding of reaction mechanism and at the same time for the characterization of all possible intermediates, by-products formed during the reaction. Such real-time analysis would help in optimizing the amount of the reactants or reagents which sometimes gets used in excess amount. This real-time in-process monitoring of the reaction could be used to develop an efficient synthetic route with better desired product yield, atom efficiency, and less E-factor. Various different types of monitoring techniques such as in-line (continuous sampling of all material) or on-line (sampling of representative aliquots) are available for the proper monitoring of a reaction.

Principle 12: *"Substances and the form of a substance used in chemical processes should be chosen to minimize the potential for chemical accidents, including releases, explosions, and fire."* [2].

It is considered to be one of the most important principles of the green chemistry which discuss about the selection of chemicals while designing a synthetic methodology or process. It highlights very vital point that the selection of chemicals which could be used as reactant, reagents and solvents or for any other purposes, should be done in such a way that would minimize the risk factor involved in storage, handling and disposal at the end of their use. It is mainly related to the safety of the workers, society, and environment. A very famous quote of Trevor Kletz who laid the concept of inherently safer process (ISP) says *"What you don't have can't leak"* [35]. It clearly means that while selecting the chemicals if we have not selected a chemical or any form of chemical (solid, gas, liquid) which is hazardous or highly toxic, then risk of probable accidents taking place due to these chemicals would be very low. For example, if for bromination step, *N*-bromosuccinimide is being used instead of bromine, then definitely there would not be risk associated with the leakage or handling of bromine which is liquid and which may cause serious health problem. Similarly, for cynation instead of using HCN gas which is highly poisonous, other alternative such as trimethylsilylcynanide could be used to avoid the risk associated with the leakage of these highly toxic chemicals. In the history of the chemicals industries, various accidents have been reported which occurred mainly due to leakage of toxic chemical, explosion, or fire. Some of them have been discussed below.

2 Principles of Green Chemistry

Scheme 28 Preparation of Carbaryl using MIC

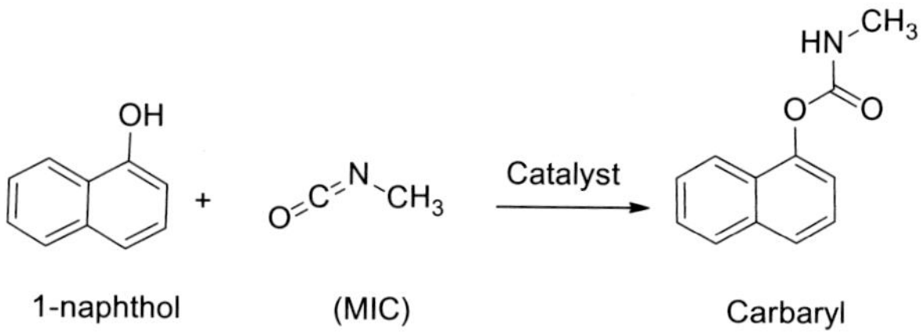

1-naphthol   (MIC)   Catalyst   Carbaryl

(a) Bhopal Gas Tragedy

Bhopal gas tragedy is one of the most disastrous accidents in the history that occurred in the Bhopal, Madhya Pradesh, India, at the chemical plant of Union Carbide India Limited (UCIL) which was jointly owned by union carbide corporation (UCC) and UCIL. It occurred due to the leakage of highly hazardous methyl isocyanate (MIC) gas which actually began in the night at around 11:00 PM on December 2, 1984, and later at around 01:00 AM on December 3, the major blast occurred due to the failure of safety valve which led to the leakage of huge amount (*approx.* 40 tons) of the MIC into the air [36]. MIC is an intermediate which is required for the further reaction with α-naphthol to produce carbaryl (naphthalen-1-yl methylcarbamate or 1-Naphthyl-N-methylcarbamate) which is a very popular pesticide sold under the trade name *Sevin* (Scheme 28).

This accident led to the death of more than 3000–10,000 people and continued affecting the lives of human as well as animal in nearby area. Several premature deaths and other health related complications are still being reported. In fact, several reports are frequently coming which suggests that despite of gap of more than 37 years, the adverse effect of this accident is still being observed.

(b) Flixborough accident

Similarly, another example of accident that occurred due to the massive explosion at one of the Nypro (UK) plant in Flixborough, UK, on June 1, 1974, led to the death of 28 workers and 36 were injured on site and several others injury were reported from nearby places [37]. Flixborough process is used for the production of the mixture of cyclohexanol and cyclohexanone through the partial oxidation of cyclohexane (Scheme 29). Cyclohaxnone is one of the main feedstocks in the synthesis of caprolactum which is an intermediate required for the preparation of

Scheme 29 Oxidation of cyclohexane to the mixture of cyclohexanol and ketone

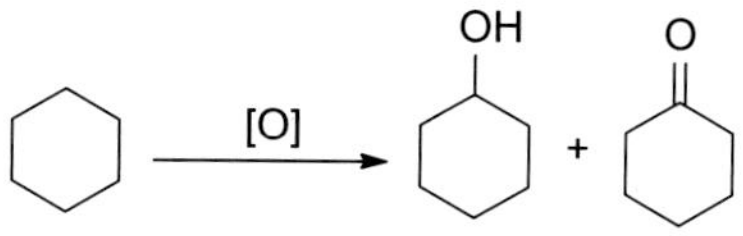

[O]

Nylon 6 [38]. The leakage in the pipe circulating cyclohexane was considered as major cause of the accident.

Various guidelines have been suggested by Anastas and Hammond to minimize or avoid the risks such as avoids using toxic or hazardous substance while designing the chemical process [39]. If at all a chemical which is required is toxic could be synthesized *on-demand* rather than transporting it or storing it for a longer period which would attract accidents. Apart from that, this principle encourages to use computational methods to understand the stability, toxicology, or probable hazard which may occur due to a new chemical.

3   Principle of Inherently Safer Design (ISD)

In order to avoid the potential hazardous effect of a chemical or process, the concept of inherently safer design was proposed which have been discussed below. The concept of inherently safer design [39] highlights about the permanent elimination or reduction of the hazard in order to make the process safer and to avoid and reduce the risk factor associated with the hazard. Inherent could be defined as permanent or undividable characteristic of something. For example, Na catches fire when it comes in direct contact with water. Similarly, HCN is highly poisonous these are their inherent nature. This principle emphasizes on the fact that instead of controlling the hazard, it is better to avoid using such chemicals which have inherent nature of being hazardous or under certain set of conditions which could have disastrous consequences. This principle underlines the complete elimination of hazardous substance, i.e., a substance which has potential to cause harm to the life, properties, and environment like HCN, nitroglycerine, NaCN, etc. or replace these chemicals with safe or less hazardous alternatives. This approach will help in decreasing the chances of probable accidents which could occur due to various reasons including human as well as instrumental errors. Basically, it is a philosophy which is applicable to all the aspects such as manufacture, storage, use, and disposal of a chemical or process. According to Trevor Kletz, *"what you don't have can't leak."* This phrase encompasses very important points that if we do not use any hazardous substance, then definitely we do not need to worry about their disastrous effect or accidents which may occur. For example, for bromination reaction instead of using bromine, *N*-bromosuccinimide (NBS) could be used which will decrease the risk associated with the leakage or spilling of bromine as NBS is solid. Similarly, instead of directly using HCN gas for cynation, NaCN or other cynating agent which is solid could be used. It will definitely avoid the risk associated with the leakage of highly poisonous gas. Similarly, organic solvents which are highly flammable and some are carcinogenic could be replaced with water or other green alternative such as ionic liquid, PEG.

Four different approaches [40] have been suggested to meet the requirement of ISD which are as follows:

Inherent: It simply talks about the complete elimination of hazardous material or process conditions. For example, replace the flammable and harmful organic solvents especially volatile organic compounds (VOC) with water. A synthetic method or process involving very high temperature could be replaced with the improved process which would occur at room temperature. Some of the reagents such as $LiAlH_4$ or *tert*-butyllithium catch fire in presence of moisture or water. Therefore, method could be developed where use of such chemicals could be replaced with safe reagents.

Passive: This approach emphasizes on the design of process or equipment which will reduce the hazard without using any highly sensitive controlling devices. For example, performing a chemical reaction in robust pressure vessel which is capable of bearing the more pressure such as 10 bar than that is expected from the reaction which is 5 bar. So, in that case, there is no need of installing any sensor which will be monitoring the pressure change. Similarly, performing a reaction of 10 kg in a reaction vessel which is capable in handling 20 kg of reaction mixture would minimize the potential risk. Therefore, this approach is considered as *passive*.

Active: Unlike passive approach, it emphasizes on the use of safety instrumented system (SIS) or control system which will ensure the efficient use of a chemical or process safely. For example, sensors could be installed to monitor the leakage of a gas, to monitor pressure or temperature change. Installation of efficient fire protection system that will allow the use of flammable substance, also.

Procedural: It highlights about the use of standard operating procedure (SOP) in conducting any reaction or in developing a process. For example, there are certain sets of rules for storage, use, and disposal of a hazardous substance. So, by following those accidents could be avoided. Similarly, by having proper emergency plan to handle any untoward accidents or incidents any loss of life, property or environment could be avoided or minimized. Apart from that by providing proper training to the employee or worker dealing with such hazardous substance or reaction conditions, its disastrous effect could be avoided or minimized. For example, if there is sudden rise in the temperature of a reactor, then in that condition, an operator who is monitoring that reaction should be trained enough to apply emergency cooling to avoid any major accident.

4 Various Barriers in the Implementation of Green Chemistry

The implementations of the various principles formulated for the green chemistry and discussed earlier are essential to achieve the goal of sustainable development and also to maintain a healthy and safe environment but certain barriers [41] and issues have been identified which have to be addressed to achieve the target. For example, Matus et al. performed one study in USA that was mainly based on the interviews of the leaders in the area of green chemistry from various sectors like industry, NGO,

32   1 Green Chemistry: Introduction to the Basic Principles

academia, etc. and identified six major barriers that need to be addressed for the successful implementation of green chemistry [42].

(a) **Economic and Financial Barriers**

In general, the aim of any business is to make profit; otherwise, it would be tough to run the same. The same is also applicable in chemical industry as well. During the development and production of a new product or existing product as per the principles of green chemistry, the present methods of production and formulation would require additional modification or in some cases replacement of existing set up could be a costlier process. Apart from the product, the nature of waste material may also change which need different ways to treat. In addition to the cost, shuttering the previous production plant and establishing new one would also require time as well as extra efforts; therefore, in general, chemical enterprises are reluctant to implement these green chemistry principles until unless the new process is highly profitable compare to the existing one. For example, Pfizer received Presidential Green Chemistry in 2002 for developing a greener method for the synthesis of sertraline (Zoloft) (Scheme 1) which is among one of the most prescribed drug in the USA. This greener method involving one-pot transformation and replacing four different solvents such as DCM, THF, toluene, and hexane used in different steps of the synthesis by single and environmentally benign greener alternative ethanol allowed the production to be doubled using new method. In addition to that, the amount of raw material used also reduced upto 60%. More importantly, the amount of hazardous waste material including titanium dioxide also got reduced by approximately 1.8 million pound. Therefore, in such case where using newer method allowed led to double the yield, drastically reduced generation of hazardous waste material, and replaced mixture of solvents with single solvent, implementation of greener method would be welcomed; otherwise, it would not be easy to replace the existing method.

(b) **Regulatory Barriers**

It is important to mention that every industry has to follow certain rules and regulations devised by different national and international agencies. For example, a pharmaceutical industry in USA has to receive approval from US Food and Drug Administration (FDA) if they wish to make any change in the existing method of production of the product which is already in the market. This kind of recertification process is generally time as well as money consuming process, and therefore, despite of having sound understanding of green chemistry, the implementation of green chemistry principle gets obstructed. Similarly, if some industry wishes to adopt a greener way for the production of a pesticide replacing the existing set up or replacing carcinogens with greener alternative, it has to go for the registration with Environmental Protection Agency (EPA) and receive approval from Federal Insecticide, Fungicide, and Rodenticide Act (FIFRA) and Toxic Substances Control Act of 1976 (TSCA), respectively, which is again not very smooth process. Therefore without receiving any incentive, companies are reluctant to implement greener methods.

4 Various Barriers in the Implementation of Green Chemistry

(c) **Technical Barriers**

Apart from the other aspects in the implementation of green chemical approach, one important issue is the non-availability of greener alternative for all types of chemical transformations. For example, performing an organometallic reaction using sensitive catalyst and organometallic reagent generally requires dry organic solvents like NMP, DMF, DMA, toluene, etc. Other important point is the readily availability of the information related to developed greener alternative as some research groups in industry hesitates to share it due to certain competitive reasons. Therefore, lack of comprehensive database and references related to the developed greener alternatives hinders their ready use. Another important barrier is the lack of expertise in diverse disciplines which is essential for the production of a new material or technology as per the need of green chemistry principles. For example, an organic chemist who is involved in the synthesis of a new compound of material or medicinal importance may not have knowledge about the toxic effect of the product on the ecosystem or environment. Similarly, a biologist involved in drug discovery may not be having enough understanding about the intricacies involved in the development of a medicinally important compound as per the need of green chemistry. For example, BASF Company while developing biobased polymer observed those scientists who were involved in this work had insufficient expertise in handling of the feedstock. In addition to that, the lack of complete information including the toxicology, hazardous nature about several compounds is also another barrier in the development of environmentally benign methods.

(d) **Organizational Barriers**

Several times, the different divisions of an organization may not have similar approach or views about adopting of greener approach; in that case, one part of the organization could be reluctant in accepting that newly developed method and may stick to the older one. In addition to that, sometimes the scientists who are involved in the development of greener methods or process may not receive proper support from the higher authority in the organization so in that case again, it would not be easy to implement the principles of green chemistry. So the differences in the views about the implementation of principles of green chemistry at the different levels as well as in the different part of the organization have also been realized as one of the major barrier.

(e) **Cultural Barriers**

Awareness and willingness are two important driving forces behind the implementation of an approach which may require rectification of the existing process. In general, the lack of awareness about the impact of adopting different principles of green chemistry in achieving the goal of sustainable development and protecting the environment as well as human lives is one of the major barriers in this direction. In general in some of the academic institutions, green chemistry itself is not a part of curriculum. Sometimes despite of having a course on green chemistry, its impact on transforming an existing method and developing an environmentally benign one requires proper discussion to make the audience realize about the benefits of this concept. The same is true for industry also where some of the scientist may be having good understanding of this subject, but their collaborators or other groups may not be aware of its business potential so implementation becomes difficult. Sometimes, some having misconception that these are simply a part of environmental agenda which may not be profitable on the ground or practically may not be very useful also discourage the scientists to implement it. In addition to that, the difference in the research interest between research groups from academics and industry is also one main obstacle in this direction as both are having different goals. For example, an environmentally method developed in some academic institutions at milligram scale may not be readily applicable and profitable at the industrial level. Sometimes, the target of a research group in academics may be a high impact journal which may not be applicable in industry so these differences in the research interest and their own target also work as a barrier.

(f) **Barriers from Disagreements Regarding Definitions and Metrics**

Similarly, the disagreement in the understanding and definitions of different parameters and metrics to defining a process greener is also considered an obstacle in adopting and implementation of green chemistry approach. For example, the developed method may not be following all the 12 principles of the green chemistry so in that case, there could be different views in considering it as greener method. Unlike different branches of the chemistry such as physical, inorganic, and organic wherein some clear demarcation exist, in that sense it is difficult to find separate boundaries in green chemistry finds as it requires multidisciplinary knowledge and understanding.

References

1. Anastas, P.T., Warner, J.C.: Green chemistry: theory and practice. Green Chem. Theory Pract. Oxford Univ. Press, New York (1998)
2. Anastas, P., Eghbali, N.: Green chemistry: principles and practice. Chem. Soc. Rev. **39**(1), 301–312 (2010)
3. Sheldon, R.A.: The E factor 25 years on: the rise of green chemistry and sustainability. Green Chem. **19**(1), 18–43 (2017)
4. Sheldon, R.A.: The E factor: fifteen years on. Green Chem. **9**(12), 1273–1283 (2007)
5. Dunn, P.J., Galvin, S., Hettenbach, K.: The development of an environmentally benign synthesis of sildenafil citrate (Viagra^{TM}) and its assessment by green chemistry metrics. Green Chem. **6**(1), 43–48 (2004)
6. Rouhi, A.M.: Fine chemicals. Chem. Eng. News Arch. **80**(29), 45–62 (2002)
7. Bruggink, A., Roos, E.C., de Vroom, E.: Penicillin acylase in the industrial production of β-Lactam antibiotics. Org. Process Res. Dev. **2**(2), 128–133 (1998)
8. Cho, H., Madden, R., Nisanci, B., Török, B.: The Paal–Knorr reaction revisited. A catalyst and solvent-free synthesis of underivatized and N-Substituted Pyrroles. Green Chem. **17**(2), 1088–1099 (2015)
9. Rao, M.L.N., Awasthi, D.K., Banerjee, D.: Microwave-mediated solvent free rap-stoermer reaction for efficient synthesis of benzofurans. Tetrahedron Lett. **48**(3), 431–434 (2007)

References

10. Li, C.-J., Trost, B.M.: Green chemistry for chemical synthesis. Proc. Natl. Acad. Sci. **105**(36), 13197–13202 (2008)
11. Nicolaou, K.C., Snyder, S.A., Montagnon, T., Vassilikogiannakis, G.: The Diels-Alder reaction in total synthesis. Angewandte Chemie - International Edition **41**, 1668–1698 (2002)
12. Agee, B.M., Mullins, G., Swartling, D.J.: Progress towards a more sustainable synthetic pathway to ibuprofen through the use of solar heating. Sustain. Chem. Process. **4**(1), 1–9 (2016)
13. Ott, D., Kralisch, D., Denčić, I., Hessel, V., Laribi, Y., Perrichon, P.D., Berguerand, C., Kiwi-Minsker, L., Loeb, P.: Life cycle analysis within pharmaceutical process optimization and intensification: case study of active pharmaceutical ingredient production. Chemsuschem **7**(12), 3521–3533 (2014)
14. Baral, E.R., Lee, J.H., Kim, J.G.: Diphenyl carbonate: a highly reactive and green carbonyl source for the synthesis of cyclic carbonates. J. Org. Chem. **83**(19), 11768–11776 (2018)
15. Memoli, S., Selva, M., Tundo, P.: Dimethylcarbonate for eco-friendly methylation reactions. Chemosphere **43**(1), 115–121 (2001)
16. Yadav, M.S., Singh, A.S., Agrahari, A.K., Mishra, N., Tiwari, V.K.: Silicon industry waste polymethylhydrosiloxane-mediated benzotriazole ring cleavage: a practical and green synthesis of diverse benzothiazoles. ACS Omega **4**(4), 6681–6689 (2019)
17. Ali, N.M., McKillop, A., Mitchell, M.B., Rebelo, R.A., Wallbank, P.J.: Palladium-catalysed cross-coupling reactions of arylboronic acids with π-deficient heteroaryl chlorides. Tetrahedron **48**(37), 8117–8126 (1992)
18. Yanagisawa, S., Ueda, K., Taniguchi, T., Itami, K.: Potassium T-butoxide alone can promote the biaryl coupling of electron-deficient nitrogen heterocycles and haloarenes. Org. Lett. **10**(20), 4673–4676 (2008)
19. Ando, N., Fukazawa, A., Kushida, T., Shiota, Y., Itoyama, S., Yoshizawa, K., Matsui, Y., Kuramoto, Y., Ikeda, H., Yamaguchi, S.: Photochemical intramolecular C−H addition of dimesityl (Hetero) arylboranes through a [1, 6]-sigmatropic rearrangement. Angew. Chemie Int. Ed. **56**(40), 12210–12214 (2017)
20. Liang, K., Li, N., Zhang, Y., Li, T., Xia, C.: Transition-metal-free α-arylation of oxindoles: vi a visible-light-promoted electron transfer. Chem. Sci. **10**(10), 3049–3053 (2019)
21. Daina, A., Michielin, O., Zoete, V.: SwissADME: a free web tool to evaluate pharmacokinetics, drug-likeness and medicinal chemistry friendliness of small molecules. Sci. Rep. **7** (2017)
22. Kim, J.H., Scialli, A.R.: Thalidomide: the tragedy of birth defects and the effective treatment of disease. Toxicol. Sci. **122**(1), 1–6 (2011)
23. Ansari, A.J., Sharma, S., Pathare, R.S., Gopal, K., Sawant, D.M., Pardasani, R.T.: Solvent-free multicomponent synthesis of biologically-active fused–imidazo heterocycles catalyzed by reusable Yb (OTf) 3 under microwave irradiation. ChemistrySelect **1**(5), 1016–1021 (2016)
24. Fiorito, S., Epifano, F., Taddeo, V.A., Genovese, S.: Ytterbium triflate promoted coupling of phenols and propiolic acids: synthesis of coumarins. Tetrahedron Lett. **57**(26), 2939–2942 (2016)
25. Li, C.-J., Zhang, W.-C.: Unexpected Barbier−Grignard allylation of aldehydes with magnesium in water. J. Am. Chem. Soc. **120**(35), 9102–9103 (1998)
26. Cicco, L., Sblendorio, S., Mansueto, R., Perna, F.M., Salomone, A., Florio, S., Capriati, V.: Water opens the door to organolithiums and grignard reagents: exploring and comparing the reactivity of highly polar organometallic compounds in unconventional reaction media towards the synthesis of tetrahydrofurans. Chem. Sci. **7**(2), 1192–1199 (2016)
27. More, S.R., Yadav, G.D.: Effect of supercritical CO2 as reaction medium for selective hydrogenation of acetophenone to 1-phenylethanol. ACS Omega **3**(6), 7124–7132 (2018)
28. Kühlborn, J., Groß, J., Opatz, T.: Making natural products from renewable feedstocks: back to the roots? Nat. Prod. Rep. **37**(3), 380–424 (2020)
29. Baran, P.S., Maimone, T.J., Richter, J.M.: Total synthesis of marine natural products without using protecting groups. Nature **446**(7134), 404–408 (2007)
30. Du, X., et al.: Oxidation of Sulfur Dioxide over V2O5/TiO2 Catalyst with Low Vanadium Loading: A Theoretical Study. J Phys Chem C **122**, 4517–4523 (2018)

31. Krištofíková, D., Modrocká, V., Mečiarová, M., Šebesta, R.: Green asymmetric organocatalysis. Chemsuschem **13**(11), 2828–2858 (2020)
32. Wu, S., Snajdrova, R., Moore, J.C., Baldenius, K., Bornscheuer, U.T.: Biocatalysis: enzymatic synthesis for industrial applications. Angew. Chemie—Int. Ed. **60**(1), 88–119 (2021)
33. Oaks, J.L., Gilbert, M., Virani, M.Z., Watson, R.T., Meteyer, C.U., Rideout, B.A., Shivaprasad, H.L., Ahmed, S., Chaudhry, M.J.I., Arshad, M.: Diclofenac residues as the cause of vulture population decline in Pakistan. Nature **427**(6975), 630–633 (2004)
34. Gross, R.A., Kalra, B.: Biodegradable polymers for the environment. Science (80) **297**(5582), 803–807 (2002)
35. Vaughen, B.K., Klein, J.A.: What you don't manage will leak: a tribute to Trevor Kletz. Process Saf. Environ. Prot. **90**(5), 411–418 (2012)
36. Adams, J.P., Brown, M.J.B., Diaz-Rodriguez, A., Lloyd, R.C., Roiban, G.: Biocatalysis: a pharma perspective. Adv. Synth. Catal. **361**(11), 2421–2432 (2019)
37. Okoh, P., Haugen, S.: Maintenance-related major accidents: classification of causes and case study. J. Loss Prev. Process Ind. **26**(6), 1060–1070 (2013)
38. Zong, B., Sun, B., Cheng, S., Mu, X., Yang, K., Zhao, J., Zhang, X., Wu, W.: Green production technology of the monomer of Nylon-6: caprolactam. Engineering **3**(3), 379–384 (2017)
39. Anastas, P.T., Hammond, D.G.: Inherent safety at chemical sites. Elsevier (2016)
40. Hendershot, D.C.: Inherently safer design: an overview of key elements. Prof. Saf. **56**(02), 48–55 (2011)
41. Poliakoff, M., Fitzpatrick, J.M., Farren, T.R., Anastas, P.T.: Green chemistry: science and politics of change. Science (80) **297**(5582), 807–810 (2002)
42. Matus, K.J.M., Clark, W.C., Anastas, P.T., Zimmerman, J.B.: Barriers to the Implementation of green chemistry in the United States. Environ. Sci. Technol. **46**(20), 10892–10899 (2012)

Chapter 2
Energy-Efficient Process in Organic Synthesis

1 Introduction

In the view of achieving the goal of the sustainable growth, the development of the energy-efficient processes and technologies is one of the most sought-after areas of research. Especially in the present scenario where the sources of non-renewable energy such as coal, petroleum, etc. are depleting fast, there is need of the judicious use of the energy. Therefore, energy has been considered as one of the parameters to evaluate the greenness of a method and it has been included the widely accepted twelve principles of green chemistry established by the Paul Anastas. It states that *"Energy requirements of chemical processes should be recognized for their environmental and economic impacts and should be minimized. If possible, synthetic methods should be conducted at ambient temperature and pressure"* [1].

It is important to highlight that apart from the environmental impact, the imprudent use of the energy has great economic impact also. Although the reaction developed at room temperature would ideally be considered as greener one but most of the reactions generally requires additional energy for the chemical activation. Conventionally, energy supplied through the heating the reaction mixture using oil bath is one of the most widely used method but due to the shortcomings such as wastage of heat energy, long reaction time, and in some cases formation of additional by-products, etc. encouraged the scientific community to explore various other greener and energy efficient modes of chemical activation [2].

Therefore, a paradigm shift has been observed in the area of organic synthesis where the focus has been diverted from achieving the best yield to the development of greener and energy efficient methods. In order to ensure the judicious consumption of energy in performing various chemical transformations various greener approaches for the chemical activation have been explored using *microwave, sound wave, light wave*, and *ball mill method,* etc. as energy efficient alternative to the widely employed conventional heating methods. This chapter will be highlighting the organic reactions developed mainly through four different energy-efficient approaches (Fig. 1) [3].

© The Author(s), under exclusive license to Springer Nature Singapore Pte Ltd. 2022
V. K. Tiwari et al., *Green Chemistry*,
https://doi.org/10.1007/978-981-19-2734-8_2

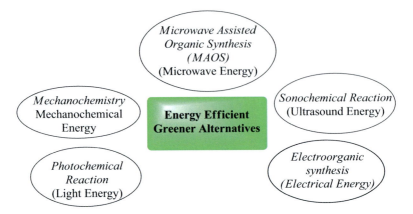

Fig. 1 Greener and energy-efficient approaches in chemical synthesis

2 Microwave-Assisted Organic Synthesis

The source of energy for a chemical reaction has kept on changing over the period of time. Bunsen burner invented by Robert Bunsen in 1855 remained one of the favorite sources of heat energy [2]. In fact, still in few undergraduate and postgraduate laboratories it could be seen frequently. Later heating mantle, hot plates and oil bath have been explored as safer alternative as these avoid the use of open flame which could be unsafe in chemistry laboratory in particular in presence of organic and flammable solvents and chemicals. Similarly, the vast application of microwave which is an electromagnetic wave of 1 mm–1 m wavelength and 0.3–300 GHz frequency, as heat source in diverse areas such as processing of food, diathermy, polymer industries, etc. encouraged scientists to employ it as greener alternative of conventional heating method to achieve different types of chemical reactions. In case of the conventional heating methods using oil bath or heating mantle, *conductance* is the prime mode of heat transfer from the source to the reaction vessels. In this case the direction of heat transfer is from the source which is placed outside to the reaction mixture present in the reaction vessel, the efficiency and the rate of heat transfer also depend on the nature of material of reaction vessel. Whereas the microwave irradiation directly interacts with the dipoles such as polar solvents as well with ions which generally happen to be the part of the reaction mixture, through two different mechanisms known as dipolar polarization and ionic conduction respectively and triggers the heating process. Therefore compared to the conventional heating method, reaction mixture attains temperature very fast in case of microwave-assisted reaction as the direction of flow of heat is from inside to outside so the wastage of thermal energy through conductance could effectively be minimized in microwave-assisted reaction process. However, it is important to mention here that all the solvents used to carry out chemical reactions are not equally efficient in absorbing the microwave and consequently the heating ability also vary by changing the solvents. Generally, the

2 Microwave-Assisted Organic Synthesis

Scheme 1
Microwave-assisted
hydrolysis of benzamide

Reaction conditions: (a) Thermal: 1 h, reflux; Yield: 90%
(b) Microwave (MW): 10 min.
sealed vessel; Yield: 99%

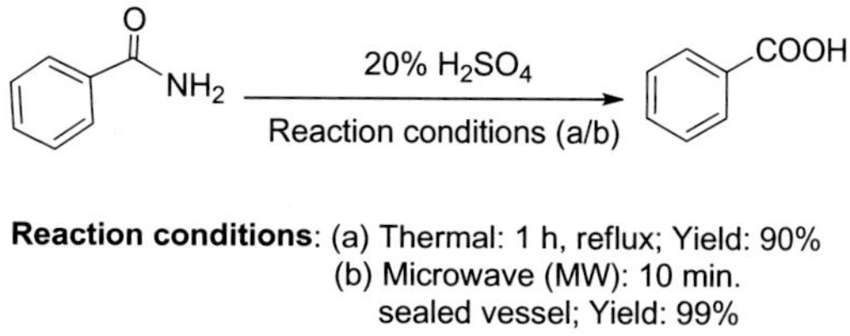

operational frequency of a microwave reactors used to carry out chemical reactions is around 2.45 Hz.

Although the application of microwave energy in the hydrogenation of alkene and hydrocracking of bitumen, etc. were already in practice, the application of microwave in the routinely used organic transformation only gained attention when Gedye et al. demonstrated its application in achieving some of the simpler reactions such as hydrolysis, oxidation, esterification, etc. For example, the hydrolysis of benzamide under microwave-assisted condition was achieved within 10 min with 99% yield compared to thermal condition which required 1 h and lesser yield of benzoic acid. It displayed the potential application of microwave in performing chemical reaction (Scheme 1) [4].

2.1 Microwave-Assisted Coupling Reactions

The applications of the microwave energy as greener alternative of conventional heating methods were demonstrated in achieving vast varieties of organic transformations such as cross-coupling reactions, multicomponent reactions (MCR), cyclization reactions, etc. [5, 6]. Among such reactions, carbon–carbon (C–C) as well as carbon-heteroatoms (C–X; X = O, N, P, S, etc.) bond-forming reactions in particular cross-coupling reactions have been one of the most productive and applied area of research in last few decades. Specifically, transition metal-catalyzed cross-coupling methods such has Suzuki, Negishi, Hiyama, Stille, etc. coupling reactions have been immensely applied as one of the key steps in the synthesis of natural products which are of medicinal and material interest [7]. Considering the vast synthetic application of metal-catalyzed cross-coupling reactions, Prof. Suzuki, Prof. Negishi, and Prof. Heck were awarded Nobel Prize in 2010 [8].

Various other methods for the generation of new C–C as well as C–X bonds including C–H activations, oxidative coupling, and reductive coupling along with the transition metal-free synthetic coupling methods have also been developed in past few years. Despite of broad substrate scope and applicability, most of these methods suffer from some of the serious drawbacks and use high temperature is one of them. Therefore, in order to minimize the energy loss through conventional heating methods, some of these coupling reactions were also performed under microwave irradiation. For example, Len and Herve demonstrated the application of microwave-assisted

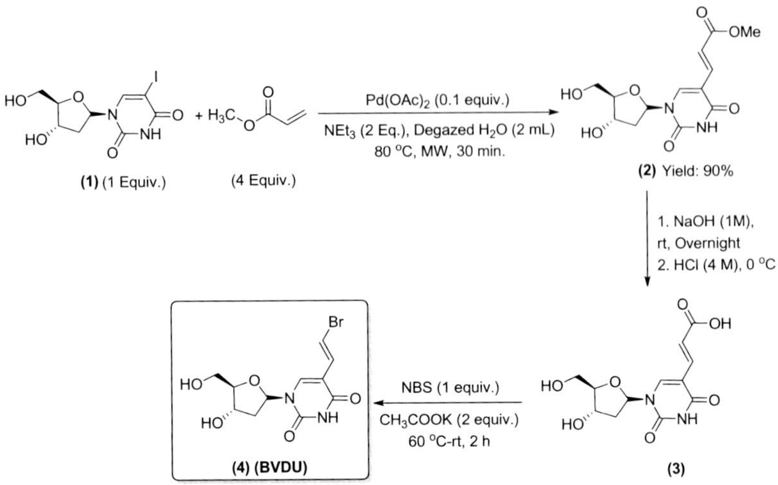

Scheme 2 Synthesis of BVDU (**4**) involving microwave-assisted Heck cross-coupling reaction of 5-iodo-2'-deoxyuridine and methyl acrylate

Heck coupling reaction using 5-iodo-2'-deoxyuridine **1** and different acrylates as coupling partner to obtain of a library of substituted nucleoside with antiviral properties [9]. For example, Heck coupling between iodouridine **1** with methyl acrylate furnished the coupled product **2** which were further converted into the acid derivative **3** and then finally into the target compound (*E*)-5-(2-bromovinyl)-2'-deoxyuridine **4** (BVDU, Zostex) which is a nucleoside based antiviral drug effective against herpes simplex virus type 1 (HSV-1) and varicella-zoster virus (VZV) (Scheme 2). The coupled product was obtained in 90% yield within 30 min using water as green media.

Similarly, Glasnov et al. demonstrated the efficacy of the microwave irradiation in achieving a the vinylation of 3-bromo-1-methyl-4-phenylquinolin-2(1*H*)-one **5** using ethyl acrylate following the Pd-catalyzed Heck reaction protocol to achieve the synthesis of bio-active 3-(quinolin-3-yl)acrylate **6** [10]. Reaction performed using the Pd(PPh$_3$)$_4$ (0.03 equiv.), Et$_3$N (3 equiv.) and DMF as a solvent at 150 °C for 45 min (Scheme 3).

Reaction was found to furnish 81% of the desired acrylate derivative along with minor amount of the debrominated quinolinone. Du and Yang have also developed a Heck coupling reaction protocol under solvent and ligand free reaction condition. Coupling reaction between aryl halides and substituted alkene performed under microwave radiation (300 W) afforded alkene derivatives in yield up to 98% within 25 min reaction period (Scheme 4) [11].

2 Microwave-Assisted Organic Synthesis

41

Scheme 3 Microwave-assisted Heck coupling of 3-bromo-1-methyl-4-phenylquinolin-2(1H)-one **5** and ethylacrylate

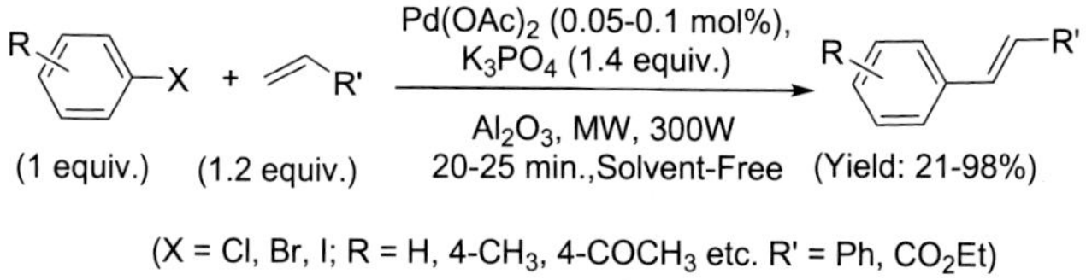

Scheme 4 An example of Microwave-assisted Heck coupling under solvent-free condition

2.2 Microwave-Assisted Heterocyclic Synthesis

Likewise, several other such Heck coupling protocols have been successfully developed under the microwave-assisted reaction condition [12]. The heterocycles especially *N*-heterocycles are very important class of biological actives compounds. A vast library of such heterocyclic compounds is known with diverse range of biological activities such as anticancerous, antibacterial, antiviral, anti-inflammatory, etc. [13]. In general, multicomponent approach is one of the primary ways to achieve the synthesis of these heterocycles [14]. The synthesis of these medicinally important heterocyclic scaffolds has also been explored under microwave irradiation. Few examples have been discussed here.

For example, Manjashetty and coworkers synthesized Isoniazid analogues under the microwave radiation using water as a solvent. Isoniazid which is chemically known as isonicotinohydrazide is an antibacterial drug used to treat tuberculosis. The synthesis of Isoniazid analog **7** was achieved under microwave-assisted reaction protocol consisting of the p-dodecylbenzene sulfonic acid (DBSA)-catalyzed multicomponent reaction using isoniazid, 5,5-dimethylcyclohexane-1,3-dione (dimedone), and functionalized benzaldehyde as reactants. Reaction afforded up to 96% yield of the isozonide analogues (Scheme 5) [15].

Mahindra and Jain developed a microwave-assisted method for the regioselective C-5arylation of protected L-histidine using aryl iodides and bromides as one of the coupling partners catalyzed under the Pd-catalyzed reaction protocol (Scheme 6) [16].

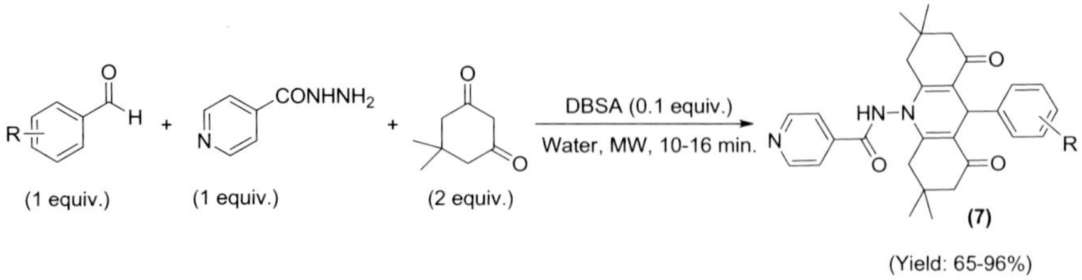

Scheme 5 Microwave-assisted synthesis of Isozonid analogues

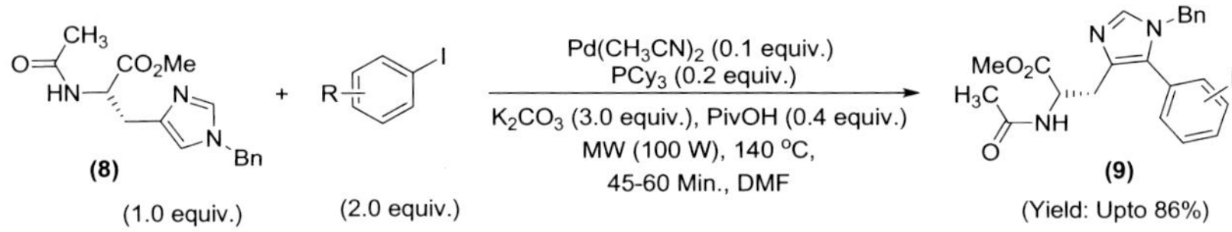

(Yield: 65–96%)

Scheme 6 Microwave-assisted regioselective C-5 arylation of protected L-histidine

(Yield: Upto 86%)

The C-2 arylation of the histidine **8** under the optimized reaction condition consisting of Pd $(CH_3CN)_2$ (0.1 equiv.), PCy_3 (0.2 equiv.), PivOH (0.4 equiv.), K_2CO_3 (3.0 equiv.) as base and DMF as a solvent provided the arylated product **9** in 86% yield. Reaction worked at 140 °C for 45–60 min under MW irradiation.

The same group also developed a microwave-assisted method for the regioselective C-2 arylation of protected L-histidine using aryl iodides and bromides as one of the coupling partners catalyzed by bimetallic system. The C-2 arylation of the histidine under the optimized reaction condition consisting of Pd $(OAc)_2$ (0.2 equiv.), CuI (2 equiv.), P(n-Bu)(1-adamantyl)$_2$ (0.4 equiv.), PivOH (0.4 equiv.), t-BuOK (3.0 equiv.) as base and NMP as a solvent provided the arylated product **10** in 36–78% yield. Reaction worked at 140 °C for 45 min under MW irradiation (Scheme 7) [17].

Likewise pyrazole derivatives are also known to be a part various drugs available in the market as well as of different bioactive compounds such as celecoxib, ruxolitinib, lonazolac, etc. [18]. Generally, the pyrazole synthesis is achieved following the popular Knorr pyrazole synthesis method that involves cyclocondensation reaction between 1,3-dicarbonyl compounds and hydrazine [19]. Similarly, the one-pot

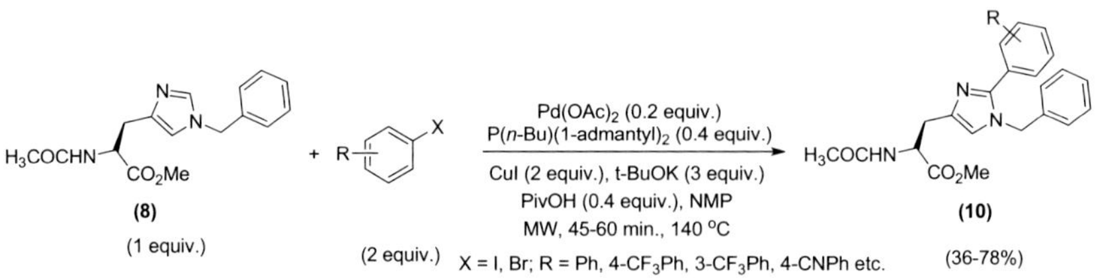

Scheme 7 Microwave-assisted regioselective C-2 arylation of protected histidine

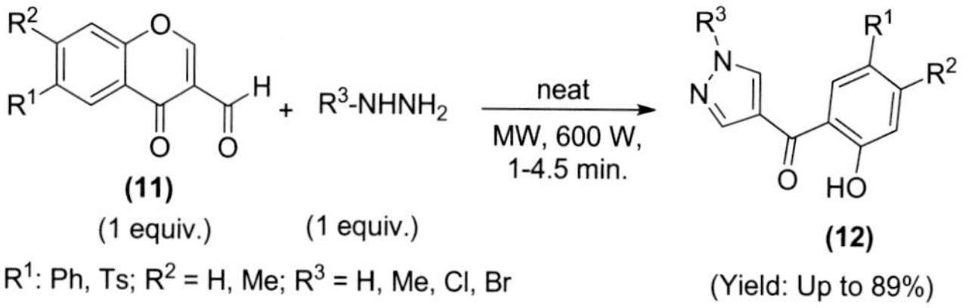

2 Microwave-Assisted Organic Synthesis

Scheme 8 Microwave-assisted solvent-free synthesis of substituted pyrazoles

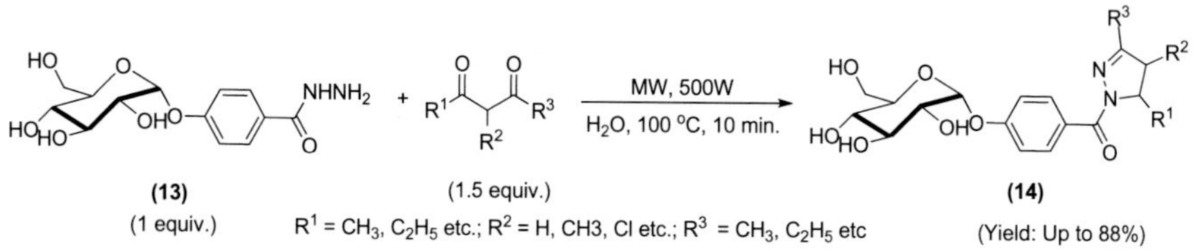

Scheme 9 Microwave-assisted synthesis of glucose-based pyrazole

method developed by Kumar and Rao provided an efficient method for the generation of 3,4-diarylpyrazoles [20]. Sabitha et al. developed a microwave-assisted reaction protocol for the synthesis of various pyrazole derivatives using aryl hydrazine and 3-formylchromone **11** as substrates. The reaction performed under microwave radiation and solvent-free condition led to the formation of several functionalized 4-(2-hydroxybenzoyl)pyrazoles **12** within very short period of time of 1–4.5 min and provided 89% yield of the product (Scheme 8) [21].

Du and coworker synthesized a sugar-based pyrazoles having anticancerous properties under the microwave-assisted reaction condition within the 10 min of reaction period. The reaction between glucose-based phenyl hydrazide **13** and substituted 1,3-dicarbonyl compounds under water radiation and water as a solvent led to the formation of sugar-based pyrazoles **14** (Scheme 9) [22].

Under this reaction condition, the alkyl substituted dicarbonyl compounds worked well and furnished good yield of pyrazoles; but no product was observed when the alkyl group present on dicarbonyl skeleton was replaced with the phenyl or electron withdrawing substituents such as trifluoromethyl group.

Similarly, fused heterocycles such as imidazo[1,2-*a*]pyridines (IPs) are also known to exhibit broad range of biological activity [23]. For example, drugs such as minodronic acids, zolpidem, alpidem, used to treat osteoporosis, anxiety, and insomnia, respectively, are examples of such fused heterocyclic compound with imidazo[1,2-*a*]pyridines as core scaffold [24].

Generally, the synthesis of imidazo[1,2-a]pyridines is achieved following a Lewis-acid-catalyzed multicomponent reaction protocol where reaction mixture consists of aldehyde, 2-aminopyrimidine, and isonitrile that undergo cyclocondensation reaction to furnish this fused heterocycle. This reaction is popularly known as Groebke–Bienayme´–Blackburn (G–B–B) reaction. Ansari et al. established a microwave-assisted

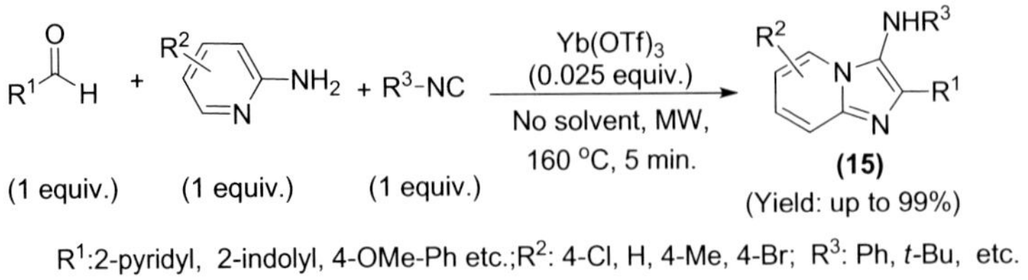

Scheme 10 Yb(OTf)$_3$-catalyzed solvent-free synthesis of imidazo[1,2-*a*]pyridines

Yb(OTf)$_3$-catalyzed Groebke–Bienaymeʹ–Blackburn (G–B–B) reaction protocol under solvent-free reaction condition where equimolar amount of all three components including aldehyde, amine, and isonitrile underwent cyclocondensation reaction to furnish up to 99% of the desired imidazopyridines **15** within 5 min (Scheme 10) [25].

Likewise, functionalized quinolones especially some of the fluoroquinolones are very popular antibacterial drugs. For example, ciprofloxacin, ofloxacin, norfloxacin are examples of commercially available drugs with quinolone scaffolds which are very frequently prescribed to treat bacterial infections [26]. Gould–Jacobs reaction [27] and the Lewis-acid-catalyzed cyclocondensation of *N*-acetyl anthralinic acids are among few popular methods to obtain functionalized quinolinones [28]. These reactions have also been explored under microwave radiation.

For example, Dave et al. established a microwave-assisted Gould–Jacobs reaction protocol which involves the cyclocondensation reaction between using aromatic amines and diethyl 2-(ethoxymethylene)malonate **16** under microwave irradiation to provide functionalized quinolin-4-ones **17** in appreciable yield within short reaction period of 2–14 min (Scheme 11) [29].

Similarly, Jia et al. explored a microwave-assisted method under cerium chloride (CeCl$_3$.7H$_2$O)-catalyzed reaction between aminophenyl ketone **18** and ester condition for the preparation of 3,4-disubstituted quinolone **19**. The formation of quinolone ring was achieved under solvent-free condition within 4–6 min of reaction time to provide good-to-excellent yields of the product (Scheme 12) [30].

Quinazoline analogues are also biologically active and medicinally important heterocyclic motif which has been found as a part of various natural product as well as drug candidates [31]. Various synthetic methods have been developed using different starting materials such as 2-aminobenzoic acid, anthranilonitrile, *N*-arylnitrilium

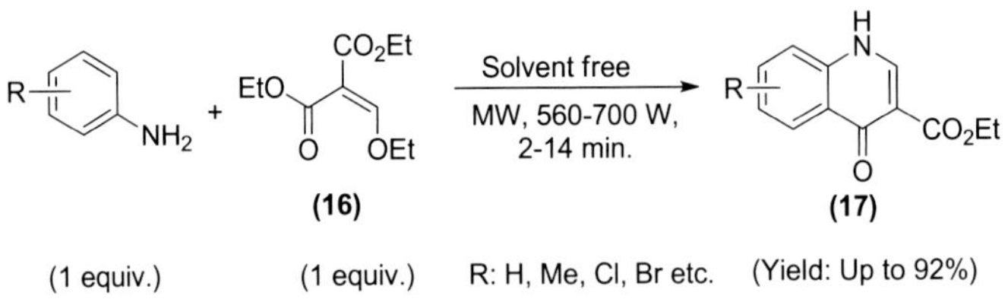

Scheme 11 Solvent-free Gould–Jacobs reaction under microwave reaction condition

2 Microwave-Assisted Organic Synthesis

Scheme 12
Microwave-assisted
CeCl$_3$.7H$_2$O-catalyzed
synthesis of 2-quinolones
under solvent-free condition

(18)
(1 equiv.)   (1 equiv.)

(19)
(Yield: Up to 91%)

R^1: 4-FPh, CH$_3$, ; R^2: CN,PhCO, CH$_3$CO etc. R^3: CH$_3$, C$_2$H$_5$

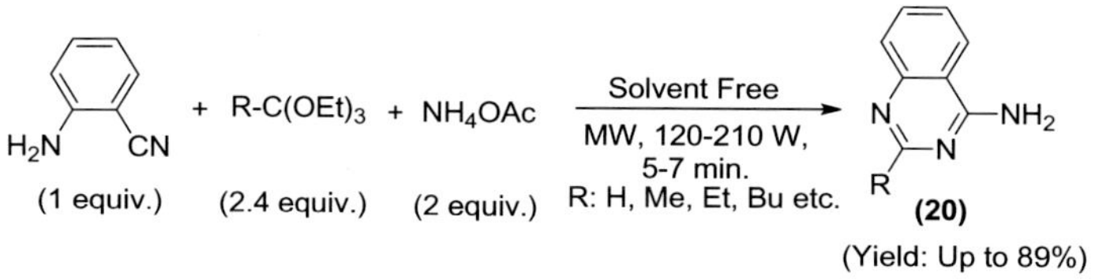

(1 equiv.)   (2.4 equiv.)   (2 equiv.)   R: H, Me, Et, Bu etc.   **(20)**
(Yield: Up to 89%)

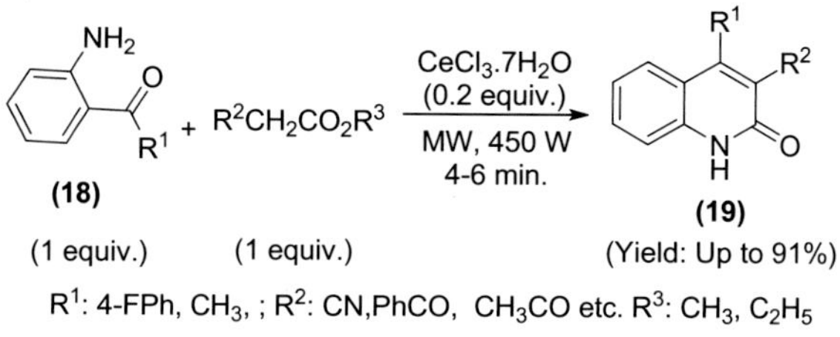

Scheme 13 Synthesis of 2-methylquinazolin-4-amine under solvent-free condition

salts, etc. for the formation of quinazoline nucleus. Some of the synthetic methods have also been developed under microwave-assisted reaction condition [32]. For example, Rad-Moghadam and coworkers synthesized quinazoline derivatives under microwave-assisted solvent-free reaction condition. Multicomponent reaction among 2-aminobenzonitrile, orthoester, and ammonium acetate led to the formation of functionalized quinazolines **20** in yield up to 89% in a short period of time (Scheme 13) [33].

Similar to the *N*-heterocycles, other heterocycles having two different heteroatoms such as Benzoxazoles constitutes core motifs of various natural products as well as drugs. Various condensation methods starting with the different substrates such as 2-hydroxyacetophenone, ortho-aminophenols are known in the literature for the synthesis of benzoxazole. Apart from the conventional heating method, these condensation methods have also been explored under microwave-assisted condition. For example, Seijas et al. prepared phenyl-substituted benzoxazole in only one minute through microwave-assisted condensation of *o*-aminophenol and thiobenzoic acid in presence of Lawesson's reagent in 83% yield (Scheme 14) [34].

Benzodiazepines are another important class of *N*-heterocyclic compound which have exhibited broad range of biological activities such as antimalarial, antidepressive, etc. [35]. A drug such as alprazolam is very popular antihypertensive drug with benzodiazepines ring as core skeleton. Mwande-Maguene et al. reported a faster

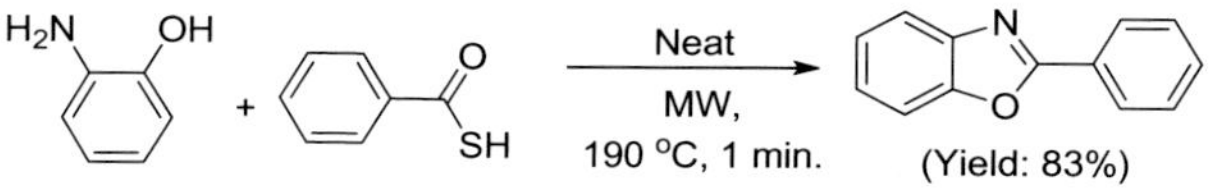

(Yield: 83%)

Scheme 14 Microwave-assisted solvent-free synthesis of 2-phenylbenzo[*d*]oxazole

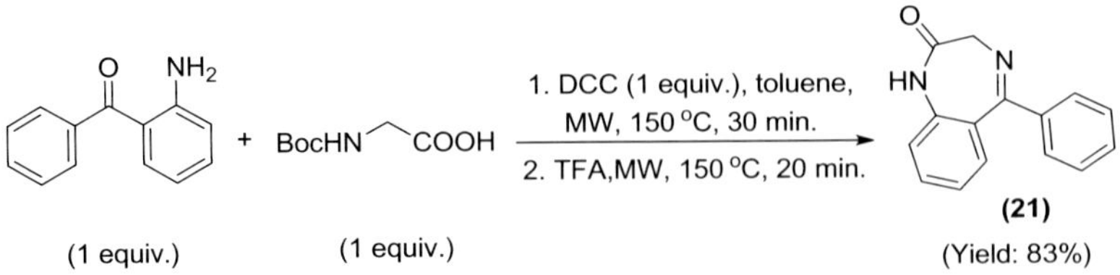

46   2 Energy-Efficient Process in Organic Synthesis

Scheme 15 Microwave-assisted synthesis of benzodiazepines

method for the preparation of this ring under microwave-assisted reaction condition. Reaction involved microwave-assisted coupling reaction in the first step using 2-aminobenzophenone and glycine followed by TFA-mediated cyclization to generate benzodiazepines **21**. Both the steps were jointly completed within 50 min of reaction time to produce 83% of the desired product (Scheme 15) [35].

Similarly, the oxygen heterocycles such as coumarin, isocoumarins, benzofuran, etc. also forms core skeletons in vast varieties of natural products as well as bioactive compounds [36]. Therefore, various methods have been explored to obtain such compounds. For example, Pechmann condensation which involves the condensation reaction between phenol and β-carbonyl containing ester or acids is one of the popular ways for obtaining the coumarin ring [37]. Therefore, this condensation reaction has also been explored under different energy efficient way. For example, Vahabi et al. developed a solvent-free and FeF_3-catalyzed Pechmann condensation reaction of phenols with ethyl 3-oxobutanoate to prepare 4-methylcoumarin using microwave radiation as greener energy source. Reaction afforded functionalized 4-methylcoumarins in yield up to 98% within 6–9 min (Scheme 16) [38].

Similarly, the synthesis of 4-arylcoumarin has also been reported under microwave-induced reaction condition. For example, Crecento-campo et al. developed a microwave-assisted condensation to prepare 4-arylcoumarins. 1,8-Diazabicyclo [5.4.0]undec-7-ene (DBU)-mediated condensation reaction between o-hydroxybenzophenone and alkyl malonate led to the formation of 4-arylcoumarin in considerably good yield within 7 min reaction time (Scheme 17) [39].

Apart from the conventional condensation method, the metal-catalyzed cyclization reaction toward the formation of coumarin has also been attempted under microwave-assisted reaction condition. For example, Fiorito et al. prepared coumarin

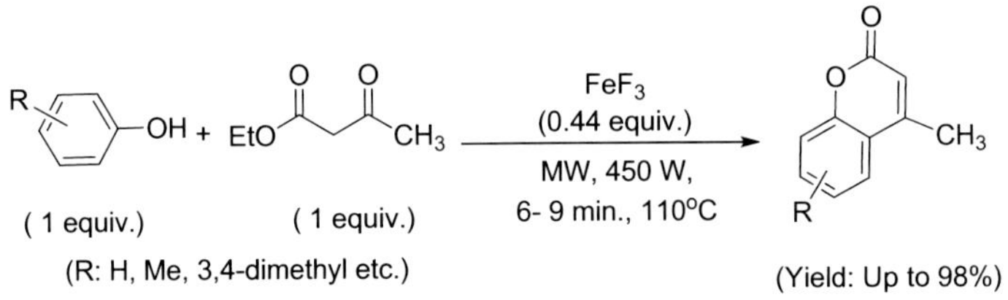

Scheme 16 Microwave-assisted FeF_3-catalyzed Pechmann condensation

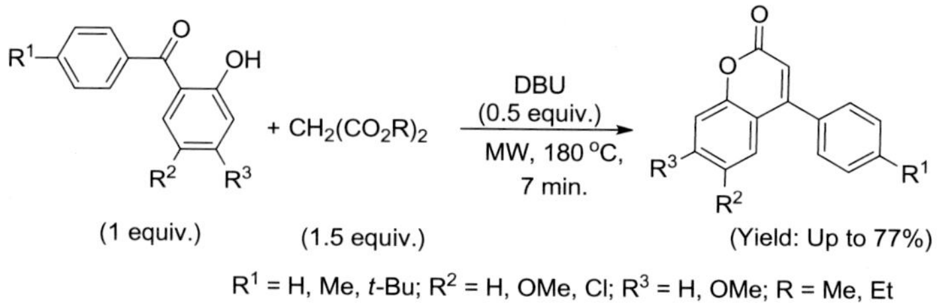

2 Microwave-Assisted Organic Synthesis 47

Scheme 17 Microwave-assisted preparation of 4-arylcoumarin

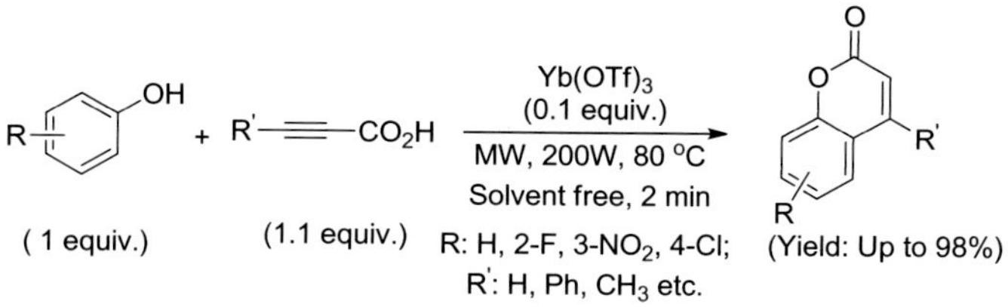

R^1 = H, Me, t-Bu; R^2 = H, OMe, Cl; R^3 = H, OMe; R = Me, Et

Scheme 18 Microwave-assisted synthesis of coumarin under Yb(OTf)$_3$-catalyzed reaction condition

ring through the microwave-assisted reaction between functionalized phenol and propiolic acid under Yb(OTf)$_3$-catalyzed reaction condition (Scheme 18). Under microwave condition, reaction completed very fast within 2 min to furnish functionalized coumarin in yield up to 98% [40].

The chromone ring formation has conventionally been prepared following famous name reactions such as Allan–Robinson reaction, Baker–Venkataraman rearrangement. It has also been attempted under the microwave-assisted reaction condition. For example, Balakrishna et al. developed a propylphosphonic anhydride (T3P®)-mediated one-pot reaction condition employing microwave as green energy source. Reaction completed within 20 min time compared to the conventional heating method that requires 24 h for the complete conversion. Reaction afforded upto 95% yield of the chromones (Scheme 19) [41].

Similarly, benzofuran which is also a very important heterocyclic skeleton found in a variety of natural products as well as medicinally important compounds have also be synthesized using microwave radiation [42]. Rap–Stoermer reaction is one of the methods for the synthesis of highly substituted benzofuran. Rao et al. developed microwave-assisted method for the Rap–Stoermer reaction to prepare functionalized

Scheme 19
Microwave-assisted one-pot synthesis of
4H-chromen-4-one

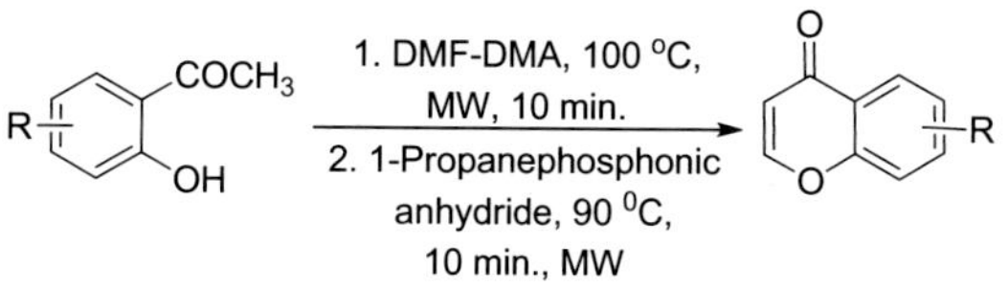

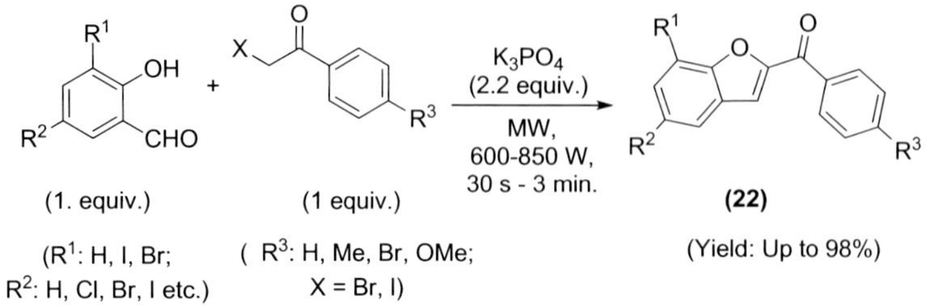

Scheme 20 Microwave-assisted solvent-free synthesis of benzofuran

benzofurans **22** using substituted salicylaldehyde, α-haloketones as substrates and K_3PO_4 as base (Scheme 20) [43].

3 Ball Milling Method for Organic Synthesis

In addition to the application of sound and various parts of the electromagnetic waves, chemical transformations have also been explored using mechanical energy and separate term m*echanochemistry* is being used to denote chemical transformation achieved using mechanical energy. IUPAC has defined the mechanochemical reaction as "*a chemical reaction that is induced by the direct absorption of mechanical energy*" [44]. The generation of mechanical energy through *grinding* process using mortar and pestle and its application in achieving chemical transformation a very old chemistry but due to the lack of proper control over the various parameters such as speed, strength of grinding, frequency, etc. involved in the manual grinding process, there was a problem in the reproducibility of the result in terms of the % conversion, yield, selectivity, etc. Therefore, a more robust and reliable technique known as *milling* has been developed where the reactants and reagents are mixed and ground with the help of balls made up of different materials such as steel inside a mill. Different types of mills such as planetary mill (PM), mixer mill (MM), high-speed ball mills (HSBMs), etc. are used for milling process leading to chemical transformation. In ball mill method, the outcome of the reaction mainly depends on various parameters such as materials of the ball and beaker, number, and the size of the ball along with milling frequency, timing, etc. [45, 46]. Various different types of reactions have been explored under ball mill method which will be discussed in the following sections.

3.1 Application of Ball Mill Method in Synthesis of Heterocycles

Zhu et al. developed a solvent-free reaction condition for the synthesis of triarylpyrazoline **23** through the cyclocondensation reaction between chalcone and phenylhydrazone in presence of $NaSO_4$. H_2O as a catalyst (Scheme 21) [47]. Reaction performed using high-speed ball mills (HSBMs) method provided good-to-excellent yield of the product under solvent-free reaction condition within 5 min of the reaction time.

Sharma et al. synthesized 1,2-disubstituted benzimidazoles ionic liquid (IL) (**24**)-coated ZnO nanoparticles (ZnO-NPs) as a recyclable catalyst under the ball milling method. Reaction between o-phenylenediamine and benzaldehyde under ball milling method provided excellent yield of the benzimidazole **25** (Scheme 22) [48].

Thorwirth et al. have developed a solvent-free method for the 1,3-dipolar cycloaddition reaction of alkyne and azide **26** to form triazole derivative **27** using ball mill technique (Scheme 23) [49]. Excellent yield of the substituted triazole was formed within 10 min of reaction time without any noticeable side product formation.

Maleki et al. synthesized 3-aminoimidazo[1,2-a]pyridine **28** through well-established Ugi-multicomponent reaction among 2-aminopyridine, benzaldehyde, and isonitrile under ball mill reaction condition (Scheme 24) [50].

This solvent-free reaction method displayed excellent efficacy in the formation of various functionalized imidazo[1,2-a]pyridines. Similarly, Wang et al. synthesized 2-aryl-imidazo[1,2-a]pyridines **30** through the iodine promoted cyclocondensation

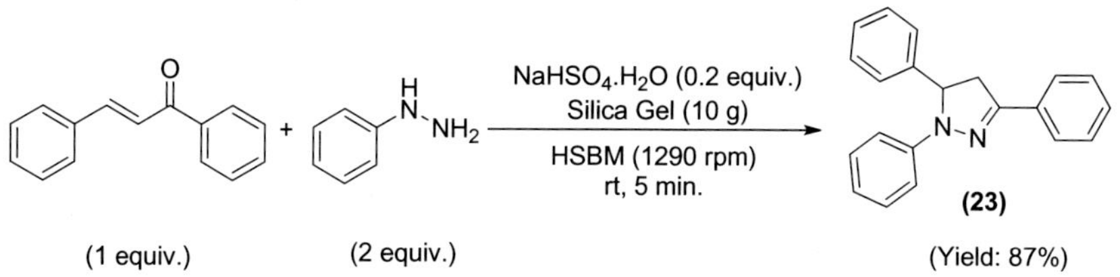

Scheme 21 Synthesis of triarylpyrazolines using HSBM method

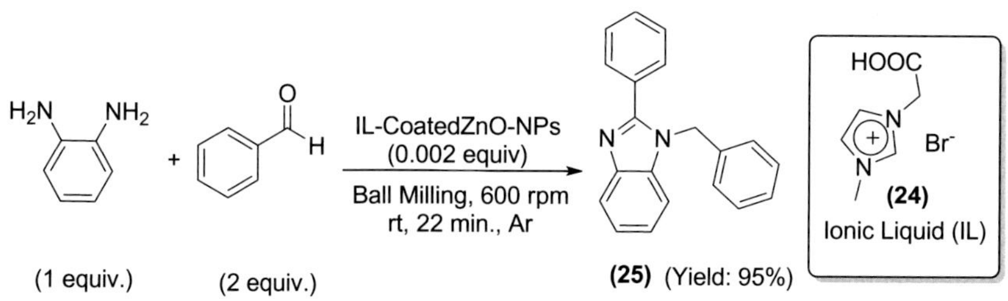

Scheme 22 Ball milling technique for the synthesis of benzimidazole

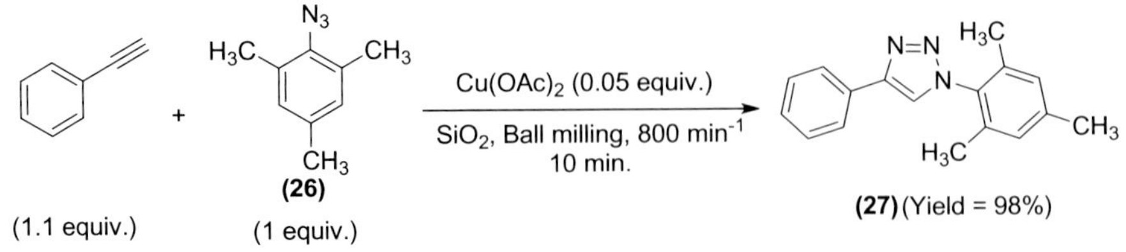

Scheme 23 1,3-Dipolar cycloadditions reaction under ball milling method

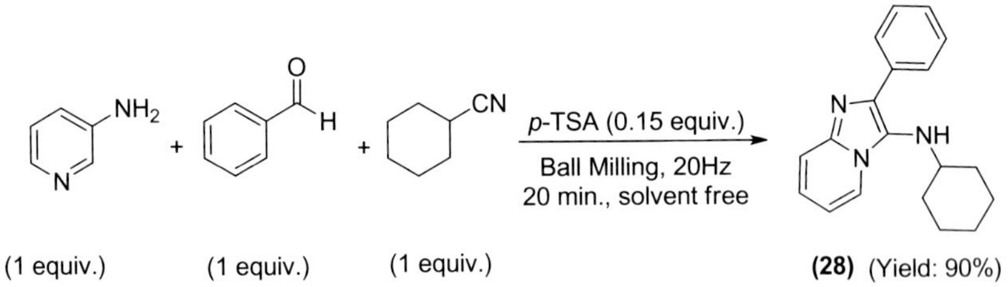

Scheme 24 Ugi-multicomponent reaction for the formation of imidazo[1,2-a]pyridine in ball mill condition

reaction of 2-aminopyridines and acetophenones **29** in presence of DMAP as a base under high-speed ball milling reaction condition (Scheme 25) [51]. The developed reaction protocol was also utilized in the synthesis of zolimidine which is a drug used for the treatment of the peptic ulcer.

Kaupp synthesized quinaxoline **31** under ball mill method through the cyclocondensation of o-phenylenediamine and benzil (Scheme 26) [52].

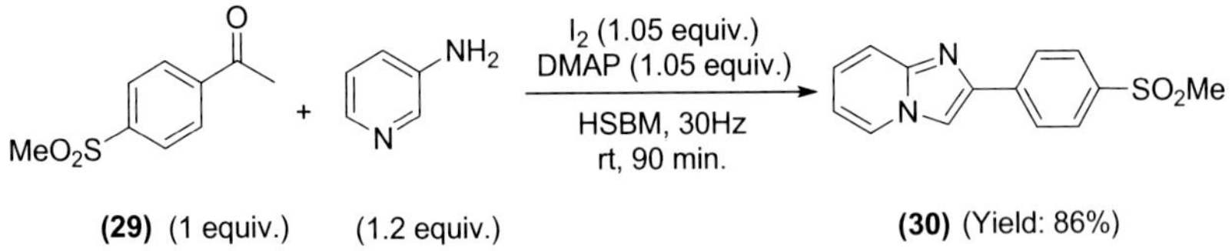

Scheme 25 Synthesis of zolimidine under ball mill method

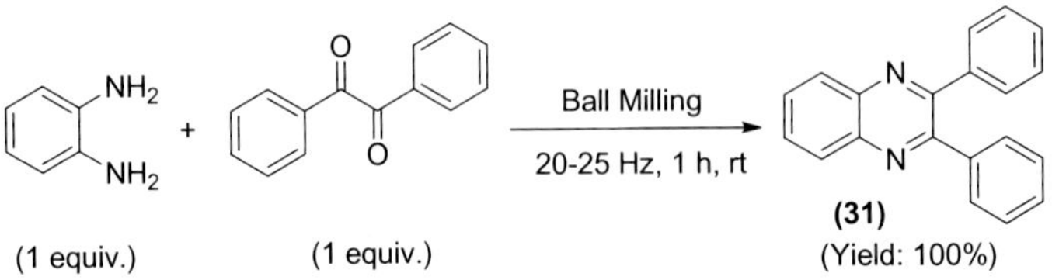

Scheme 26 Synthesis of quinoxaline under ball mill method

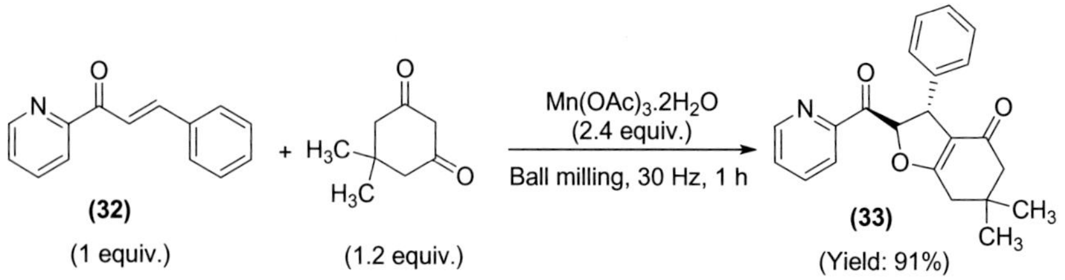

3 Ball Milling Method for Organic Synthesis 51

Scheme 27 Mn(OAc)$_3$.2H$_2$O-mediated cycloaddition reaction for the synthesis of tetrahydrofuranone

Scheme 28 Conversion of aziridine to oxazolidinone under ball mill method

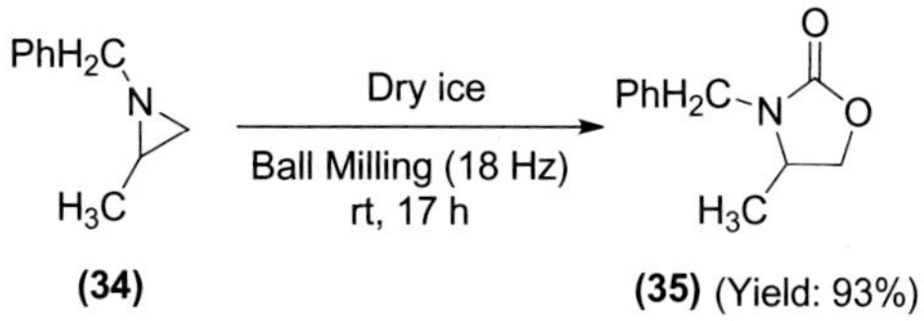

Wang et al. developed a Mn(OAc)$_3$.2H$_2$O-mediated reaction protocol using ball mill method for the synthesis of functionalized tetrahydrobenzofuran-4(5*H*)-one **33** through the Mn(III) mediated cycloaddition reaction of cyclohexane-1,3-dione and enones **32**. Reaction afforded good-to-excellent yield of various tetrahydrobenzofuran-4(5*H*)-one (Scheme 27) [53].

Phung et al. developed a ball mill reaction protocol for the synthesis of oxazolidinone **35** starting with aziridine **34**. The conversion to the oxazolidinone under ball mill reaction method worked without using any catalyst or solvent and only required dry ice (Scheme 28) [54].

3.2 Asymmetric Synthesis Under Ball Mill Method Using Organocatalysts

Ball milling technique also been employed to achieve asymmetric synthesis. Different types of such reactions have successfully been explored employing ball mill method [55]. For example, Rodrguez et al. carried out an extensive study to compare the (*S*)-Proline-catalyzed aldol condensation under conventional stirring and ball mill method. For example, the aldol condensation reaction between 3-nitrobenzaldehyde and cyclohexanone using (*S*)-Proline as organocatalyst afforded yield of 99% of the product **36** with 94% enantioselectivity within 5.5 h reaction time which is much less compared to the reaction performed under conventional magnetic stirring that requires 96 h to provide 98% of the product yield with same enantioselectivity (Scheme 29) [56].

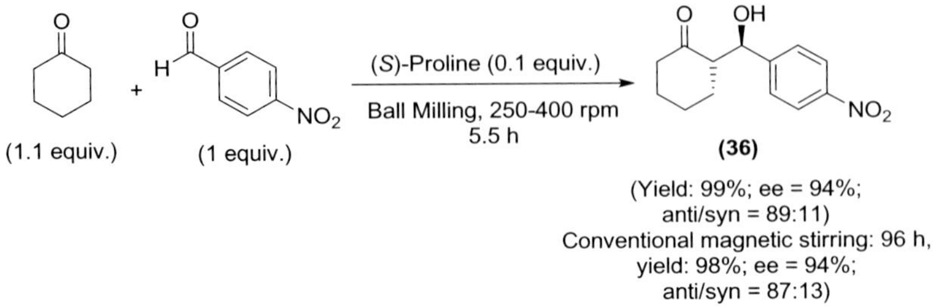

Scheme 29 (*S*)-Proline-catalyzed aldol condensation under ball milling condensation

Similarly, Wang et al. reported the asymmetric Michael addition reaction between 1,3-dicarbonyl compound with various nitroolefins using squaramide derivative **37** as a chiral H-bonding organocatalyst [56]. Interestingly, this addition reaction performed under ball mill condition worked under very low catalyst loading of 0.005 equiv. to furnish the addition product in excellent yield and very high enantioselectivity. For example, squaramide derivative **37**-catalyzed Michael addition reaction between pentane-2,4-dione and β-nitrostyrene under ball mill condition furnished 95% yield of the addition product **38** with 99% enantioselectivity (Scheme 30). Surprisingly, reaction completed within only 5 min of reaction time which was much less compared to the conventional stirring method that needed 8 h of reaction time and also under solvent-free reaction condition.

Rantanen et al. disclosed a ball mill method for the quinidine mediated asymmetric ring opening of mesoanhydride under solvent-free reaction condition to afford

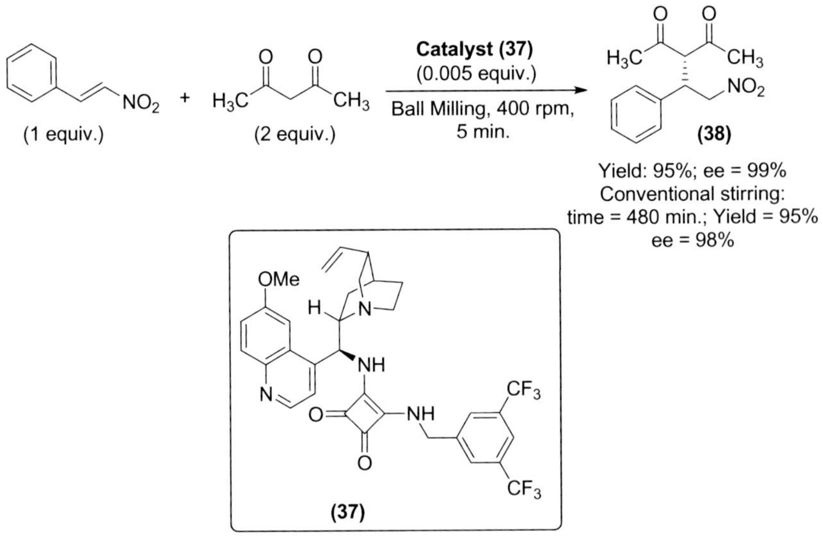

Scheme 30 Squaramide derivative-catalyzed Michael addition under ball mill method

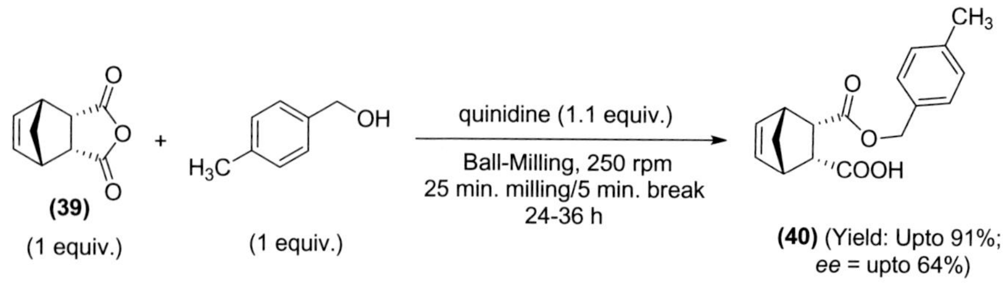

3 Ball Milling Method for Organic Synthesis

Scheme 31 Quinidine-mediated opening of anhydride in ball mill method

various optically active esters [57]. Such ring opening reaction attempted under ball mill condition provided the esters with high yield and maximum value of enantiomeric excess (*ee*) of the product obtained was 64%. For example, reaction between mesoanhydride (39) and *p*-methylbenzyl alcohol afforded the mono ester product **40** in 91% yield with 61% of *ee* (Scheme 31).

3.3 Coupling Reaction Under Ball Mill Condition

Ball mill method has also been employed to achieve the C–C bond formation through Suzuki–Miyaura coupling. For example, Seo et al. developed a Pd-catalyzed cross-coupling between aryl halide and organoboron method using 1,5-cod as an olefin additive under ball mill method. For example, coupling reaction between 4-bromo-1,1′-biphenyl and (4-(dimethylamino)phenyl)boronic acid under developed reaction protocol afforded 91% of the cross-coupled product **41** under ball mill reaction condition (Scheme 32) [58].

Olefin was assumed to prevent the aggregation of Pd-nanoparticle and stabilizes the monomeric Pd(0) species. In general, the catalytic reaction performed under ball mill reaction protocol requires the addition of catalyst in the powder form similar to the conventional solution state chemistry and this type of catalysis under mechanochemical reaction condition is known as *indirect mechanocatalysis*. But in past few years, scientist has explored the use of the metals present in the ball, itself as catalyst which is termed as *direct mechanocatalysis*. Fulmer et al. first demonstrated the application of high-speed ball milling method to perform Sonogashira coupling between aryl halides and substituted acetylenes under solvent-free condition using

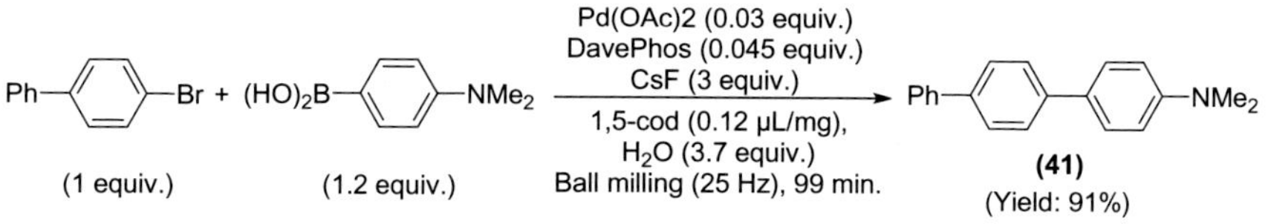

Scheme 32 Suzuki cross-coupling under ball mill method

Scheme 33 Sonogashira coupling under solvent-free and ball mill condition

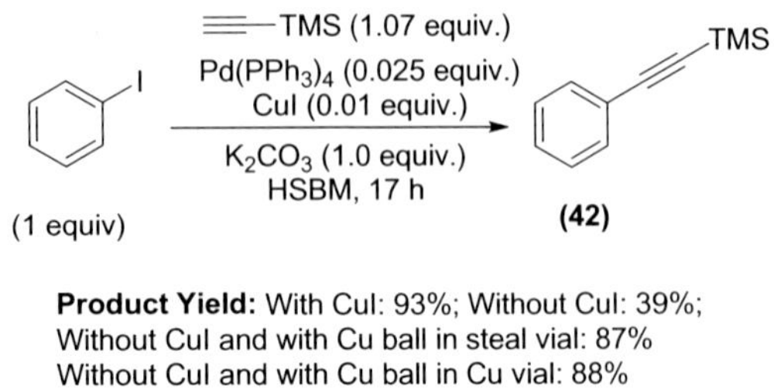

Product Yield: With CuI: 93%; Without CuI: 39%; Without CuI and with Cu ball in steal vial: 87% Without CuI and with Cu ball in Cu vial: 88%

copper ball as a cocatalyst along with Pd-catalyst. For example, reaction between idobenzene and trimethyl silylacetylene coupled under Pd/Cu-catalyzed reaction condition to afford coupled product **42** in yield up to 93% (Scheme 33) [59]. The role of CuI in performing Sonogashira reaction is well established but instead of direct addition of CuI as a source of copper, they disclosed the role of Cu present as ball and vial material as catalyst in performing this coupling. The reaction afforded good-to-excellent yield of the coupling product under this optimized protocol even without adding CuI.

Similarly, Su et al. employed HSBM technique to achieve cross-dehydrogenative coupling (CDC) reaction protocols between tetrahydroisoquinolines **43** and different pronucleophiles such as indoles, alkynes, nitroalkane in presence of DDQ as base to produce the desired coupled products **44–46** (Scheme 34) [60]. This coupling reaction afforded good-to-excellent yield of the cross-coupling product within maximum time period of 40 min. Coupling using indole and alkynes with tetrahydroisoquinolines worked efficiently in presence of copper balls only as catalyst. So this is also an example of direct mechanocatalysis.

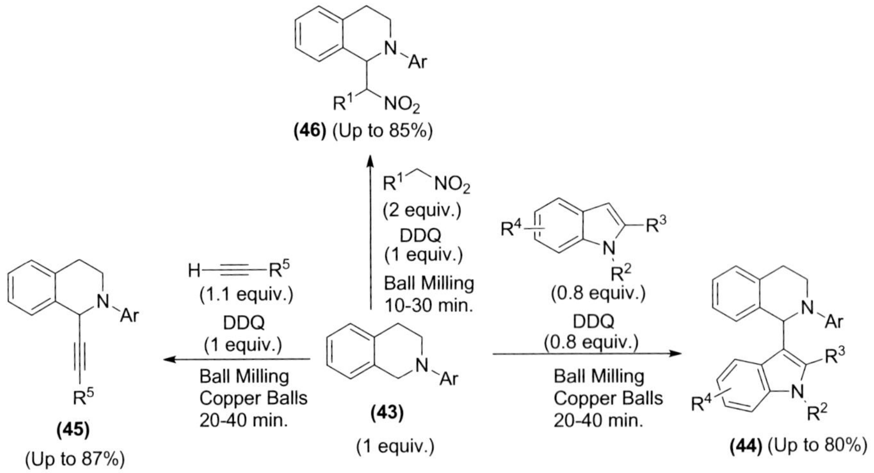

Scheme 34 Cross-dehydrogenative coupling (CDC) reaction protocols between tetrahydroisoquinolines(43) and different pronucleophiles using ball mill method

4 Ultrasonic Methods in Organic Synthesis

Similar to the application of microwave in organic synthesis, sound wave in particular ultrasound has also been explored as greener energy source. It is known to affect the reaction rate and reactivity of a reaction by imploding of microsized bubbles which forms and grows when a pressure wave of suitable intensity propagates through a reaction medium. Due to the release of energy during the collapse of bubbles, the local temperature and pressure gets increased up to 5000 K and to 1000 bar respectively, which improves the mass transfer and also enables the chemical reactions to progress in forward direction by affecting various other parameters of the reaction [61, 62]. A vast varieties of chemical reactions have been successfully established under the ultrasound-assisted reaction condition.

For example, Li et al. developed an ultrasound-assisted reaction protocol for the preparation of 1,3,5-triaryl-2-pyrazolines at mild temperature. Cyclocondensation reaction between functionalized chalcones and phenylhydrazine hydrochloride in presence of sodium acetate as base and the mixture of water and acetic acid as solvent led to the formation of the desired product **47** in yield up to 96% yield within 1.5–3 h reaction time under the ultrasound-assisted reaction condition (Scheme 35) [63].

K. F. Shelke et al. developed an ultrasound-assisted method for the synthesis of triarylimidazole **48** through the ammonium acetate mediated reaction between benzil and aromatic aldehyde in presence of ceric ammonium nitrate (CAN) as a catalyst (Scheme 36) [64].

Reaction afforded good to excellent yield of the desired imidazoles in 30–70 min time. Under the same reaction condition just by increasing the reaction time, the

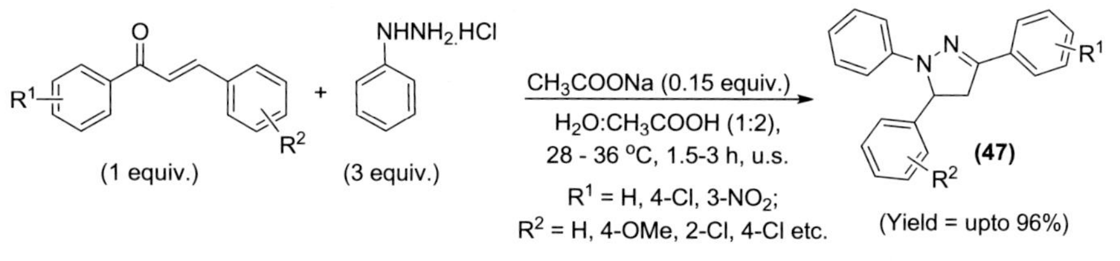

Scheme 35 Ultrasound-assisted synthesis of 1,3,5-triaryl-2-pyrazolines

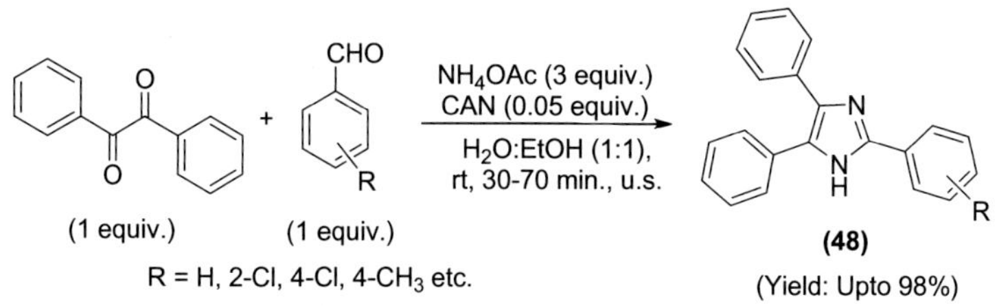

Scheme 36 Ultrasound-assisted synthesis of triarylimidazole

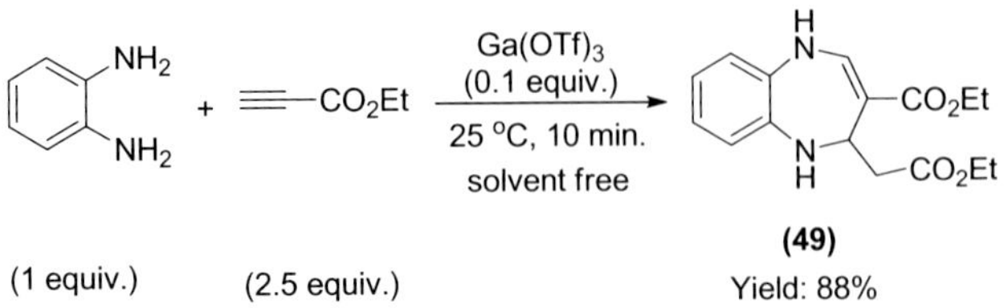

Scheme 37 Ultrasound-assisted Ga(OTf)$_3$-catalyzed formation of 3,4-Disubstituted-1,5-benzodiazepines under solvent-free condition

formation of triarylimidazole was also achieved using benzoin in place of benzyl in 85–94% yield. Similarly, Jiang et al. developed a Ga(III)-catalyzed method for the synthesis of substituted 1,5-benzodiazepines **49** involving [4+2+1] cycloaddition reaction between o-phenylenediamines and ethyl propiolate under ultrasound-assisted reaction condition (Scheme 37) [65]. This transformation was achieved in absence of any solvent to provide 88% yield of the desired product.

Guzen et al. prepared 1,5-benzodiazepines **50** under the ultrasonication method at room temperature following the p-toluene sulfonic acid (pTSA)-catalyzed cyclocondensation reaction between o-phenylenediamines and pentane-2,4-dione (Scheme 38) [66]. This ultrasound-assisted method furnished the good yield of 1,5-benzodiazepines **50**.

Braibante et al. prepared aminopyrazoles under ultrasound-assisted reaction protocol in two steps. In the first step, the dichloromethane solution of O-Ethyl 3-oxo-3-phenylpropanethioate **51** dispersed on montmorillonite K-10 was reacted with the primary amine under ultrasound-condition that led to the formation of α-oxoketene O,N-acetals **52** in 22 h. It was then reacted with hydrazine hydrate (80%) to produce substituted pyrazoles **53** (Scheme 39) [67].

Venigalla et al. developed an ultrasound-assisted synthesis of triazole in one-pot operation. The reaction between aldehyde and semicarbazide under ultrasound-assisted reaction condition leads to the formation of triazole **54** through the cyclocondensation reaction (Scheme 40) [68].

Castillo et al. prepared a Cu(I) complex using bis(pyrazolyl)methane (**L$_1$**) as ligand and demonstrated its efficacy as a catalyst in performing the click reaction using a combination of alkyl halide with terminal alkyne and sodium azide under the ultrasonic reaction condition. This ultrasonication method led to the formation of

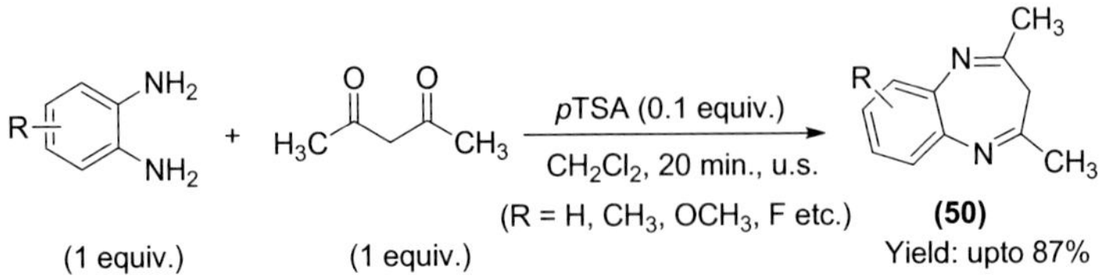

Scheme 38 Ultrasound-assisted synthesis of 1,5-benzodiazepines using p-TSA as a catalyst

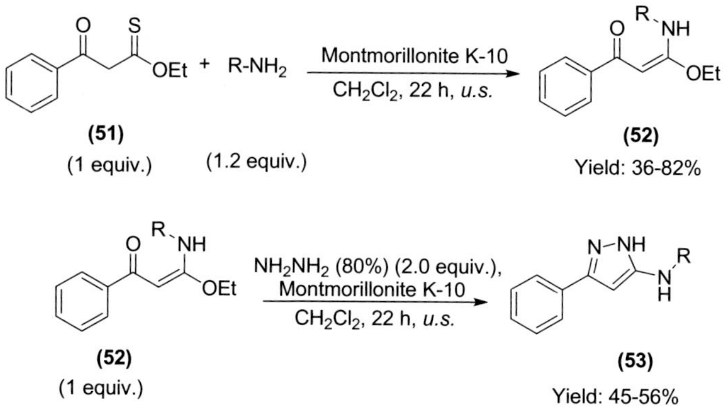

4 Ultrasonic Methods in Organic Synthesis

Scheme 39 Ultrasound-assisted pyrazole synthesis

Scheme 40
Ultrasound-assisted
synthesis of triazole
derivatives in one-pot
operation

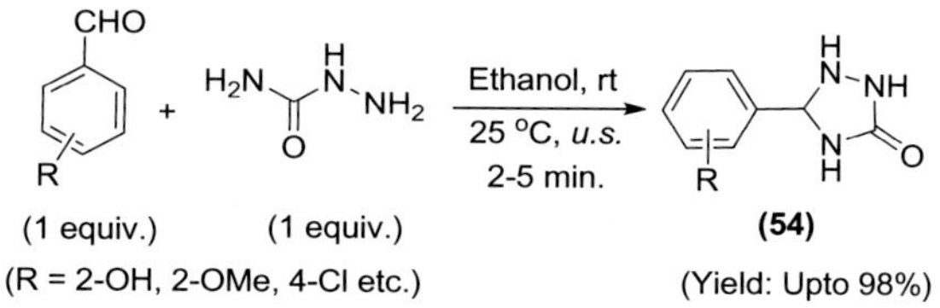

various functionalized 1,2,3-triazoles **55** within 30 min reaction time and in presence of water as a solvent (Scheme 41) [69]. Reaction afforded good to excellent yield of the desired triazoles.

Pereira et al. prepared benzotriazole **56** following a reaction between *o*-phenylenediamine and sodium nitrite in acetic acid as solvent under the ultrasound irradiation. This conversion to benzotriazole completed within 10–15 min (Scheme 42) [70].

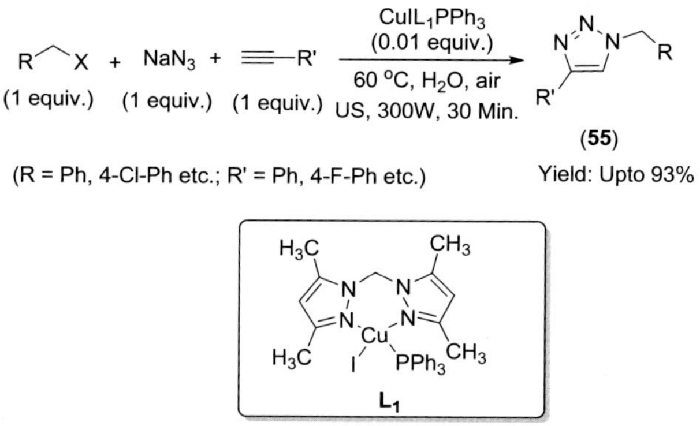

Scheme 41 Ultrasound-assisted Cu(I)-complex-catalyzed click reaction

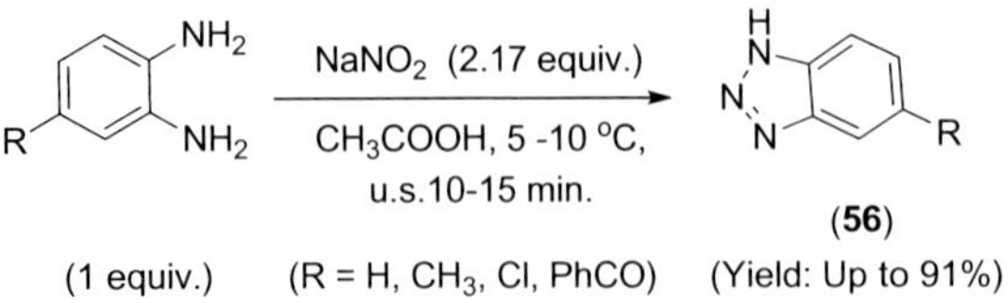

Scheme 42
Ultrasound-assisted
synthesis of
1,2,3-benzotriazole starting
with *o*-Phenylenediamine

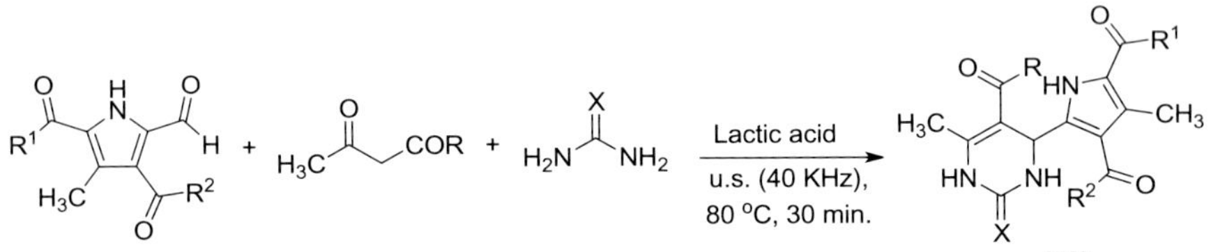

(1 equiv.)   (R = H, CH₃, Cl, PhCO)   (Yield: Up to 91%)

Scheme 43 Synthesis of pyrrole dihydropyrimidinones under ultrasound-assisted condition using lactic acid as solvent

Li et al. developed an ultrasound-assisted reaction protocol for the Biginelli reaction wherein the efficient and green method has been developed for the synthesis of various pyrrole dihydropyrimidinones through Biginelli reaction. For example, the multicomponent reaction among pyrrole derivative **57**, 1,3,-dicarbonyl compound and urea/thiourea led to the formation of pyrrole dihydropyrimidinones **58** under the sonochemical condition using lactic acid as green solvent (Scheme 43) [71].

5   Photo-Induced Organic Transformations

Photo-induced chemical transformation in nature could be considered as one of the oldest chemical reaction. Absorption of light energy by photosystem mainly in the region of 400–700 nm to induce the formation of glucose using CO_2 and H_2O as reacting substrates through photosynthesis is well-established chemical reaction that has been occurring in nature and oxygen produced thereby has allowed us to survive [72]. Apart from this natural phenomenon, some of the very significant observations made by Joseph Priestley, J. W. Dobeveiner, and Hermarin Trommsdorff in different periods made us to realize the importance of light in carrying out chemical transformations. Photo-induced chemical changes observed by Hermarin Trommsdorff in 1834 in *santonin*, a naturally occurring anthelminthic compound, is one of the oldest example known in the solid-state photochemistry [73]. Since then diverse varieties of chemical transformations have been developed exploiting light as a green energy source. The proceeding sections will be highlighting some of such recent examples in this area.

5 Photo-Induced Organic Transformations

5.1 Photo-Catalyst Catalyzed C–C and C–X Bond Forming Reaction

The arylation reaction is one of the most prevalent reactions which is generally achieved via the well-established cross-coupling methods such as Suzuki, Negishi coupling reactions using different organometallic reagents. In addition to that various new approaches such as arylation using C–H activation or oxidative/reductive coupling, etc. are being developed for the same which generally requires conventional heating methods. Therefore, new approaches exploiting light as a green alternative in presence of photo-catalyst is being exploited to achieve diverse varieties of arylated compounds. For example, the visible light mediated reaction has also been used to achieve the direct C–H arylation of isoquinoline using aryldiazonium salt as arylating agent and [Ru(byp)$_3$Cl$_2$].6H$_2$O as photo-catalyst. The [Ru(byp)$_3$Cl$_2$].6H$_2$O complex is one of the most commonly used photo-redox catalyst that works by reversibly changing its nature as oxidant or reductant in presence of a suitable quencher and therefore considered as very efficient photo-redox catalyst [74] (Fig. 2).

On irradiation of the light of a suitable wavelength ($\lambda = 452$ nm), the Ru(byp)$_3$Cl$_2$ complex moved to excited state Ru(bpy)$_3^{2+*}$ that in presence of a suitable quencher gets reduced or oxidized to Ru(bpy)$_3^+$ or Ru(bpy)$_3^{3+}$, respectively. This visible-light-induced photo-redox nature of this Ru-complex has extensively been exploited in the organic synthesis [74]. For example, Cano-Yelo and Deronzier exploited this photo-redox nature of Ru(byp)$_3$Cl$_2$ and synthesized functionalized phenanthrenes **60** from diazonium salt **59** under the irradiation of visible light (Scheme 44) [75].

This is the very first example of Ru-complex-catalyzed Pschorr reaction under the photo-catalytic condition. Later on, several such reactions have been developed under visible light reaction condition. For example, Zhang et al. developed a visible-light-induced method for the arylation of isoquinoline using aryldiazonium salt as aryl donor and Ru(byp)$_3$Cl$_2$ as a photo-redox catalyst. This mild reaction condition was applicable in the preparation of a library of arylated isoquinolines **61** (Scheme 45) [76].

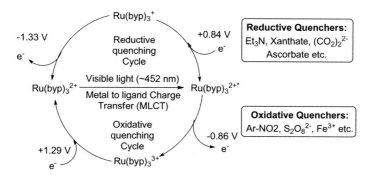

Fig. 2 Photo-redox cycles of [Ru(byp)$_3$Cl$_2$]

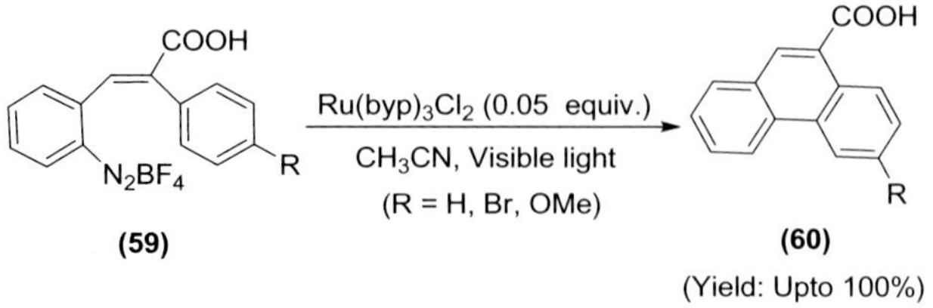

Scheme 44 Visible-light-induced Pschorr reaction for phenanthrene formation

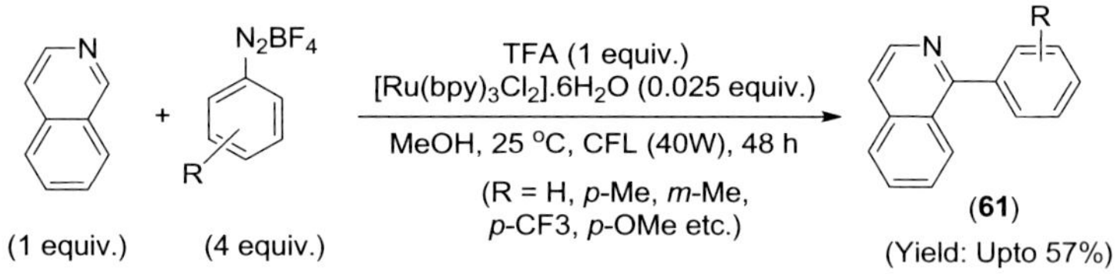

Scheme 45 Visible-light-mediated C-H arylation of isoquinoline

They further demonstrated the efficacy of the developed direct C–H arylation protocol as one of the key step in the synthesis of Menisporphine. Reaction involves arylation of isoquinolone derivative **62** to produce arylated product **63** which was further converted into Menisporphine(**64**) in two steps. Menisporphines are examples of alkaloids with isoquinoline as core skeleton (Scheme 46) [76].

Similarly, under visible light, Xue et al. reported the direct arylation method of heteroarenes such as pyridine, quinoline, imidazole, etc. using aryldiazonium salt as arylating agent and $Ru(byp)_3Cl_2$ as photo sensitizer. The reaction protocol for the direct arylation with water as a solvent worked effectively at room temperature to furnish considerable yield of the arylated product. For example, the arylation of

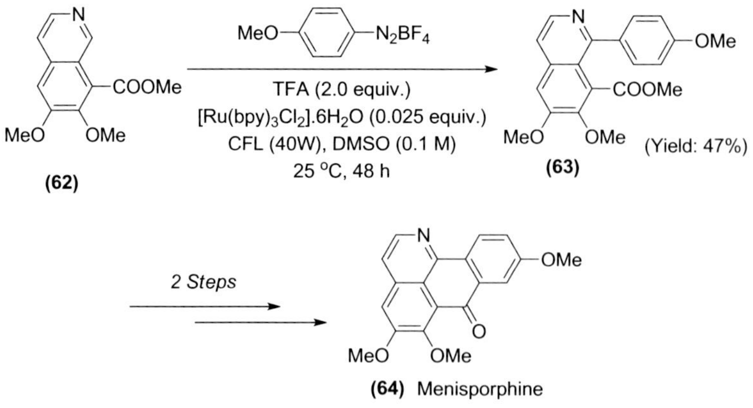

Scheme 46 Synthesis of Menisporphine via photo-induced direct C–H arylation of isoquinoline

5 Photo-Induced Organic Transformations

pyridine hydrochloride with aryldiazonium salt led to the formation of arylpyridine in yield up to 93% (Scheme 47) [77].

Although for the arylation of caffeine under similar condition, 88% of aqueous formic acid was found to be a superior solvent as low yield of the arylated product **65** was obtained in water medium probably due to the low solubility of the caffeine in water compared to the 70% of yield obtained using formic acid as solvent (Scheme 48) [77].

Li et al. developed a mild reaction protocol for the synthesis of functionalized benzimidazole using consisting of o-phenylenediamine and aldehydes in presence of fluorescein which is an inexpensive dye, as a photo-catalyst under visible light (Scheme 49) [78]. Reaction was found to be effective with electronically diverse

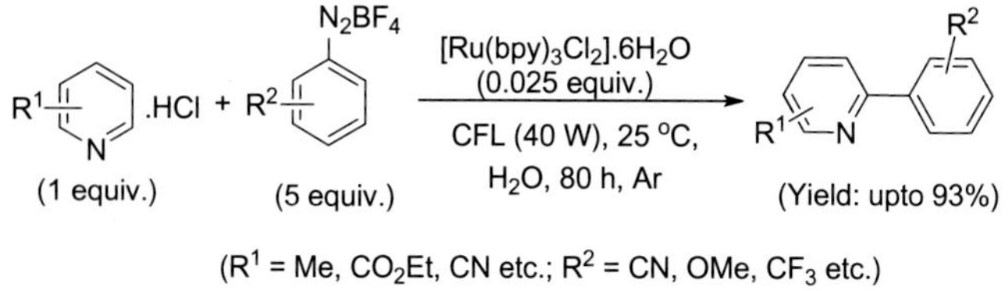

Scheme 47 Visible-light-mediated arylation of aryldiazonium salt in water

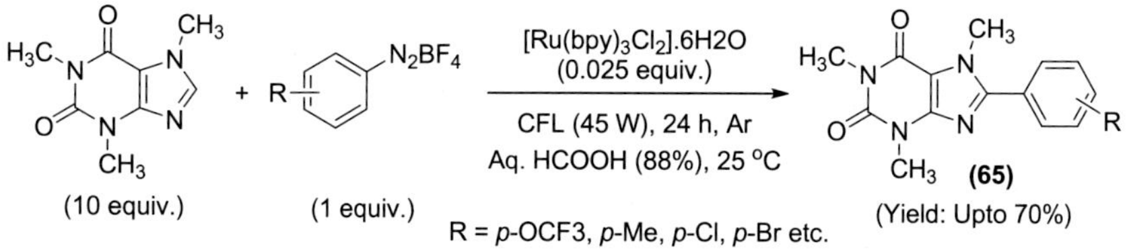

Scheme 48 Visible-light-mediated arylation of caffeine aryldiazonium salt in water

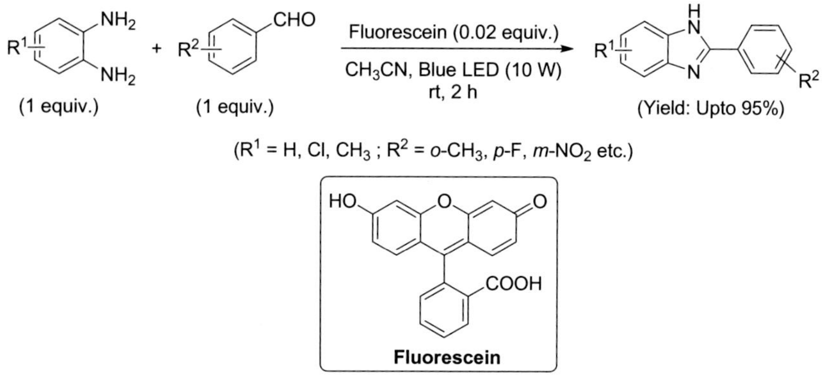

Scheme 49 Light-induced synthesis of benzimidazoles

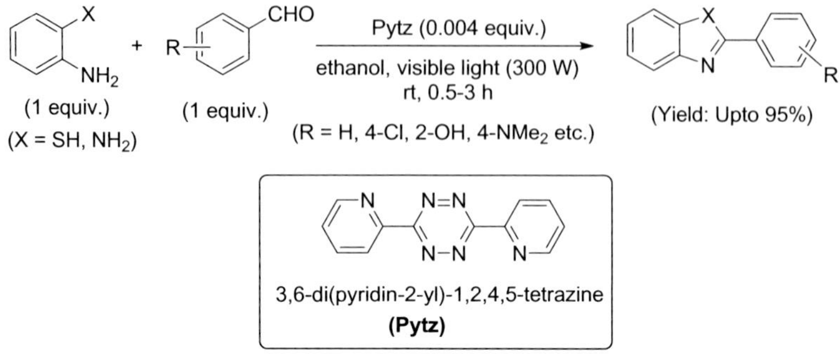

Scheme 50 Visible-light-induced synthesis of 2-arylbenzimidazole and benzothiazole

aldehydes as well as phenylenediamine.

For example, in case of aromatic, aliphatic as well as heteroaromatic aldehyde as one of the substrates, reaction afforded good-to-excellent yield of the benzimidazole. Under the standard reaction condition, 2-aminobenzamide instead of diamine, successfully provided 2-phenylquinazolin-4(3H)-one in 55% yield. The utility of other organic compounds as photo-catalyst$s$-tetrazine has also been explored. For example, Samanta et al. synthesized 3,6-di(pyridin-2-yl)-1,2,4,5-tetrazine (pytz) and demonstrated its role as a photo-catalyst to achieve the synthesis of 2-alkyl and aryl substituted benzimidazole and benzothiazole using o-phenylenediamine and o-aminothiophenol respectively along with the different aldehydes (Scheme 50) [79]. Reaction afforded good to excellent yield of the alkyl and aryl-substituted benzimidazole and benzothiazole. The ferrocenecarboxaldehyde under the standard condition also provided ferrocene substituted benzimidazole in 80% yield. Reaction worked well in presence of both electron withdrawing and electron-donating substituents.

The presence of reactive functional group such as –OH on the aldehyde doesn't seem to affect the reactivity as excellent yield of the benzimidazole was obtained in this case also. The reaction using terephthalaldehyde with o-phenylenediamine and o-aminothiophenol furnished phenyl 1,4-dibenzimidazole and 4-benzothiazole-yl-benzaldehyde, respectively. Although by changing the solvent to THF, reaction provided phenyl-1,4-dibenzothiazole exclusively.

Similarly various methods such as photocyclization, [80] cycloisomerization [81], and benzannulation [82] of stilbene, arynes and biaryl, respectively, are known for the synthesis of phenanthrene derivatives. Xiao et al. utilized light energy to achieve the synthesis of phenanthrene *via* [4+2] benzannulation of alkynes with diazonium salt in presence of eosin-Y which is an inexpensive yellowish-red colored xanthene dye, as photo-redox catalyst (Scheme 51) [83]. The use of eosin Y-catalyzed method provides a metal-free and visible light-induced greener alternative to the existing methods that generally involves the use of transition metal under conventional heating condition.

Niu et al. developed a selectfluor promoted oxidative coupling method to achieve α sp^3 C–H arylation of alcohol (Scheme 52). Selectfluor is a popular oxidizing as well as fluorination agent. Under the irradiation of blue light, the reaction begins with the

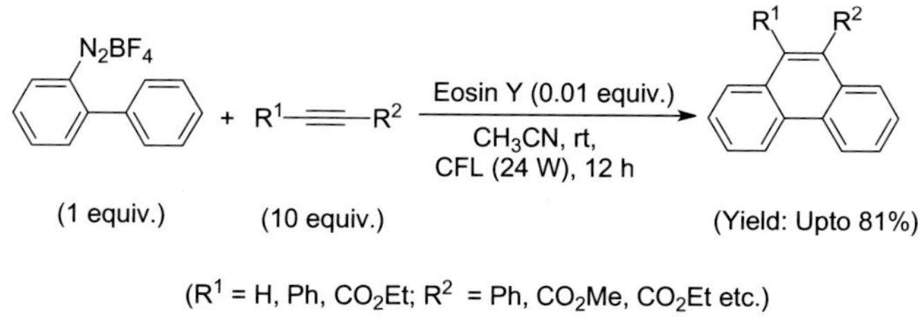

5 Photo-Induced Organic Transformations

Scheme 51 Visible-light-induced [4+2] benzannulation of biaryldiazonium salts with alkynes

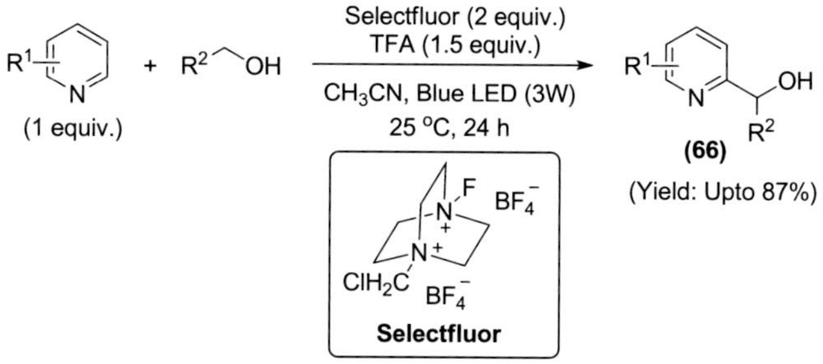

$(R^1 = H, Ph, CO_2Et; R^2 = Ph, CO_2Me, CO_2Et etc.)$

Scheme 52 Selectfluor-promoted α C–H arylation of alcohols

activation of N–F bond to form radical cation which later combines with alcohol to produce hydroxyalkyl radical. It reacts with the heteroarenes that undergo oxidative aromatization to produce the α C–H arylated alcohol **69**. Reaction using heteroarenes such as isoquinolines, quinoline, benzothiazole produced good-to-excellent yield of the desired product.

For example, Liang et al. developed a metal-free arylation method of the oxindole **67** under visible light using aryl halides (Scheme 53) [84]. The reaction afforded good-to-excellent yield of the desired arylated product **68** under the reaction condition consisting of CsOH as base and acetonitrile as a solvent in presence of white LED light of 23 W. The developed methods were found to work efficiently with the aryl halides in particularly with aryl bromide and iodides having electronically diverse substituents (Scheme 53).

Peptide coupling is one of the most useful reactions in the organic synthesis which allow the synthesis of a diverse range of peptide based natural products as well as

Scheme 53 Photo-catalyzed transition metal-free arylation of oxindoles

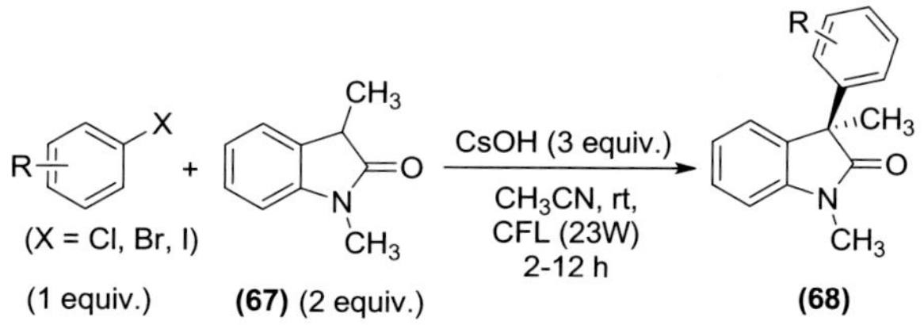

64   2 Energy-Efficient Process in Organic Synthesis

bioactive compounds and drugs. In general, peptide bond formation is achieved using carbodiimide-based reagents. For example, *N,N'*-dicyclohexylcarbodiimide (DCC), 1-ethyl-3-(3-dimethylaminopropyl) carbodiimide (EDC) are some of the routinely used coupling reagents. Similarly, organic salts of hexafluorophosphates such as benzotriazol-1-yloxy)-tripyrrolidino-phosphoniumhexafluoro phosphate (PyBOP), (O-(benzotriazol-1-yl)-1,1,3,3-tetramethyluronium hexafluorophosphate (HBTU), etc. are other expensive coupling reagents. Although the peptide synthesis using all these reagents is advantageous in terms of the yield and selectivity, the generation of large amount of waste materials from these reagents is one major concern that demands the discovery of novel coupling approach.

5.2 Photo-Induced Peptide Coupling Reaction

In that direction, Mishra et al. developed an environmentally benign coupling method under solar light where the iminium salt (**69** and **70**, Fig. 3) as novel coupling reagents was generated in situ, using DMAP and BrCCl$_3$ [85]. These two reagents form charge–transfer complex DMAP···BrCCl$_3$ that undergoes light-assisted transformation to the iminium salt at the later stage. This reaction worked under solar light and dry solvent was not mandatory for this coupling step. Optimized reaction condition tolerated the presence of different protecting group such as Boc, Cbz, Fmoc, etc. and successfully delivered the peptides in good-to-excellent yield of the peptides (Scheme 54).

Fig. 3 In situ formation of iminium salt as coupling reagents

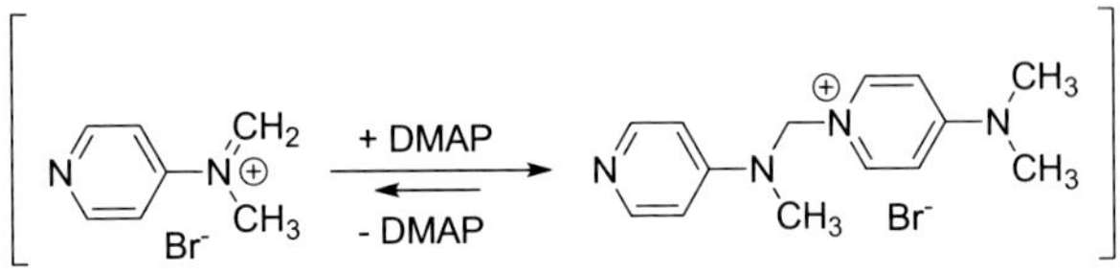

(69)   (70)

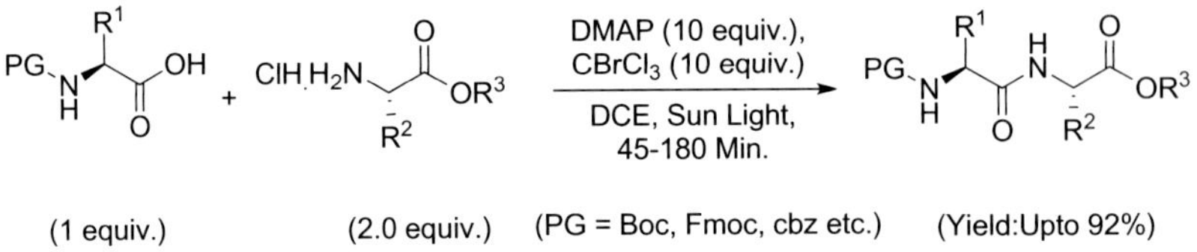

(1 equiv.)   (2.0 equiv.)   (PG = Boc, Fmoc, cbz etc.)   (Yield:Upto 92%)

Scheme 54 Peptide coupling under solar light

5.3 Photo-Induced Decarboxylative Coupling

In last few decades, the decarboxylative coupling reactions which generally involve the use of carboxylic acids and their derivatives (ester and salt) have emerged as very attractive methods for the construction of carbon–carbon and carbon–heteroatom bonds [86]. The use of generally inexpensive, stable carboxylic acid or its derivatives which could also easily be accessed through simple organic transformation methods are few important advantages of decarboxylative coupling method [87]. As compared to the conventional thermal heating, the photo-induced methods have been considered as better approach as it reduces the energy loss [88]. In addition to that, the photo-induced methods appear to be operationally simple and relatively safe. Several such photo-induced decarboxylative reaction protocols have been developed in recent years. In general, the decarboxylative coupling requires high temperature using conventional heating which impedes its wider application. In past few years, several photo-induced decarboxylative coupling reaction protocols employing redox-active ester such as N-(acyloxy)phthalimide as one of the coupling partner have been developed [89].

For example, Fu and coworker developed a transition metal-free decarboxylative coupling method for the alkylation of silyl enol ether which under the developed reaction condition led to the formation of functionalized ketones [90]. They demonstrated the synthetic importance of the combination of triphenyl phosphine and sodium iodide as an efficient catalytic system upon irradiation of blue LED (456 nm) in catalyzing the alkylation of silylenol ether **71** using N-(acyloxy)phthalimides (NPhth) as redox-active coupling partner (Scheme 55). However, the application of sodium iodide in the reduction of aryl halide and pseudohalide such as triflate to form aryl radical is known in the literature but only under the irradiation of high energy-UV radiation. The presence of triphenylphosphine facilitates the electron transfer process from sodium iodide to N-(acyloxy)phthalimides (NPhth) by forming complex and generation of R_3P-I• radical via intermolecular charge transfer with it which overall makes this process exergonic. Controlled experiments revealed that the combination of all three components NaI, PPh$_3$ and the presence of light are crucial for the successful transformation to product as no to trace product formation was observed in absence of any of these. The efficacy of the established reaction protocol in the formation of diverse varieties of ketone involving decarboxylative alkylation was successfully demonstrated in the reaction of a wide range of silylenol ether (Scheme 56). The reaction protocol was found to be compatible with different

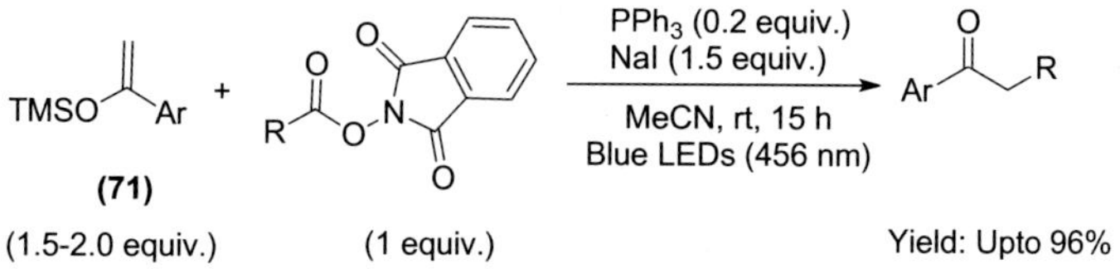

Scheme 55 Photo-induced synthesis of alkylated ketones involving decarboxyaltive coupling

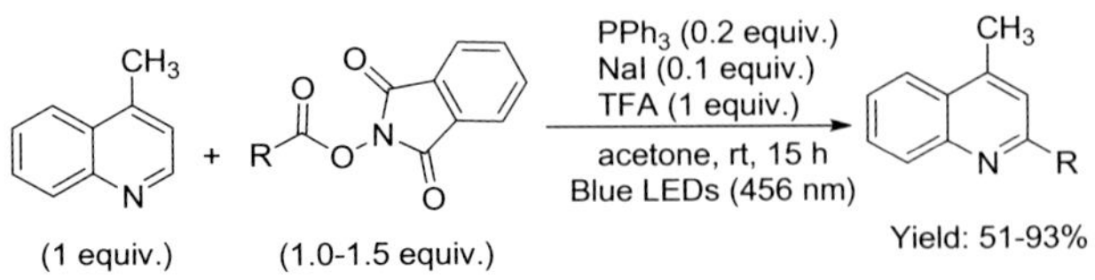

Scheme 56 Photo-induced Minisci-type alkylation of 4-methylquinoline

reactive functional groups such as alkene, alkyne, ester, Boc, halides, etc. Even in the presence of the strong electron withdrawing group such as CF_3, good yield of the ketone was obtained. The use of NaI in super stoichiometric amount (1.2. equivalent) was vindicated due to its dual role in photo-induced catalysis as well as in the succeeding step, i.e., desilylation. With slight modification in the reaction condition, i.e., by addition of trifluoroacetic acid (TFA) in stoichiometric amount (1 equiv.), the Minisci-type alkylation of N-heteroarenes was also achieved to provide functionalized 2-alkyl quinolines in good-to-excellent yield (Scheme 56) [91].

Such methods are synthetically viable as the reaction was performed under *metal-free* condition avoiding use of any conventional organometallic nucleophiles such as organoboron, organozinc, etc. and uses the ester of N-hydroxyphthalimide which is synthetically easier to achieve. In addition to that, the other key feature which attracts attention is that it minimizes or completely purges the possibility of β-hydride elimination which is generally observed and most troublesome during the alkylation using organometallic reagents. In another example, Jin and coworkers demonstrated the photo-induced alkylating ability of N-hydroxyphthalimide ester to attain regioselective alkylation of coumarin at C-3 position which allowed the synthesis of 3-alkyl coumarins in good-to-excellent yield (Scheme 57) [92]. The controlled experiments revealed that reaction proceeds via radical mechanism and both visible light and Ir(ppy)$_3$ as photo-catalyst are essential to achieve the desired transformation. In addition to that the use of triflic acid as an additive was essential as by reducing the amount of it from 0.2 equiv. the yield of the desired alkylated product got reduced significantly. The broad synthetic applicability of this environmentally benign condition was also demonstrated in regioselective chain alkylation of coumarin (Scheme 57). Reaction condition was found to be compatible with alkene, halide, ester, Boc, etc. The desired transformation was completely hampered by using

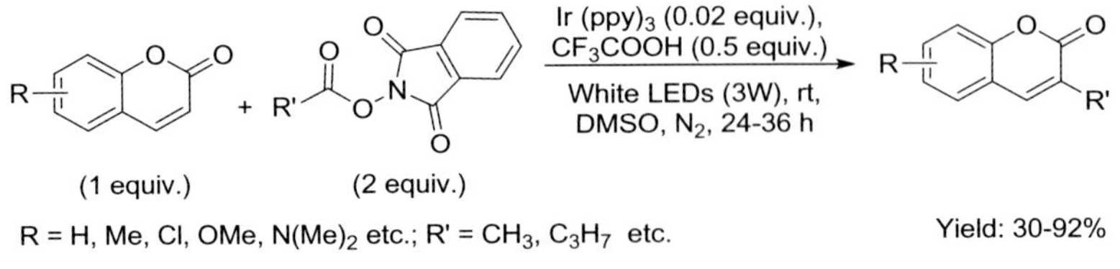

Scheme 57 Visible-light-induced regioselective alkylation of coumarins via decarboxylative coupling with N-hydroxyphthalimide esters

5 Photo-Induced Organic Transformations

TEMPO (2,2,6,6-tetramethyl-1-piperidinyloxy) (3.0 equiv) and 2,6-di-tert-butyl-4-methyphenol (BHT) as radical scavengers in the controlled experiments which indicated the formation of radical species in the mechanistic pathway. Similarly, other controlled experiments supported the formation of alkyl radical during the progress of the reaction.

Besides the C–C bond forming transformation, the application of light-induced photo-redox coupling was also employed to accomplish C–S bond formation involving decarboxylation step as one of the key steps. For example, Jin and coworkers developed an operationally simple and archetypal photo-catalyst and transition-metal-free reaction protocols to realize arylthiation *via* visible-light-mediated decarboxylative coupling of ester of N-hydroxyphthalimide under (Scheme 58) [93]. Excellent reactivity with wide substrate scope was demonstrated under developed reaction condition.

The capability of *N*-hydroxyphthalimide as redox-active alkylating agent was also demonstrated in the synthesis of alkyl substituted styrene derivatives. The coupling reaction leading to the formation of alkyl substituted styrene derivatives are uncommon and synthetically exigent due to the difficult initial oxidative step. The other key factor which restricts the use of alkyl halide as common electrophile similar to their aryl analogues is the fast decomposition of alkyl palladium(II) species to alkene via fast β-elimination step.

For example, Koy and coworkers developed a base-free and light-induced green protocol for the preparation of styrene derivative (Scheme 59) [93]. It involves the photo-activation of Pd(PPh$_3$)$_4$ followed by reduction of *N*-hydroxyphthalimide ester via single electron transfer (SET) to generate alkyl radical which further reacts with styrenes to produce (*E*)-substituted olefin as exclusive coupled product (Scheme 60). The developed method was found to be tolerating reactive functional groups like alkyne, ester, OH, nitrile, etc. This reaction protocol offers a green alternative to

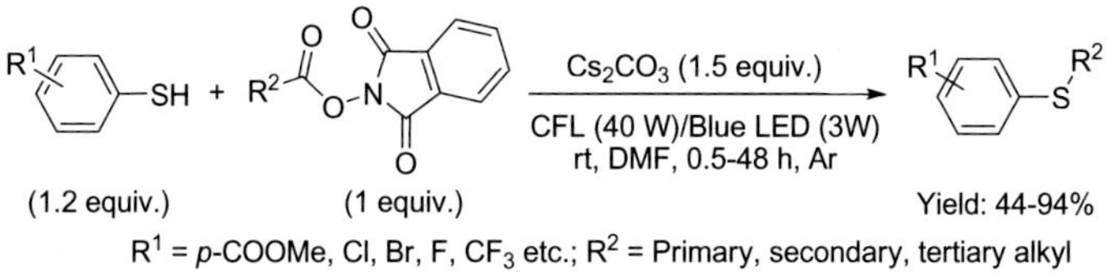

(1.2 equiv.) (1 equiv.) Yield: 44–94%

R^1 = *p*-COOMe, Cl, Br, F, CF$_3$ etc.; R^2 = Primary, secondary, tertiary alkyl

Scheme 58 Visible-light-induced photo-redox arylthiation involving decarboxylative coupling

(1.5 equiv.) (1 equiv.) Yield: 23–86%

R^1 = H, *p*-Ph, Cl, OMe, F, OAc etc.;R^2 = H, Ph; R^3 = Primary, secondary, tertiary alkyl

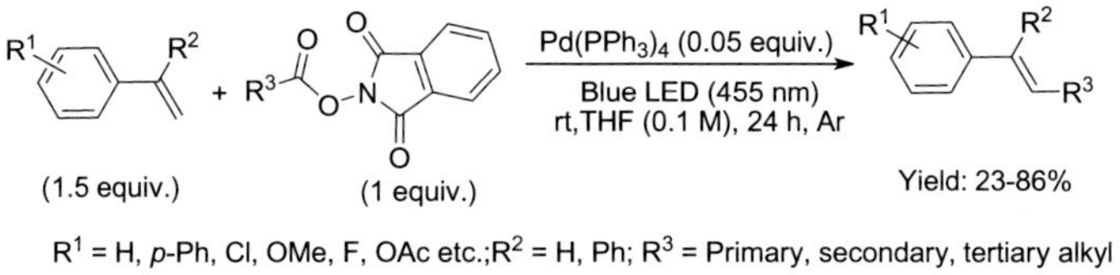

Scheme 59 Visible-light-mediated decarboxylative Heck coupling for the synthesis of (*E*)-substituted olefins

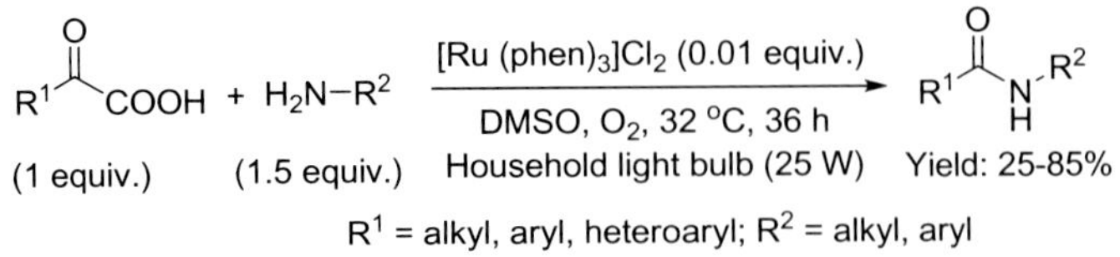

Scheme 60 Visible-light-mediated oxidative decarboxylative amidation

the popular Heck coupling as it is a base-free and light-induced methods which work at room temperature. Similarly, the visible-light-promoted C–N bond formation was also achieved involving oxidative decarboxylative coupling process. Liu and coworkers developed a visible-light-promoted oxidative decarboxylative amidation method using α-keto acids and amines (Scheme 60) [94]. This method was found to be compatible with both electronically diverse α-keto acids and amines for the synthesis of aryl, heteroaryl, and alkyl substituted amides in moderate to good yield under such mild and environmentally benign reaction condition. Alkyl-substituted 2-oxopropanoic acid and propargylamines were found to provide lower product yield under optimized reaction condition. In case of primary and secondary amines as coupling partner, excess amount (5–10 equiv.) was used to achieve the good yield of amides.

Wang and coworkers developed a reaction protocol comprised of 9-mesityl-10 methyl-acridinium perchlorate ([Acr+-Mes]ClO$_4$), oxygen and Na$_2$HPO$_4$ as photo-catalyst, oxidant, and base, respectively to achieve the C2-alkylated benzothiazole in presence of blue LED light (400–415 nm, 1.5 W) at room temperature (Scheme 61) [95]. Controlled experiments revealed that the presence of all three components, i.e., photo-catalyst, visible-light, and base were indispensable as in absence of any one of them, no product formation was observed. Similarly, the absence of product formation under nitrogen atmosphere confirmed the necessity of oxygen. The 1:1 combination of acetonitrile and water was also essential as low conversion of product was observed when either water (35%) or acetonitrile (58%) was used alone as solvent. The developed reaction conditions were found to be very efficient in achieving the C2-alkylation of benzothiazole through the reaction of electronically diverse benzothiazole and secondary and tertiary carboxylic acids. Sterically hindered acid counterpart, such as adamantane-1-carboxylic acid also exhibited excellent reactivity under established protocol.

Although reaction worked effectively with secondary and tertiary carboxylic acid but with primary carboxylic acid as acid counterpart, the C2-alkylation could not

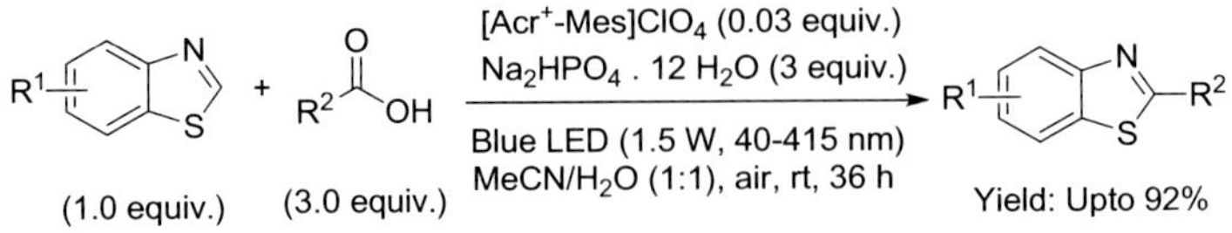

Scheme 61 Light-promoted metal-free decarboxylative C-2 alkylation of benzothiazole

be achieved. During the progress of reaction, the generation of alkyl radical was confirmed as upon the addition of 2,2,6,6-tetramethyl-1-piperidinyloxy (TEMPO) as radical scavenger, no desired product was formed.

6 Electrochemical Approach for Organic Synthesis

Similar to the previously discussed environmentally benign approaches adopted in achieving different types of organic transformations, electrochemical transformations have also been explored as greener alternative to the traditional synthetic methods which generally require the supply of energy through conventional heating and also the use of non-renewably accessible petroleum derived solvents and organic compounds [96]. Interestingly, electrochemical transformation relied on the use of electrical energy which could be generated by various renewable means such as through hydro-thermal, wind, solar, etc. electric current emerged as renewable and greener alternatives to the toxic and conventional heavy metal-based oxidizing and reducing agents. Similar to the simple, electrochemical cell where two-half reactions, i.e., oxidation and reduction take place at anode and cathode respectively, electro-organic synthesis also require similar setup. The electrode which could either is anode or cathode where reaction with the substrate takes place known as *working electrode* and another one which is required to complete the circuit is known as *counter electrode.* Both the electrodes could be placed in the same chamber (*undivided cells*) or could be divided with the help of salt bridge and partially permeable membrane (*divided cells*) to avoid the premature reduction or oxidation of intermediate. Electrosynthesis could be achieved by using constant current supply (*galvanostatic*) or by fixing the potential (*potentiostatic*). Although the electrical energy to achieve the organic transformations has been successfully employed as early as in 1849 when Hermann Kolbe, a German chemist who explained that in the previously developed method for the electrolysis of acetic acid by Michael Faraday in 1834, ethane is also produced along with carbon dioxide as a by-product [97]. Later, the Kolbe electrolysis of aliphatic acid has emerged as a popular method for the formation of hydrocarbons involving decarboxylation. It generally involves the use of carboxylate salt which leads to the generation of radical involving decarboxylation step under the electrochemical reaction condition. The radical further dimerizes to generate higher alkane through the formation of new C–C bond. It is one of the oldest and widely used electrochemical method for the generation of higher alkanes [98]. The application of electrochemical approach in organic synthesis has been highlighted with the help of some of the examples. For example, C–C bond was created under the electrochemical method using solid-supported base and aliphatic acid **72** as substrate. Reaction furnished the highly substituted dimer **73** as product involving Kolbe electrolysis of the acid (Scheme 62) [99].

Similarly, the application of electrochemical methods has also been demonstrated in the synthesis of adiponitrile form the glutamic acid 5-methyl ester **74** [100]. Importantly, glutamic acid could be obtained from the renewable sources. Under

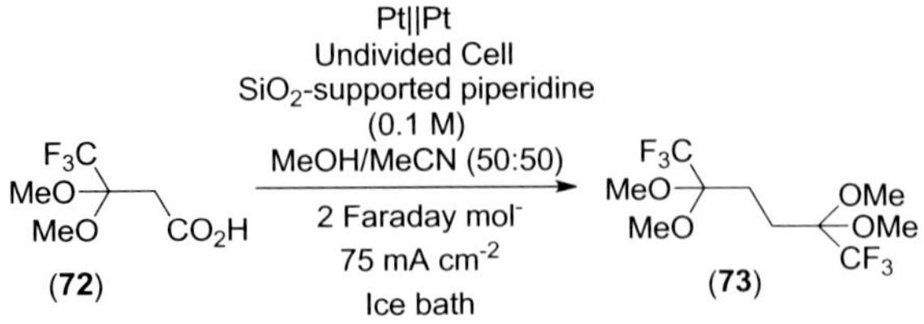

Scheme 62 C–C bond formation involving Kolbe electrolysis using solid-supported base

the electrochemical reaction condition, the ester of glutamic acid **74** converts into 3-cyanopropanoic acid methyl ester **76** through the generation of an intermediate **75** involving oxidative decarboxylation step. The ester **76**, under the Kolbe electrolysis method, leads to the formation of adiponitrile (Scheme 63). Importantly, adiponitrile is an important precursor to the hexamethylenediamine which further reacts with adipic acid to produce nylon 6,6 which is one of the most widely used polymer [101].

The application of electrochemical methods has also been demonstrated in the synthesis of various medicinally important heterocyclic compounds. For example, the synthesis of fused tetracyclic benzimidazole **79** was achieved under the electrochemical method using heteroarene **77** as a starting material. Reaction involved the formation of amidinyl radical **78** via N–H cleavage under electrochemical reaction condition. It further cyclizes and re-aromatized to produce tetracyclic benzimidazole **79** (Scheme 64) [102]. Conventionally, such transformation requires transition

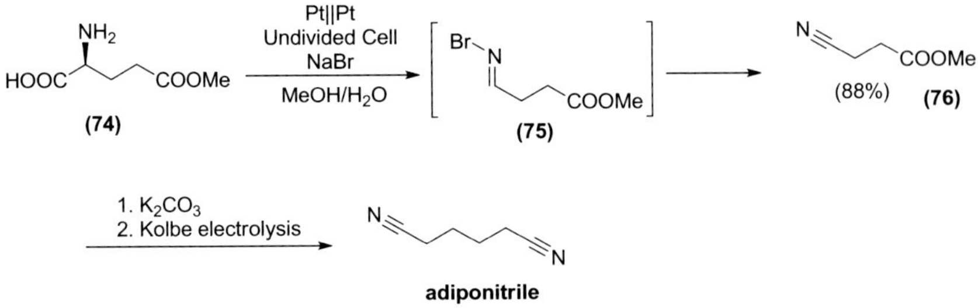

Scheme 63 Formation of adiponitrile from glutamate ester using electrochemical method

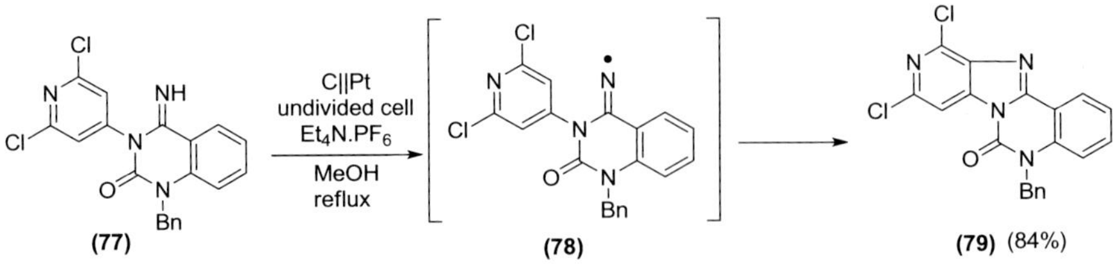

Scheme 64 Synthesis of tetracyclic *N*-heterocycle using electrochemical method

6 Electrochemical Approach for Organic Synthesis

metal-based catalyst which has successfully been avoided under the aforementioned protocol using electrochemical method.

Likewise, Shono-oxidation is an important electrical energy-driven method for the α-functionalization of cyclic and acyclic amines such as amides and carbamates through the formation of N-acyliminium ion via anodic oxidation of these amine derivatives [103]. The ion formed reacts with the alcohol which was mainly used as alcoholic solvents. This approach was also utilized widely in the synthesis of alkoxy mainly methoxy substituted functionalized amines, carbamates, and heterocycles. For example, the α-methoxylation of piperidine **80** was successfully carried out under Shono-oxidation reaction protocol to form methoxylated product **81** (Scheme 65).

Recently, Vitale, Grimaud, and coworkers reported a very first Ugi-type convenient and chemoselective oxidative isocyanide-based multicomponent reactions for an easy access of amides as biologically important molecular scaffold starting from readily available and stable substrates such as carboxylic acids, amines, isocyanides, and alcohols under environmental benign and mild reaction conditions (Scheme 66). This MCR reaction was catalyzed by TEMPO-catalyzed electro-oxidation process in absence of any supporting electrolyte.

Electrochemical approaches have also been adopted to achieve various other different types of organic transformation which generally require the use of transition metals [104]. The electrochemical approach can also be used for the chemoselective oxidation of secondary alcohols [105]. A number of biobased reagents were also explored under electrosynthesis in diverse range of carbohydrates as feedstock [106, 107].

Scheme 65 α-methoxylation of piperidine derivative

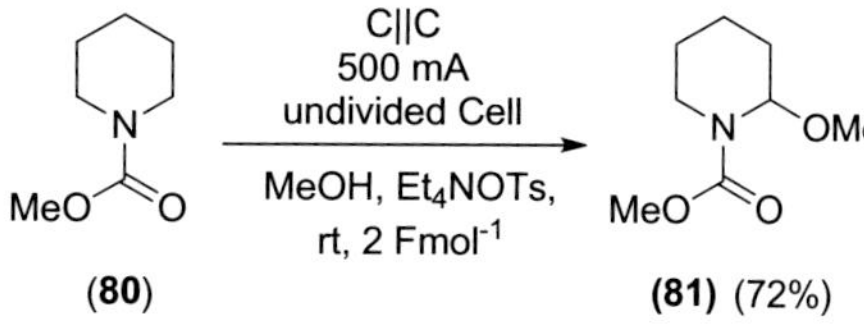

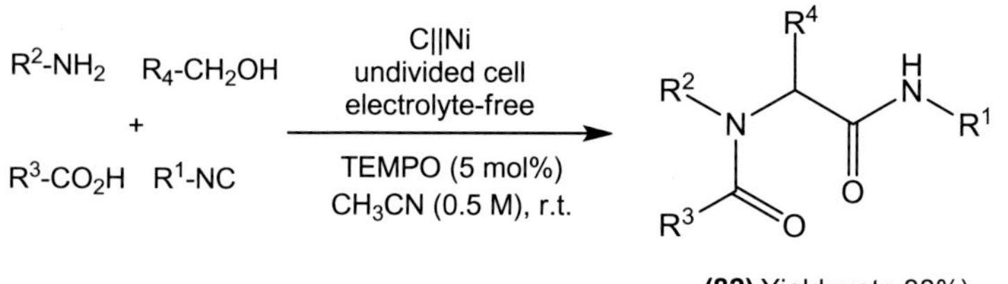

Scheme 66 TEMPO-catalyzed Ugi-type oxidative isocyanide-based multicomponent reactions under electrochemical condition

7 Conclusions

This chapter includes different environmentally benign and energy-efficient alternatives of the conventional heating methods. Reaction protocols developed under microwave, ultrasound, and visible light-assisted conditions have been discussed with the help of appropriate examples. In addition to that, reaction conditions developed under the ball milling and electrochemical approach have also been included in this chapter. Overall, it will provide a broad understanding and applications of various energy-efficient and green ways in achieving the different types of organic transformations.

References

1. Anastas, P., Eghbali, N.: Green chemistry: principles and practice. Chem. Soc. Rev. **39**(1), 301–312 (2010)
2. Kappe, C.O., Dallinger, D.: The impact of microwave synthesis on drug discovery. Nat. Rev. Drug. Discov. **5**(1), 51–63 (2006)
3. Mandal, B.: Alternate energy sources for sustainable organic synthesis. ChemistrySelect **4**(28), 8301–8310 (2019)
4. Gedye, R., Smith, F., Westaway, K., Ali, H., Baldisera, L., Laberge, L., Rousell, J.: The use of microwave ovens for rapid organic synthesis. Tetrahedron Lett. **27**(3), 279–282 (1986)
5. Salih, K.S.M., Baqi, Y.: Microwave-assisted palladium-catalyzed cross-coupling reactions: generation of carbon-carbon bond. Catalysts **10**(1), 4 (2020)
6. Fairoosa, J., Saranya, S., Radhika, S., Anilkumar, G.: Recent advances in microwave assisted multicomponent reactions. ChemistrySelect **5**(17), 5180–5197 (2020)
7. Nicolaou, K.C., Bulger, P.G., Sarlah, D.: Palladium-catalyzed cross-coupling reactions in total synthesis. Angew. Chemie Int. Ed. **44**(29), 4442–4489 (2005)
8. Johansson Seechurn, C.C.C., Kitching, M.O., Colacot, T.J., Snieckus, V.: Palladium-catalyzed cross-coupling: a historical contextual perspective to the 2010 nobel prize. Angew. Chemie Int. Ed. **51**(21), 5062–5085 (2012)
9. Hervé, G., Len, C.: First ligand-free, microwave-assisted, heck cross-coupling reaction in pure water on a nucleoside-application to the synthesis of antiviral BVDU. RSC Adv. **4**(87), 46926–46929 (2014)
10. Glasnov, T.N., Stadlbauer, W., Kappe, C.O.: Microwave-assisted multistep synthesis of functionalized 4-arylquinolin-2(1 H)-ones using palladium-catalyzed cross-coupling chemistry. J. Org. Chem. **70**(10), 3864–3870 (2005)
11. Du, L., Wang, Y.: Microwave-promoted heck reaction using Pd(OAc)$_2$ as catalyst under ligand-free and solvent-free conditions. Synth. Commun. **37**(2), 217–222 (2007)
12. Karu, R., Gedu, S.: Microwave assisted domino heck cyclization and alkynylation: synthesis of alkyne substituted dihydrobenzofurans. Green Chem. **20**(2), 369–374 (2018)
13. Heravi, M.M., Zadsirjan, V.: Prescribed drugs containing nitrogen heterocycles: an overview. RSC Adv. **10**(72), 44247–44311 (2020)
14. Jiang, B., Rajale, T., Wever, W., Tu, S., Li, G.: Multicomponent reactions for the synthesis of heterocycles. Chem. Asian J. **5**(11), 2318–2335 (2010)
15. Manjashetty, T.H., Yogeeswari, P., Sriram, D.: Microwave assisted one-pot synthesis of highly potent novel isoniazid analogues. Bioorg. Med. Chem. Lett. **21**(7), 2125–2128 (2011)
16. Mahindra, A., Bagra, N., Jain, R.: Palladium-catalyzed regioselective C-5 arylation of protected L-histidine: microwave-assisted C–H activation adjacent to donor arm. J. Org. Chem. **78**(21), 10954–10959 (2013)

References

17. Mahindra, A., Jain, R.: Regiocontrolled palladium-catalyzed and copper-mediated C–H bond functionalization of protected L-histidine. Org. Biomol. Chem. **12**(23), 3792–3796 (2014)
18. Tripathi, G., Singh, A.K., Kumar, A.: Arylpyrazoles: heterocyclic scaffold of immense therapeutic application. Curr. Org. Chem. **24**(14), 1555–1581 (2020)
19. Flood, D.T., Hintzen, J.C.J., Bird, M.J., Cistrone, P.A., Chen, J.S., Dawson, P.E.: Leveraging the Knorr Pyrazole synthesis for the facile generation of thioester surrogates for use in native chemical ligation. Angew. Chemie **130**(36), 11808–11813 (2018)
20. Kumar, A., Rao, M.L.N.: Pot-economic synthesis of diarylpyrazoles and pyrimidines involving Pd-catalyzed cross-coupling of 3-trifloxychromone and triarylbismuth. J. Chem. Sci. **130**(12), 1–11 (2018)
21. Sabitha, G., SatheeshBabu, R., Yadav, J.S.: One pot synthesis of 4-(2-hydroxybenzoyl)-pyrazoles from 3-formylchromones under microwave irradiation in solvent-free conditions. Synth. Commun. **28**(24), 4571–4576 (1998)
22. Du, K., Xia, C., Wei, M., Chen, X., Zhang, P.: Microwave-assisted rapid synthesis of sugar-based pyrazole derivatives with anticancer activity in water. RSC Adv. **6**(71), 66803–66806 (2016)
23. Molina, P., Tárraga, A., Otón, F.: Imidazole derivatives: a comprehensive survey of their recognition properties. Org. Biomol. Chem. **10**(9), 1711–1724 (2012)
24. Attanasi, O.A., Bianchi, L., Campisi, L.A., De Crescentini, L., Favi, G., Mantellini, F.: A novel solvent-free approach to imidazole containing nitrogen-bridgehead heterocycles. Org. Lett. **15**(14), 3646–3649 (2013)
25. Ansari, A.J., Sharma, S., Pathare, R.S., Gopal, K., Sawant, D.M., Pardasani, R.T.: Solvent-free multicomponent synthesis of biologically-active fused–imidazo heterocycles catalyzed by reusable Yb(OTf)$_3$ under microwave irradiation. ChemistrySelect **1**(5), 1016–1021 (2016)
26. Aldred, K.J., Kerns, R.J., Osheroff, N.: Mechanism of quinolone action and resistance. Biochemistry **53**(10), 1565–1574 (2014)
27. Lengyel, L.C., Sipos, G., Sipocz, T., Vago, T., Dormán, G., Gerencser, J., Makara, G., Darvas, F.: Synthesis of condensed heterocycles by the Gould-Jacobs reaction in a novel three-mode pyrolysis reactor. Org. Process Res. Dev. **19**(3), 399–409 (2015)
28. Zewge, D., Chen, C., Deer, C., Dormer, P.G., Hughes, D.L.: A mild and efficient synthesis of 4-quinolones and quinolone heterocycles. J. Org. Chem. **72**(11), 4276–4279 (2007)
29. Dave, C.G., Joshipura, H.M.: Microwave assisted Gould-Jacob reaction: synthesis of 4-quinolones under solvent-free conditions. Ind. J. Chem. B **41**(3), 650–652 (2002)
30. Jia, C.-S., Dong, Y.-W., Tu, S.-J., Wang, G.-W.: Microwave-assisted solvent-free synthesis of substituted 2-quinolones. Tetrahedron **63**(4), 892–897 (2007)
31. Michael, J.P.: Quinoline, quinazoline and acridone alkaloids. Nat. Prod. Rep. **25**(1), 166–187 (2008)
32. Mohammadkhani, L., Heravi, M.M.: Microwave-assisted synthesis of quinazolines and quinazolinones: an overview. Front Chem. **8**, 921 (2020)
33. Rad-Moghadam, K., Samavi, L.: One-pot three-component synthesis of 2-substituted 4-aminoquinazolines. J. Heterocycl. Chem. **43**(4), 913–916 (2006)
34. Seijas, J.A., Vázquez-Tato, M.P., Carballido-Reboredo, M.R., Crecente-Campo, J., Romar-Lopez, L.: Lawesson's reagent and microwaves: a new efficient access to benzoxazoles and benzothiazoles from carboxylic acids under solvent-free conditions. Synlett **2**, 313–317 (2007)
35. Mwande-Maguene, G., Jakhlal, J., Lekana-Douki, J.-B., Mouray, E., Bousquet, T., Pellegrini, S., Grellier, P., Ndouo, F.S.T., Lebibi, J., Pelinski, L.: One-pot microwave-assisted synthesis and antimalarial activity of ferrocenyl benzodiazepines. New J. Chem. **35**(11), 2412–2415 (2011)
36. Rao, M.L.N., Kumar, A.: Pd-catalyzed chemo-selective mono-arylations and bis-arylations of functionalized 4-chlorocoumarins with triarylbismuths as threefold arylating reagents. Tetrahedron **70**(39), 6995–7005 (2014)
37. Rajitha, B., Kumar, V.N., Someshwar, P., Madhav, J.V., Reddy, P.N., Reddy, Y.T.: Dipyridine copper chloride catalyzed coumarin synthesis via Pechmann condensation under conventional heating and microwave irradiation. ARKIVOC **12**, 23–27 (2006)

38. Vahabi, V., Hatamjafari, F.: Microwave assisted convenient one-pot synthesis of coumarin derivatives via Pechmann condensation catalyzed by FeF$_3$ under solvent-free conditions and antimicrobial activities of the products. Molecules 19(9), 13093–13103 (2014)
39. Crecente-Campo, J., Vazquez-Tato, M.P., Seijas, J.A.: Microwave-promoted, one-pot, solvent-free synthesis of 4-arylcoumarins from 2-hydroxybenzophenones. Eur. J. Org. Chem. 21, 4130–4135 (2010)
40. Fiorito, S., Epifano, F., Taddeo, V.A., Genovese, S.: Ytterbium triflate promoted coupling of phenols and propiolic acids: synthesis of coumarins. Tetrahedron Lett. 57(26), 2939–2942 (2016)
41. Balakrishna, C., Kandula, V., Gudipati, R., Yennam, S., Devi, P.U., Behera, M.: An efficient microwave-assisted propylphosphonic anhydride (T3P®)-mediated one-pot chromone synthesis via enaminones. Synlett 29(8), 1087–1091 (2018)
42. Heravi, M.M., Zadsirjan, V., Hamidi, H., Amiri, P.H.T.: Total synthesis of natural products containing benzofuran rings. RSC Adv. 7(39), 24470–24521 (2017)
43. Rao, M.L.N., Awasthi, D.K., Banerjee, D.: Microwave-mediated solvent-free rap-stoermer reaction for efficient synthesis of benzofurans. Tetrahedron Lett. 48(3), 431–434 (2007)
44. McNaught, A.D., Wilkinson, A.: IUPAC. Compendium of Chemical Terminology, 2nd ed. (the "Gold Book") (1997)
45. Stolle, A., Szuppa, T., Leonhardt, S.E.S., Ondruschka, B.: Ball milling in organic synthesis: solutions and challenges. Chem. Soc. Rev. 40(5), 2317–2329 (2011)
46. Avila-Ortiz, C.G., Juaristi, E.: Novel methodologies for chemical activation in organic synthesis under solvent-free reaction conditions. Molecules 25(16), 3579 (2020)
47. Zhu, X., Li, Z., Jin, C., Xu, L., Wu, Q., Su, W.: Mechanically activated synthesis of 1,3,5-triaryl-2-pyrazolines by high speed ball milling. Green Chem. 11(2), 163–165 (2009)
48. Sharma, H., Kaur, N., Singh, N., Jang, D.O.: Synergetic catalytic effect of ionic liquids and ZnO nanoparticles on the selective synthesis of 1,2-disubstituted benzimidazoles using a ball-milling technique. Green Chem. 17(8), 4263–4270 (2015)
49. Thorwirth, R., Stolle, A., Ondruschka, B., Wild, A., Schubert, U.S.: Fast, ligand-and solvent-free copper-catalyzed click reactions in a ball mill. Chem. Commun. 47(15), 4370–4372 (2011)
50. Maleki, A., Javanshir, S., Naimabadi, M.: Facile synthesis of imidazo[1,2-a]pyridines via a one-pot three-component reaction under solvent-free mechanochemical ball-milling conditions. RSC Adv. 4(57), 30229–30232 (2014)
51. Wang, F.-J., Xu, H., Xin, M., Zhang, Z.: I 2-mediated amination/cyclization of ketones with 2-aminopyridines under high-speed ball milling: solvent-and metal-free synthesis of 2,3-substituted imidazo[1,2-a]pyridines and zolimidine. Mol. Divers 20(3), 659–666 (2016)
52. Kaupp, G., Naimi-Jamal, M.R.: Quantitative cascade condensations between o-phenylenediamines and 1,2-dicarbonyl compounds without production of wastes. Eur. J. Org. Chem. 2002(8), 1368–1373 (2002)
53. Wang, G.-W., Dong, Y.-W., Wu, P., Yuan, T.-T., Shen, Y.-B.: Unexpected solvent-free cycloadditions of 1,3-cyclohexanediones to 1-(pyridin-2-yl)-enones mediated by manganese (III) acetate in a ball mill. J. Org. Chem. 73(18), 7088–7095 (2008)
54. Nathaniel, T.G.: The solvent-free and catalyst-free conversion of an aziridine to an oxazolidinone using only carbon dioxide. Green Chem. 13(11), 3224–3229 (2011)
55. Egorov, I.N., Santra, S., Kopchuk, D.S., Kovalev, I.S., Zyryanov, G.V., Majee, A., Ranu, B.C., Rusinov, V.L., Chupakhin, O.N.: Ball milling: an efficient and green approach for asymmetric organic syntheses. Green Chem. 22(2), 302–315 (2020)
56. Wang, Y.-F., Chen, R.-X., Wang, K., Zhang, B.-B., Li, Z.-B., Xu, D.-Q.: Fast, solvent-free and hydrogen-bonding-mediated asymmetric michael addition in a ball mill. Green Chem. 14(4), 893–895 (2012)
57. Rantanen, T., Schiffers, I., Bolm, C.: Solvent-free asymmetric anhydride opening in a ball mill. Org. Process Res. Dev. 11(3), 592–597 (2007)
58. Seo, T., Ishiyama, T., Kubota, K., Ito, H.: Solid-state Suzuki-Miyaura cross-coupling reactions: olefin-accelerated C–C coupling using mechanochemistry. Chem. Sci. 10(35), 8202–8210 (2019)

References

59. Fulmer, D.A., Shearouse, W.C., Medonza, S.T., Mack, J.: Solvent-free Sonogashira coupling reaction via high speed ball milling. Green Chem. **11**, 1821–1825 (2009)
60. Su, W., Yu, J., Li, Z., Jiang, Z.: Solvent-free cross-dehydrogenative coupling reactions under high speed ball-milling conditions applied to the synthesis of functionalized tetrahydroisoquinolines. J. Org. Chem. **76**(21), 9144–9150 (2011)
61. Draye, M., Chatel, G., Duwald, R.: Ultrasound for drug synthesis: a green approach. Pharmaceuticals **13**(2), 23 (2020)
62. Chatel, G., Leclerc, L., Naffrechoux, E., Bas, C., Kardos, N., Goux-Henry, C., Andrioletti, B., Draye, M.: Ultrasonic properties of hydrophobic bis(trifluoromethylsulfonyl)imide-based ionic liquids. J. Chem. Eng. Data **57**(12), 3385–3390 (2012)
63. Li, J.-T., Zhang, X.-H., Lin, Z.-P.: An improved synthesis of 1,3,5-triaryl-2-pyrazolines in acetic acid aqueous solution under ultrasound irradiation. Beilstein J. Org. Chem. **3**(1), 13 (2007)
64. Shelke, K.F., Sapkal, S.B., Shingare, M.S.: Ultrasound-assisted one-pot synthesis of 2,4,5-triarylimidazole derivatives catalyzed by ceric (IV) ammonium nitrate in aqueous media. Chin. Chem. Lett. **20**(3), 283–287 (2009)
65. Jiang, Y.-J., Cai, J.-J., Zou, J.-P., Zhang, W.: Gallium (III) triflate-catalyzed [4+2+1] cycloadditions for the synthesis of novel 3,4-disubstituted-1,5-benzodiazepines. Tetrahedron Lett. **51**(3), 471–474 (2010)
66. Guzen, K.P., Cella, R., Stefani, H.A.: Ultrasound enhanced synthesis of 1,5-benzodiazepinic heterocyclic rings. Tetrahedron Lett. **47**(46), 8133–8136 (2006)
67. Braibante, M.E.F., Braibante, H.T.S., Da Roza, J.K., Henriques, D.M., de Carvalho Tavares, L.: Synthesis of aminopyrazoles from α-oxoketene O, N-acetals using Montmorillonite K-10/ultrasound. Synthesis **8**, 1160–1162 (2003)
68. Venigalla, L.S., Maddila, S., Jonnalagadda, S.B.: Facile, efficient, catalyst-free, ultrasound-assisted one-pot green synthesis of triazole derivatives. J. Iran. Chem. Soc. **17**(7), 1539–1544 (2020)
69. Castillo, J.-C., Bravo, N.-F., Tamayo, L.-V., Mestizo, P.-D., Hurtado, J., Macías, M., Portilla, J.: Water-compatible synthesis of 1,2,3-triazoles under ultrasonic conditions by a Cu(I) complex-mediated click reaction. ACS Omega **5**(46), 30148–30159 (2020)
70. Pereira, C.M.P., Stefani, H.A., Guzen, K.P., Orfao, A.T.G.: Improved synthesis of benzotriazoles and 1-acylbenzotriazoles by ultrasound irradiation. Lett. Org. Chem. **4**(1), 43–46 (2007)
71. Li, Y., Wang, L.T., Wang, Z., Yuan, S., Wu, S., Wang, S.: Ultrasound-assisted synthesis of novel pyrrole dihydropyrimidinones in lactic acid. ChemistrySelect **1**(21), 6855–6858 (2016)
72. Liu, J., van Iersel, M.W.: Photosynthetic physiology of blue, green, and red light: light intensity effects and underlying mechanisms. Front Plant Sci. **12**, 328 (2021)
73. Roth, H.D.: The beginnings of organic photochemistry. Angew. Chemie Int. Ed. Eng. **28**(9), 1193–1207 (1989)
74. Narayanam, J.M.R., Stephenson, C.R.J.: Visible light photoredox catalysis: applications in organic synthesis. Chem. Soc. Rev. **40**(1), 102–113 (2011)
75. Cano-Yelo, H., Deronzier, A.: Photocatalysis of the Pschorr reaction by tris-(2,2′-bipyridyl)ruthenium (II) in the phenanthrene series. J. Chem. Soc. Perkin Trans. **2**(6), 1093–1098 (1984)
76. Zhang, J., Chen, J., Zhang, X., Lei, X.: Total syntheses of menisporphine and daurioxoisoporphine C enabled by photoredox-catalyzed direct C–H arylation of isoquinoline with aryldiazonium salt. J. Org. Chem. **79**(21), 10682–10688 (2014)
77. Xue, D., Jia, Z., Zhao, C., Zhang, Y., Wang, C., Xiao, J.: Direct arylation of N-heteroarenes with aryldiazonium salts by photoredox catalysis in water. Chem. Eur. J. **20**, 2960–2965 (2014)
78. Li, Z., Song, H., Guo, R., Zuo, M., Hou, C., Sun, S., He, X., Sun, Z., Chu, W.: Visible-light-induced condensation cyclization to synthesize benzimidazoles using fluorescein as a photocatalyst. Green Chem. **21**(13), 3602–3605 (2019)
79. Samanta, S., Das, S., Biswas, P.: Photocatalysis by 3,6-disubstituted-s-tetrazine: visible-light driven metal-free green synthesis of 2-substituted benzimidazole and benzothiazole. J. Org. Chem. **78**(22), 11184–11193 (2013)

80. Almeida, J.F., Castedo, L., Fernández, D., Neo, A.G., Romero, V., Tojo, G.: Base-induced photocyclization of 1,2-diaryl-1-tosylethenes. A mechanistically novel approach to phenanthrenes and phenanthrenoids. Org. Lett. **5**(26), 4939–4941 (2003)
81. Liu, Y.-L., Liang, Y., Pi, S.-F., Huang, X.-C., Li, J.-H.: Palladium-catalyzed cocyclotrimerization of allenes with arynes: selective synthesis of phenanthrenes. J. Org. Chem. **74**(8), 3199–3202 (2009)
82. Seganish, W.M., DeShong, P.: Application of aryl siloxane cross-coupling to the synthesis of allocolchicinoids. Org. Lett. **8**(18), 3951–3954 (2006)
83. Xiao, T., Dong, X., Tang, Y., Zhou, L.: Phenanthrene synthesis by Eosin Y-catalyzed, visible light-induced [4+2] benzannulation of biaryldiazonium salts with alkynes. Adv. Synth. Catal. **354**(17), 3195–3199 (2012)
84. Liang, K., Li, N., Zhang, Y., Li, T., Xia, C.: Transition-metal-free α-arylation of oxindoles via visible-light-promoted electron transfer. Chem. Sci. **10**(10), 3049–3053 (2019)
85. Mishra, A.K., Parvari, G., Santra, S.K., Bazylevich, A., Dorfman, O., Rahamim, J., Eichen, Y., Szpilman, A.M.: Solar and visible light assisted peptide coupling. Angew. Chemie Int. Ed. **60**(22), 12406–12412 (2021)
86. Zhang, T., Wang, N.-X., Xing, Y.: Advances in decarboxylative oxidative coupling reaction. J. Org. Chem. **83**(15), 7559–7565 (2018)
87. Gooßen, L.J., Deng, G., Levy, L.M.: Synthesis of biaryls via catalytic decarboxylative coupling. Scienc **313**(5787), 662–664 (2006)
88. Reischauer, S., Pieber, B.: Emerging concepts in photocatalytic organic synthesis. I Sci. 102209 (2021)
89. Murarka, S.: N-(acyloxy) phthalimides as redox-active esters in cross-coupling reactions. Adv. Synth. Catal. **360**(9), 1735–1753 (2018)
90. Fu, M.-C., Shang, R., Zhao, B., Wang, B., Fu, Y.: Photocatalytic decarboxylative alkylations mediated by triphenylphosphine and sodium iodide. Science **363**(6434), 1429–1434 (2019)
91. Li, G.-X., Hu, X., He, G., Chen, G.: Photoredox-mediated minisci-type alkylation of N-heteroarenes with alkanes with high methylene selectivity. ACS Catal. **8**(12), 11847–11853 (2018)
92. Jin, C., Yan, Z., Sun, B., Yang, J.: Visible-light-induced regioselective alkylation of coumarins via decarboxylative coupling with N-hydroxyphthalimide esters. Org. Lett. **21**(7), 2064–2068 (2019)
93. Jin, Y., Yang, H., Fu, H.: An N-(acetoxy) phthalimide motif as a visible-light pro-photosensitizer in photoredox decarboxylative arylthiation. Chem. Commun. **52**(87), 12909–12912 (2016)
94. Liu, J., Liu, Q., Yi, H., Qin, C., Bai, R., Qi, X., Lan, Y., Lei, A.: Visible-light-mediated decarboxylation/oxidative amidation of α-keto acids with amines under mild reaction conditions using O_2. Angew. Chemie Int. Ed. **53**(2), 502–506 (2014)
95. Wang, B., Li, P., Miao, T., Zou, L., Wang, L.: Visible-light induced decarboxylative C2-alkylation of benzothiazoles with carboxylic acids under metal-free conditions. Org. Biomol. Chem. **17**(1), 115–121 (2019)
96. Yan, M., Kawamata, Y., Baran, P.S.: Synthetic organic electrochemical methods since 2000: on the verge of a renaissance. Chem. Rev. **117**(21), 13230–13319 (2017)
97. Holzhäuser, F.J., Mensah, J.B., Palkovits, R.: (Non-)Kolbe electrolysis in biomass valorization—a discussion of potential applications. Green Chem. **22**(2), 286–301 (2020)
98. Kolbe, H.: Untersuchungen Über Die Elektrolyse Organischer Verbindungen. Justus Liebigs Ann Chem **69**(3), 257–294 (1849)
99. Kurihara, H., Fuchigami, T., Tajima, T.: Kolbe carbon–carbon coupling electrosynthesis using solid-supported bases. J. Org. Chem. **73**(17), 6888–6890 (2008)
100. Dai, J., Huang, Y., Fang, C., Guo, Q., Fu, Y.: Electrochemical synthesis of adiponitrile from the renewable raw material glutamic acid. Chemsuschem **5**(4), 617–620 (2012)
101. Blanco, D.E., Dookhith, A.Z., Modestino, M.A.: Enhancing selectivity and efficiency in the electrochemical synthesis of adiponitrile. React. Chem. Eng. **4**(1), 8–16 (2019)

References

102. Zhao, H., Hou, Z., Liu, Z., Zhou, Z., Song, J., Xu, H.: Amidinyl radical formation through anodic $N-H$ bond cleavage and its application in aromatic $C-H$ bond functionalization. Angew. Chemie Int. Ed. **56**(2), 587–590 (2017)
103. Shono, T., Hamaguchi, H., Matsumura, Y.: Electroorganic chemistry. XX. Anodic oxidation of carbamates. J. Am. Chem. Soc. **97**(15), 4264–4268 (1975)
104. Kärkäs, M.D.: Electrochemical strategies for C–H functionalization and C–N bond formation. Chem. Soc. Rev. **47**(15), 5786–5865 (2018)
105. Sommer, F., Kappe, C.O., Cantillo, D.: Chemoselective electrochemical oxidation of secondary alcohols using a recyclable chloride-based medoator. Synlett **33**, 166–170 (2022)
106. Vedovato, V., Vanbroekhoven, K., Pant, D., Helsen, J.: Elecrtosynthesis of biobased chemicals using carbohydrates as a feedstock. Molecules **25**, 3712 (2020)
107. Pan, N., Lee, M.X., Bunel, L., Grimaud, L., Vitale, M.R.: Electrochemical TEMPO-catalyzed oxidative Ugi-type reaction. ACS Org. Inorg. Au **1**(1), 18–22 (2021)

Chapter 3
Green Solvents: Application in Organic Synthesis

1 Introduction

Solvents play very important role in a chemical reaction by providing a medium to solubilize the reacting components, reagents and also facilitate mass and heat transfer processes [1]. It also affects the reaction kinetics and the stability of various reacting species and intermediates. Organic solvents mainly hydrocarbon and their derivatives have been a choice of solvents particularly in organic synthesis. Although solvents are essential component of the reaction, it is important to mention that it also contributes immensely in the total waste generation and increase e-factor value particularly in case of fine chemical and pharmaceutical industries [2]. In addition to that, low boiling point, toxicity, flammability, etc., in case of most of the organic solvents, some of the important issues which different types of safety, health, and environmental issues are also associated with their use. Therefore, various greener strategies have been explored. For example, the development of *solventless processes* is one of such attractive strategy [3, 4]. Diverse types of chemical transformations such as multicomponent reactions and cross-coupling reactions have been developed for the synthesis of vast varieties of heterocyclic skeletons and other medicinally important compounds. One limitation is that all types of reaction may not be performed under solvent-free reaction condition. Therefore in order to minimize and the environmental impact as well as impact on health that are known to occur because of organic solvents [5], several greener alternatives have been explored. Green solvents have potential to minimize and replace the use of conventional organic solvents some of which are considered to be hazardous and have severe adverse effect on the environment [6]. In addition to that one important aspect which demands the discovery of other greener alternative is its procurement from the non-renewable petroleum sources [7]. Therefore, the application of different greener alternatives of the conventional solvents has been explored which are eco-friendly, biodegradable and have minimal environmental impact [8].

© The Author(s), under exclusive license to Springer Nature Singapore Pte Ltd. 2022
V. K. Tiwari et al., *Green Chemistry*,
https://doi.org/10.1007/978-981-19-2734-8_3

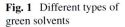

Fig. 1 Different types of green solvents

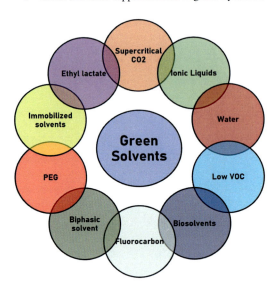

2 Types of Green Solvents

In order to categorize a solvent a green one, various properties such as its source, biodegradability, impact on three safety, health and environment (SHE) have been considered. In addition to that various guidelines have also been proposed to which have been discussed in the subsequent section. Figure 1 represents some of the green solvents which will be discussed in this chapter [9].

3 Solvent Selection Guides

In order to classify a solvent as green, various guidelines have been formulated by different academic and industrial institutions. For example, GlaxoSmithKline (GSK) has developed a *solvent selection tool* to identify the appropriate non-aqueous solvents which could be used in leather industry [10, 11]. Similarly, solvent selection guide developed by Sanofi helps in selecting a sustainable solvent which would have broader environmental acceptability [12]. Based on that solvents have been placed in four categories with different color codes (Table 1).

Likewise various other agencies such as *"Registration, Evaluation, Authorisation and Restriction of Chemicals (REACH)"* in Europe guides and control the usage of different varieties of chemicals including solvents. So while using any chemicals the conditions and regulations established in REACH have to be followed strictly otherwise these chemicals or materials containing such chemicals could be removed immediately [13]. Although there are few differences in classifying a solvent in a particular category and it may vary according to different guidelines established by

Table 1 Four classes of solvents as per Sanofi's solvent selection guide

S. No	Color code	Class	Example
1	Green	Recommended solvent	Ethyl acetate, Methyl-THF
2	Yellow	Substitution advisable	THF, Diisopropyl ether
3	Red	Substitution requested	Dioxane, Diglyme
4	Brown	Banned solvent	Diethyl ether, Benzene, n-pentane

different agencies, the aim of all is to classify the solvents mainly according to their impact on *Health, Safety and Environment* (HSE).

4   Green Solvents and Their Application

4.1   *Water as Solvent*

Water is a universal solvent with high dielectric constant, so high polarity and is present in abundant amount in human beings as well as in our biosphere. Its abundance, least cost and least hazard potential makes water a great green solvent. Also, its interesting behavior like specialized selectivity, three-dimensional hydrogen-bonding potential gives it tendency of hydrophobic effect, high surface tension, and a large cohesive energy density, pH values that can be modifiable, peculiar salting in-and-out effect and also can be used in biphasic reaction systems. The property of hydrogen bonding makes water a solvent that has capability to accelerate some reactions both under normal atmospheric and also at high pressures [6]. Reactions at high-pressure condition with high temperatures are often accepted for the "sub-critical water region" that is liquid water phase above the boiling point at ambient pressure (100 °C) and below the critical point (374 °C, 22.1 MPa). The chemical and physical properties of subcritical water display a noteworthy dependency on the particular temperature and pressure, so that solvent properties are strongly prejudiced and adaptable. The thick wall glass containers are used in low-pressure condition. The special supercritical water is very promising solvent for biosolvents and fuel obtained by processing biomass without using any hazardous chemicals [14]. Exploiting this peculiar and interesting behavior of water are seen in various chemical reactions in plethora of literature at temperatures between 150 and 250 °C [15].

A huge range of reactions has been successfully explored in water [6]. For example, reaction performed between cyclopentadiene and butenone in water was found to be 700 faster in waster compared to other organic solvents like 2,2,4-trimethyl pentane along with high endo/exo ratio (Scheme 1) [16]. Similarly, the considerable enhancement in the reaction rate was also observed in case of the cycloaddition reaction between cyclopentadiene and acrylonitrile (Scheme 1).

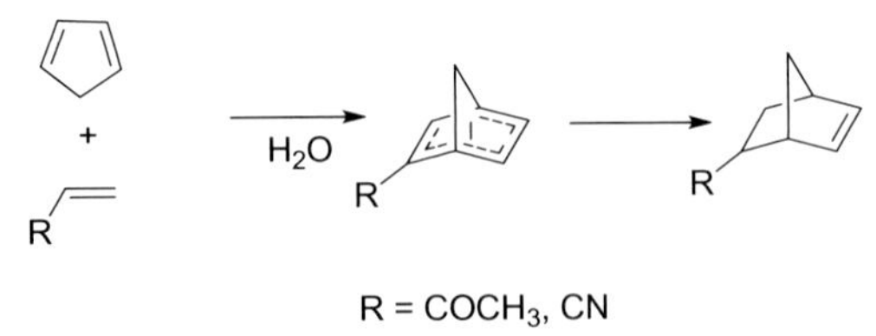

Scheme 1 DielsAlder reaction between cyclopentadiene with butanone (R=COCH$_3$) and acrylonitrile (R=CN)

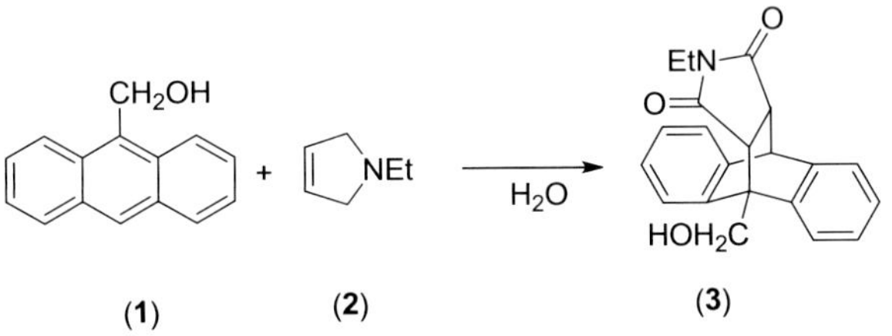

R = COCH$_3$, CN

Scheme 2 Diels–Alder reaction between anthracene-9-carbinol and N-ethylmaleimide

(1)　　　(2)　　　(3)

An exceptional effect of the use of water as a solvent was observed in case of the Diels–Alder reaction between anthracene-9-carbinol **1** and N-ethylmaleimide **2** for the formation of cycloaddition product **3** (Scheme 2) [16]. Generally, the rate of this reaction was found to be faster in case of nonpolar hydrocarbons as solvent compared to the polar one where slower reaction rate was observed. But due to hydrophobic effect, the reaction rate in water was found to be exceptionally high in this case.

Effect of water as solvent was also observed in Claisen rearrangements reaction [17] where the reactivity and selectivity were increased in water compared to other organic solvents. For example, the Claisen rearrangement of ether **4** to naphthol derivative **5** was also found to be much faster in case of water comparing to other organic solvents. This rearrangement reaction completed at room temperature within five days using aqueous suspension (Scheme 3) [18].

Along with Diels–Alder cycloadditions and Claisen rearrangements, other reactions such as lactone formation also show drastic improvement in the yield when reaction was performed in water. For example, cyclization of acyclic ester **6** to lactone **7** was also achieved with better yield in water (Scheme 4) [19].

Similarly, the cycloaddition reaction of quadricyclane **8** and azodicarboxylate **9** displayed faster reactivity in the presence of water as solvent. Reaction completes within 10 min. compared to other organic solvents such as toluene, ethyl acetate which require more than 120 h for completion (Scheme 5) [18].

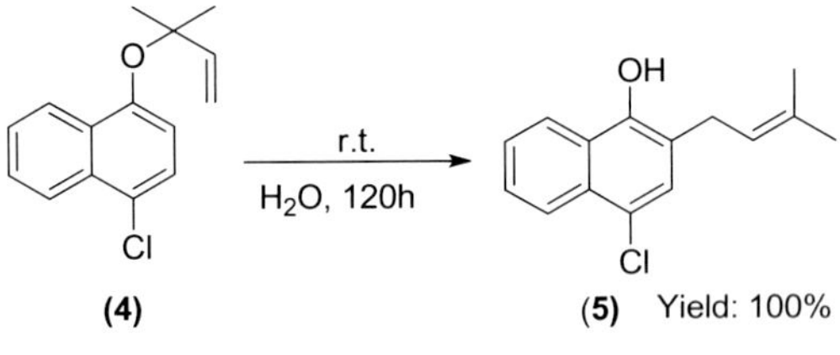

Scheme 3 An example of Claisen rearrangement in water

(4)　　　(5)　Yield: 100%

4 Green Solvents and Their Application

Scheme 4 Lactone formation through iodine transfer cyclization in water

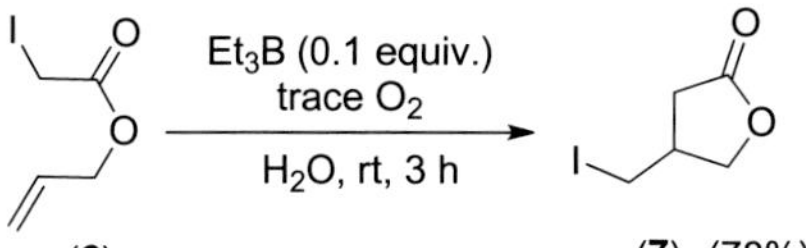

(6) → **(7) (78%)**

Et$_3$B (0.1 equiv.) trace O$_2$
H$_2$O, rt, 3 h

Scheme 5 Cycloaddition reaction of quadricyclane and azodicarboxylate

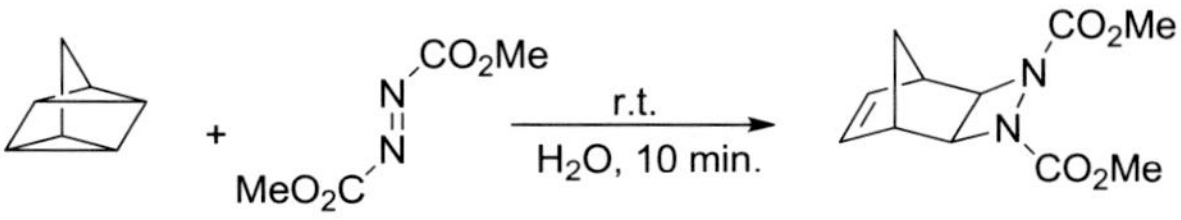

(8) + **(9)** → **(10)**

r.t.
H$_2$O, 10 min.

Reactions of carbanion equivalent in water

Along with pericyclic reactions, other reactions like carbanion-based reactions that involve organometallic reagents and reactions are also showed the enhancement of rate because these reactions involve complete prohibition of moisture and oxygen and often require low temperature. Metal mediators like indium, tin, and zinc for reactions of carbonyl compounds and imines with allyl halides in water. Phenylation and acylation in water and under an atmosphere of air by using metal in catalytic amount can be carried out in high yields through Barbier–Grignard type reaction (Schemes 6 and 7) [20, 21]. Modifications like the replacement of methyl group with halogens can cause enhancement in reactivity.

Scheme 6 Rh-catalyzed carbonyl addition to form alcohol in water

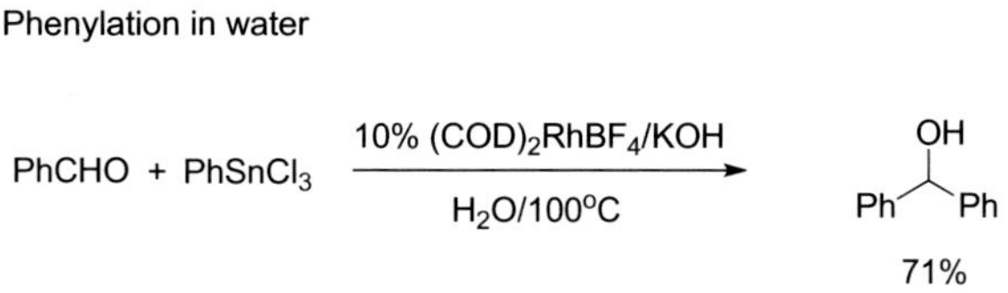

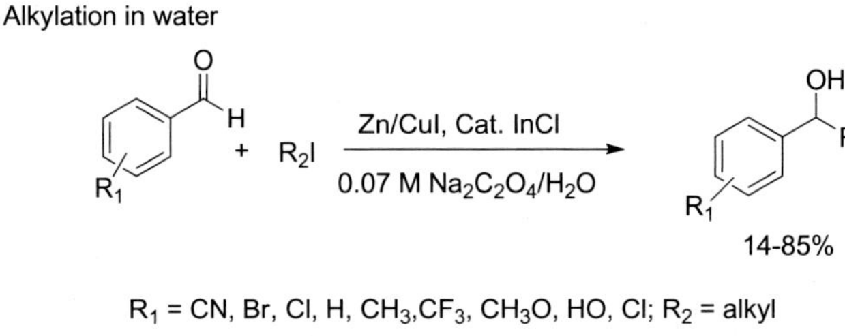

Phenylation in water

$$PhCHO + PhSnCl_3 \xrightarrow[\text{H}_2\text{O}/100^\circ\text{C}]{10\% \ (\text{COD})_2\text{RhBF}_4/\text{KOH}}$$

71%

Alkylation in water

+ R$_2$I

Zn/CuI, Cat. InCl
0.07 M Na$_2$C$_2$O$_4$/H$_2$O

14-85%

R$_1$ = CN, Br, Cl, H, CH$_3$,CF$_3$, CH$_3$O, HO, Cl; R$_2$ = alkyl

Scheme 7 Alkylation reaction in water

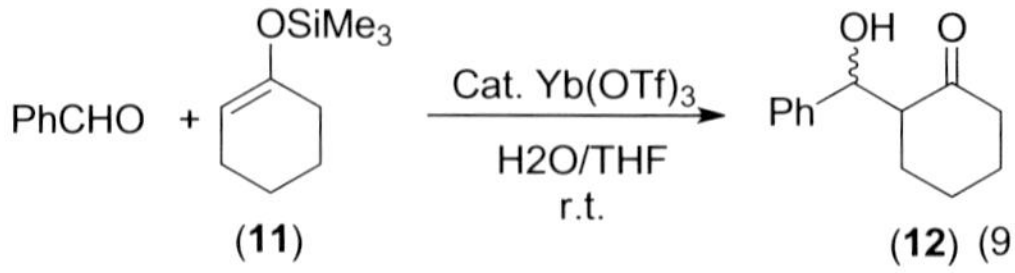

Scheme 8 Yb(OTf)$_3$-catalyzed Mukaiyama aldol reaction of benzaldehyde

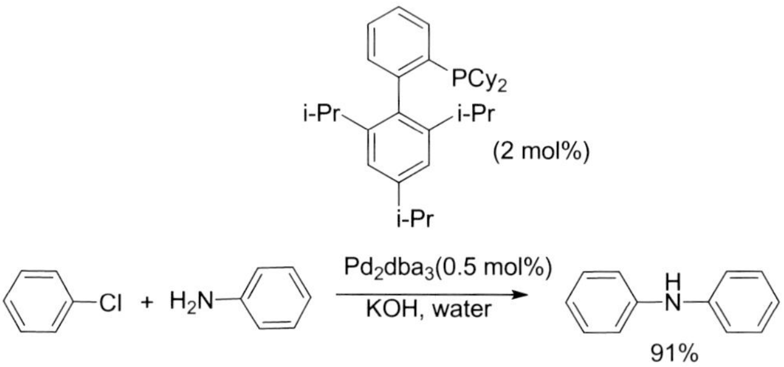

Scheme 9 Pd-catalyzed C–N bond formation in water

Reactions of carbocation equivalent in water

Along with reactions of carbanion equivalent, a diverse variety of protonic and Lewis acids have been seen to exhibit catalytic activities in water. Mukaiyama aldol reaction of benzaldehyde with silyl enol ether **11** was performed using Yb(OTf)$_3$ as an active catalyst in water to produce β-hydroxyketone **12** as aldol condensation product in 91% yield (Scheme 8) [22].

The C–N bond formation through palladium-catalyzed reactions can also take place in water. For example, reaction between chlorobenzene and aniline under Pd-catalyzed reaction condition led to the formation of diphenylamine (Scheme 9).

Transition metal-catalyzed asymmetric reaction can also be performed in water as a solvent. For example using copper as catalyst and pybox as ligand, the propargylamine **14** could be prepared by using terminal alkynes and imines **13** in water (Scheme 10) [23].

Oxidation and reduction in water

There are diverse oxidation and reduction reactions in water which are used nowadays that involve H$_2$O$_2$ and O$_2$ as water compatible oxidants. As primary and secondary alcohols into aldehyde and ketones using stable and recyclable O$_2$ (O$_2$/N$_2$ mixture was utilized together with a pressure of 10–30 bar) and a water-soluble palladium(II) bathophenanthroline complex [24]. Also in refluxing water, palladium nanoparticles (ARP-Pd) have been used with 1 atm of O$_2$ in refluxing water [25]. Hydrogen peroxide has been used with a tungstate complex and quaternary ammonium-hydrogen sulfate as an acidic phase transfer catalyst in aqueous biphasic systems for the oxidation of alcohols, alkenes, and sulfides [26]. Hydrogen peroxide in a biphasic reaction system has also been utilized for the oxidation of alcohols with a polyoxometalate (Na$_{12}$[WZn$_3$(H$_2$O)$_2$(ZnW$_9$O$_{34}$)$_2$]) prepared in situ [27]. Importantly, the catalyst could be reused with little loss of catalytic activity.

4 Green Solvents and Their Application

Scheme 10 Cu(I)-catalyzed
formation of propargylamine
in water

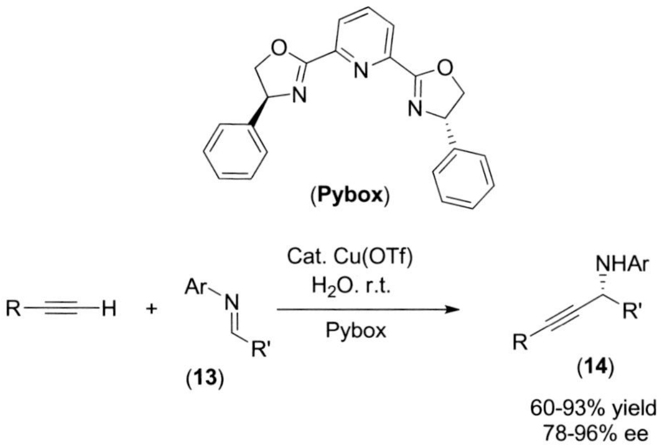

As reducing agents are incompatible with water so reduction reactions are rarely carried out in water while ketones and aldehydes can be reduced using lithium/sodium borohydride in water as they are compatible in water or using heterogeneous (THF) reaction conditions, and good enantioselectivities were observed. Reductive aminations occur using R-picoline-borane in water in the presence of small amounts of ethanoic acid. This sodium borohydride is also used amphiphilic dendrimers that are derived from polyamidoamine (PAMAM) and D-gluconolactone for the reduction of prochiral ketones.

Water in Carbohydrate chemistry

When water is used in this type of reaction, the yield can be increased as the protection and deprotection processes can be eliminated. With commercially available D-glucose and pentane-2,4-dione as starting substrates, β-C-glycosidic ketone **15** is obtained in one step with almost a quantitative yield in aqueous $NaHCO_3$ solution (Scheme 11). In a sharp contrast, in organic solvents the β-C-glycosidic is prepared from benzylated D-glucose, a protected sugar, in seven steps in an overall low yield [28].

Disadvantage of water as solvent in chemical reactions

Its temperature-restricted range of liquid state and especially its significant heat capacity, which makes distillation processes extremely energy-consuming is one important drawback in using water as a solvent. As like dissolves like, so the major

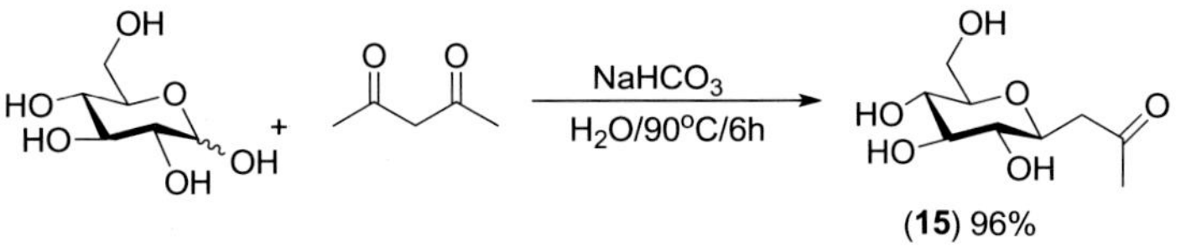

Scheme 11 Synthesis of β-C-glycosidic ketone in water in one-pot operation

drawback with water is the less solubility of organic compounds in water that can be overcome by using surfactants. One of the major drawbacks of using water as a solvent is the low solubility of most organic substrates in water. It could be overcome by the use of surfactants, which solubilize organic substances in water. Also, the same things happen when the property of evaporation and reverse osmosis is considered which are necessary when water must be separated from a non-volatile reaction product or impurities [17]. In addition to that, various reactions such as reactions involving reactive organometallic reagents such as organolithium, organozinc, and Grignard reagents are there which require exclusively dry reaction condition. Similarly in various others cases, the presence of reactive metals such as Na and other pyrophoric reagents such as $LiAlH_4$ limit the use of waster as a solvent.

4.2 Ionic Liquids

Mixture of molten salts, cations and anions forms ionic liquids. They are best solvent and a good alternative to convenient solvents in organic syntheses as their melting points can reach to 100 °C. Unlike other organic solvents, ionic liquids have a negligible vapor pressure and flammability so they can be considered as green solvents [29]. Many ionic liquids are supposed to be very stable in many redox reactions; moreover, it is impressive to observe its electrochemical stability. Apart from its use as solvent, ionic liquid could be used for diverse applications such as for the preparation vast varieties of materials such as plasticizers, battery, and lubricants [30].

Also, they can work in high-vacuum systems and exclude possible contaminants because they are non-volatile and non-flammable. They produce an anhydrous polar alternate option for two-phase systems as do not form a homogeneous mixture with many organic solvents. When they are hydrophobic, they may provide immiscible polar phases with water.

Among so many classes of ionic liquids, one containing N,N'-dialkyl-substituted imidazolium cations combined with a convenient counterion is the most widespread class. A number of significant steps are involved in its synthesis along with many non-green approaches [31]. Instead of having an acceptable toxicity, for cleavage of C–C bonds, there is no enzyme. Because of non-biodegradability of cations, they do not fall under the category of green solvent [32]. But there are two categories of ionic liquids according to their application which is greener than usual ionic liquids.

Category first belongs to ionic liquids that involve cations that are either found in the metabolism of living things or their derivative example choline, betaine, carnitine, or amino acids [33]. Choline is somewhat synthetic but its cheap and its availability is also in large quantities. Example as green solvent ionic liquids, Choline hexanoate dissolves suberin that is a natural polyester, the main component of cork, and it is synthesized by the reaction of choline hydroxide and hexanoic acid in a one-step process. Second category consists of ethylene oxide (EO) groups. These ionic liquids have a benefits of high entropy contributions as they have several EO groups, so have

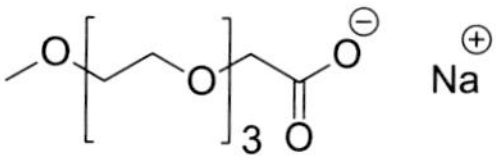

4 Green Solvents and Their Application

Fig. 2 [Na][TOTO] as the example of an easy-to-make low toxic IL

crystallization properties in comparison with other ionic liquids (ILs) which are energetically hindered to crystallize. Because of these special feature of Eos, these ions show high flexibility and thus favor the liquid state. Example sodium 2,5,8,11-tetraoxatridecan-13-oate ([Na][TOTO]) [34], which is shown in Fig. 2. These TOTO compounds are already commercialized by Kao Chemicals under the trade name Akypo. Again, there are cheap, with low toxicity and available in large quantities in comparison with choline so can be used in cosmetics. Along with that they convince with high thermal and electrochemical stability. Only thing is that although it is cheap and simple, in the synthesis of this ILs, the use of "non-green" reactants cannot be avoided. These green solvents should follow all the twelve criteria of green solvents so that they can beneficially replace conventional solvents. The enhancement in performance for this greener solvent in comparison with conventional one should also be there that is achieved by application of an IL, as compared to the usual solvent.

There are several reactions that can be performed in these green ILs that have been discussed in the next chapter of this book. Their disadvantages include a whole bunch of chemically very different solvent and liquid classes with extremely different characteristics, relatively high price, especially for reasonably pure ILs, their often high to very high viscosity and the high effort required for their synthesis, only best thing is that they are all salts and liquid at low to moderate temperature.

4.3 Switchable Solvents

For achieving the need of solvent, mainly distillation is used as a common practice but the distillation of volatile solvent causes damage to environment mainly through the formation of smog. In Canada, 4400 tons per year of hexane are emitted to the atmosphere, one-third of that from oilseed processing. In addition to the emission of solvent vapours in atmosphere, the need of very high energy input to perform the process of the distillation is another important issues associated with the use of volatile organic solvents. Therefore it is good to search for a new non-distillation route for the separation of solvents from products, so that volatile solvents would not have to be used. Here comes the introduction of switchable solvent [35] that can remove solvent from a hydrophobic product that means this can be achieved if there were a solvent that could be reversibly switched from being hydrophobic to hydrophilic, that's why it is a switchable solvent [36]. Basically, there is an urge of a solvent that shows the following desired phase behavior with water, very little miscibility of the solvent with water in the absence of CO_2 but complete miscibility with water in the presence of CO_2. Because of non-toxic, benign, inexpensive property of this solvent it can be easily removed and is preferred as the trigger for the switching process. Actually secondary amine such as DBU and guanidine reacts with alkyl carbonate which its self form through the reaction between alcohol and CO_2 to form salt which is polar in nature and after reaction by removing CO_2 through heating, the salt changes into nonpolar form and thereby gets separated. Similarly instead of alkyl carbonate, amine reacts with carbonic acid and attains polar form by forming salt. Amines which include another functional group, especially oxygen-containing groups, are less hazardous than alkyl amines. Secondary amines can have improved switching speeds relative to tertiary amines. A switchable-hydrophilicity solvent (SHS), meaning a solvent that can reversibly switch from having poor miscibility with water to having very good miscibility with water, has been identified. CO_2 at atmospheric pressure is used to switch the solvent to its hydrophilic form, and air and/or heat can be used to switch it back again. This solvent can be used to extract and isolate organic materials, such as soybean oil, without the need for a distillation step. Because distillation is not required, there is no longer a need to use a volatile organic solvent. The variety of switchable-hydrophilicity solvent (SHSs) identified suggests that amine SHSs can be designed to have ideal properties for a given application [37]. This reaction is based on protonation of amines and is exothermic. As shown in Fig. 3, the ammonium bicarbonate salt formed through the reaction between tertiary amine with carbonic acid in the presence of CO_2 attains polar form of SHS. With the removal of CO_2 by heating the solvent or by passing nitrogen gas (N_2) or with basic treatment, the salt again converts into a polar form, i.e., amine.

SHS has been used for different purposes such as for the extraction of natural products such as vegetable oil and phenolic compounds as well as also for the removal of pesticides, insecticides, dyes which are present in trace amount. Tremendous advantages of these switchable solvents are the uncomplicated isolation of the product and the simple recovery of the organic solvent from water by removing carbonate from the mixture. In this way, an energy-consuming distillation can be completely avoided. Instead of using CO_2 as a trigger, temperature can be used for this purpose, as shown by Khono et al. in water/IL mixtures. Tetrapentylammonium bromide (Pe4NBr) is nearly insoluble in water at room temperature. With a melting point close to 100 C, it is at the limit of what is usually called IL. But what matters here is its remarkable phase behavior when mixed with water. Above approximately 30 C, it becomes highly water-soluble, but at the same time, and as otherwise only known

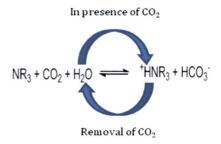

Fig. 3 Preparation and formation mechanism of polar–nonpolar SHS

from polymer solutions, it separates into two phases, building a water-rich and a second water-poor phase in equilibrium. The latter behaves like an organic phase. As a result, organic molecules can be dissolved in this phase during a chemical reaction. At the end, the hydrophobic phase can easily be recycled simply by decreasing the temperature, because then the phase vanishes and the salt precipitates. As such, it even does not pollute the water phase. A liquid organic product phase separates from the aqueous phase and can easily be isolated. As in the case of Jessop's switchable solvents, a distillation process can be avoided. However, one disadvantage should be mentioned: the salt contains bromide, which is not always desirable in industrial processes.

4.4 Supercritical CO_2 as Solvent

In general, all three different phases of a substance could be interchanged by modifying temperature and pressure. Similarly, a gas below or at its critical temperature could be converted into its liquid phase by applying certain amount of pressure. Likewise carbon dioxide remains in gaseous phase at STP. But under the effect of enhanced pressure and temperature, near its critical point, it attains a state where its properties lie somewhere between liquid and gaseous phases and this is known as supercritical CO_2. It has also been explored as an environmentally green alternative to conventional solvent, and different types of reactions have also been explored using it as reaction medium [38]. Moderate critical temperature and pressure of 31 °C and 74 bar, respectively, for CO_2 allow its conversion into supercritical state using relatively small amount of energy. Apart from that, its non-flammable, relatively non-toxic and chemically inert nature makes $scCO_2$ as a greener alternative to existing solvents obtained from non-renewable petroleum sources. Interestingly by simple depressurization, this could also be recovered after completion of the reaction. Similarly, physical properties such as polarity could also be manipulated simply by adjusting pressure and temperature.

Application of Supercritical CO_2

Extraction of natural products: One of the main applications of $scCO_2$ is in the decaffeination of coffee [39]. By using $scCO_2$, use of chlorinated hydrocarbon solvent could be avoided. Although in some cases small amount of organic solvents is also required as cosolvent, for example, higher concentration of phenolic compounds was obtained in case of $scCO_2$ compared to ethanol (~400 ppm as compared to ~200 ppm for ethanol extraction). However in this case, small amount of ethanol was also used as cosolvent [40]. Similarly, the extraction of thymol from thyme was also achieved using $scCO_2$ [41]. Likewise, diterpenoid with antitumor properties has also been extracted from the *Pteris semipinnata* with high selectivity and efficiency using $scCO_2$ [42].

Application in organic synthesis: Due to the greater miscibility of gases in CO_2, several reactions involving gases could also be performed in $scCO_2$. For example, the conversion of CO_2 to methanol and formic acid was efficiently through the hydrogenation reaction using ruthenium(II)-phosphine complex as catalyst and $scCO_2$ as solvent as well as reactant also [43]. Reaction was found to be highly effective in $scCO_2$ due to better miscibility of hydrogen gas and CO_2. Similarly, the conversion to of esters and amides using the corresponding alcohols and amines have also been achieved [44, 45]. Asymmetric hydrogenation of alkenes using different transition metals such as Rh and Ir along with appropriate ligands has also been performed in $scCO_2$. For example, Rh-catalyzed hydrogenation of olefins **16** using suitable ligands such as DuPhos **18** has been achieved to obtain amino acid derivatives **17** with *ee* up to 82% which is higher compared to that obtained in hexane (*ee*: 70%) (Scheme 12a) [46]. Similarly, the hydrogenation of tiglic acid **19** was also achieved under Ru-catalyzed reaction condition to obtain chiral acid **20** in $scCO_2$ with 89% *ee* (Scheme 12b). However, in this case another fluorinated alcohol was also used as cosolvent [47].

Likewise, various other examples such as hydrogenation of dimethyl itaconate **22** to ester **23** could be achieved using a ruthenium and a fluorinated BINAP-derived ligand **24** with only 73% *ee* in scCO2, although in this case selectivity was more in methanol (95% *ee*) (Scheme 13a) [48].

Likewise, chiral amines **26** could also be prepared starting from imines **25** (Scheme 13b) through hydrogenation and enantioselectivity was comparable to organic solvents (up to 90% *ee* for $scCO_2$ and up to 87% *ee* for dichloromethane) but rate was much faster in $scCO_2$ as reaction completes within 50 min as compared to the reaction in dichloromethane which requires 19 h [49]. Various other such transformations are reported in $scCO_2$ such as synthesis of chiral alcohol from ketones, hydroformylations from alkene, functionalization of alkane, formation of triazole using Click Chemistry approach, etc. [50–52].

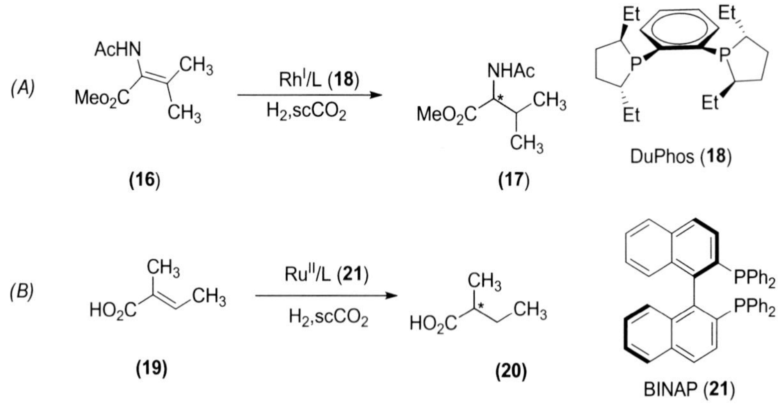

Scheme 12 Stereoselective hydrogenation of olefins in $scCO_2$

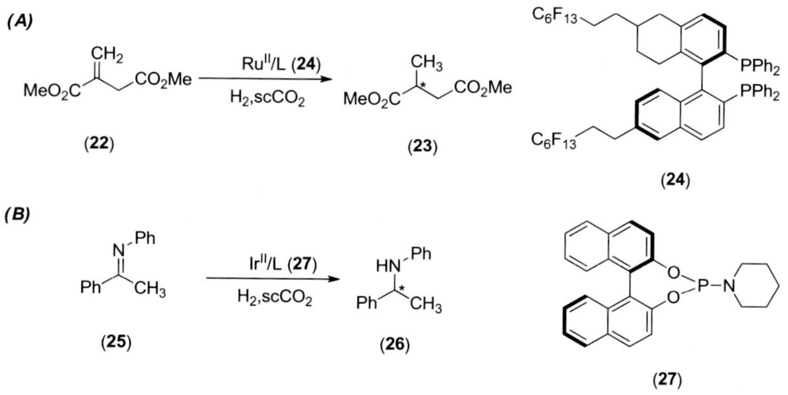

4 Green Solvents and Their Application

Scheme 13 Rh(II) and Ir(III) catalyzed asymmetric hydrogenation of alkene and imine

Application of scCO$_2$ in Materials Processing

Supercritical CO$_2$ has also been used extensively for the preparation and processing of polymers [53]. Different varieties of polymerization processes including step and chain growth polymerizations have also been performed for the synthesis of polymers with different functional groups such as polyesters, polyamides, polyureas, and polycarbonates. Although there are various types of polymerization processes, radical polymerization is the most commonly observed in supercritical CO$_2$ as a solvent [54]. Current research is focused on the synthesis of CO$_2$-philic surfactants which are necessary for emulsion polymerizations, for example, by employing vinyl trifluorobutyrate as the monomer instead of vinyl acetate [54]. Furthermore, scCO$_2$ has been currently used to process porous organic polymers [55] to enhance their total pore volume and specific surface area, as a drying medium for aerogel formation [56] and for unidirectional foaming of PLA/graphene oxide foams [57]. Synthesis and processing of bulk materials, aerogels, thin films, coatings, particle suspensions, powders, and nanoparticles can also be conducted in scCO$_2$, and materials that have been processed in this solvent have often superior properties due to the very effective wetting of surfaces, facilitated by the fact that CO$_2$ has no surface tension. This way, chemical reactions can easily happen on substrate surfaces, resulting in materials with higher surface areas and better-defined nanostructures. The scCO$_2$ has also been used for the dispersion of metal nanoparticles such as Pd nanoparticles on grapheme to produce an efficient electrocatalyst that could be used for the oxidation of methanol and formic acid for their application in different types of fuels cells such as *methanol fuel cell* (DMFC) and *direct formic acid fuel cell* (DFAFC) [58].

Similarly, this technique was used for efficient deposition of Pt on carbon nanotubes which thereby enhanced the catalytic efficiency in achieving electro-catalytic reduction of carbon dioxide (CO$_2$) to the mixture of organic compounds which mainly includes formic acid (up to 89%), methane (up to 33%) along with the formation of CO and methanol in minor proportion. for catalytic CO$_2$ reduction, with formic acid and methane as the main products [59]. Likewise, the application of scCO$_2$ has also been demonstrated in the dyeing of different types of textiles of natural and synthetic origin using hydrophobic nonpolar dyes which have displayed good coloration behavior [60–63]. In addition to the fastening of the dyeing process, the use of scCO$_2$ also reduces the operational cost by up to 50%.

Disadvantage using scCO$_2$

One of the major challenges in using scCO$_2$ as a solvent is the need of high working pressure which requires additional safety. It also adds to the additional energy requirement to carry on the process using scCO$_2$ as a solvent. Therefore, its industrial use is not very feasible like other conventional solvents. Apart from that, requirement of costly equipment and setup is also a major challenge to deal with [64].

4.5 Solvents from the Renewable Bio-Based Feedstock

In addition to the conventional and other greener alternatives which have been discussed above, various other unconventional solvents derived from the plant as feedstock is also being explored as an environmentally benign and renewable alternatives [7]. For example, various agro-based products derived from crops such as sugarcane, corn, and Miscanthus (*Miscanthus sinensis*) which is also known as elephant grass are being explored as sources of solvents and fuels [65]. Importantly, an additional advantage is also there as the use of biofuels would promote the cultivation of trees and other crops which consequently would improve the air quality and support and reducing pollution also but it would also require proper management of land [66]. For example, *bioethanol,* which is produced through the fermentation of starch and sugar extracted from the sugarcane and corn, is an example of first-generation biofuel. Apart from that, ethanol which is considered as an organic solvent with relatively lower toxicity could also be produced from bioethanol. Similarly, biofuels produced from the grass, agricultural, and forestry residues which are high in concentration of biopolymers such as lignin, cellulose, and hemicellulose are examples of second generation of biofuel [67]. The continuous supply of these bio-based solvents at the industrial scale with lower price is one of the major challenges in using these solvents on regular basis in academic institution and industries.

Solvents Derived from Cellulose and Starch

Cellulose and starch are among the most abundant polysaccharide that could easily be obtained from plant. Glucose is the monomer in both the cases which are joined through α-1,4 and β-1,4 connections in case of cellulose and starch, respectively [68, 69]. Therefore, the hydrolysis of these biopolymers achieved through chemically or in the presence of enzyme leads to the production of glucose which is used as precursor of several biofuels [70]. Some of the solvents which are derived from plant biomass have also been explored extensively are discussed below [71].

4 Green Solvents and Their Application

2-Methyltetrahydrofuran (MeTHF): Methyltetrahydrofuran (*mp*: $-136\,°C$, *bp*: 80.2 °C) which is commercially available solvent is being explored as a greener and renewable substitute to THF (*mp*: $-108.4\,°C$, *bp*: 66 °C). The synthesis of 2-MeTHF could be achieved using levulinic acid and furfural as precursor which could be synthesized from xylose and glucose (Fig. 4) [67]. It displays lower toxicity compared to tetrahydrofuran (THF). Therefore, it has been used as greener alternative to THF and has also been approved for its use in the pharmaceutical industries [72].

γ-Valerolactone (GVL): Similarly, γ-Valerolactone (GVL) (Fig. 4b) could also be used as renewable solvent and prepared starting from lignocellulosic biomass (Fig. 5) [73].

Cyrene: Similarly, cyrene (Fig. 4c) is another example of biofuels which could be manufactured from cellulose through the formation of levoglucosenone (LGO) which could further be subjected to catalytic hydrogenation to produce cyrene (Fig. 5) [74]. Cyrene could be used as an alternative to polar aprotic solvents like NMP, DMF, and DMAc and has also been explored in achieving different types of organic transformation such as for the synthesis of MOF [75].

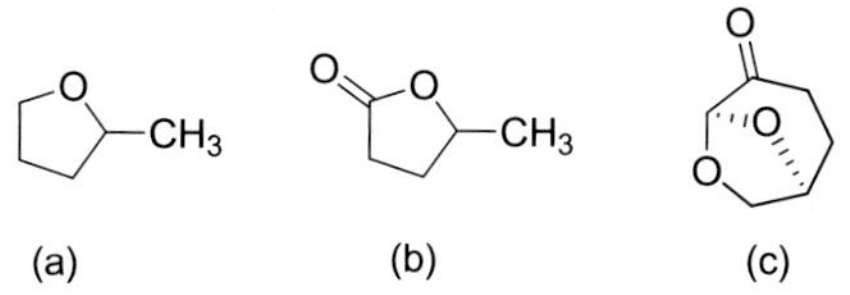

Fig. 4 **a** 2-MeTHF, **b** γ-Valerolactone (GVL), **c** Dihydrolevoglucosenone (Cyrene)

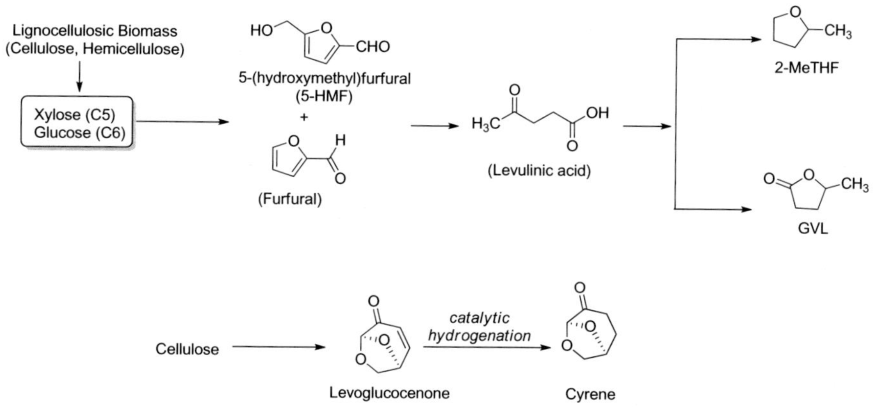

Fig. 5 Synthesis of 2-methyltetrahydrofuran (2-MeTHF), γ-valerolactone (GVL), and cyrene from renewable sources

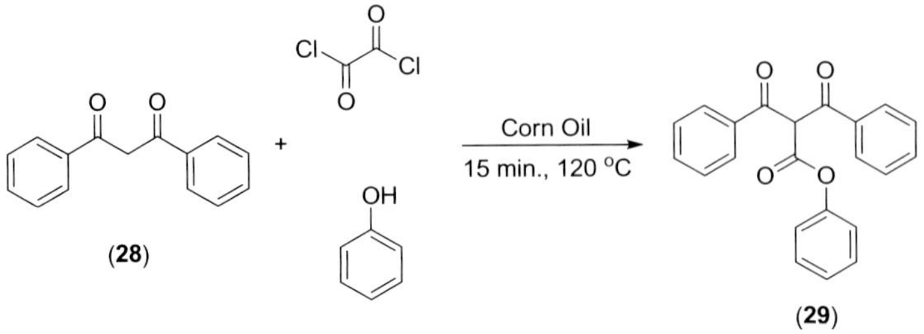

The acylation of a mixture of dibenzoylmethane and phenol was performed using corn oil as solvent.

Scheme 14 Acylation reaction using corn oil as solvent

4.6 *Water Extract of Agrowaste Ash (AWEs)*

Similar to the biofuels and solvents, water extract of *agro-waste ash* (AWEs) is also been explored as an environmentally benign reaction medium to achieve various types of transformations which generally require conventional organic solvents. Water extract of the peels of various fruits and vegetable such as banana, papaya, and pomegranate, and also, the extract of straw of crops have been investigated for their use as solvent. For example, *water extract of papaya bark ash* (WEPBA), *water extract of banana* (WEB) peel ash, *Water Extract of Rice Straw Ash* (WERSA) have been used as reaction medium to achieve different types of reaction such as Suzuki coupling, Sonogashira coupling, and oxidation [76]. Compared to the conventional volatile organic compounds (VOC) as organic solvents, these are non-toxic and biodegradable as well as cheap also.

Vegetable oil as a green solvent

Other renewable, biodegradable, and green alternative to organic solvent includes oils extracted from seeds of different types of plants and fruits [77]. Various organic compounds which mainly include triglyceride (95–98% which is the ester of glycerol and different types of long-chain fatty acids including saturated fatty acids such as lauric, palmitic acid, and other unsaturated fatty acids like oleic acid and linoleic acid. In addition to that, other minor component includes sterol, squalene, polyphenols, etc. Various types of reactions have also been attempted using oil as reaction medium. For example, the acylation of a mixture of dibenzoylmethane **28**, and phenol with oxalyl chloride was performed for the formation of acylated product **29** using corn oil as reaction medium at 120 °C (Scheme 14) [78].

4.7 *Glycerol as Green Solvent*

Similarly, glycerol which is 1,2,3-propanetriol is emerging as a popular choice of solvents [78]. It is abundant in the form of triglycerides which is a major component of oil. It is obtained as a by-product of various processes such soap formation which

4 Green Solvents and Their Application

requires hydrolysis of fatty acids. Similarly, it could also be produced as by-product during the formation of biodiesel. It could also be prepared using propylene oxide [79].

Glycerol is a viscous, colorless and odorless and non-toxic, biodegradable liquid (LD 50 (oral rat) = 12,600 mg/Kg) liquid which is completely miscible in water and short-chain alcohols [80]. Apart from dissolving organic compounds, it also has ability to dissolve various inorganic salts, bases, acids, etc. In addition to that, it is immiscible in hydrocarbons, ethers, and other such nonpolar solvents. Unlike volatile organic compound (VOC), it is non-volatile (b.p. 290 °C). Therefore, compatibility with organic as well as inorganic compounds, high boiling point, non-toxicity, biodegradable nature, low cost, etc. makes it as eco-friendly solvent. Therefore, various different types of reactions have been attempted using glycerol as solvent.

Different types of organic transformations in glycerol

Several different types of reactions have been performed using glycerol as solvent. For example, the *Aza-Michael reaction* using *p*-anisidine and *n*-butyl acrylate as reactants was successfully performed using glycerol as solvent under catalyst-free reaction condition to furnish 82% of aniline derivative **30** (Scheme 15) [81]. Interestingly reaction afforded high yield in both pure and technical grade glycerol, whereas other organic solvent like toluene, 1,2-dichloroethane, and DMSO were found to be ineffective.

Similarly, Michael addition reaction between indole and nitrostyrene was achieved under catalyst-free condition to afford indole derivative **31** in 80% yield (Scheme 16) [82]. The advantage of using glycerol as solvent is its easy separation from the

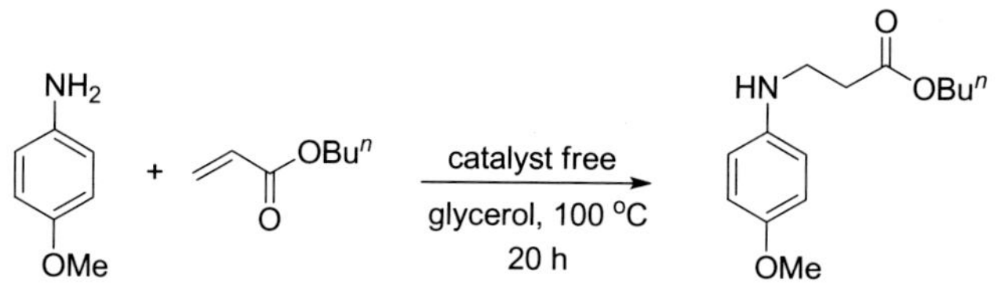

Scheme 15 Aza-Michael reaction using glycerol as solvent

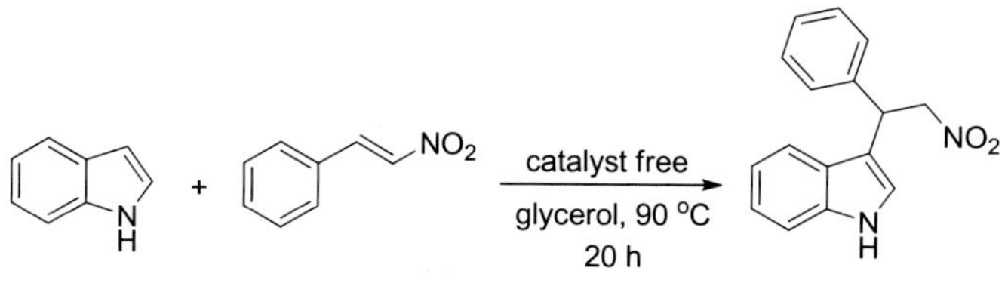

Scheme 16 Michael reaction of indole using glycerol as solvent

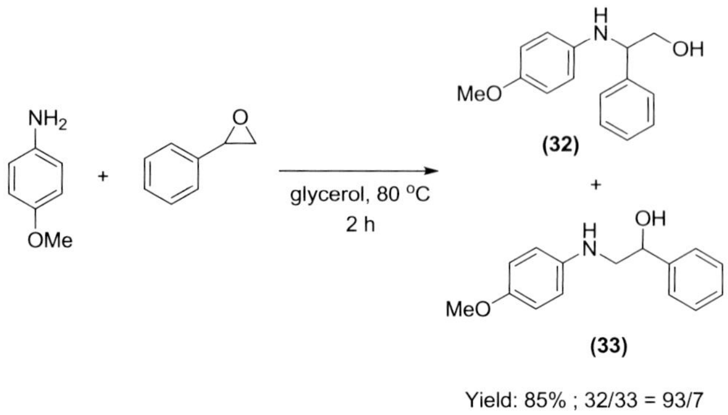

Scheme 17 Epoxide ring opening in glycerol

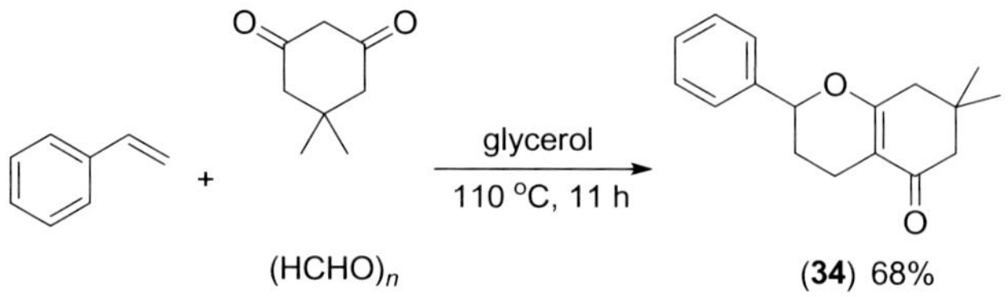

Scheme 18 Multicomponent reaction in glycerol solvent

reaction mixture by means of liquid–liquid phase extractions with other solvents such as ethyl acetate.

Glycerol as a solvent for increasing reaction selectivity

Similarly, the ring opening of epoxide with higher selectivity could also be achieved in glycerol under catalyst-free condition. For example, the reaction between p-anisidine and styrene oxide leads to the formation of alcohol involving ring opening of epoxide to produce products **32** as major product in glycerol (Scheme 17) [81]. Minor amount of other isomer **33** was also obtained. Generally, this type of reaction requires the use of Lewis or Brønsted acids as catalyst.

Similarly, multicomponent reaction for the formation of heterocycles ring has also been achieved. For example, reaction among styrene, dimedone and paraformaldehyde in glycerol led to the formation of compound containing tetrahydropyran ring **34** through tandem Knoevenagel and hetero-Diels–Alder (Scheme 18) [83]. Reaction led to the formation of desired product in 68% yield.

Similarly, the cyclocondensation reaction arylhydrazines and β-ketone esters **35** in glycerol at 110 °C for 4 h leads to the formation of pyrazolone derivative **36** in good yield (Scheme 19) [84].

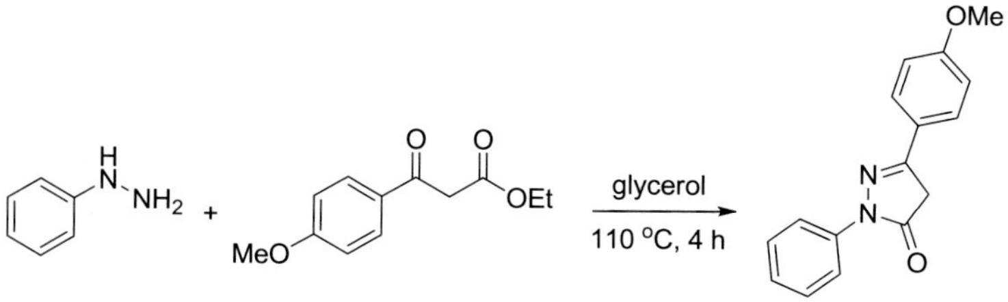

4 Green Solvents and Their Application

Scheme 19 Preparation of pyrazolone in glycerol

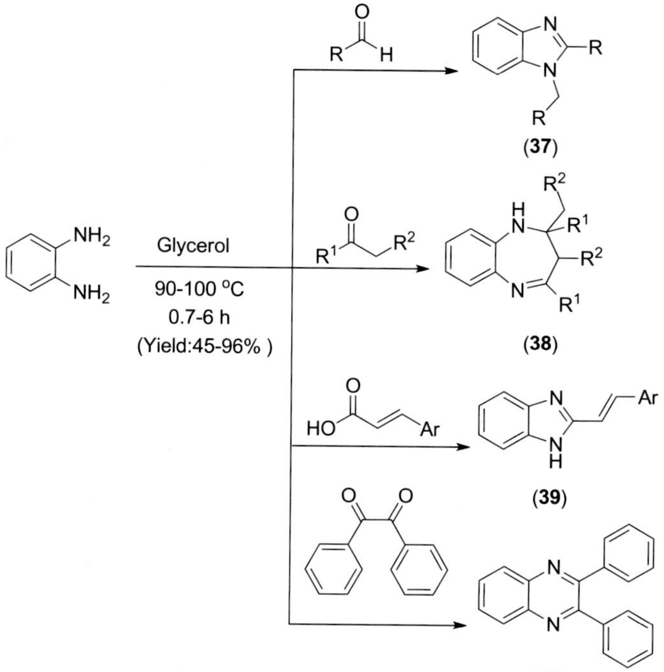

Scheme 20 Catalyst-free synthesis of *N*-heterocycles using glycerol as solvent

Glycerol as a solvent for catalyst-free synthesis

Likewise, glycerol has also been used for the condensation reaction between *o*-phenylenediamine with aldehyde, ketone, cinnamic acid derivatives and 1,3-diketone led to the formation of benzimidazoles **37**, benzodiazepines **38**, 2-styrylbenzimidazoles **39**, and quinaxoline **40**, respectively, in good-to-high yields (Scheme 20) [85–87]. In case of quinaxoline formation, mixture of glycerol and water was used as solvent at 90 °C.

Scheme 21 Asymmetric reduction of ketone under Baker's yeast catalyzed condition in glycerol

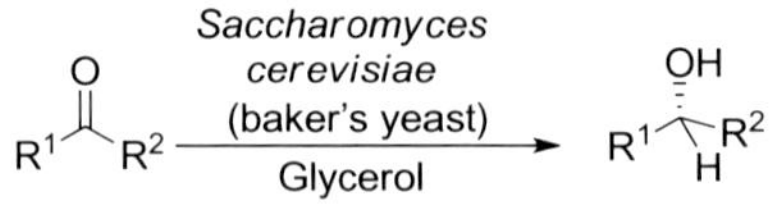

Scheme 22 Bioreduction of chloroacetophenone in glycerol

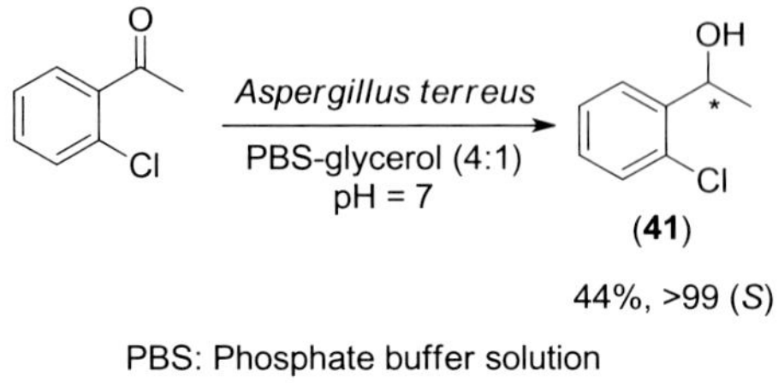

44%, >99 (*S*)

PBS: Phosphate buffer solution

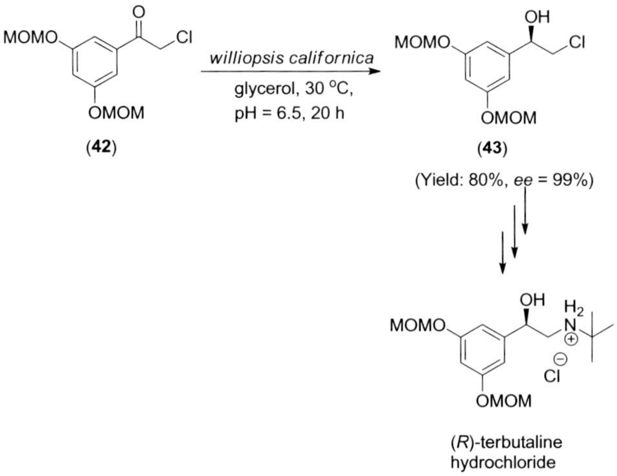

(**42**)

(**43**)

(Yield: 80%, *ee* = 99%)

(*R*)-terbutaline hydrochloride

Scheme 23 Asymmetric synthesis of (*R*)-terbutaline involving biocatalytic asymmetric reduction of ketone to alcohol

Glycerol as a solvent for biocatalysis

Glycerol has also been used as solvent to perform biocatalyst catalyzed transformations. For example, the asymmetric reduction of ketone to chiral alcohol was achieved using Baker's yeast catalyzed reaction condition in glycerol (Scheme 21) [88].

Similarly, bioreduction of 2-chloroacetophenone to benzyl alcohol **41** was also carried out using *Aspergillus terreus* as biocatalyst and in the presence of glycerol as a cosolvent (Scheme 22) [89].

Likewise, the asymmetric reduction of the α-chloroacetophenone **42** to substituted benzyl alcohol **43** under *Williopsis californica* catalyzed reaction condition was also involved as one of the important steps in the preparation of (*R*)-terbutaline hydrochloride which is a potent 2-adrenoceptorstimulating agent (Scheme 23) [90].

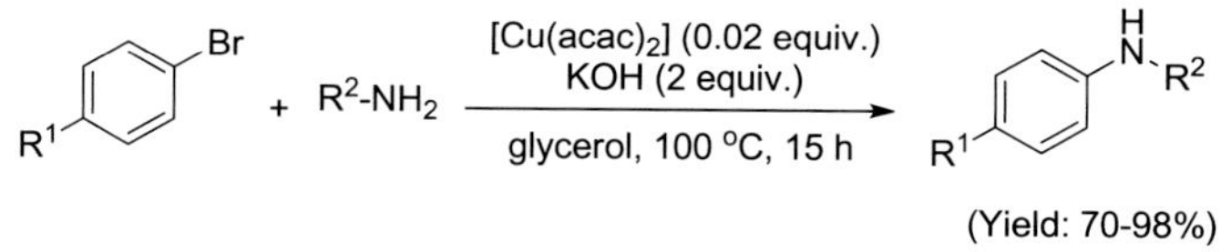

4 Green Solvents and Their Application

Scheme 24 Cu-catalyzed C–N bond formation in glycerol

Glycerol as a solvent in metal-catalyzed transformation

The application of glycerol as solvent has also been demonstrated in achieving the metal-catalyzed C–C and C–N bond-forming reactions. In general, such reaction requires the use of toxic solvents like DMSO, DMF, NMP, etc. For example, reaction C–N cross-coupling reaction between aniline and bromobenzene was achieved using glycerol as green solvent in the presence of under $Cu(acac)_2$ as catalyst [91]. Reaction afforded yield upto 98% (Scheme 24). After completion of the reaction, glycerol and catalyst could be recycled for another reaction simply by separating the reaction mixture containing product using diethyl ether. Interestingly, the formation of homocoupled product was not observed under given reaction condition.

Glycerol as a solvent for catalyst design and recycling

The important advantage in using the glycerol as solvent is the recycling of solvent and catalytic system after the reaction. For example, in the Ce(III)-catalyzed synthesis of bis(indolyl)methanes **44**, the reaction completes within 1.5 h to furnish yield up to 96% (Scheme 25) [92]. After the reaction, the glycerol and $CeCl_3$ mixture was extracted through liquid-phase extraction with another solvent ethyl acetate.

Similarly, the β,β-diarylation of alkene has also been achieved through the reaction between aryl iodide and acrylates using palladium nanoparticles that was stabilized over the sugar aminopolysaccharide (AP)-based surfactant (Pd/AP). Under the established reaction condition symmetrically β,β-diarylated alkene **45** was achieved at using 2 equiv. of ArI at 120 °C. Reaction afforded good yield of the diarylated product (Scheme 26) [93].

Interestingly, the synthesis of unsymmetrical diarylated product using two different aryl iodides was also achieved just by modifying the reaction temperature. Monoarylated product **46** was achieved at 90 °C and later in the same pot after adding 2 equiv. of other aryl iodide and increasing the reaction temperature to 120 °C

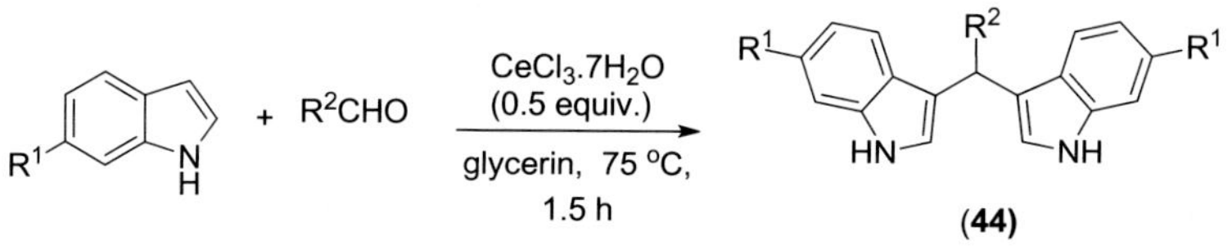

R^1 = H, Br; R^2 = Ph, 4-NO$_2$Ph, 4-ClPh etc.

Scheme 25 Ce(III)-catalyzed synthesis of bis(indolyl)methanes

Scheme 26 Diarylation of acrylates in glycerol

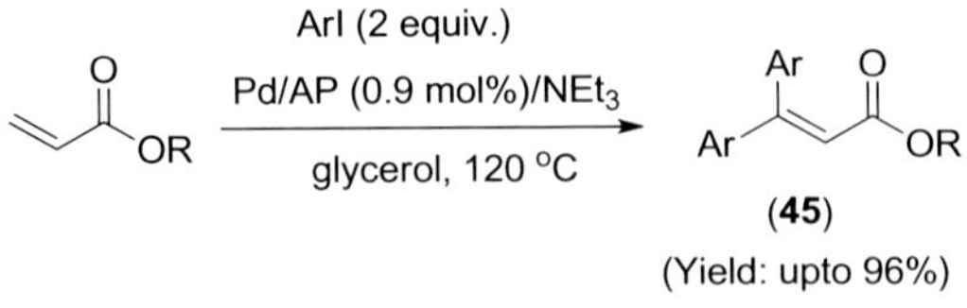

(**45**)
(Yield: upto 96%)

R = Butyl, cyclohexyl, Ar = Ph, napthyl etc.

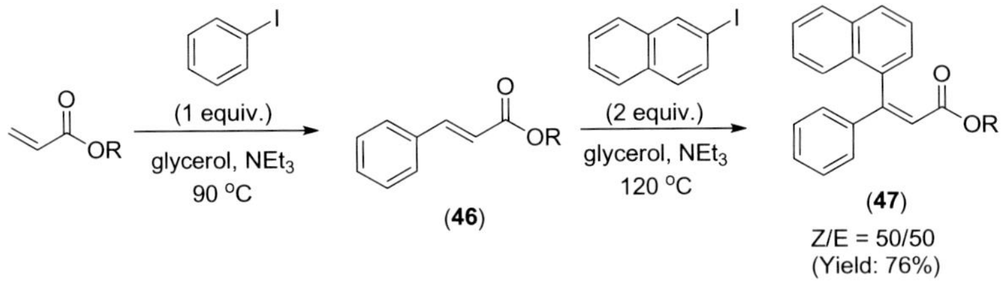

Scheme 27 Unsymmetrical diarylation of acrylate in glycerol

unsymmetrical diarylated product **47** was obtained in overall 76% yield (Scheme 27) [93].

Preparation of disulfide using glycerol as solvent

Glycerol has also been used a reaction medium to prepare disulfide bond. For example, thiol reacted to form disulfide under microwave-assisted reaction condition in the presence of Na_2CO_3. Reaction successfully delivered the corresponding disulfides using different types of thiols including alkyl, aryl, heteroaryl-substituted thiols under the optimized reaction condition in glycerol (Scheme 28) [94]. After extracting the reaction mixture in the mixture of ethyl acetate and hexane, the recovered glycerol could also be used for the subsequent reactions.

Reduction of carboxylic acids to primary alcohols

The reduction of carboxylic acid directly to the primary alcohol could also be achieved through transfer hydrogenation process in the presence of catalytic amount of $COCl_2 6H_2O$ and KOH (Scheme 29) [95]. Although reaction requires high temper-

Scheme 28 Base-promoted synthesis of disulfide in glycerol

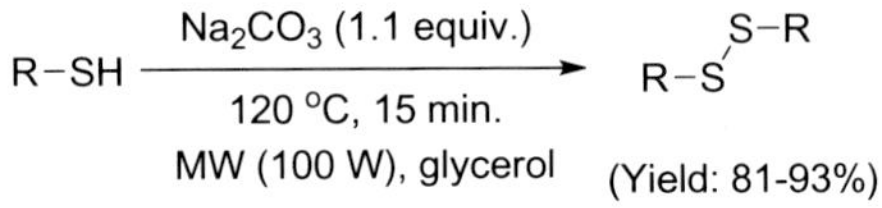

(Yield: 81-93%)

Scheme 29 Direct reduction of carboxylic acid under Co-catalyzed condition in glycerol

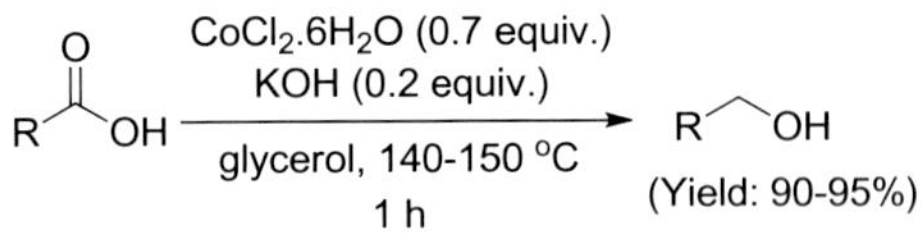

(Yield: 90-95%)

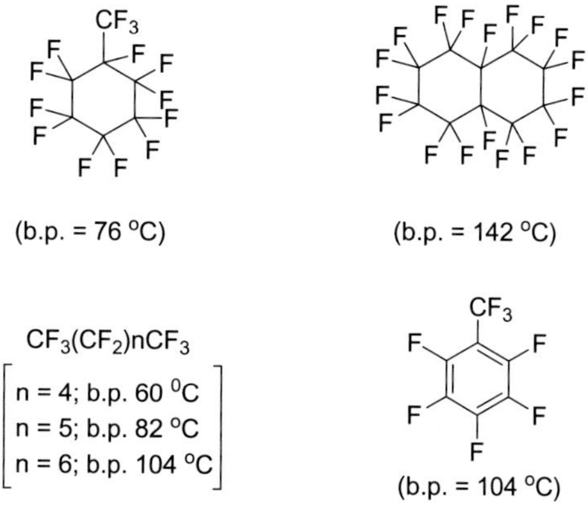

4 Green Solvents and Their Application

Fig. 6 Examples of fluorinated solvents

(b.p. = 76 °C) (b.p. = 142 °C)

$CF_3(CF_2)nCF_3$

$$\begin{bmatrix} n = 4;\ \text{b.p. } 60\ ^\circ C \\ n = 5;\ \text{b.p. } 82\ ^\circ C \\ n = 6;\ \text{b.p. } 104\ ^\circ C \end{bmatrix}$$

(b.p. = 104 °C)

ature (140–150 °C) compared to the other reducing agent such as H_2 and other borohydrides like $NaBH_4$ but it offers a greener method where glycerol acts as H-donor in presence of the combination of the catalyst and KOH. Various such reports are also available in the literature.

4.8 Fluorous Biphasic Solvents

The application of liquid–liquid biphasic system has also been explored for their use as environmentally benign alternative to the conventional organic solvents. The fluorous biphasic system which consist of two different solvent phases one of them is fluorous one. As name itself indicate the *fluorous phase* which is analogous to *aqueous phase* basically consists of perfluorinated alkanes, amines, ether, etc., and other such fluorinated hydrocarbons [96]. A large number of fluorous solvents are available commercially although it is expensive compared to the common organic solvents (Fig. 6) [8].

Whereas other layer could consist of other solvents with very poor solubility into it. With increasing in the temperature, these two solvents become miscible that allows the reacting components to interact and react to form product. Their reaction could be designed in such a way that product after formation could reside in one phase whereas other reagents and catalysts may get separated as these will stay in different phases. The fluorinated compounds having suitable physical properties such as melting point and boiling point could also be used as reaction medium to achieve various types of organic transformation. The properties of these fluoroderivatives also vary with its composition and the amount of fluorine. In order to achieve catalysis in fluorous biphasic system and to solubilize the catalyst in this phase, the metal complexes with fluorophilic ligands are preferred as the presence of fluorine help in solubilizing the catalyst in the fluorous layer [97].

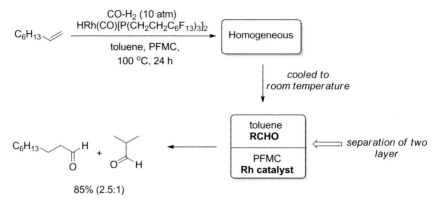

Fig. 7 Rh-catalyzed hydroformylation reaction in fluorous biphasic system

Reactions in Fluorous Biphasic System (FBS)

Hydroformylation of alkene to aldehyde

For example, the hydroformylation of octene to aldehyde was achieved under Rh-catalyzed reaction condition using fluorous biphasic system containing mixture of toluene and perfluoro(methylcyclohexane) (PFMC) (Fig. 7) [97, 98]. These solvents upon heating change into one phase where reaction occurs, and after reaction upon cooling, these two solvents get separated into two different layers. The fluorous ligand-containing Rh catalyst remains in the fluorous layer, whereas the product moves to toluene layer which allows the separation and recycling of the catalyst.

Grubb's Metathesis Reaction

Similarly, using Grubb's second-generation catalyst having fluorous phosphine ligands, the metathesis reaction has also been achieved.

For example, the cyclization reaction of alkene **48** using Rh-complex **49** as catalyst reaction condition leads to the formation of cyclopentene **50** in 70% (Scheme 30) [99]. Compared to the reaction in monophasic system using dichloromethane as single solvent system, the yield was found to be better in case of biphasic system.

Allylation of 1,3-diketone

The allylation of 1,3-diketone using allyl alcohol and lanthanide metal complex as catalyst has also been achieved successfully under biphasic approach. For example, reaction between allylic alcohol **51** and acetylacetone using Yb(III)-catalyst afforded excellent yield of the allylated diketone **52** using biphasic fluorous solvent consists of nitromethane and perfluorodecalin (PFD) (Scheme 31) [100]. After the completion of the reaction, the catalyst was recovered in fluorous phase and was further used for the next reaction.

4 Green Solvents and Their Application

Electrophilic Fluorination of Grignard Reagents

Similarly, fluorous biphasic system was found to be a better yielding reaction medium in case of the electrophilic fluorination of Grignard reagents. For example, fluorination reaction of heteroaryl Grignard reagent 53 with *N*-fluorobenzenesulfonimide (NFSI) as fluorinating agent performed in monophasic system with single solvent such as dichloromethane, THF, ether led to the lower yield of fluorinated pyridine compared to that of biphasic system consisting of DCM:PFD (4:1) (Scheme 32) [101].

Although various reactions have been attempted in using fluorous solvent, these solvents are not very frequently and commercially available. In addition to that, its high cost and higher half-lives limits their frequent use.

4.9 Polyethylene Glycol [PEG] as a Solvent

Polyethylene glycol (PEG) (Fig. 8), which is a polymer of ethylene glycol containing ether functional group, has also been used as environmentally benign alternative to the other conventional organic solvent.

The physical properties and application of PEG depend on their average molecular weight. For example, the number of monomer requires for their formation and on their molecular weight. For example, PEG with molecular weight <700 is viscous liquid at room temperature, whereas those with molecular weight lying between 1000 and 2000 are soft solid and waxy solid. Their biocompatibility, non-toxic nature, low flammability, non-volatility, amphiphilicity, etc., are some of the important features which make it suitable to be used as green alternative to other conventional organic solvent. It has also been added in the list of food additives and placed in the compounds generally recognized as safe (GRAS) list approved by USFDA [102]. In addition to that, the polar nature of polyethylene glycol makes it a solvent to be used for microwave-assisted reaction.

Scheme 30 Grubb's metathesis reaction in fluorous biphasic solvent

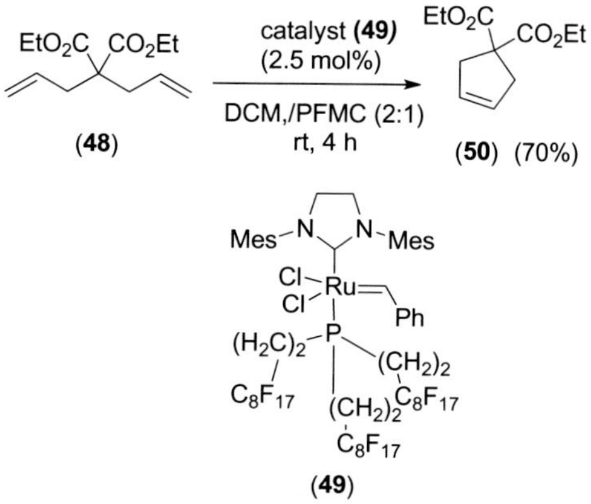

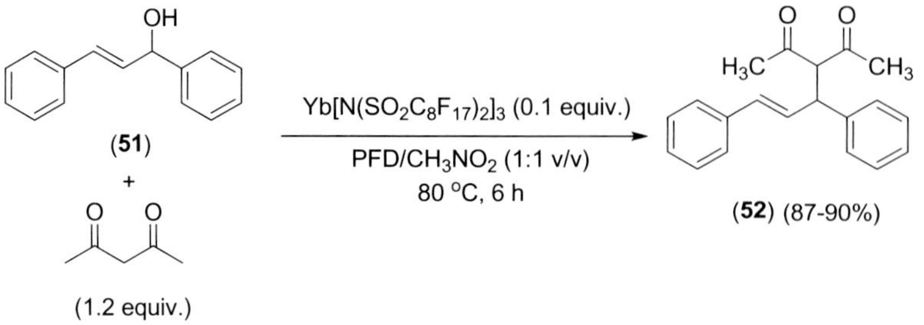

Scheme 31 Yb(III)-catalyzed alylation reaction under fluorous biphasic system

Application of Polyethylene Glycol in Organic Synthesis

Microwave-assisted Extraction using PEG: It has been used as an environmentally benign solvent used for the *microwave-assisted extraction* (MAE) of flavonoids such as flavone and coumarin derivatives from the medicinally important plant [102]. In addition to that, diverse varieties of organic reactions have been performed using PEG a single solvent or one of the cosolvents.

PEG as solvent for C–C bond-forming coupling Reactions: Polyethylene glycol has been widely been used as reaction medium to achieve metal-catalyzed cross-coupling reactions [103]. For the formation of C–C bond, for example, the microwave-assisted Suzuki coupling reaction between aryl halides and organoboron compounds has been achieved under Pd-catalyzed reaction condition (Scheme 33)

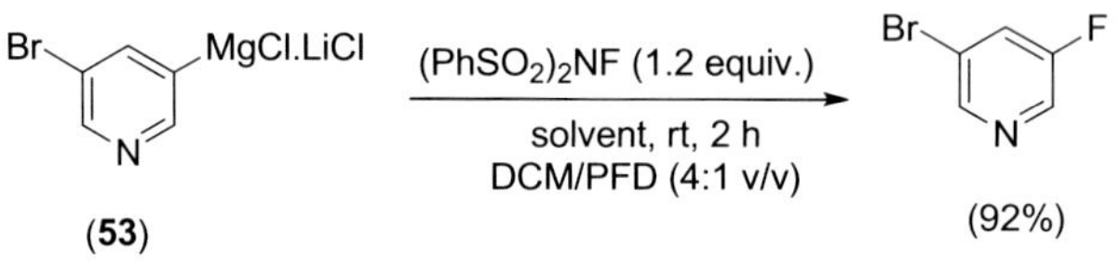

Scheme 32 Electrophilic fluorination of Grignard reagent in fluorous biphasic system

Fig. 8 Structure of polyethylene glycol

Polyethylene Glycol (PEG)

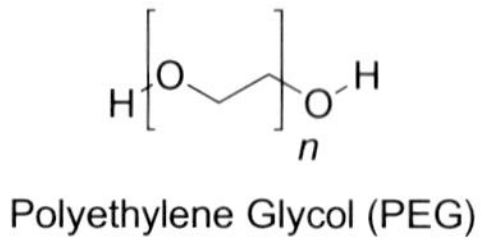

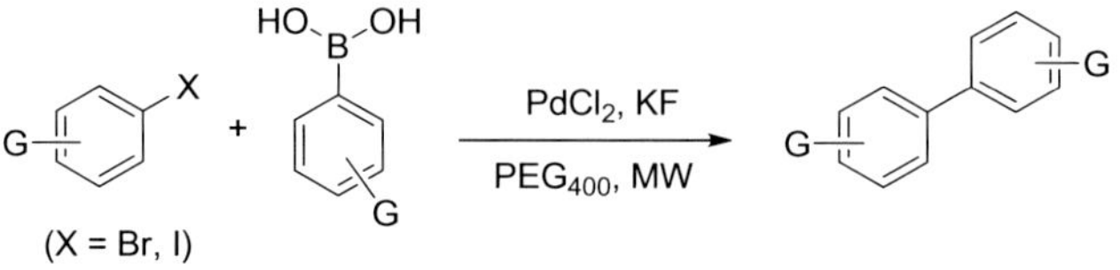

Scheme 33 Suzuki coupling using PEG$_{400}$ as solvent

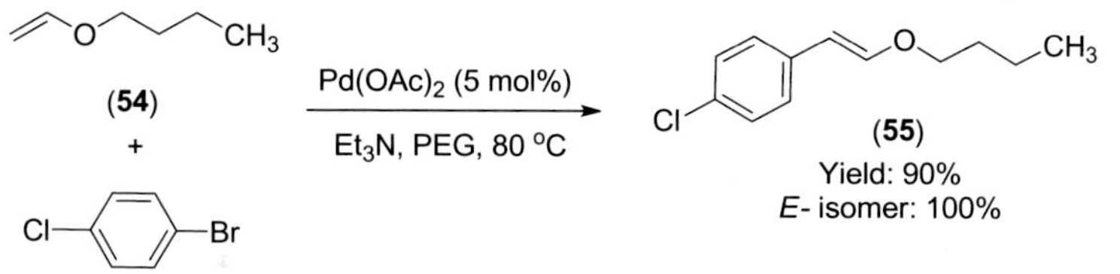

4 Green Solvents and Their Application

Scheme 34 An example of Heck coupling reaction in PEG$_{2000}$

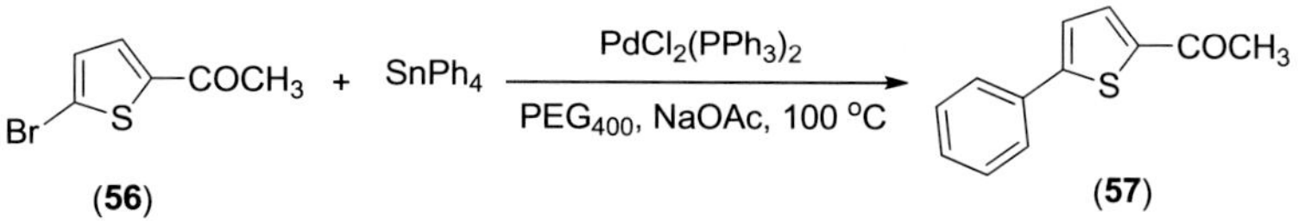

Scheme 35 An example of Stille coupling in PEG$_{400}$

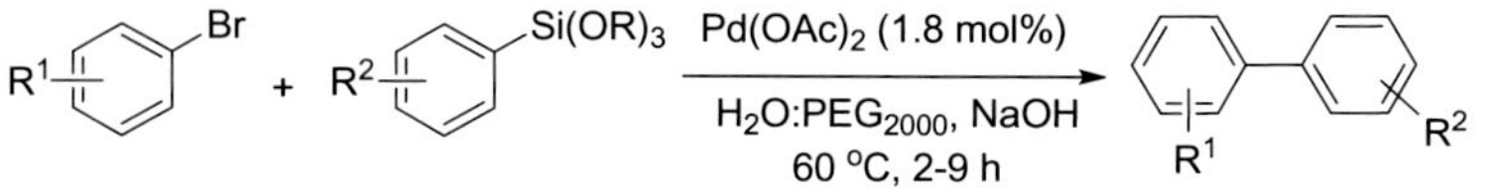

Scheme 36 Hiyama coupling reaction in water:PEG$_{2000}$

[104].

Similarly, Pd-catalyzed Mizoroki–Heck reaction between ether **54** and aryl halides and triflates has also been demonstrated using PEG as a green solvent [105]. Coupling reaction performed at 80 °C using PEG2000 as a reaction medium, led to the formation of single regioisomer of the coupled alkene **55** with high diastereoselectivity and in some cases reaction affords only *E*-diastereomer (Scheme 34).

Likewise, Stille coupling reaction between organostannanes with organic halides has also been demonstrated using PEG$_{400}$ as a greener alternative to conventional solvent. For example, Pd-catalyzed cross-coupling between substituted bromothiophene **56** and tetraphenyl tin reagent led to the formation of phenyl-substituted thiophene **57** as cross-coupled product (Scheme 35) [106].

Hiyama coupling

Similarly, the cross-coupling reaction coupling between organosilanes and organic halides which are known as Hiyama coupling has also been performed using aqueous PEG as solvent (Scheme 36) [107].

Sonogashira coupling reaction between a terminal alkyne and aryl halide has also been achieved using PEG$_{6000}$ as a reaction medium at 150 °C using PEG-PdL as catalyst. Sonogashira reaction between 4-bromoacetophenone and phenylacetylene using PEG-PdL as catalyst and PEG$_{6000}$ as solvent furnished internal alkyne **58** (Scheme 37) [108].

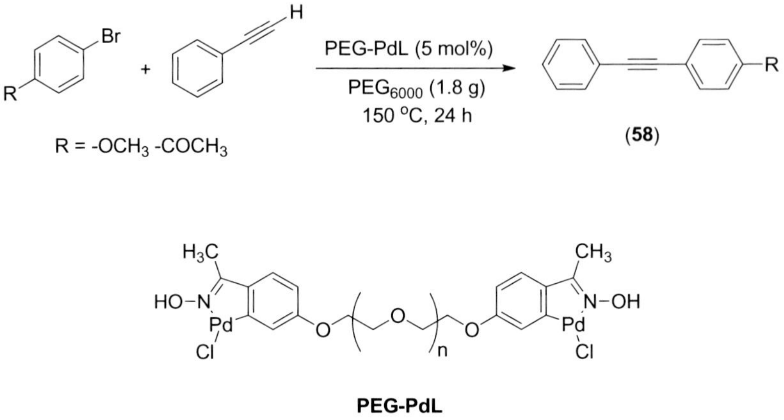

Scheme 37 Sonogashira coupling reaction in PEG$_{6000}$

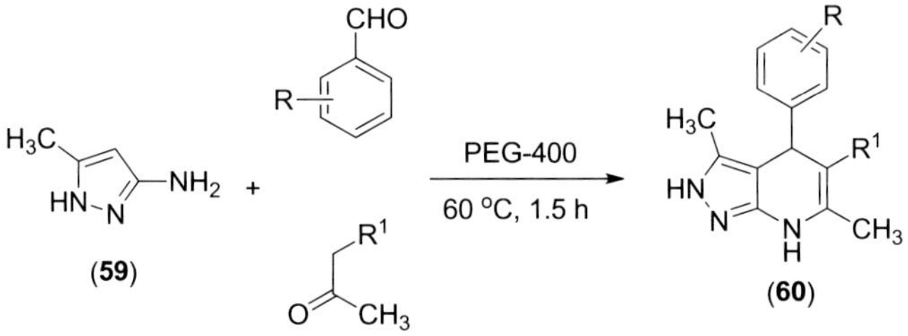

Scheme 38 Synthesis of pyrazolo[3,4-b]pyridines using PEG-400 as solvent

Polyethylene Glycol as a solvent for Heterocyclic synthesis

Heterocyclic compounds are important scaffold present in the diverse varieties of biologically active compounds and drugs to treat a diverse range of physiological disorders and diseases ranging from anticancerous, antiviral, anti-inflammatory, etc. Polyethylene glycol has also been used as a reaction medium to perform multicomponent reaction for the formation of N-heterocyclic ring. For example, reaction among amino pyrazole **59**, aldehyde, and ketone in PEG-400 leads to the formation of dihydro-2H-pyrazolo[3,4-b]pyridine derivatives **60** in 90–96% within 1.5 h (Scheme 38) [109].

Similarly, using PEG-400 as a solvent and (diacetoxyiodo)benzene (PIDA) as an oxidant, the formation of diarylmethanes **61** was achieved involving the coupling between sp3 and sp2-hybridized carbons under metal-free conditions (Scheme 39) [110].

Scheme 39 Dimerization of fused heterocycles using PIDA as an oxidant in PEG-400

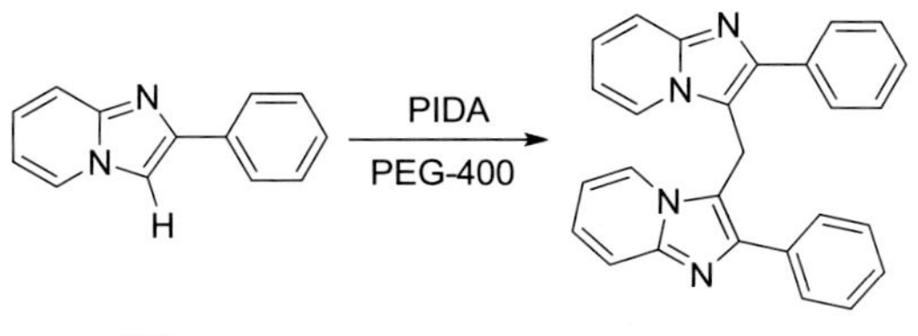

(61)   (62)

5 Conclusions

Overall this chapter dealt with the scope and applications of different types of environmentally benign alternatives of the conventional organic solvents. Mainly elaborate discussion on the various types of organic transformations achieved in green solvents such as water, supercritical carbon dioxide ($scCO_2$), ionic liquid, biosolvents, and polyethylene glycol (PEG) will provide an extensive knowledge and understanding about the different types of environmentally benign solvents which could be used to perform various types of chemical reactions.

References

1. Reichardt, C.: Solvents and solvent effects: an introduction. Org. Process Res. Dev. **11**(1), 105–113 (2007)
2. Sheldon, R.A.: The E factor 25 years on: the rise of green chemistry and sustainability. Green Chem. **19**(1), 18–43 (2017)
3. Rodríguez, B., Rantanen, T., Bolm, C.: Solvent-free asymmetric organocatalysis in a ball mill. Angew. Chemie **118**(41), 7078–7080 (2006)
4. Tripathi, G., Kumar, A., Rajkhowa, S., Tiwari, V.K.: Synthesis of biologically relevant heterocyclic skeletons under solvent-free condition. In: Green Synthetic Approaches for Biologically Relevant Heterocycles, pp. 421–459. Elsevier (2021)
5. Tobiszewski, M., Namieśnik, J., Pena-Pereira, F.: Environmental risk-based ranking of solvents using the combination of a multimedia model and multi-criteria decision analysis. Green Chem. **19**(4), 1034–1042 (2017)
6. Simon, M.-O., Li, C.-J.: Green chemistry oriented organic synthesis in water. Chem. Soc. Rev. **41**(4), 1415–1427 (2012)
7. Clark, J.H., Farmer, T.J., Hunt, A.J., Sherwood, J.: Opportunities for bio-based solvents created as petrochemical and fuel products transition towards renewable resources. Int. J. Mol. Sci. **16**(8), 17101–17159 (2015)
8. Sheldon, R.A.: Green solvents for sustainable organic synthesis: state of the art. Green Chem. **7**(5), 267–278 (2005)
9. Yilmaz, E., Soylak, M.: Type of green solvents used in separation and preconcentration methods. In: New Generation Green Solvents for Separation and Preconcentration of Organic and Inorganic Species (2020). https://doi.org/10.1016/b978-0-12-818569-8.00005-x
10. Sathish, M., Silambarasan, S., Madhan, B., Rao, J.R.: Exploration of GSK's solvent selection guide in leather industry: a CSIR-CLRI tool for sustainable leather manufacturing. Green Chem. **18**(21), 5806–5813 (2016)

11. Alder, C.M., Hayler, J.D., Henderson, R.K., Redman, A.M., Shukla, L., Shuster, L.E., Sneddon, H.F.: Updating and further expanding GSK's solvent sustainability guide. Green Chem. **18**(13), 3879–3890 (2016)
12. Prat, D., Pardigon, O., Flemming, H.-W., Letestu, S., Ducandas, V., Isnard, P., Guntrum, E., Senac, T., Ruisseau, S., Cruciani, P.: Sanofi's solvent selection guide: a step toward more sustainable processes. Org. Process Res. Dev. **17**(12), 1517–1525 (2013)
13. Byrne, F.P., Jin, S., Paggiola, G., Petchey, T.H.M., Clark, J.H., Farmer, T.J., Hunt, A.J., McElroy, C.R., Sherwood, J.: Tools and techniques for solvent selection: green solvent selection guides. Sustain. Chem. Process **4**(1), 1–24 (2016)
14. Bembenic, M.A.H., Clifford, C.E.B.: Subcritical water reactions of a hardwood derived organosolv lignin with nitrogen, hydrogen, carbon monoxide, and carbon dioxide gases. Energ. Fuels **26**(7), 4540–4549 (2012)
15. Avola, S., Goettmann, F., Antonietti, M., Kunz, W.: Organic reactivity of alcohols in superheated aqueous salt solutions: an overview. New J. Chem. **36**(8), 1568–1573 (2012)
16. Rideout, D.C., Breslow, R.: Hydrophobic acceleration of diels-alder reactions. J. Am. Chem. Soc. **102**(26), 7816–7817 (1980)
17. Chao-Jun Li, L.C.: Organic chemistry in water. Chem. Soc. Rev. **35**, 68–82 (2006)
18. Narayan, S., Muldoon, J., Finn, M.G., Fokin, V.V., Kolb, H.C., Sharpless, K.B.: "On Water": unique reactivity of organic compounds in aqueous suspension. Angew. Chemie Int. Ed. **44**(21), 3275–3279 (2005)
19. Yorimitsu, H., Nakamura, T., Shinokubo, H., Oshima, K., Omoto, K., Fujimoto, H.: Powerful solvent effect of water in radical reaction: triethylborane-induced atom-transfer radical cyclization in water. J. Am. Chem. Soc. **122**(45), 11041–11047 (2000)
20. Huang, T., Meng, Y., Venkatraman, S., Wang, D., Li, C.-J.: Remarkable electronic effect on rhodium-catalyzed carbonyl additions and conjugated additions with arylmetallic reagents. J. Am. Chem. Soc. **123**(30), 7451–7452 (2001)
21. Keh, C.C.K., Wei, C., Li, C.-J.: The Barbier−Grignard-type carbonyl alkylation using unactivated alkyl halides in water. J. Am. Chem. Soc. **125**(14), 4062–4063 (2003)
22. Kobayashi, S.: Lanthanide trifluoromethanesulfonates as stable Lewis acids in aqueous media. Yb(OTf)$_3$ catalyzed hydroxymethylation reaction of silyl enol ethers with commercial formaldehyde solution. Chem. Lett. **20**(12), 2187–2190 (1991)
23. Wei, C., Li, C.-J.: Enantioselective direct-addition of terminal alkynes to imines catalyzed by copper (I) pybox complex in water and in toluene. J. Am. Chem. Soc. **124**(20), 5638–5639 (2002)
24. ten Brink, G.-J., Arends, I.W.C.E., Sheldon, R.A.: Green, catalytic oxidation of alcohols in water. Science **287**(5458), 1636–1639 (2000)
25. Uozumi, Y., Nakao, R.: Catalytic oxidation of alcohols in water under atmospheric oxygen by use of an amphiphilic resin-dispersion of a nanopalladium catalyst. Angew. Chemie **115**(2), 204–207 (2003)
26. Noyori, R., Aoki, M., Sato, K.: Green oxidation with aqueous hydrogen peroxide. Chem. Commun. **16**, 1977–1986 (2003)
27. Sloboda-Rozner, D., Alsters, P.L., Neumann, R.: A water-soluble and "self-assembled" polyoxometalate as a recyclable catalyst for oxidation of alcohols in water with hydrogen peroxide. J. Am. Chem. Soc. **125**(18), 5280–5281 (2003)
28. Rodrigues, F., Canac, Y., Lubineau, A.: A convenient, one-step, synthesis of β-C-glycosidic ketones in aqueous media. Chem. Commun. **20**, 2049–2050 (2000)
29. Welton, T.: Ionic liquids: a brief history. Biophys. Rev. **10**(3), 691–706 (2018)
30. Keskin, S., Kayrak-Talay, D., Akman, U., Hortaçsu, Ö.: A review of ionic liquids towards supercritical fluid applications. J. Supercrit. Fluids **43**(1), 150–180 (2007)
31. Jessop, P.G.: Searching for green solvents. Green Chem. **13**(6), 1391–1398 (2011)
32. Keskin, S., Kayrak-Talay, D., Akman, U., Hortaçsu, O.: A review of ionic liquids towards supercritical fluid applications. J. Supercrit. Fluids **43**, 150–180 (2007)
33. Yang, B., Zhang, Q., Fei, Y., Zhou, F., Wang, P., Deng, Y.: Biodegradable betaine-based aprotic task-specific ionic liquids and their application in efficient SO_2 absorption. Green Chem. **17**(7), 3798–3805 (2015)

References

34. Eilmes, A., Kubisiak, P.: Quantum-chemical and molecular dynamics study of $M^+[TOTO]^-$ (M = Li, Na, K) ionic liquids. J. Phys. Chem. B **117**(41), 12583–12592 (2013)
35. Plaumann, H.: Switchable polarity solvents: are they green? Phys. Sci. Rev. **2**(3), 27–30 (2017)
36. Phan, L., Brown, H., White, J., Hodgson, A., Jessop, P.G.: Soybean oil extraction and separation using switchable or expanded solvents. Green Chem. **11**(1), 53–59 (2009)
37. Vanderveen, J.R., Durelle, J., Jessop, P.G.: Design and evaluation of switchable-hydrophilicity solvents. Green Chem. **16**(3), 1187–1197 (2014)
38. Peach, J., Eastoe, J.: Supercritical carbon dioxide: a solvent like no other. Beilstein J. Org. Chem. **10**(1), 1878–1895 (2014)
39. De Marco, I., Riemma, S., Iannone, R.: Supercritical carbon dioxide decaffeination process: a life cycle assessment study. Chem. Eng. Trans. **57**, 1699–1704 (2017)
40. Pinelo, M., Ruiz-Rodríguez, A., Sineiro, J., Señoráns, F.J., Reglero, G., Núñez, M.J.: Supercritical fluid and solid-liquid extraction of phenolic antioxidants from grape pomace: a comparative study. Eur. Food Res. Technol. **226**(1), 199–205 (2007)
41. Villanueva Bermejo, D., Angelov, I., Vicente, G., Stateva, R.P., Rodriguez García-Risco, M., Reglero, G., Ibañez, E., Fornari, T.: Extraction of thymol from different varieties of thyme plants using green solvents. J. Sci. Food Agric. **95**(14), 2901–2907 (2015)
42. Lu, Y., Mu, B., Zhu, B., Wu, K., Gou, Z., Li, L., Cui, L., Liang, N.: Comparison of supercritical fluid extraction and liquid solvent extraction on antitumor diterpenoid from *Pteris semipinnata L.*. Sep. Sci. Technol. **47**(16), 2436–2443 (2012)
43. Jessop, P.G., Ikariya, T., Noyori, R.: Homogeneous catalytic hydrogenation of supercritical carbon dioxide. Nature **368**(6468), 231–233 (1994)
44. Jessop, P.G., Hsiao, Y., Ikariya, T., Noyori, R.: Catalytic production of dimethylformamide from supercritical carbon dioxide. J. Am. Chem. Soc. **116**(19), 8851–8852 (1994)
45. Jessop, P.G., Hsiao, Y., Ikariya, T., Noyori, R.: Homogeneous catalysis in supercritical fluids: hydrogenation of supercritical carbon dioxide to formic acid, alkyl formates, and formamides. J. Am. Chem. Soc. **118**(2), 344–355 (1996)
46. Burk, M.J., Feng, S., Gross, M.F., Tumas, W.: Asymmetric catalytic hydrogenation reactions in supercritical carbon dioxide. J. Am. Chem. Soc. **117**(31), 8277–8278 (1995)
47. Xiao, J., Nefkens, S.C.A., Jessop, P.G., Ikariya, T., Noyori, R.: Asymmetric hydrogenation of α,β-unsaturated carboxylic acids in supercritical carbon dioxide. Tetrahedron Lett. **37**(16), 2813–2816 (1996)
48. Hu, Y., Birdsall, D.J., Stuart, A.M., Hope, E.G., Xiao, J.: Ruthenium-catalysed asymmetric hydrogenation with fluoroalkylated binap ligands in supercritical CO_2. J. Mol. Catal. A Chem. **219**(1), 57–60 (2004)
49. Lyubimov, S.E., Rastorguev, E.A., Petrovskii, P.V., Kelbysheva, E.S., Loim, N.M., Davankov, V.A.: Iridium-catalyzed asymmetric hydrogenation of imines in supercritical carbon dioxide using phosphite-type ligands. Tetrahedron Lett. **52**(12), 1395–1397 (2011)
50. Berthod, M., Mignani, G., Lemaire, M.: New perfluoroalkylated BINAP usable as a ligand in homogeneous and supercritical carbon dioxide asymmetric hydrogenation. Tetrahedron Asym. **15**(7), 1121–1126 (2004)
51. Gava, R., Olmos, A., Noverges, B., Varea, T., Álvarez, E., Belderrain, T.R., Caballero, A., Asensio, G., Pérez, P.J.: Discovering copper for methane C–H bond functionalization. ACS Catal. **5**(6), 3726–3730 (2015)
52. Zhang, W., He, X., Ren, B., Jiang, Y., Hu, Z.: $Cu(OAc)_2 \cdot H_2O$—an efficient catalyst for Huisgen-click reaction in supercritical carbon dioxide. Tetrahedron Lett. **56**(19), 2472–2475 (2015)
53. López-Periago, A.M., Vega, A., Subra, P., Argemí, A., Saurina, J., García-González, C.A., Domingo, C.: Supercritical CO_2 processing of polymers for the production of materials with applications in tissue engineering and drug delivery. J. Mater. Sci. **43**(6), 1939–1947 (2008)
54. Liu, X., Coutelier, O., Harrisson, S., Tassaing, T., Marty, J.-D., Destarac, M.: Enhanced solubility of polyvinyl esters in $ScCO_2$ by means of vinyl trifluorobutyrate monomer. ACS Macro Lett. **4**(1), 89–93 (2015)

55. Chakraborty, S., Colón, Y.J., Snurr, R.Q., Nguyen, S.T.: Hierarchically porous organic polymers: highly enhanced gas uptake and transport through templated synthesis. Chem. Sci. 6(1), 384–389 (2015)
56. Maleki, H., Durães, L., Portugal, A.: Synthesis of mechanically reinforced silica aerogels via surface-initiated reversible addition-fragmentation chain transfer (RAFT) polymerization. J. Mater. Chem. A 3(4), 1594–1600 (2015)
57. Kuang, T.-R., Mi, H.-Y., Fu, D.-J., Jing, X., Chen, B., Mou, W.-J., Peng, X.-F.: Fabrication of poly (lactic acid)/graphene oxide foams with highly oriented and elongated cell structure via unidirectional foaming using supercritical carbon dioxide. Ind. Eng. Chem. Res. 54(2), 758–768 (2015)
58. Zhao, J., Liu, Z., Li, H., Hu, W., Zhao, C., Zhao, P., Shi, D.: Development of a highly active electrocatalyst via ultrafine Pd nanoparticles dispersed on pristine graphene. Langmuir 31(8), 2576–2583 (2015)
59. Jiménez, C., Garcia, J., Camarillo, R., Martínez, F., Rincón, J.: Electrochemical CO_2 reduction to fuels using Pt/CNT catalysts synthesized in supercritical medium. Energ. Fuels 31(3), 3038–3046 (2017)
60. Cid, M.V.F., Van Spronsen, J., Van der Kraan, M., Veugelers, W.J.T., Woerlee, G.F., Witkamp, G.J.: Excellent dye fixation on cotton dyed in supercritical carbon dioxide using fluorotriazine reactive dyes. Green Chem. 7(8), 609–616 (2005)
61. Abou Elmaaty, T., Abd El-Aziz, E.: Supercritical carbon dioxide as a green media in textile dyeing: a review. Text. Res. J. 88(10), 1184–1212 (2018)
62. Xiao, H., Zhao, T., Li, C.-H., Li, M.-Y.: Eco-friendly approaches for dyeing multiple type of fabrics with cationic reactive dyes. J. Clean. Prod. 165, 1499–1507 (2017)
63. Zhang, Y.-Q., Wei, X.-C., Long, J.-J.: Ecofriendly synthesis and application of special disperse reactive dyes in waterless coloration of wool with supercritical carbon dioxide. J. Clean. Prod. 133, 746–756 (2016)
64. DeSimone, J.M., Tumas, W.: Green Chemistry Using Liquid and Supercritical Carbon Dioxide. Oxford University Press (2003)
65. Lee, W., Kuan, W.: Miscanthus as cellulosic biomass for bioethanol production. Biotechnol. J. 10(6), 840–854 (2015)
66. Beringer, T.I.M., Lucht, W., Schaphoff, S.: Bioenergy production potential of global biomass plantations under environmental and agricultural constraints. GCB Bioenerg. 3(4), 299–312 (2011)
67. Pace, V., Hoyos, P., Castoldi, L., Dominguez de Maria, P., Alcántara, A.R.: 2-methyltetrahydrofuran (2-MeTHF): a biomass-derived solvent with broad application in organic chemistry. ChemSusChem 5(8), 1369–1379 (2012)
68. Brandt, A., Gräsvik, J., Hallett, J.P., Welton, T.: Deconstruction of lignocellulosic biomass with ionic liquids. Green Chem. 15(3), 550–583 (2013)
69. Farrán, A., Cai, C., Sandoval, M., Xu, Y., Liu, J., Hernáiz, M.J., Linhardt, R.J.: Green solvents in carbohydrate chemistry: from raw materials to fine chemicals. Chem. Rev. 115(14), 6811–6853 (2015)
70. Onda, A., Ochi, T., Yanagisawa, K.: Selective hydrolysis of cellulose into glucose over solid acid catalysts. Green Chem. 10, 1033–1037 (2008)
71. Ciriminna, R., Lomeli-Rodriguez, M., Cara, P.D., Lopez-Sanchez, J.A., Pagliaro, M.: Limonene: a versatile chemical of the bioeconomy. Chem. Commun. 50(97), 15288–15296 (2014)
72. Antonucci, V., Coleman, J., Ferry, J.B., Johnson, N., Mathe, M., Scott, J.P., Xu, J.: Toxicological assessment of 2-methyltetrahydrofuran and cyclopentyl methyl ether in support of their use in pharmaceutical chemical process development. Org. Process Res. Dev. 15(4), 939–941 (2011)
73. Liguori, F., Moreno-Marrodan, C., Barbaro, P.: Environmentally friendly synthesis of γ-valerolactone by direct catalytic conversion of renewable sources. ACS Catal. 5(3), 1882–1894 (2015)

References

74. Sherwood, J., Constantinou, A., Moity, L., McElroy, C.R., Farmer, T.J., Duncan, T., Raverty, W., Hunt, A.J., Clark, J.H.: Dihydrolevoglucosenone (Cyrene) as a bio-based alternative for dipolar aprotic solvents. Chem. Commun. **50**(68), 9650–9652 (2014)
75. Botella, L., Nájera, C.: Controlled mono and double heck reactions in water catalyzed by an oxime-derived palladacycle. Tetrahedron Lett. **45**(9), 1833–1836 (2004)
76. Sarmah, M., Mondal, M., Bora, U.: Agro-waste extract based solvents: emergence of novel green solvent for the design of sustainable processes in catalysis and organic chemistry. ChemistrySelect **2**(18), 5180–5188 (2017)
77. Yara-Varón, E., Li, Y., Balcells, M., Canela-Garayoa, R., Fabiano-Tixier, A.-S., Chemat, F.: Vegetable oils as alternative solvents for green oleo-extraction, purification and formulation of food and natural products. Molecules **22**(9), 1474 (2017)
78. Menges, N., Şahin, E.: Metal-and base-free combinatorial reaction for C-acylation of 1,3-diketo compounds in vegetable oil: the effect of natural oil. ACS Sustain. Chem. Eng. **2**(2), 226–230 (2014)
79. García, J.I., García-Marín, H., Pires, E.: Glycerol based solvents: synthesis, properties and applications. Green Chem. **16**(3), 1007–1033 (2014)
80. Becker, L.C., Bergfeld, W.F., Belsito, D.V., Hill, R.A., Klaassen, C.D., Liebler, D.C., Marks, J.G., Jr., Shank, R.C., Slaga, T.J., Snyder, P.W.: Safety assessment of glycerin as used in cosmetics. Int. J. Toxicol. **38**(3), 6S-22S (2019)
81. Gu, Y., Barrault, J., Jerome, F.: Glycerol as an efficient promoting medium for organic reactions. Adv. Synth. Catal. **350**(13), 2007–2012 (2008)
82. Gu, Y., Jérôme, F.: Glycerol as a sustainable solvent for green chemistry. Green Chem. **12**(7), 1127–1138 (2010)
83. Li, M., Chen, C., He, F., Gu, Y.: Multicomponent reactions of 1,3-cyclohexanediones and formaldehyde in glycerol: stabilization of paraformaldehyde in glycerol resulted from using dimedone as substrate. Adv. Synth. Catal. **352**(2–3), 519–530 (2010)
84. Tan, J.-N., Li, M., Gu, Y.: Multicomponent reactions of 1,3-disubstituted 5-pyrazolones and formaldehyde in environmentally benign solvent systems and their variations with more fundamental substrates. Green Chem. **12**(5), 908–914 (2010)
85. Radatz, C.S., Silva, R.B., Perin, G., Lenardão, E.J., Jacob, R.G., Alves, D.: Catalyst-free synthesis of benzodiazepines and benzimidazoles using glycerol as recyclable solvent. Tetrahedron Lett. **52**(32), 4132–4136 (2011)
86. Kumar, T.A., Devi, B.R., Dubey, P.K.: Simple, facile and complete green synthesis of N-alkyl-2-styrylbenzimidazoles using glycerol and PEG-600 as green solvents. Der. Chem. Sin. **4**, 116–121 (2013)
87. Bachhav, H.M., Bhagat, S.B., Telvekar, V.N.: Efficient protocol for the synthesis of quinoxaline, benzoxazole and benzimidazole derivatives using glycerol as green solvent. Tetrahedron Lett. **52**(43), 5697–5701 (2011)
88. Wolfson, A., Dlugy, C., Tavor, D., Blumenfeld, J., Shotland, Y.: Baker's yeast catalyzed asymmetric reduction in glycerol. Tetrahedron Asym. **17**(14), 2043–2045 (2006)
89. Andrade, L.H., Piovan, L., Pasquini, M.D.: Improving the enantioselective bioreduction of aromatic ketones mediated by aspergillus Terreus and Rhizopus Oryzae: The role of glycerol as a co-solvent. Tetrahedron Asym. **20**(13), 1521–1525 (2009)
90. Taketomi, S., Asano, M., Higashi, T., Shoji, M., Sugai, T.: Chemo-enzymatic route for (R)-terbutaline hydrochloride based on microbial asymmetric reduction of a substituted α-chloroacetophenone derivative. J. Mol. Catal. B Enzym. **84**, 83–88 (2012)
91. Khatri, P.K., Jain, S.L.: Glycerol ingrained copper: an efficient recyclable catalyst for the N-arylation of amines with aryl halides. Tetrahedron Lett. **54**(21), 2740–2743 (2013)
92. Silveira, C.C., Mendes, S.R., Líbero, F.M., Lenardão, E.J., Perin, G.: Glycerin and CeCl$_3$ · 7H$_2$O: a new and efficient recyclable medium for the synthesis of bis(indoly) methanes. Tetrahedron Lett. **50**(44), 6060–6063 (2009)
93. Delample, M., Villandier, N., Douliez, J.-P., Camy, S., Condoret, J.-S., Pouilloux, Y., Barrault, J., Jérôme, F.: Glycerol as a cheap, safe and sustainable solvent for the catalytic and regioselective β,β-diarylation of acrylates over palladium nanoparticles. Green Chem. **12**(5), 804–808 (2010)

94. Cabrera, D.M.L., Libero, F.M., Alves, D., Perin, G., Lenardao, E.J., Jacob, R.G.: Glycerol as a recyclable solvent in a microwave-assisted synthesis of disulfides. Green Chem. Lett. Rev. **5**(3), 329–336 (2012)
95. Chung, W.J., Baskar, C., Chung, D.G., Han, M.D., Lee, C.H.: Catalytic transfer hydrogenation of carboxylic acids to their corresponding alcohols by using glycerol as hydrogen donor. Repub Korean Kongkae Taeho Kongbo (2012)
96. Gladysz, J.A., Curran, D.P., Horváth, I.T.: Handbook of Fluorous Chemistry. Wiley (2006)
97. Horváth, I.T., Kiss, G., Cook, R.A., Bond, J.E., Stevens, P.A., Rábai, J., Mozeleski, E.J.: Molecular engineering in homogeneous catalysis: one-phase catalysis coupled with biphase catalyst separation. The fluorous-soluble HRh(CO){P[CH$_2$CH$_2$(CF$_2$)5CF$_3$]$_3$}$_3$ hydroformylation system. J. Am. Chem. Soc. **120**(13), 3133–3143 (1998)
98. Horváth, I.T., Rábai, J.: Facile catalyst separation without water: fluorous biphase hydroformylation of olefins. Science **266**(5182), 72–75 (1994)
99. da Costa, R.C., Gladysz, J.A.: Syntheses and reactivity of analogues of Grubbs' second generation metathesis catalyst with fluorous phosphines: a new phase-transfer strategy for catalyst activation. Adv. Synth. Catal. **349**(1–2), 243–254 (2007)
100. Shen, M.-G., Cai, C., Yi, W.-B.: Yb[N(SO$_2$C8F17)$_2$]$_3$-catalyzed allylation of 1,3-dicarbonyl compounds with allylic alcohols in a fluorous biphase system. J. Fluor. Chem. **130**(6), 595–599 (2009)
101. Yamada, S., Gavryushin, A., Knochel, P.: Convenient electrophilic fluorination of functionalized aryl and heteroaryl magnesium reagents. Angew. Chemie **122**(12), 2261–2264 (2010)
102. Zhou, T., Xiao, X., Li, G., Cai, Z.: Study of polyethylene glycol as a green solvent in the microwave-assisted extraction of flavone and coumarin compounds from medicinal plants. J. Chromatogr. A **1218**(23), 3608–3615 (2011)
103. Corma, A., García, H., Leyva, A.: Polyethyleneglycol as scaffold and solvent for reusable CC coupling homogeneous Pd catalysts. J. Catal. **240**(2), 87–99 (2006)
104. Namboodiri, V.V., Varma, R.S.: Microwave-accelerated Suzuki cross-coupling reaction in polyethylene glycol (PEG). Green Chem. **3**(3), 146–148 (2001)
105. Chandrasekhar, S., Narsihmulu, C., Sultana, S.S., Reddy, N.R.: Poly(ethylene glycol) (PEG) as a reusable solvent medium for organic synthesis. Application in the heck reaction. Org. Lett. **4**(25), 4399–4401(2002)
106. Zhou, W., Wang, K., Wang, J.: Atom-efficient, palladium-catalyzed Stille coupling reactions of tetraphenylstannane with aryl iodides or aryl bromides in polyethylene glycol 400 (PEG-400). Adv. Synth. Catal. **351**(9), 1378–1382 (2009)
107. Shi, S., Zhang, Y.: Pd(OAc)$_2$-catalyzed fluoride-free cross-coupling reactions of arylsiloxanes with aryl bromides in aqueous medium. J. Org. Chem. **72**(15), 5927–5930 (2007)
108. Corma, A., García, H., Leyva, A.: Comparison between polyethylenglycol and imidazolium ionic liquids as solvents for developing a homogeneous and reusable palladium catalytic system for the Suzuki and Sonogashira coupling. Tetrahedron **61**(41), 9848–9854 (2005)
109. Kerru, N., Gummidi, L., Maddila, S., Jonnalagadda, S.B.: Polyethylene glycol (PEG-400) mediated one-pot green synthesis of 4,7-dihydro-2H-pyrazolo[3,4-b]pyridines under catalyst-free conditions. ChemistrySelect **5**(40), 12407–12410 (2020)
110. Kumar, R., Rawat, D., Adimurthy, S.: Polyethylene glycol (PEG-400) as methylene spacer and green solvent for the synthesis of heterodiarylmethanes under metal-free conditions. Eur. J. Org. Chem. **2020**(23), 3499–3507 (2020)

Chapter 4
Growing Impact of Ionic Liquids in Heterocyclic Chemistry

1 Introduction

In the quest for replacing the conventional toxic organic solvents, a continual effort has been made in fabricating a solvent that is versatile, compatible with reaction condition and at the same time environmentally benign too. The search reached to a satisfactory end with the discovery of ionic liquids (ILs) [1]. Although, molten metal halide salts have been in very much application since 1830s with Faraday's electrolysis processes, the requirement of high temperature practically restricted their applicability. Therefore, ILs started gaining the attention of researchers in the 1980s. Those initial years of ILs involved in developing novel inorganic salt mixtures (more often than not) that could serve as versatile solvents in analytical and synthetic chemistry. Later, with the development of room-temperature ionic liquids (RTILs), the research in this domain gains the momentum as the energy consumption and has been mollified as they provide a media to carry out reactions at moderate temperature. ILs are also termed "designer solvent" based on the concept that there are hundreds of cation–anion combinations that will form ILs and the easy tenability of ILs makes them an ideal system for any particular application can be selected [1, 2]. They were first used in chemical synthesis during the late 1990s and since then they have extensively used in variety of applications including enzymatic synthesis. The wide range of applications lies in their versatile nature such as low vapor pressure at room temperature, high solubility and thermal stability, hydrophilic/hydrophobic properties with biocompatibility, and moreover the environmentally benign nature. ILs is overall a suitable candidate for chemical, biological, and material sciences due to their easy and tailor-made fabrication as per the demand, and therefore, they are better known as "task-specific ILs" [3]. These ILs have been studied for their potential applications in various fields ranging from material science to medicinal chemistry and catalysis. Probably the most extensive use was the application of ILs to "clean" and/or "green" technologies and specifically the growing green chemistry revolution. The ionic nature of ILs provides unique solvation environments compared to the conventional molecular organic/inorganic solvents, and this is exploited in a variety

© The Author(s), under exclusive license to Springer Nature Singapore Pte Ltd. 2022
V. K. Tiwari et al., *Green Chemistry*,
https://doi.org/10.1007/978-981-19-2734-8_4

of different synthetic reactions, materials processing/extraction, and gas separation. Most importantly, their ability to dissolve materials that are usually insoluble in common organic solvents is highly advantageous, while their ionic nature helps in stabilizing nanoparticle dispersions during the synthesis. They are chemically mixtures of molten salts or molten oxides having ionic-covalent crystalline structures with melting points not more than 100 °C [2]. The presence of coulombic and van der Waals interactions as well as the hydrogen bonds among the cations and anions imparts special properties to ILs [4]. With the synthesis of ILs that can remain liquid at or below room temperature, termed as room-temperature ionic liquids (RTILs) has gained immense popularity due to their vital role in organic synthesis for the past few decades [4–15]. RTILs possess a number of inherent properties which are also tunable according to fulfill the requirements of a specific task such as melting point, viscosity, density, refractive index, polarity, as well as physical properties such as non-volatility, non-flammability, thermal stability, solvability, and recyclability [16]. The rapid expansion of the field of ILs includes bench-scale and industrial scale research and applications. The commercial use of ILs has been under development since the late 1990s. It would be noteworthy that the ILs has potential applications as battery electrolytes for energy generation and storage after the discovery of the chloroaluminate ILs in the latter half of the twentieth century. The advantages of ILs go beyond simple non-flammability to include wide electrochemical windows, stability to the various electrode materials, separation agents for azeotropic mixtures, application in pharmaceutical industries, fabricating nanomaterials, and polymers of wide utilities.

In addition, thermoregulated ILs due to their notable features have displayed great applications in catalysis and other emerging areas [17]. ILs were used for energy storage [18] and also for the extraction and separation processes for a various bioactive candidates [19]. With an easy separation and recyclability features, they have identified as "immobilized catalyst" or "biphasic catalyst" [19]. ILs serve as a promising catalysts, solvents, and/or reagents in various of oxidation reactions, for example, Baeyer–Villiger oxidation of ketones to esters, oxidation of alkanes, alcohols and sulfides, and epoxidation of alkenes [20]. Fascinatingly, a novel API-IL concept is explored to develop traditional drug candidates for their effective use [21]. A large number of pharmacologically active molecules as well approved drug candidates contains the heterocyclic skeletons [22–25]. They are part of our life saving drugs, and these structural units were found in hundreds of bioactive natural products. This chapter discusses all these aspects of RTILs, their challenges and future perspectives in emerging issues in science and technology in special reference to their impact in heterocyclic synthesis.

2 Structure and Types of RTILs

ILs are commonly a mixture of ions, where the cations are organic moieties having a delocalized charge that prevents stable crystal lattice formation, and the anions

2 Structure and Types of RTILs

could be either organic or inorganic moieties. Figure 1 shows a variety of cations and anions that when combined together can resulted into the desired ILs of versatile utilities in synthesis.

The first report on RTIL for the preparation of ethylammonium nitrate, [EtNH$_3$][NO$_3$], was in 1914 by Paul Walden, better known as the father of ILs [26]. The first chloroaluminate anion-based RTIL was introduced by the middle of twentieth century [27, 28]. With the progress in development of novel combinations of cations and anions or replacement of functional groups, ILs have been dominating the synthesis field in terms of suitable solvents and/or catalysts over the years. Thus, ILs can be categorized in three distinct generations (1–3) on the basis of such developments and their applications as shown in Table 1.

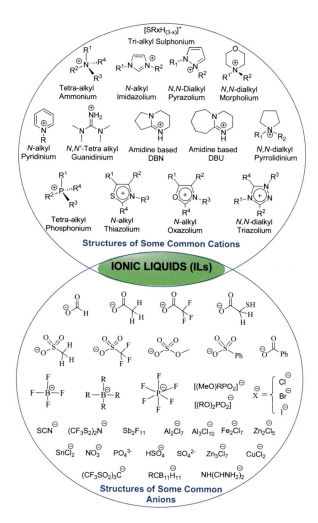

Fig. 1 Structure of some common cations and anions used in the preparation of RTILs

Table 1 Illustration for the three generations of ILs

Generation	Structure of ILs	Era and property	References
First generation	**1**	1980s Chloroaluminate IL	[29]
Second generation	**2**	1990s Air- and moisture-stable IL	[30]
Third generation	**3**	2000s Task-specific IL	[31]

The notable feature of ILs as green media drive for the gradual advancement in this growing field. In 1978, Osteryoung and coworkers developed a molten salt system through quaternization of pyridine mixing with $AlCl_3$ [32]. Followed by, Wilkes and Hussey evaluated the effect of cations that resulted to discovery of dialkylimidazolium-based ILs [33]. These ILs were very sensitive to moisture and not suitable to be use in open-air condition. Latter, Wilkes and Zaworotko successfully developed a series of water- and air-stable dialkylimidazolium-based ILs [34]. The authors invented an anion exchange framework with more hydrolytically stable anions such as acetate, BF_4^-, PF_6^-, NO_3^- or SO_4^{2-} [35].

Nevertheless, the aim of attempting different cation–anion combinations is to endow the overall structure of ILs with various physicochemical properties, and to modulate their characteristics depending on the conditions. This kind of fabrication allows the user to customize the IL with appropriate density, viscosity, hydrophobicity, solubility, and melting point in accordance with process's specifications. Moreover, ILs serve as better solvents than the conventional organic solvents in as much as ILs are more suitable in product separation, synthesis, catalysis, electrochemistry, nanotechnology, and biotechnology. Thus, it is utmost necessary to design novel class of ILs equipped with essential physicochemical properties required for the sustainable and green synthetic processes. A good number of reports have been found on application of ILs in extraction and synthesis that advocate their advantageous properties over conventional organic solvents [36, 37]. Various types of ILs can be summarized in Fig. 2.

Environmental impact, cost-effectiveness, and recyclability are few major concerns related to ILs. Thus, a number of inexpensive, biodegradable ILs extracted from renewable sources are being synthesized in recent times and developed as promising biomaterials in several emerging areas [38]. A representative example is choline chloride having deep eutectic mixtures (**4**) (Fig. 3) [39]. Likewise, choline cations (**5**) are designed as a cost-effective ILs by direct neutralization of choline hydroxide with various carboxylic acids. A great interest on protic ILs grows due to their high potential for direct proton transfer from a Bronsted acid to a base, an essential property required for application in fuel cell technologies.

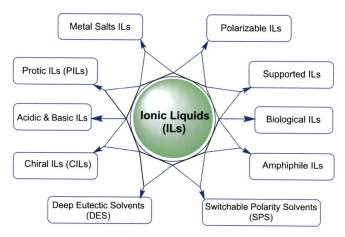

Fig. 2 A general classification of various types of ionic liquids (ILs)

Biodegradable ILs (4) **Cost effective ILs (5)**

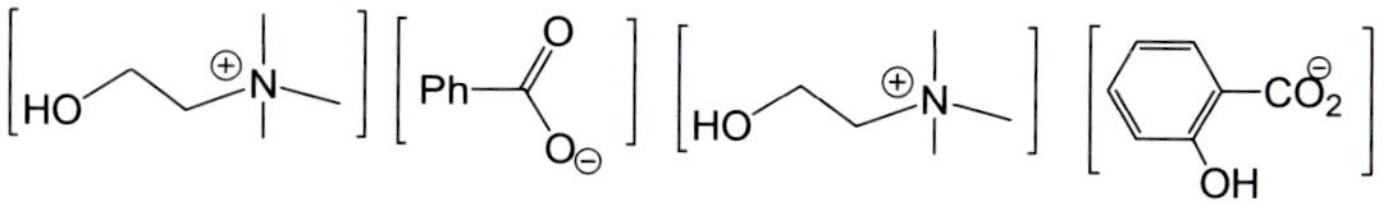

Fig. 3 Chemical structure of a biodegradable and cost-effective ILs

When acidic or basic functional groups are present on either side of anion or cation in ILs (Fig. 4), they are known as acidic ILs and basic ILs [40]. ILs containing polynuclear metallic anions, such as chloroaluminates, exhibit Lewis acidity, and super acidity in presence of protons is extensively studied for their physical and chemical significance.

The chiral ILs can easily be obtained by deriving from readily available chiral biomolecules including carbohydrates, amino acids, and other renewable sources [41]. Chiral center can be generated on either of the cation or anion or both (Fig. 5). Toward this end, a diverse range of chiral ILs composed of chiral amino acids as anions and imidazolium, ammonium, phosphonium, and other cations were developed for their wide applications [42, 43]. Chiral ligand-supported ILs is generally supported on inorganic materials and they are equally useful in various field. For example, a highly ordered mesoporous functional organosilica appended chiral camphorsulfonamide **7** was developed from chiral imidazolium-tetraethoxysilane salt using a hydrolysis–polycondensation route [44]. Interestingly, switchable polar solvents that are neutral liquids, when exposed to carbon dioxide, can be reversibly converted to the required polar ILs.

Importantly, the stability of ILs during the course of any reaction could be considered as a prime factor for their selection as environmentally benign organic media

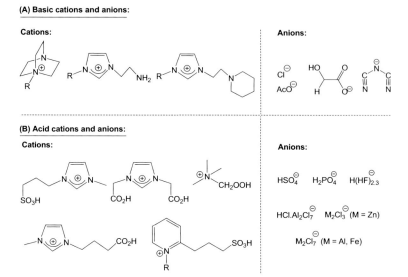

Fig. 4 Chemical structure of some common acidic and basic ILs

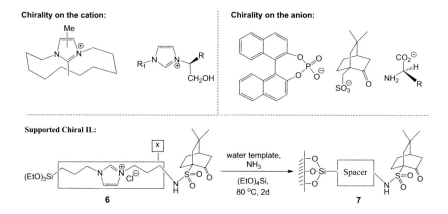

Fig. 5 Structure of some chiral ILs and silica-supported chiral IL **7**

and/or catalytic system for the required conversion steps or synthesis of biologically relevant scaffolds. A number of successful attempts have been made to achieve chemically stable ILs; N-heterocyclic carbene is a notable example. Thus, the C(2)-H activation of imidazolium heterocycle under basic condition was widely explored for the synthesis of required N-heterocyclic carbene as a stable IL of wide synthetic applications [45]. Another important example is the synthesis of useful methylimidazole-based IL **8** (Fig. 6), which could be obtained in good yield by replacing C(2)-H with a alkylthiol residue [46]. Likewise, biological ILs also play a significant role in synthetic and medicinal fields [47]. Interestingly, replacement of hydrazine with

2 Structure and Types of RTILs

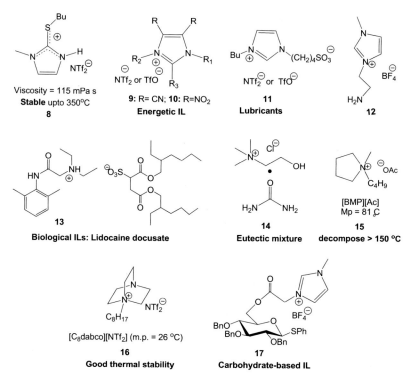

Fig. 6 Structure of some widely explored functional ILs

imidazolium-based ILs containing dicyanamide anions (**9**) in hypergolic fuels is expected to improve the energy content and physical parameters [48]. IL cations containing amine bases can be used for capturing of carbon dioxide [49, 50]. The basicity of specially fabricated ILs, e.g., **IL 12** and **IL 14**, may be adjusted by adding amino groups to either side of cation or anion in the IL structure. Furthermore, IL derived from DABCO **IL-16** containing N-octyl cation and [NTf$_2$]$^-$ anion exhibit substantial thermal stability with a low melting point [51, 52]. A few basic ILs derived from naturally occurring amino acids display interesting properties such as biodegradability [53]. Structures of some widely explored functional ILs (ILs **8–17**) are shown in Fig. 6.

In recent years, a great number of ILs were derived from chiral substrates including natural amino acids, carbohydrate derivatives, and other optically active molecules acting as a chiral pool. In addition to the use in chiral resolution techniques, these chiral ILs were widely explored as suitable media for several enantioselective reactions [54]. In addition to carbohydrate-based chiral ILs, a number of stable and well-defined three-dimensional dissymmetric architectures, e.g., imidazolium salts with cyclophane-type planar chirality (**18–20**), are well documented in literature (Fig. 7) [55].

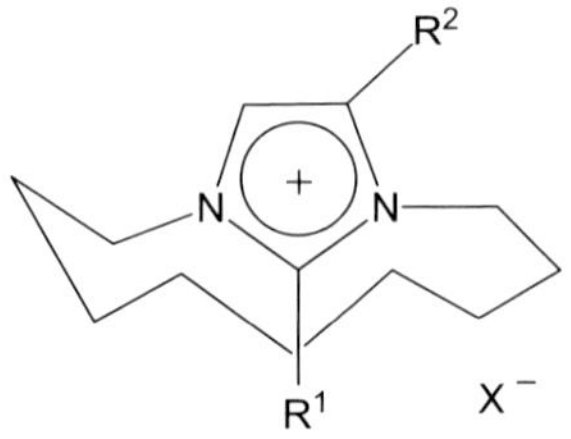

Fig. 7 Some cyclophane-type imidazolium salts (**18–20**) as promising ILs

18: $R^1 = R^2 = H$; **19:** $R^1 = Me$, $R^2 = H$; **20:** $R^1 = R^2 = Me$

(a) X = Br ; **(b)** X = $(CF_3SO_2)_2\bar{N}$
(c) X = $(C_2F_5SO_2)_2\bar{N}$
(d) X = (1S)-(+)-10-camphorsulfonate

Chiral ILs **20a–d** contain a C(4) methyl group, which is required for the induction of planar chirality, while a C(2) methyl group is essential for the suppression of a rope-skipping procedure. Therefore, it is also significant for the racemization of imidazolium-based planar cyclophane architecture. Chiral IL **20b** has a low melting point (42–45 °C) in comparison to analogues **18** and **19b**. Interestingly, chiral IL **20d** comprising of camphor sulfonate that helps to form diastereomeric salts can be used as a suitable chiral medium in asymmetric synthesis and also in the resolution of enantiomers.

With the ever-increasing applications of RTILs, they have received much attention in various disciplines of science and technology. ILs have an extended application as deep eutectic solvents, switchable polarity solvents (SPS), polymer-supported ILs, amphiphile ILs, polarizable ILs, protic ILs, metal salts ILs, chiral ILs, and biological ILs. Due to synergic chemical properties, multifunctional ILs is another popular branch of ILs. Carbohydrate-based chiral ILs have attained much consideration as biological ILs in the fields of medicinal chemistry, catalysis in several important reactions, value-added products, green reaction media, supramolecular chemistry, and materials science [4–20]. A growing impact of ionic liquids in various areas of chemistry during the last twenty years can be witnessed through a simple search in SciFinder (Chemical Abstract Service) using "ionic liquid" as a keyword that result amazing number of hits (Fig. 8) [56].

Ionic liquids are basically one of the most important motifs having a wide range of applications in industry in comparison to the volatile organic solvents. Noticeably, organic solvents have been well known from centuries, but the current natural demands prefer the ionic liquids in industry. Several properties of ionic liquids were discovered and found to be dominant over organic solvents. Some of the properties of organic solvents and ionic liquids are compared in Table 2 [37].

3 Properties of ILs

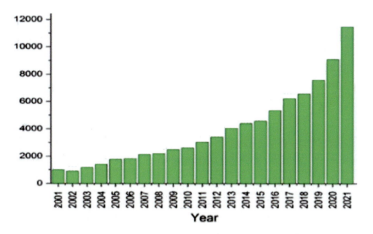

Fig. 8 Number of contributions per year dealing to ILs in organic synthesis

Table 2 Comparison of common organic solvents and ionic liquid (ILs) [37]

Property	Organic solvents	Ionic liquids
Applicability in a given process	Single function	Multifunction
Number of solvents	>1000	>10^6
Flammability	Usually flammable	Usually, non-flammable
Vapor pressure	Volatile in nature	Negligible vapor pressure under normal condition
Tuneability	Limited range of solvents available	Virtually unlimited range means "designer solvents"
Catalytic ability	Rare	Common and tunable
Chirality	Rare	Common and tunable

3 Properties of ILs

ILs remain in liquid state up to a temperature slightly more than 400 K having infinitesimal vapor pressure and higher density than water; thus, they are well suited for catalysis and also used as solvents in both organic and inorganic chemoenzymatic synthesis. The key role of ILs in synthetic chemistry is attributed to the fact that they are recyclable with ease and less reaction time to produce high yields and also fulfill the protocols of green chemistry principles [3]. There is a set of guidelines documented to maintain sustainable development with synthesis progress [57, 58]. They are summarized in six emphasized as given below:

(i) The process could be made economy by minimizing waste generation,
(ii) The increased atom economy by transforming as many reagents used as possible into the final product,

(iii)   Maintaining process safety by using non-toxic, non-inflammable solvents or reagents,

(iv)   Enhancement in process efficiency either by material recycling or following catalytic route for energy conservation,

(v)   Environmental compatibility by using chemicals derived from renewable sources,

(vi)   and lastly the generation of biodegradable products as far as possible.

For any synthesis, safe and chemically non-toxic processes are always preferred therefore, use of green solvents like supercritical CO_2 and ILs are used over the corrosive aromatic and halogenated solvents both in laboratory and industrial processes. The unique physicochemical properties of ILs makes them suitable candidate in solvent-based and absorption-based extraction processes. The interaction of ILs with organic and inorganic species is via strong $\pi-\pi$ interactions, hydrogen-bonding, ion-exchange, dispersion, and/or hydrophobic–hydrophilic interactions [59]. Biomolecules like carbohydrates having unique electrostatic nature can act as electrostatic dipoles as they possess a number of hydroxyl (–OH) groups in their structure and thus can readily interact with ILs. Although, other factors, viz. hydrogen bonding, polarizability, steric interactions, and dispersion forces also contribute to the overall free energy of system, their effect is less significant than the electrostatic interaction between RTIL and biomolecules [60].

In addition, negative Gibbs free energies, large entropy changes, and small lattice enthalpies associated with the large size and conformational flexibility of ions make the interaction thermodynamically favorable which enables the mixture of salts to remain in liquid state [61].

In general, RTILs are composed of large asymmetric organic cations and organic/inorganic anions. The structure and nature of constituent moieties and the interaction among them in the melt determine the chemical properties of RTILs, whereas the physical properties depend on the level of purity of ILs. Physical properties can be altered by the presence of impurities such as water or chloride ion [62, 63]. Thus, determination of water and halide content must be carried out by any reliable methods (scanning of ^{1}H NMR spectra) prior to evaluation of the thermophysical properties of ILs. But, the wide combination of constituent ions with variety of applications makes it difficult to generalize their properties. Nevertheless, the required RTILs equipped with necessary physicochemical properties can be designed either by making slight changes in cationic side chain (R groups) or proper choice of anionic moiety as shown Fig. 1. In this chapter, a systematic discussion on of the common properties of pure RTILs is presented.

3.1 Melting Point (m. p.)

RTILs have melting point below 100 °C. Various factors that determine the melting point of ILs are (i) symmetry of ions, (ii) van der Waal interactions, (iii) H-bonding capability, (iv) charge distribution on constituent ions. Replacement of small inorganic cations with large asymmetric organic cations can significantly reduce the m. p. of the mixture; e.g., m. p. of NaCl is 803 °C, while that of 1-propyl-3-methylimidazolium chloride (**21**) is 60 °C. Also, imidazolium-based halides can greatly influence the m. p. of RTILs (Fig. 9) [14, 31].

There are several other factors that influence the m. p. of ILs listed below:

(i)   In case of imidazolium-based ILs, three important regions, i.e., the charge-rich region, symmetry-breaking region, and hydrophobic region have a cumulative effect on melting point [64].

(ii)   Alkyl chain length of the substituent: with increasing its length, there is an initial decrease in m. p. which reaches a minimum in case of octyl group, and then starts increasing from tetradecyl and higher alkyl groups. This variation can be justified on the fact that the longer the alkyl chains, the symmetry of imidazolium cation reduces and hence no effective crystal packing occurs. A slight interference in crystal packing may be reason for decrease in m. p. On the other hand, the attractive van der Waals forces of the alkyl chain are sufficiently strong to hold the ions together which reflects in the steady increase in m. p.

(iii)   Alkyl chain branching: With increasing alkyl chain branching, higher the m. p. will be for imidazoles with the same molecular mass.

(iv)   Substituted alkyl groups: Presence of partially fluorinated alkyl groups are reported to increase the m. p. as compared to their non-fluorinated counterparts [65]. The pattern of heat treatment greatly influences the m. p. of several ILs. For instance, most RTILs are supercooled above 100 K below their m. p. since they form glass at low temperatures [66, 67].

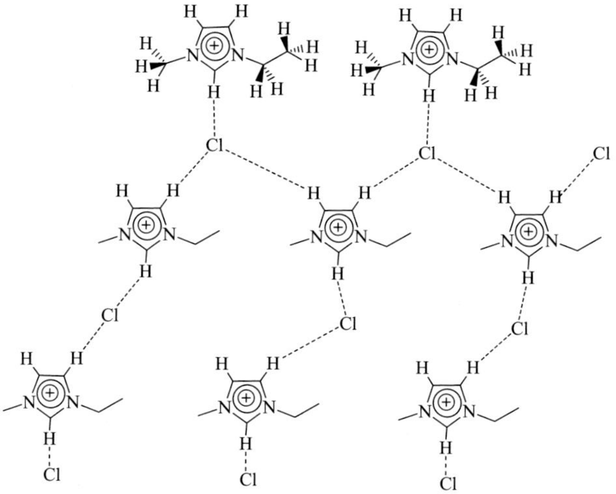

Fig. 9 Schematic representation of H-bonding in [emIm]Cl [14, 31]

3.2 Thermal Stability/Decomposition

Another interesting feature associated with RTILs is their stability and recyclability when compared with other conventional solvents. In addition, non-volatility and non-inflammability at ambient or higher temperature make them suitable to undergo fabrication for their utility in energy sectors [48, 68]. But, their thermal stability depends on the chemical nature of cation/anion, structural alterations of the cation (e.g., chain length, presence of functional groups, etc.), impurities like water, chlorides, etc., atmospheric variations, preparation method (isothermal or gradient), heating pattern, and exposure time to elevated temperatures [69]. ILs composed of pyrrolidinium and imidazolium cations are regarded as the most thermally stable whereas those with pyridinium and acyclic tetraalkylammonium cations are found to be least thermally stable, and with phosphonium, ammonium, pyrrolidinium, and piperidinium cations are considered as the intermediates. It is reported that ILs with imidazolium cation are thermally stable over 300 °C. The stability of these ILs is due to the existence of C–N bond between the imidazole nitrogen and neighboring alkyl chain carbon. Nevertheless, the anion also plays a pivotal role in determining their thermal stability. RTILs having less nucleophilic/coordinating anions as one of the component have high thermal stability. The thermal stability of ILs decreases in the order of $[(C_2F_5SO_2)_2 N]^- > [Tf_2N]^- > [TfO]^- > [PF_6]^- > [BF_4]^- > [(NC)_2N]^- > Cl^-$ anions [70]. Deprotonation at the 2-position is another route for imidazolium destruction [65]. Germinal dicationic ILs such as *bis*(trifluoromethylsulfonyl)imides possesses higher thermal stability than the common monocationic ILs. From thermal decomposition calculations, it is evident that the most of the ILs are thermally stable over 350 °C. On the contrary, ILs having thermal stability at low temperatures remains stable for a longer duration due to which they are commonly used in various catalytic processes. Since longer the alkyl chain length lower is the thermal stability of imidazolium salts thus, thermal stability of ILs is inversely proportional to the length of alkyl chain of the cationic component. This dependency on alkyl chain length was also observed in pyridinium, pyrrolidinium, piperidinium, and ammonium ILs. Siedlecka and coworkers revealed in their study that a small alteration either in the alkyl chain or in the imidazolium ring produces ILs with desired stability [71]. The chemical stabilities of ILs to stimulate concern on the potential decomposition toward its utilization need to be further considered [72].

3.3 Viscosity

Viscosity of RTILs falls within the range of ~30 to >5000 cP that corresponds to liquid-fluid phases [65]. RTILs as compared to conventional organic solvents are found to be more viscous, e.g., ethylene glycol with a viscosity of 21 cP at 20 °C. In general, viscosity of ILs is affected by two interactions (i) van der Waals forces and (ii) hydrogen bonding. The role of anion is crucial than the cation in case of

3 Properties of ILs   125

RTILs having low viscosity. In addition, temperature as well as impurities may also have a major role to play the overall density determination of system. Viscosity of dialkylimidazolium-based ILs with cyano groups is found to be decreased in the order of $[SCN]^- > [B(CN)_4]^- > [N(CN)_2]^- > [C(CN)_3]^-$ at any particular temperature eliminating any correlation of the size of anion with viscosity for such ILs. However, there might be a possibility of intermolecular interactions present in the bulk liquid for controlling the viscosity of ILs. The tetrahedral shape of a highly symmetric tetracyanoborate anion reduces the viscosity significantly as a function of number of –CN groups. On the other hand, higher viscosity of $[BF_4]$-based ILs than the –CN containing salts is attributed to a higher degree of charge distribution and lower polarizability that simultaneously contribute toward the weakening of van der Waals interactions, cohesive forces, and hydrogen bonding. Nonetheless, cyano-based ILs, especially $[C(CN)_3]^-$-based compounds, are in wide industrial applications [73]. Chernikova and coworkers observed a gradual decrease in the viscosity of ILs having $[C_4HC_1Im]^+$ cation as: $Cl^- > [PF_6]^- > [NO_3]^- > [BF_4]^- > [TfO]^- > [CF_3CO_2]^- > [Tf_2N]^- > [(NC)_2N]^- > [(NC)_3C]^-$ depending on the nature of anion. Even the cation has an insignificant role than the anion; the viscosity of a system can readily be increased by introducing long alkyl groups, bulky phenyl and/or hydroxyl-containing substituents into the cation moiety [74].

3.4 Conductivity

Since the conductivity of RTILs is inversely proportional to their viscosity, RTILs having lower viscosities display higher conductivities, and in general the increasing trend in the conductivity of RTILs is in accordance with rise in temperature. Due to their high conductive nature, RTILs are extensively used as electrolytes for SO_2 solvation, supercapacitor applications, desulphurization, and separation of azeotropic mixtures. It is also interesting to find that RTILs with highest conductivities are quite comparable with non-aqueous electrolyte systems in solvents like acetonitrile. A recent report has suggested of fabricating next-generation electrochemical sensors by tuning the electrical and electrochemical properties of several common RTILs [75]. Conductivity of ILs depends upon a number of factors such as viscosity, density, anionic charge delocalization, ion size, aggregation, and ionic mobility, etc. [76]. A decrease in conductivity may observed due to aggregation or ion-pairing in solution that reduces ion mobility. Reports have suggested that conduction in ILs is a phenomenon that takes place via Grotthuss-like hopping rather than simple transfer mechanism, e.g., solar cells in particular [77]. A lower conductivity of $[NTf_2]^-$-based ILs is believed to be due to the presence of strong ion-pair associations when compared with $[BF_4]^-$-based ILs [78]. Twenty of the most common ILs were found to have thermal conductivities in the range of $0.1–0.22$ W m^{-1} K^{-1} [74]. These conductivity values are supposed to be dependent upon the nature of anion rather than the temperature or type/structure of cation. In a recent report on aqueous [BMIm]Br solutions, the thermal conductivity is found to decrease with increasing temperature and concentration. It is assumed due to increasing molecular collisions with rise in temperature and hence, a subsequent lowering in the mean free path of molecules [79]. On the other hand, decrease in the thermal conductivity with concentration is justified with an increase in viscosity and simultaneous decrease in ionic mobility in the system at a particular temperature [79].

3.5 Density and Polarity

Density and polarity of ILs are least affected by temperature change or presence of impurities such as halides, solvents, water, etc. Common RTILs are more thick than water in texture having a density ranging from 1 to 1.6 g cm^{-3} [80]. Generally, the density of imidazolium-based RTILs decreases with alkyl chain growth, whereas with additional halide content either in alkyl chains or in anions results in an increase in density [65].

Polarity of solvent is prime factor to determine the fate of any chemical reaction [81]. There is no single parameter or any direct method that could readily determine the polarity of ILs due to lack of suitable soluble probe which would independently measure the polarity of the medium [82, 83]. With the development in ILs and their studies, a number of methods and parameters were designed that could determine measure the polarity of ILs precisely [84, 85]. However, polarity could only be predicted by varying the combination of anion and cation in their constituents. Guan et al. have introduced a simple scale of polarity that could measure the polarity of ILs on the basis of enthalpy of vaporization. Their method nullified the effect of impurities or water if present in trace amounts. A decrease in polarity with the increase in cationic alkyl chain length is observed in ILs when measured by this method [86]. In a separate study, seven-membered ring system (N-methyl-N-butylhexamethyleneiminium, [HME$_{1,4}$]$^+$)-based ILs have displayed higher polarity relative to those with five- or six-membered cyclic cations (pyrrolidinium and piperidinium). Polarity of ILs can also be enhanced by introduction of hydroxyl (–OH) groups onto the cationic alkyl chain [84]. The bulk polarity of ILs and its linear solvation energy relationships or both at the same time are two major perspectives to explicit the solvent effect of ILs to considered them as nanostructured fluids [87].

3.6 Toxicity

In recent years, green reactions and synthesis are drawing the interest of researchers in recent years [88]. A continuous effort is being made to design and develop non-toxic, non-hazardous, biodegradable, and environmentally benign ILs. A few common non-biodegradable ILs are still in practice that could have long persistence in the environment [68, 89]. Being green in nature, ILs are more favorable over traditional organic solvents by which the environmental risks can be mitigated to certain extent.

4 Chemical Synthesis of Some ILs

However, ILs can cause air pollution by forming aerosols or other volatile products on thermal decomposition. Therefore, ILs with improved properties was designed in the recent years emphasizing their industrial applications. For instance, ILS based on amino acids and food-grade imidazolium derivatives with choline cation are nontoxic biodegradable chemicals [90–92]. Nature of cations and anions present in ILs greatly influence the toxicity and biodegradability of ILs [74, 93]. In particular, tetraethylammonium halides impart ganglion blocking effect in animals (rats), i.e., a decrease in the respiratory and cardiac activity causing papillary dilation and motor disturbances with a lethal dose (LD_{50}) of 2630 mg kg^{-1} when administered orally. *Bis*-quaternary ammonium salts having similar effect which are commonly used as hypotensive and ganglion blocking drugs (benzohexonium, pentamine, pentolinium). In case of imidazolium-based ILs, toxicity to aquatic animals, snails, microorganisms, and malignant cell cultures is also influenced by the length of alkyl radical [93].

In addition, solubility of ILs in water can easily be modulated by choosing an appropriate cation and anion combination making the entire system recyclable. RTILs are widely used as organic media, solvents, catalysts, etc. since they exhibit better performances than conventional solvents [4–15, 37]. The physical and chemical properties of ILs can be varied to produce a choice to get new solvent system, achieve affordable reaction media, more efficient catalyst, easy workup procedure and pure targeted compound only by varying them in accordance to criteria.

4 Chemical Synthesis of Some ILs

ILs based on imidazolium ion are widely used organic synthesis. Commercial synthesis of ILs is widely known. For an example, imidazolium-based RTIL **24** was obtained at bulk scale and in high-yields by direct ethylation followed by trifluoroethylation of *N*-methyl imidazole **22**. Similarly, ILs based on related heterocyclic systems including pyridine can be achieved with great ease by utilizing trifluoroethylphenyliodonium *bis*-((trifluoromethyl)sulphonyl) imide or *N*-methyl *bis*((perfluoroalkyl)sulfonyl)imides under optimized conditions (Scheme 1) [94].

The systematic procedure involved in the synthesis of 1-butyl-3-methylimidazolium tetrafluoroborate (bmim[BF_4]) **26**, a room-temperature ionic liquid (RTIL) from *N*-methyl imidazole is schematically presented in Scheme 2

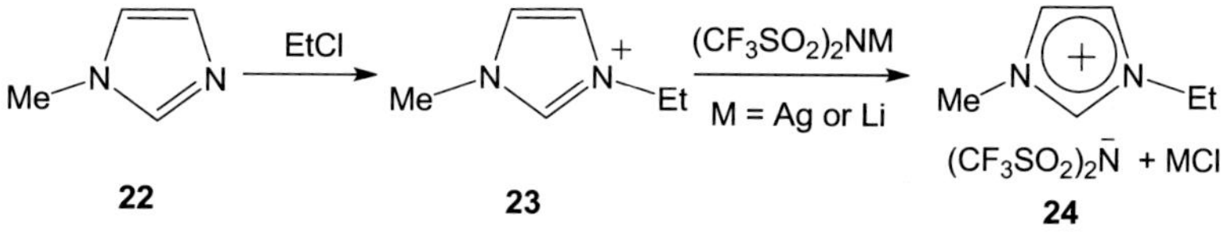

Scheme 1 Practical synthesis of imidazolium-based RTILs

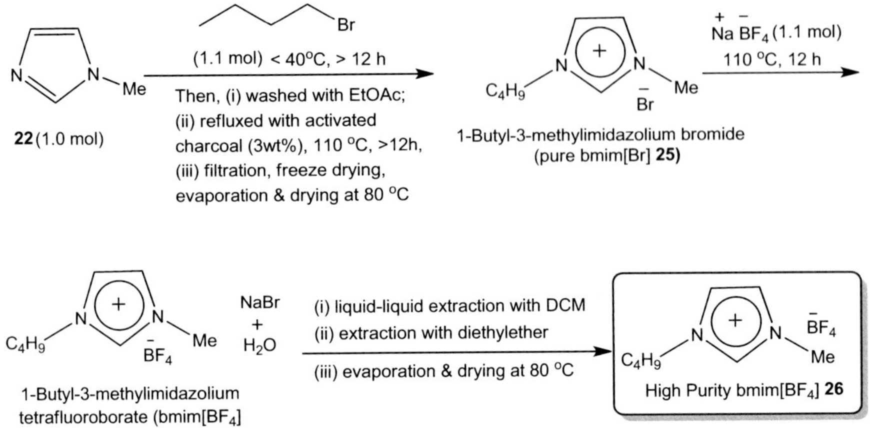

4 Growing Impact of Ionic Liquids in Heterocyclic Chemistry

22 (1.0 mol)

(1.1 mol) < 40 °C, > 12 h
Then, (i) washed with EtOAc;
(ii) refluxed with activated
charcoal (3wt%), 110 °C, >12h,
(iii) filtration, freeze drying,
evaporation & drying at 80 °C

Na BF$_4$ (1.1 mol)
110 °C, 12 h

1-Butyl-3-methylimidazolium bromide
(pure bmim[Br] **25**)

NaBr + H$_2$O

(i) liquid-liquid extraction with DCM
(ii) extraction with diethylether
(iii) evaporation & drying at 80 °C

1-Butyl-3-methylimidazolium
tetrafluoroborate (bmim[BF$_4$]

High Purity bmim[BF$_4$] **26**

Ionic Liquids	^{1}HNMR (500 MHz, CDCl$_3$ + DMSO)	^{13}C NMR (125 MHz, CDCl$_3$ + DMSO)
bmim[Br] **25**	δ = 9.71 (s, 1H), 7.31 (s, 1H), 7.45 (s, 1H), 4.29 (t, 2H), 4.06 (s, 3H), 1.87 (quint, 2H), 1.42 (sextet, 2H), 0.96 (t, 3H)	δ = 137.00, 123.56, 121.97, 49.76, 36.54, 32.00, 19.34, and 13.33 ppm
bmim[BF$_4$] **26**	δ = 9.40 (s, 1H), 7.40 (s, 1H), 7.46 (s, 1H), 4.25 (t, 2H), 4.02 (s, 3H), 1.87 (quint, 2H), 1.36 (sextet, 2H), 0.95 (t, 3H)	δ = 135.47, 123.07, 121.72, 48.90, 35.52, 31.19, 18.54 and 12.59 ppm

Scheme 2 Typical procedures for the synthesis of 1-butyl-3-methylimidazolium tetrafluoroborate bmim, (bmim[BF$_4$] **26**) [95]

[95]. In a typical experimental procedure, *n*-bromobutane (27 mL) was slowly added to 11 mL of freshly distilled 1-methylimidazole solution in a 500 mL two-neck RB flask fitted with a dropping funnel. The reaction was stirred for >48 h at <40 °C. The resultant yellowish-white solid of bmim[Br] was washed with ethylacetate (3 × 200 mL) to remove the unreacted starting materials. The decolorization of bmim[Br] was accomplished by dissolving it in deionized water (250 mL) in 1 L beaker followed by addition of activated charcoal (3–5 g) and then refluxed at 100 °C with consistent stirring for >12 h. The activated charcoal was filtered out (using a Buchner funnel containing Whatman filter paper) and clear transparent solution was then sprayed into liquid nitrogen in a 1 L vacuum flask. This was subjected to freeze-drying for overnight, evaporated at 80 °C and vacuum dried to afford white-colored solid bmim[Br] **25** in pure form. A metathesis reaction of bmim[Br] **25** (25 g) with sodium tetrafluoroborate (13 g) in deionized water (250 mL) at rt for 24 h gave the desired bmim[BF$_4$] **26**. The product was washed with diethylether (to eliminate NaBr, a byproduct generated in aqueous layer) and organic layer was evaporated under reduced pressure, dried in vacuum at 70 °C overnight to give desired bmim[BF$_4$] **26** in pure form.

4 Chemical Synthesis of Some ILs

The related imidazolium-based ILs with a low melting point and low viscosity can be obtained simply by replacement of the alkyl group on the imidazolium heterocycle with a flexible ether group (Fig. 10). Similar substitution at imidazolium cation and anion (e.g., sulfate) could also result in favorable outcome [96, 97].

Furthermore, chiral imidazolium triflates **33** were developed in enantiomerically pure form starting from the corresponding oxazoline **32** as the chiral building block (Scheme 3) [98]. A number of chiral and biorenewable ionic liquids were developed from readily available chiral starting materials like carbohydrates, for example, monosaccharide D-Fructose and D-Glucose [99–101].

Interestingly, 3,3-[disulphanylbis(hexane-1,6-diyl)]*bis*(1-methyl-1*H*-imidazol-3-ium)dichloride (**34**) on reaction with $HAuCl_4$ using $NaBH_4$ furnished respective gold nanoparticle immobilized on imidazolium cation **35** (Scheme 4) [101]. This gold nanoparticle-based IL **35** showed interesting surface phenomena that changed from hydrophilic to hydrophobic nature. Further, this could be resulted to the desired IL of notable recyclable utilities mainly in biphasic catalysis.

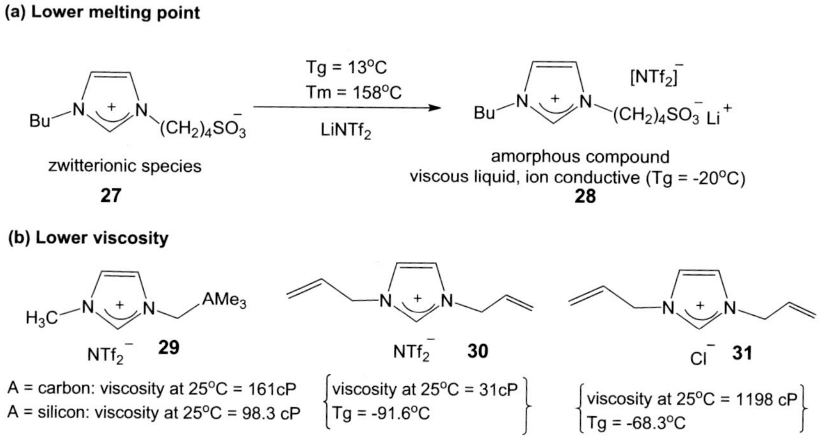

(a) Lower melting point

(b) Lower viscosity

Fig. 10 Imidazolium-based ILs (**27–31**) with notable properties (e.g., decrease in viscosity and density)

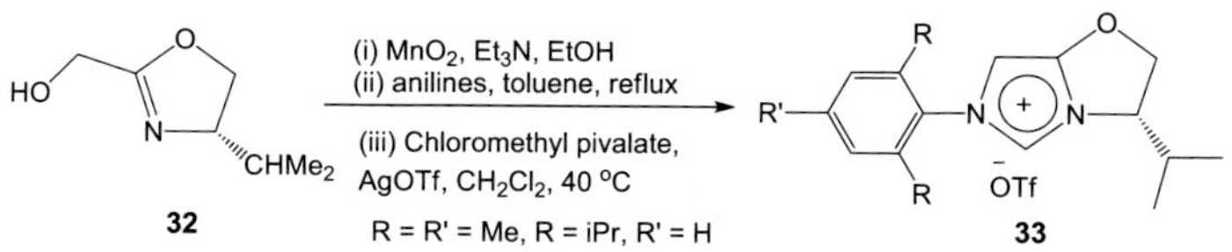

Scheme 3 Synthesis of imidazolium triflate-based chiral ILs (**33**)

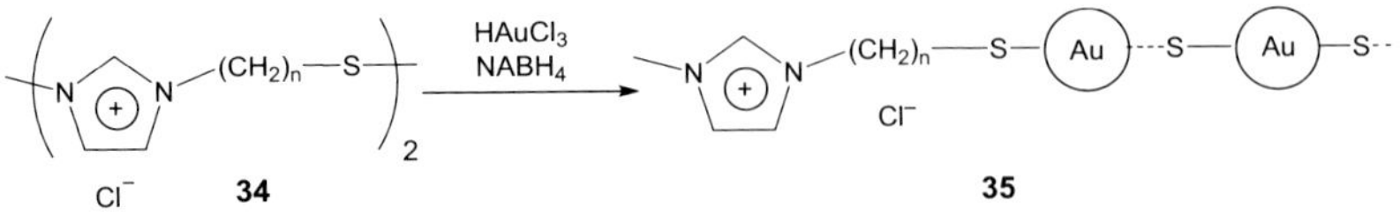

Scheme 4 Recyclable imidazolium gold nanoparticle **35** useful in biphasic catalysis [101]

5 Impact of RTILs in the Synthesis of Biologically Relevant Skeletons

Decades of effort has enlarged the field of ionic liquid in diverse application. Today, it can be used as a green media and/or as a safe catalyst for a particular reaction in organic synthesis to achieve the high-to-excellent yield of targeted product in easy and short span of time. Besides this it has found its broader application in pharmaceutical industry as solvent system, drug loading system, drug delivery system or to enhance the activity spectrum of the API. Besides this, it has application in a number of biomedical extraction procedures in greater way.

The Heck reaction can be carried out in presence of 1-*n*-butyl-3-methylimidazolium hexafluorophosphate **36**. In this reaction, the coupling product **40** and **41** can be extracted with cyclohexane after completion of reaction of aryl halides **38** and **39** separately with olefin **37**, and the byproduct triethylammonium halides can be easy washed away with water (Scheme 5) [102]. Interestingly, there is no loss of any catalytic activity even after sixth use.

The next example is the **Diels–Alder cycloaddition reaction** between diene **43** and dienophile **44** performed in the presence of 1-*n*-butyl-3-methylimidazolium trifluoromethanesulfonate **42** and the product cyclohexene derivative **45** was obtained in good yield. In this cycloaddition, the catalyst is retained in ionic liquid **42**, and the recovered ionic liquid phase containing the catalyst can be recycles and reused for several times (Scheme 6) [103]. Interestingly, there is no loss of catalytic activity even after eleventh use. They are recycled as solvents and often can be reused in a reaction with the catalyst. Because of environmental benign nature of ionic liquids, they are receiving considerable attention in a number of important reactions in academia and industry.

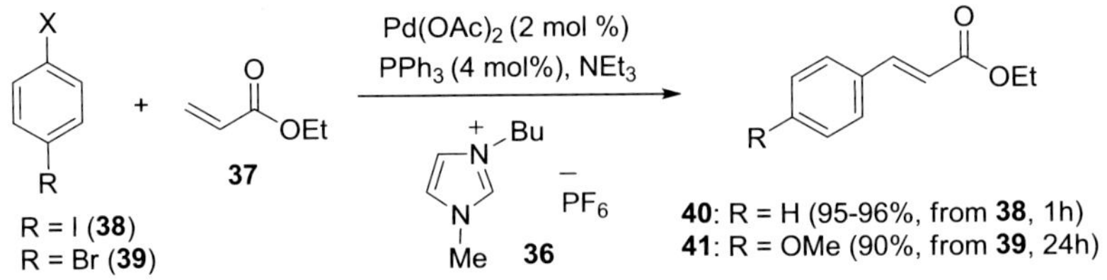

Scheme 5 IL-mediated Heck coupling

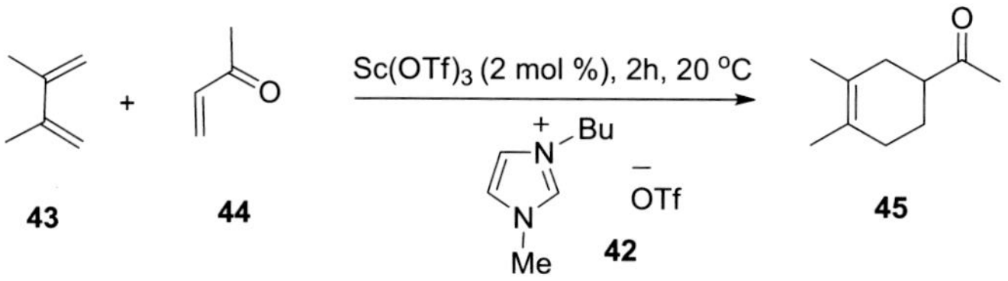

5 Impact of RTILs in the Synthesis of Biologically Relevant …

Scheme 6 IL-mediated Diels–Alder cycloaddition reaction

5.1 Impact of RTILs in the Synthesis of Biologically Relevant Heterocyclic Skeletons

Heterocyclic compounds contain some other hetero atoms (like, oxygen, nitrogen, sulfur, etc.) in a closed-ring system, constitutes a great number of biologically potent drug candidates as well significant common structural motifs of hundreds of natural products. For a notable example, ring systems like pyrrole, furan, thiophene, pyridine, indole, pyrimidine, oxazole, oxepine, azepine, azocine, oxocin, diazepin, oxazepine, thiazepine, etc. (Fig. 11), and their fused analogues are known to display a wide range of pharmacological activities. A great number of structural units of heterocyclic system were found in the commercial synthetic drugs approved during last twelve years (Table 3) [104–106].

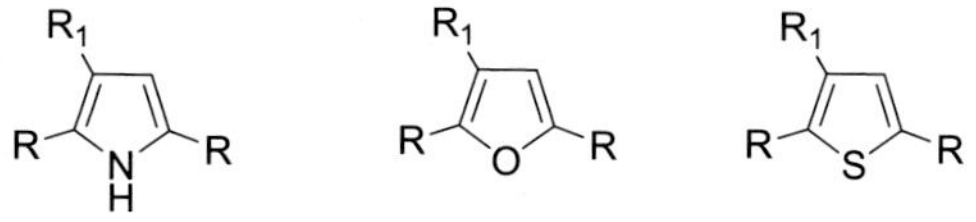

Fig. 11 Chemical structure of five-membered heterocycles having one hetero atom

Table 3 An overview of synthetic drugs approved during 2009–2019

Year	Total synthetic drug approved	Heterocyclic in origin	References
2009	21	17	[104a]
2010	15	12	[104b]
2011	26	22	[104c]
2012	27	26	[104d]
2013	28	24	[104e]
2014	24	21	[104f]
2015	29	21	[105a]
2016	19	16	[105b]
2017	31	28	[106a]
2018	39	35	[106b]
2019	40	34	[106c]

5.2 RTIL-Mediated Synthesis of Five-Membered Heterocycles

The pyrrole motif has found in a large number of drug candidates as well displayed pivotal use in agricultural and material science. Atorvastatin calcium (Lipitor), a pyrrole derivative, is widely used as cholesterol-lowering drug. A common established synthetic protocols for an easy access of pyrroles includes the classical Hantzsch pyrrole synthesis, Knorr and Paal–Knorr synthesis [107]. Besides these, a number of ionic liquid-mediated high-yielding synthesis of pyrroles are well explored in academia and industry.

Yadav and coworkers used 1-butyl-3-methylimidazolium tetrafluoroborate [bmim]BF$_4$ immobilized in 5 mol% Bi(OTf)$_3$ as reusable catalytic system in the reaction of 1,4-diketones **47** with various primary amines (**46** decene for an expeditious synthesis of N-aryl-2,3,5-substitued pyrroles **48** (Scheme 7) [108]. Kumar et al. developed a library of N-substituted benzene-fused pyrroles **51** by reacting 4-hydroxyproline **50** with isatins **49** in presence of [Bmim]BF$_4$ under microwave (*MW*) heating condition for 10–15 min [109]. This [Bmim]BF$_4$ was recycled up to six times without any notable loss of its reactivity. Shamekhi and coworkers utilized

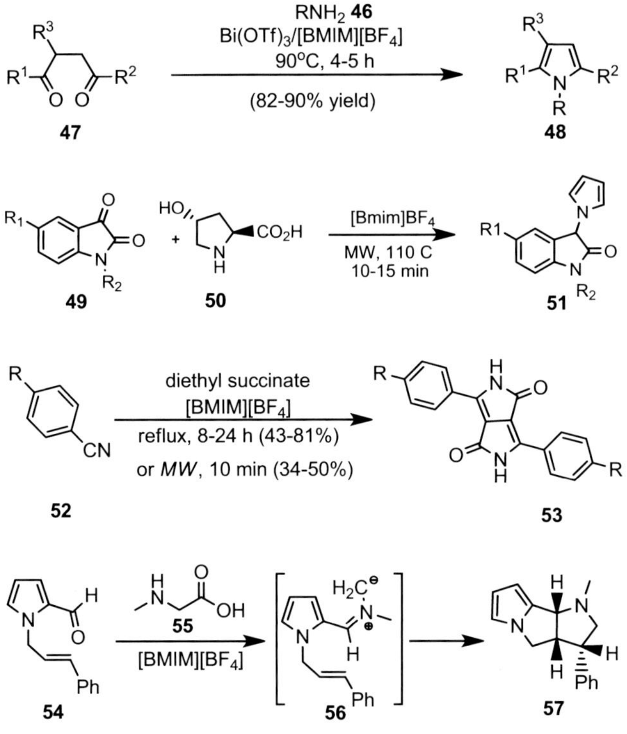

Scheme 7 RTIL-promoted synthesis of functionalized pyrrole and their analogues [108–111]

5 Impact of RTILs in the Synthesis of Biologically Relevant ... 133

[BMIM][BF$_4$] as green reaction media in the reaction of benzonitrile **52** with diethyl succinate under microwave condition to furnish high-yield of diketopyrrolopyrrole **53**, an important class of high performing pigments frequently used in paints and plastic industry [110]. Raghunathan and coworkers reported a facile route for an easy access of pyrrole fused-polycyclic heterocycles namely pyrrolopyrrolidines **57** through intramolecular 1,3-dipolar cycloaddition reaction of *N*-alkenyl carboxalde-hyde **54** with secondary amino acid **55** in presence of [BMIM][BF$_4$]. The reaction was proceeded via azomethine ylide **56** as 1,3-dipole which underwent intramolecular 1,3-dipolar cycloaddition with *N*-alkenyl group of the same molecule as dipolarophile in good yields (Scheme 7) [111].

Another important five-membered aromatic heterocycle ring with one oxygen atom is furans that possess an imperative pharmacophore and also contains core structural component in a number of biologically active alkaloids, e.g., cembranolides and kallolides. The acid-catalyzed cyclization of 1,4-dicarbonyl compound (Paal–Knorr synthesis) is widely used as for the classical synthesis of furans. Likewise pyrrole, a substituted 1,4-dicarbonyl compound **47** in presence of Bi(OTf)$_3$/[BMIM][BF$_4$] underwent cyclization and could deliver high-to-excellent yields of desired furan and their analogues **58** [108]. Further, the three-component reaction of aldehydes **59**, dimethyl acetylenedicarboxylate (DMAD), and cyclohexyl isocyanide **60** were performed in presence of [BMIM][BF$_4$] as recyclable green media afforded high-to-excellent yield of respective 2-aminofuran analogues **61** [112]. In continuation, Shaabani and coworkers reported a highly expeditious synthesis of furopyrimidine **63** via condensation reaction of various aldehydes **59** with isocyanides and *N,N'*-dimethylbarbituric acid **62** at room temperature in [BMIM][Br] as green solvent [113]. Villemin and coworkers reported a base-catalyzed facile synthesis of butenolide analogues **66** by reacting ethyl cyanoacetate **65** with α-hydroxyketones **64** using [BMIM][BF$_4$] as recyclable ionic liquid. The reaction proceed through base promoted formation of alkoxide nucleophile from the α-hydroxyketones **64** which then reacted with ester functionality of ethyl cyanoacetate **65**, followed by Knoevenagel reaction to furnished high yield of title butenolide **66** (Scheme 8) [114].

Kim and coworkers developed 5-hydroxymethyl-2-furfural (HMF, **68**) from cellulose **67**. Compound **67** in presence of metal chlorides (CrCl$_2$/RuCl$_3$) and IL [EMIM][Cl] underwent cellulolysis to afford the targeted product **68** (Scheme 9) [115]. Lin and coworkers used CrCl$_3$ · 6H$_2$O in presence of tetraethylammonium chloride (TEAC) as an ionic liquid and converted D-glucose to HMF (**68**) [116]. 1-Hydroxyethyl-3-methylimidazolium tetrafluoroborate ([C$_2$OHMIM]BF$_4$) as a recyclable catalyst was used by Huang and coworkers for the effective dehydration of glucose and fructose that furnish high yield of desired HMF **68** [117]. Likewise, Argyropoulos and coworkers utilized 3-(2-chloroethyl)-1-methylimidazolium chloride as an efficient ionic liquid for the successful conversion of fructose to 5-HMF **68** in aqueous media [118]. Interestingly, SBA-15-SO$_3$H was identified to be effective catalyst for the dehydration of fructose to HMF **68** provided [BMIM][Cl] was used as green media [119].

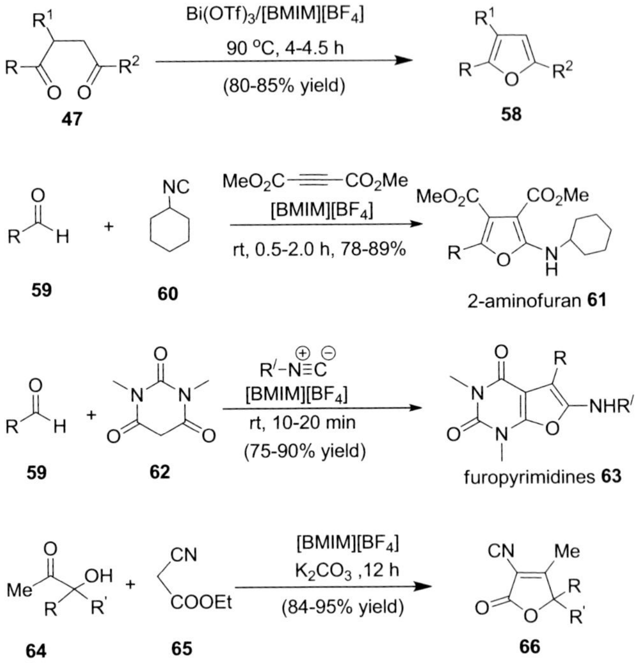

Scheme 8 IL-mediated synthesis of substituted furans and their analogues

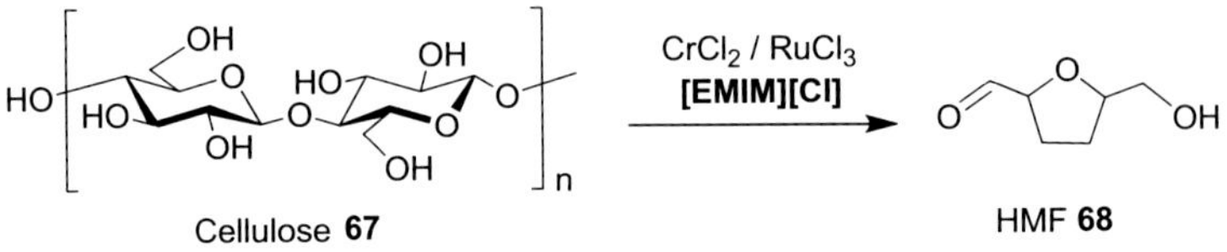

Scheme 9 Synthesis of 5-hydroxymethyl-2-furfural **68** in presence of IL

Thiophenes, a five-membered heterocycles with sulfur as a heteroatom, displayed wide range of applications ranging from early dye chemistry to modern drug development as well as biodiagnostic tools, self-assembled superstructures, conductivity-based sensors, and electro-optical devices. 1,4-Dicarbonyl compounds **47** were first reacted with Lawesson's reagent to generate 1,4-dithiocarbonyl compound (in situ) which in presence of $Bi(OTf)_3/[BMIM][BF_4]$ (5 mol%) at 90 °C underwent cyclization–condensation reaction to furnish high-to-excellent yield of functionalized thiophenes **69** under one-pot mild reaction condition (Scheme 10) [108].

In another context, Hu and coworkers reported a facile synthesis of 2-aminothiophenes **74** by utilizing the Gewald reaction. Thus, α-methylene carbonyl compound on condensation with α-cyanoacetic acid and sulfur using ionic liquid as soluble support [120]. At first, hydroxyl group of ionic liquid **70** was proceed

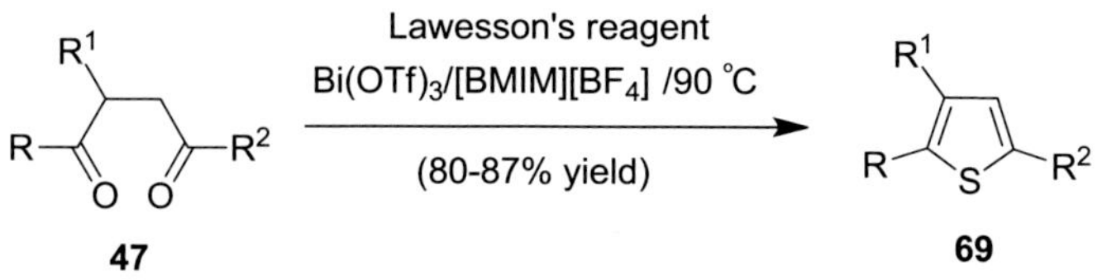

5 Impact of RTILs in the Synthesis of Biologically Relevant …

Scheme 10 [BMIM][BF$_4$] promoted expeditious synthesis of trisubstituted functionalized thiophenes **69**

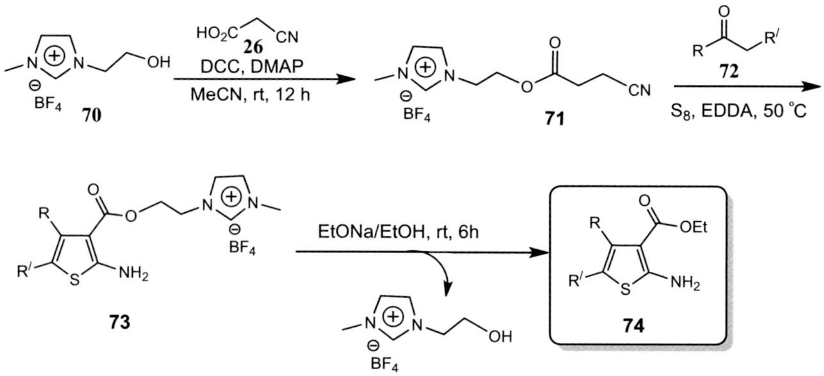

Scheme 11 Ionic liquid-mediated synthesis of diverse 2-aminothiophenes (**74**)

to acid-alcohol coupling with cyanoacetic acid and then the resulted ester **71** underwent condensation reaction with α-methylene carbonyl compound **72** in presence of S$_8$/EDDA to give desired thiophene **73**. Then after, this ionic liquid appended thiophene **73** under basic hydrolysis furnished the high yield of desired 2-aminothiophenes **74** in high purity (Scheme 11).

A amino acid-based chiral ionic liquid, e.g., [Bz-his(n-propyl)$_2$-OMe$^+$Br$^-$] **76** was developed by Kumar and coworkers [121]. The authors used imidazole-containing amino ester **75** which underwent acid–amine coupling with carboxylic acid, for example, benzoic acid using DCC/HOBt as standard coupling agent and the resulted N-benzylated amino ester on treatment with propyl bromide in acetonitrile under basic condition afforded the title ionic liquid **76**. Kumar and coworkers implemented a chiral IL [Bz-his(n-propyl)$_2$-OMe$^+$Br$^-$] **76** for the diastereoselective synthesis of dihydrothiophenes **80**. Thus, 2-arylidenemalononitrile **77** on one-pot three-component coupling reaction with 1,3-thiazolidinedione **78** and amines **79** in presence of the chiral ionic liquid **76** in water at 70 °C resulted to the formation of chiral dihydrothiophenes analogues **80** (Scheme 12) [121].

Ionic liquid-mediated practical and expeditious synthesis of a large number of biologically relevant five-membered heterocycles with two hetero atoms, such

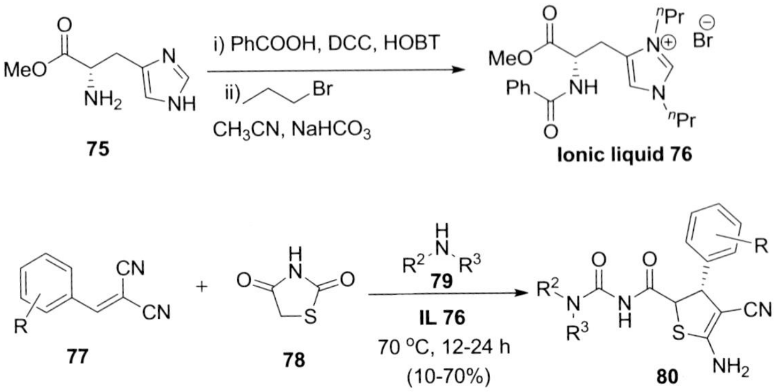

Scheme 12 Chiral IL-mediated diastereoselective synthesis of dihydrothiophenes (**80**)

as pyrazoles, imidazoles, isoxazoles, oxazoles, oxazolines, oxazolidinones, thiazoles, and thiazolidinones and or more hetero atoms e.g., 1,2,4-thiadazoles, 1,2,3-disubstituted triazoles, benzotriazoles, and tetrazoles (Fig. 12) has been well documented in literature.

The condensation of α,β-unsaturated carbonyl compounds with hydrazine or their derivatives then subsequent dehydrogenation (aromatization) was frequently explored classical route for the high-yielding synthesis of pyrazoles [122]. 1,3-Diketones on condensation with hydrazine also led to the good yield of substituted pyrazoles. Likewise, enones **81** on condensation with hydrazine in presence of

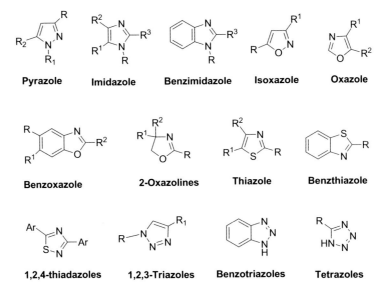

Fig. 12 Structure of five-membered heterocyclic skeletons with two or more hetero atoms

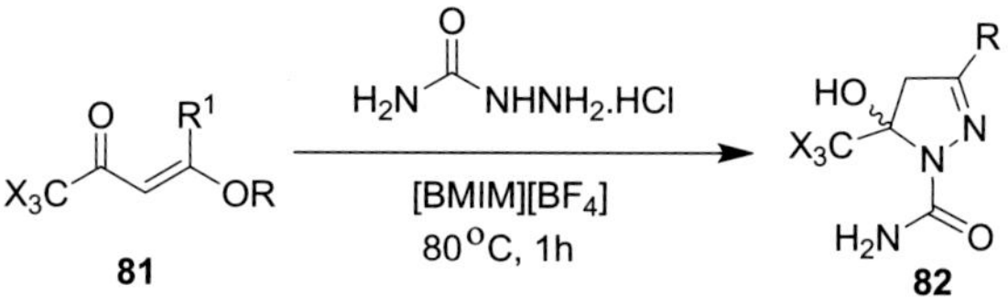

Scheme 13 [BMIM][BF$_4$]-mediated synthesis of 4,5-dihydropyrazoles

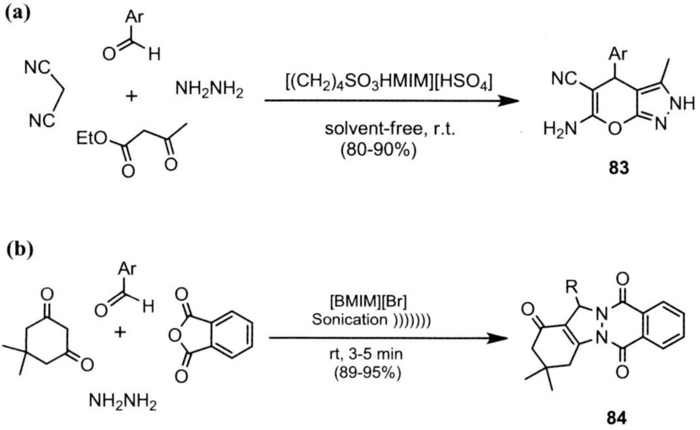

Scheme 14 Ionic liquid-mediated expeditious synthesis of fused-pyrazole skeletons

equimolar quantities of [BMIM][BF$_4$] at 80 °C for 1 h furnish high yields of title 4,5-dihydropyrazoles **82** (Scheme 13) [123].

Interestingly, four component reactions of malononitrile, aldehyde, ethyl acetoacetate, and hydrazine hydrate in presence of green Brønsted-acidic ionic liquid, e.g., [(CH$_2$)$_4$SO$_3$HMIM][HSO$_4$] under solvent-free condition afforded high-yield of desired dihydropyrano[2,3-c]pyrazoles **83** (Scheme 14a) [124]. Furthermore, similar multicomponent coupling reaction of dimedone, aldehyde, phthalic anhydride, and hydrazine hydrate in presence of [BMIM][Br] ionic liquid under sonication condition furnished high-to-excellent yield of respective 2*H*-indazolo [2,1-*b*]phthalazinetriones **84** (Scheme 14b) [125].

Safaei and coworkers reported a high-yielding regioselective synthesis of pyrazoles **86** or **87** by reacting 1,3-diketones with hydrazines or hydrazides in aqueous media in presence of multi sulfonic acid group containing Brønsted acidic ionic liquid **85** (Scheme 15). The authors claimed that the this reaction when performed in ionic liquid **85** is even more fruitful in term of reaction yield than the reaction in other common Brønsted acidic ionic liquids including [HMIM][HSO$_4$] or [MSPIM][HSO$_4$] or some other ILs, e.g., [BMIM][BF$_4$] or [BMIM][PF$_6$] [126].

1,2-Diketone on condensation with ammonium acetate (as ammonia source) and aromatic aldehyde in presence of [HBIM][BF$_4$] under one-pot refluxing condition afforded high-to-excellent yield of desired 2,4,5-triarylimidazoles **88** (Scheme 16) [127]. PEG1000-based functional dicationic acidic ionic liquid was further utilized

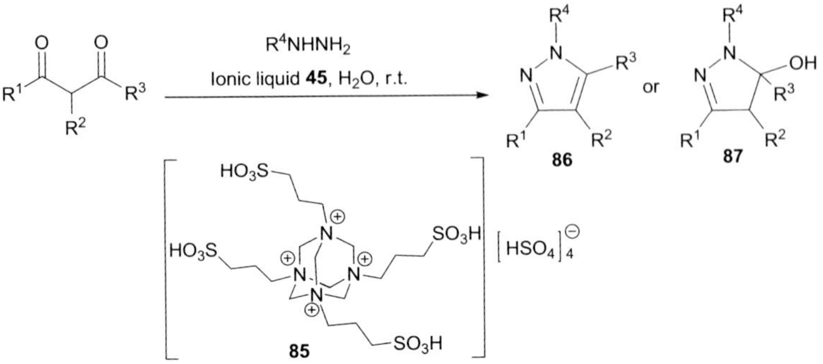

Scheme 15 Brønsted acidic IL-mediated expeditious synthesis of pyrazole heterocycle

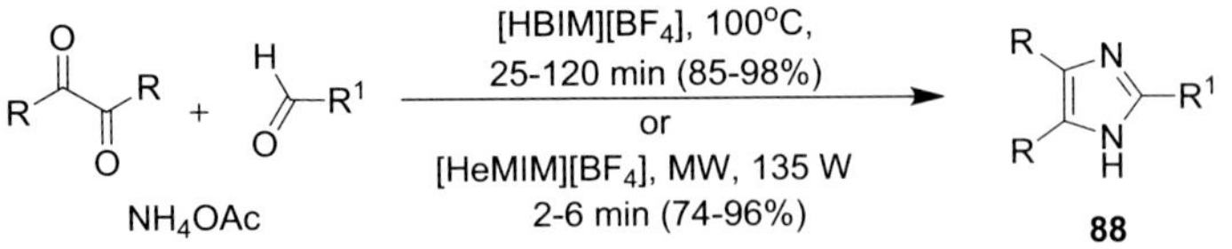

Scheme 16 IL-mediated facile synthesis of 2,4,5-triarylimidazoles

as a suitable phase-separation catalytic system for an easy access of similar 2,4,5-trisubstituted imidazoles [128].

A number of imidazole-fused heterocycle including imidazo[1,2-*a*]pyrimidines are the core structural component in several commercial drugs, e.g., zolimidine (used for the treatment of peptic ulcer) and zolpidem (used for the treatment of insomnia). Such heterocyclic systems were developed from 2-aminopyrimidine by reacting it with R-bromo-acetophenone in polar solvent [129]. In continuation, a three-component reaction of 2-aminopyridines **89**, aldehydes, and isocyanides in presence of [BMIM][Br] at room temperature afforded high-to-excellent yield of respective 3-aminoimidazo[1,2-*a*]pyridines **90** (Scheme 17a) [130]. Further, condensation of 2-aminopyrimidines **91** or 2-aminopyridines **92** separately with a α-bromoacetophenones and hydroxy(tosyloxy)iodobenzene (HTIB) in presence of n-butylpyridinium tetrafluoroborate [BPY][BF$_4$] under one-pot condition afforded high-to-excellent yield of respective imidazo[1,2-*a*] pyrimidines **93** or imidazo[1,2-*a*]pyridines **94** (Scheme 17b) [131]. Similar reaction performed in conventional solvent furnished very low yield of products (**93** and **94**) [132]. Interestingly, the condensation of α-tosyloxyketones **96** with 1-amino isoquinoline **95** in presence of [BPY][BF$_4$] as solvent at room temperature and Na$_2$CO$_3$ as non-nucleophilic base afforded good yields of desired 2-phenylimidazo[2,1-*a*]isoquinoline **97** (Scheme 17c) [132].

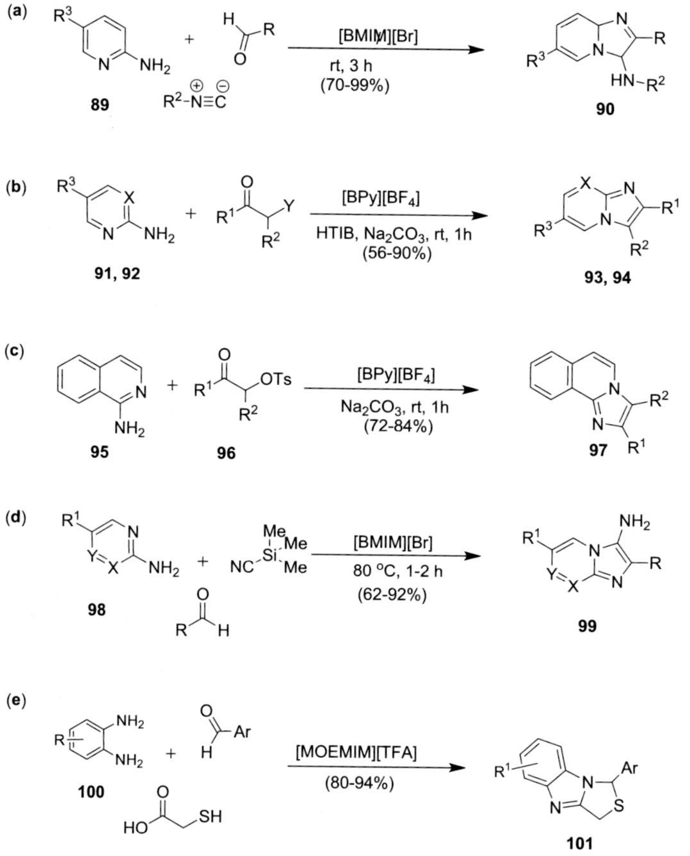

5 Impact of RTILs in the Synthesis of Biologically Relevant …

Scheme 17 ILs-mediated an expeditious synthesis of biologically relevant fused imidazoles

Likewise, condensation of 2-aminoazines **98**, aldehydes, and TMSCN in presence of [BMIM][Br] afforded high-to-excellent yield of 3-aminoimidazo[1,2-*a*]azine analogues **99** (Scheme 17d) [133]. The multicomponent reaction of 1,2-phenylenediamine **100**, 2-mercaptoacetic acid and aromatic aldehydes in presence of ionic liquid, e.g., 1-butyl-3-methyl-imidazolium tetrafluoroborate or 1-methoxyethyl-3-methylimidazolium trifluoroacetate in one-pot condition resulted to the regioselective synthesis of substituted benzimidazoles **101** (Scheme 17e) [134].

Isoxazole heterocycle occurs as a structural component on a number of commercial drugs, e.g., ibotenic acid (a COX-2 inhibitor), furoxan (a nitric oxide donor), etc. A number of methods reported for the facile synthesis of this skeleton including the cyclocondensation of nitrile oxides with alkynes [135]. Martins and coworkers reported a high-yielding practical method of trihalomethyl-substituted 4,5-dihydroisoxazoles **103** by reacting enones **102** with 1-hydroxylamine hydrochloride in presence of ionic liquid at 80 °C for 1 h (Scheme 18a) [136]. Similar reaction

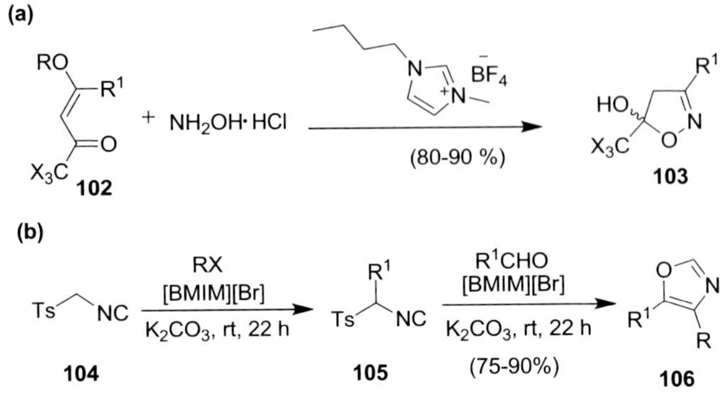

Scheme 18 IL-mediated synthesis of 4,5-dihydroisoxazoles **103** and oxazoles **106**

in refluxing methanol when carried out in absence of ionic liquid required longer reaction time (16 h). Interestingly, an alkylation of tosyl isocyanide **104** followed by in situ reaction with aldehydes in presence of [BMIM][Br] afforded good-to-excellent yield of the desired oxazoles **106** (Scheme 18b) [137].

Oxazole, oxazoline, and oxazolidinone heterocyclic skeleton were found to a part of several drug candidates. The classical Fischer oxazole synthesis includes the acid-catalyzed condensation of cyanohydrins with aldehyde. Other method includes the dehydration of 2-acylaminoketones (Robinson–Gabriel oxazole synthesis), or reaction of α-haloketone with formamide (The Bredereck reaction), reaction of aldehyde with TosMIC using strong base (Van Leusen reaction) and notably Hantzsch reaction. Fused-oxazole heterocycles displayed interesting bioactivities. Tiwari et al. explored benzotriazole ring cleavage (BtRC) methodology to afford diverse benzoxazoles by treating N-acyl benzotriazole with anhydrous $AlCl_3$ in refluxing toluene in sealed tube [138].

A number of other practical syntheses are well documented under environment benign condition. Thus, Fan and coworkers reported an expeditions synthesis of 2-substituted benzoxazoles **108** by reacting 2-aminophenols **107** with aldehydes in [BMIM][BF$_4$] as reaction media and catalytic amount of $RuCl_3 \cdot 3H_2O$ (Scheme 19) [139]. Furthermore, Srinivasan et al. utilized [HBIM][BF$_4$] ionic liquid for the reaction of 2-aminophenols **107** with benzoyl chlorides at room temperature to furnish high yield of respective 2-aryl benzoxazoles **109** [140].

In addition, 2-oxazoline heterocycles has been frequently used as suitable ligands, protecting groups in organic synthesis and chiral auxiliaries in asymmetric synthesis [141]. Kamakshi and coworkers reported a practical one-pot synthesis of oxazolines **111** through acid–amine coupling of carboxylic acids with amino alcohols **110** followed by subsequent cyclative dehydration in ionic liquid and $InCl_3$ as catalyst (Scheme 20) [142].

Thiazoles or fused-thiazoles heterocycles has been further identified for their considerable interest in medicinal chemistry. The classical Hantzsch thiazole synthesis involves the reaction of thioamide with α-haloketone. On the other hand,

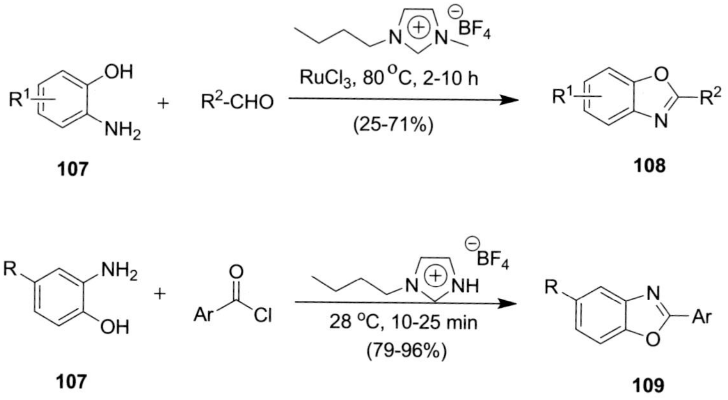

5 Impact of RTILs in the Synthesis of Biologically Relevant ...

Scheme 19 Ionic liquid-mediated synthesis of 2-substituted benzoxazoles

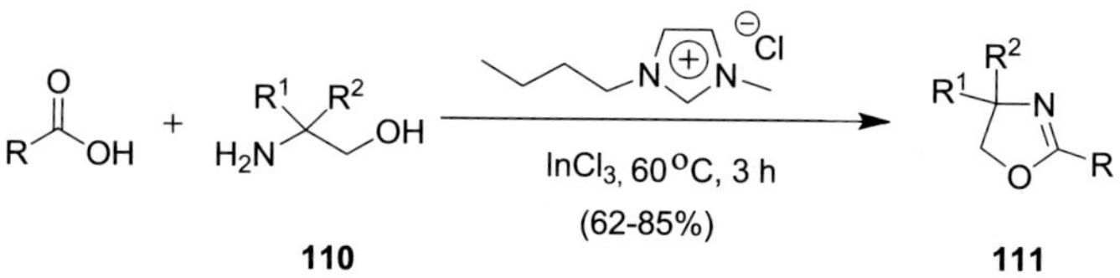

Scheme 20 IL-mediated practical synthesis of substituted 2-oxazolines **111**

Robinson–Gabriel synthesis involves the reaction of 2-acylamino-ketones with P_2S_5 [143]. Chen and coworkers reported an expeditious synthesis of diverse thiazoles **114** through cyclocondensation of α-tosyloxyketones **112** with thiobenzamide **113** in presence of [BMIM][PF$_6$] in a molar ratio of 1:10 (reactant/IL) at room temperature (Scheme 21) [144].

Tiwari's BtRC approach involves the ring cleavage of thioacyl benzotriazole under free radical condition and furnish high-to-excellent yield of diverse benzothiazoles [145, 146]. Furthermore, Le and coworkers reported a facile one-pot synthesis of 2-substituted benzothiazoles **115** through nucleophilic addition of amine to phenyl isothiocyanate to result respective thiourea which could underwent intramolecular nucleophilic aromatic substitution via its S-atom in presence of ionic liquids, e.g., [BMIM][Br$_3$] of [BMIM][BF$_4$] (Scheme 22) [147]. In another approach, Srini-

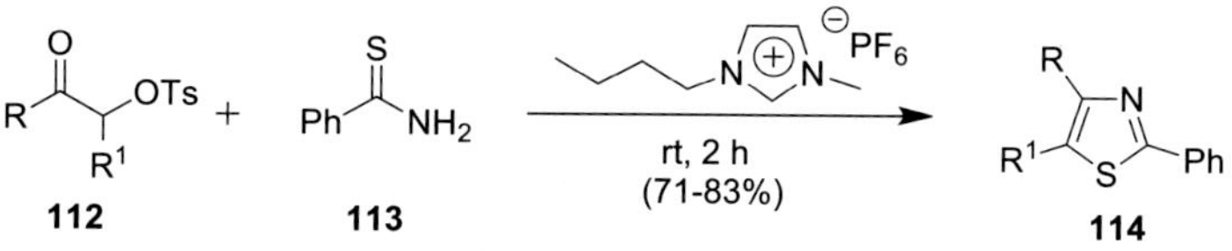

Scheme 21 Synthesis of substituted 2-thiazoles **114** using [BMIM][PF$_6$]

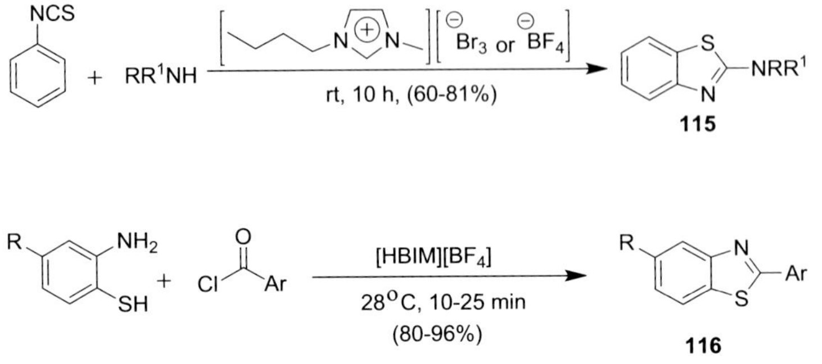

Scheme 22 IL-mediated synthesis of 2-substuituted benzothiazoles (**115, 116**)

vasan and coworkers utilized the regioselective condensation of 2-aminothiophenols with benzoyl chlorides in presence of imidazolium-based ILs, e.g., [HBIM][BF$_4$] to furnish high-to-excellent yield of desired 2-substituted benzothiazoles **116** (Scheme 22) [140].

A number of biologically relevant five-membered heterocycles with three or more hetero atoms such as thiadiazoles, triazoles, and tetrazoles has also received considerable attention due to their interesting features in chemistry, biology, medicinal chemistry, and material science. Ionic liquid-mediated practical and expeditious synthesis of a number of pyrazoles, imidazoles, isoxazoles, oxazoles, oxazolines, oxazolidinones, thiazoles, and thiazolidinones has been well documented in literature. The classical catalytic cyclative oxidative dimerization of thioamides is widely explored for an easy access of symmetrical 3,5-substituted-1,2,4-thiadiazoles. Interestingly, cyclative oxidative dimerization of thioamide **113** in catalytic amount of pentylpyridinium tribromide afforded high yield of 3,5-disubstituted 1,2,4-thiadiazoles **117** (Scheme 23) [148].

Just after the discovery of CuAAC "Click Chemistry" [149, 150], the impact of regioselective 1,4-disubstituted triazoles in various field including carbohydrate chemistry is increasing exponentially [151, 152]. For azide–alkyne cycloaddition reaction, a well-known CuI/DIPEA catalytic system has also been explored in RTILs to deliver high-to-excellent yield of desired 1,4-disubstituted triazoles **120** in short span of time (Scheme 24). This method is advantageous particularly with

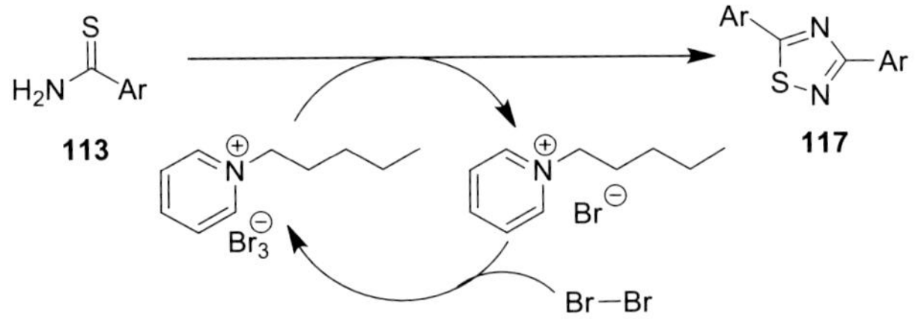

Scheme 23 Pentylpyridinium tribromide-mediated synthesis of 1,2,4-thiadiazoles **117**

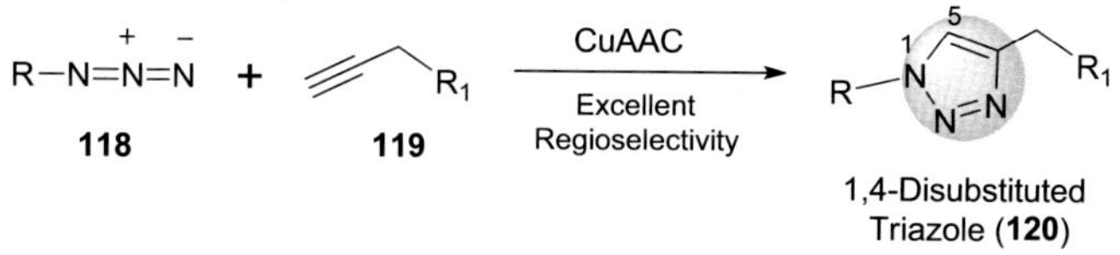

5 Impact of RTILs in the Synthesis of Biologically Relevant …

Scheme 24 CuAAC-mediated regioselective synthesis of 1,4-disubstituted triazoles **120** in ionic liquids

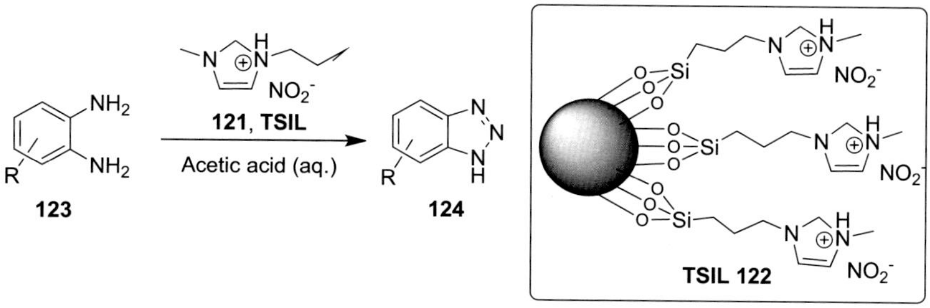

Scheme 25 Synthesis of benzotriazoles **124** in presence of IL

the substrates having some solubility issues [153]. ILs, for example, [bmim]BF$_4$ when merged with ([Cu(Im^{12})$_2$]CuCl$_2$ was considered as "environmental benign green catalyst" in the regioselective CuAAC reaction [154].

1-Butyl-3-methylimidazolium nitrite **121** and 1-(3-trimethoxysilylpropyl)-3-methylimidazolium nitrite **122**-based nanoparticles of organosilane-based nitrite ionic liquid immobilized on silica are well-known source of nitrosonium ion. Thus, Valizadeh et al. explored the reaction of 1,2-diaminobenzenes **123** with either of ILs **121** or **122** in acidic medium at room temperature to afford high yield of respective benzotriazoles **124** (Scheme 25) [155].

The [3+2] cycloaddition of nitriles with hydrazoic acid or its derivatives is well-known route for the tetrazoles synthesis. The 5-monosubstituted tetrazoles **125** and 1,5-disubstituted tetrazoles **126** were obtained by reacting nitriles with NaN$_3$ or activated nitriles with organic azides in presence of [EMIM][HSO$_4$] (Scheme 26a) [156]. Fused-tetrazole skeletons, for example, 5-(trifluoromethyl)-4,7-dihydrotetrazolo[1,5-a]pyrimidine derivatives **128** were developed in high-to-good yields through multicomponent reaction of 5-aminotetrazole **127**, aldehydes and 1,3-diketo compounds in presence of [BMIM][BF$_4$] or [BYP][BF$_4$] at 110 °C (Scheme 26b) [157].

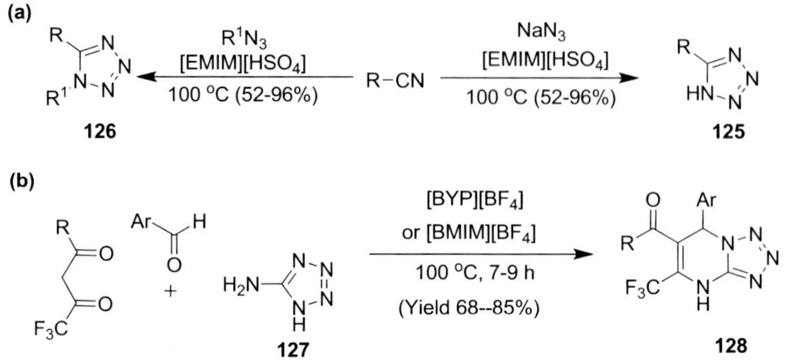

Scheme 26 IL-mediated synthesis of 5-mono- and 1,5-disubstituted tetrazoles

5.3 Ionic Liquid-Mediated Synthesis of Six-Membered Heterocycles

A number of six-membered heterocyclic systems are known for their promising application in organic synthesis. They displayed interesting biological activities and also found as structural component of a hundreds of the bioactive natural products. These heterocyclic systems (Fig. 13) have been well explored as suitable scaffold toward the development new chemical entities (NCE's) useful in drug discovery and development.

The pyridine heterocycles, because of their wide spread applications in medicinal, biological, biochemical, pharmaceutical, material science and supramolecular chemistry has been a fascinating attention over the century. Khosropour and coworkers reported an expeditious synthesis of tri-aryl-substituted pyridines **129** [158]. Thus, benzyl alcohols on reaction with 1-arylethanols, acetophenone or 2-aryl ethanols and ammonium acetate in presence of ionic liquids, e.g. [HMIM][NO$_3$] (150 mol%) and [BMIM][BF$_4$] (100 mol%) under microwave irradiation at 96 °C for 4 min afforded high-to-excellent yield of tri-aryl-substituted pyridines **129** (Scheme 27a). In this reaction, [HMIM][NO$_3$] first bring oxidation of aryl alcohols to respective aryl aldehydes which then reacted with acetophenone and ammonium acetate in the presence of catalytic amount of [BMIM][BF$_4$] to give the desired pyridine analogues **129** [158]. Interesting, Perumal and coworkers developed a practical synthesis of tetra-substituted pyridine analogues **130** by reacting 1,3-dicarbonyl compounds with alkynones and ammonium acetate in presence of [HMIM][Tfa] under one-pot solvent-free condition (Scheme 27b) [159]. The reaction proceed via in situ enaminone formation which could underwent heteroannulation under the Bohlmann–Rahtz condition to afford the title compounds **130**. In continuation, Mohammad and coworkers reported a ZrOCl$_2$ · 8H$_2$O/NaNH$_2$-mediated convenient synthesis of

5 Impact of RTILs in the Synthesis of Biologically Relevant ... 145

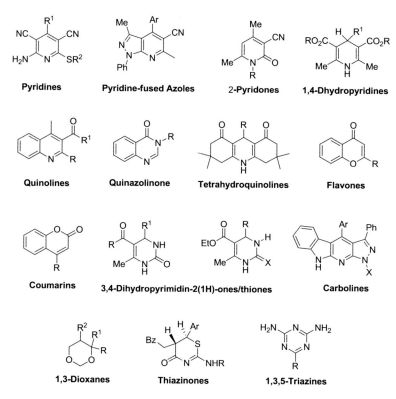

Fig. 13 Structure of some selected six-membered heterocycles of biological relevance

2-amino-6-(arylthio)-4-arylpyridine-3,5-dicarbonitriles **131** through MCR of aldehydes, arylthiols, and malononitrile in the presence of [BMIM][BF₄] under ultrasound condition at room temperature for 2–20 min (Scheme 27c) [160]. Interestingly, aldehydes on similar reaction with thiols and malononitrile in presence of 2-hydroxyethylammonium acetate (2-HEAA) as a task-specific ionic liquid at room temperature afforded high yield of respective pyridine analogues [161]. The similar condensation using [BMIM][OH] also furnished high yield of functionalized pyridines **131** [162].

Pyridine-fused skeletons have been considered as special featured systems due to their remarkable bioactivities. Toward this end, Shi and coworkers reported a facile one-pot MCR of aromatic aldehydes with 3-oxo-3-phenylpropanenitrile **132** and 3-methyl-1-phenyl-1*H*-pyrazol-5-amine **133** in [bmim]Br as green reaction media to afford high yield of respective pyridine-fused analogues **134** (Scheme 28) [163].

Aliphatic and aromatic cyanoacetamides **135** on reaction with 1,3-dicarbonyl compounds or chalcones in presence of ionic liquid, e.g., 1,1,3,3-tetramethylguanidine lactate [TMG][Lac] under solvent-free condition afforded high yield of 4,6-disubstituted-3-cyano-2-pyridones **136** (Scheme 29). The notable

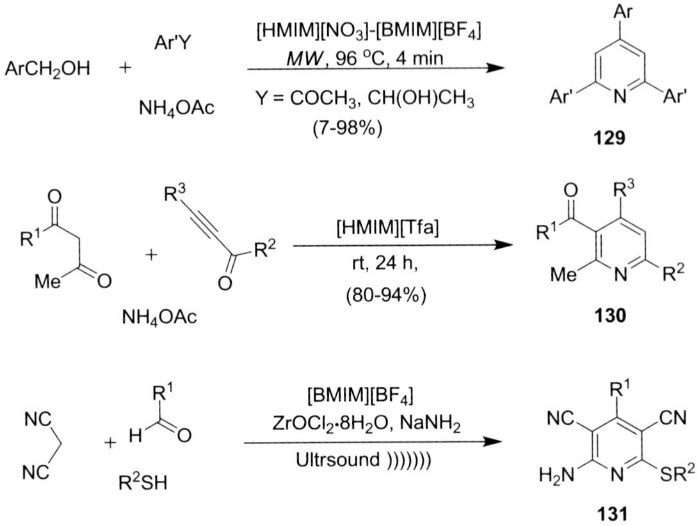

Scheme 27 ILs-mediated synthesis of 2,4,6-triarylpyridines and 2-amino-6-(arylthio)-4-arylpyridine-3,5-dicarbonitrile analogues

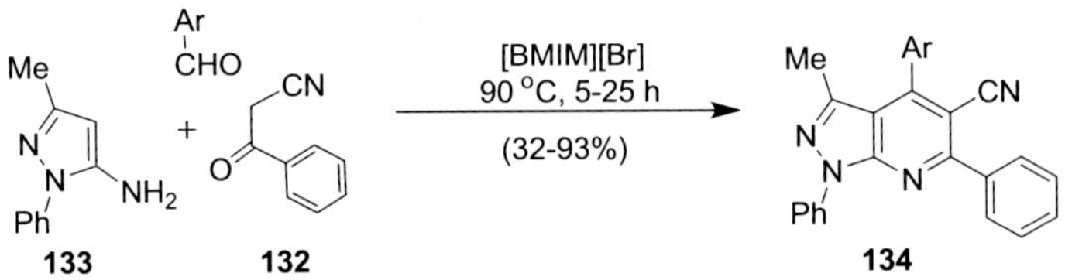

Scheme 28 Synthesis of fused-pyridine derivatives using ionic liquid [BMIM][Br]

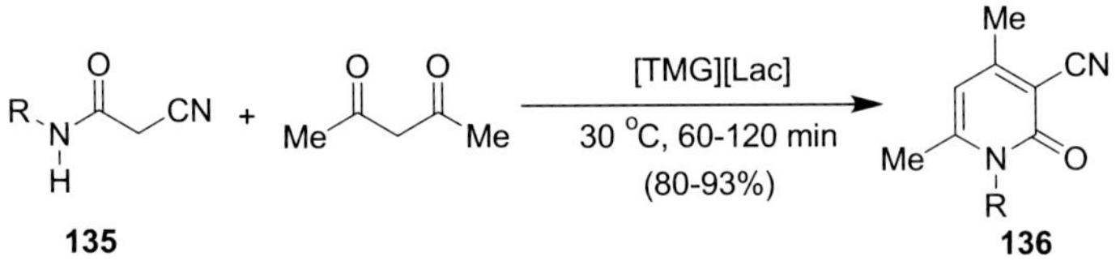

Scheme 29 [TMG][Lac]-mediated synthesis of 4,6-disubstituted-3-cyano-2-pyridones

catalytic activity of ionic liquid [TMG][Lac] can be correlated with its basicity due to intramolecular hydrogen bonding with lactic acid [164].

In continuation, Martins and coworkers reported a convenient synthesis of related pyridine system **139** by reacting benzylidene cyanoacetohydrazide **137** with 4-alkoxy-1,1,1-trifluoro-3-alken-2-ones **138** using $BF_3 \cdot OEt_2$/triethylamine/[BMIM][BF$_4$] as an efficient catalytic system (Scheme 30)

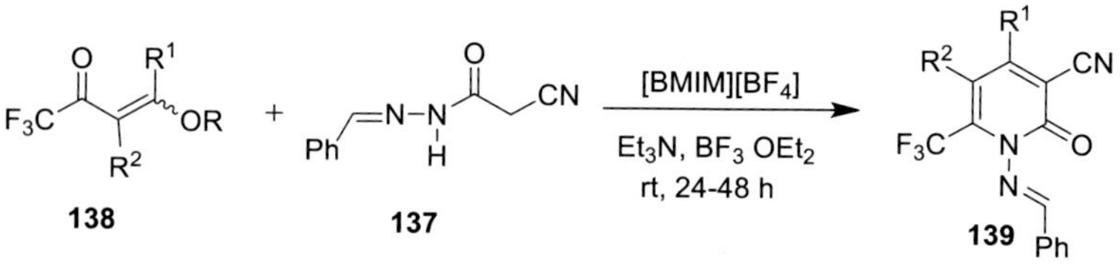

5 Impact of RTILs in the Synthesis of Biologically Relevant … 147

Scheme 30 IL-mediated synthesis of functionalized $1H$-2-pyridones **139**

[165].

Interestingly, a bislactam heterocycle **141** was obtained in high yield by reacting 6-methyl-4-hydroxypyran-2-one **140** with aldehyde and primary amine (or ammonium acetate) for 2–5 h at 95 °C in [BMIM][Br] as green media (Scheme 31) [166].

The 1,4-dihydropyridines (DHPs) is again a very important class of heterocycles known for their potent bioactivities and also for their great utilities as organocatalyst as well scaffold explored in organic synthesis. Hantzsch's route of DHP synthesis has been recognized a paramount importance since its first synthesis in the year 1882. The Hantzsch conventional MCR method includes the cyclocondensation of β-keto esters (two equivalent) with aldehydes and ammonium acetate [167]. Vijayakumar and coworkers used 10 mol% of ionic liquid, e.g., [EMIM][OAc] for the MCR of aldehydes, acetylacetone or ethylacetoacetate and ammonium acetate under sonication at room temperature to deliver high-to-excellent yield of desired 1,4-dihydropyridines **142** (Scheme 32a) [168]. In continuation, Shaabani and coworkers reported a facile route for 1,4-dihydropyridines **143** via MCR of aldehydes, β-ketoesters, and ammonium acetate in presence of 1,1,3,3-N,N,N',N'-tetramethylguanidinium trifluoroacetate (TMGT) under similar condition (Scheme 32b) [169].

The cyclocondensation reaction of enaminoesters, e.g., methyl-3-aminocrotonate with β-ketoester and aldehyde in the presence of [bmim]BF$_4$ (in 1:5 molar ratio, reactant/IL) afforded high-to-excellent yield of 1,4 dihydropyridines **144** (Scheme 32c) [170]. Dihydropyridine synthesis via similar cyclocondensation reaction using [BMIM][Cl] as green media and 3,4,5-trifluorobenzene boronic acid as catalyst is also well documented under ultrasonication condition [171].

Quinoline heterocycle found in a number of bioactive natural products and commercial pharmaceuticals. The classical Skraup, Friedlander, Conrad–Limpach, and Pfitzinger methods are most commonly employed protocol for the practical synthesis of quinoline analogues. In recent years, several RTILs were frequently

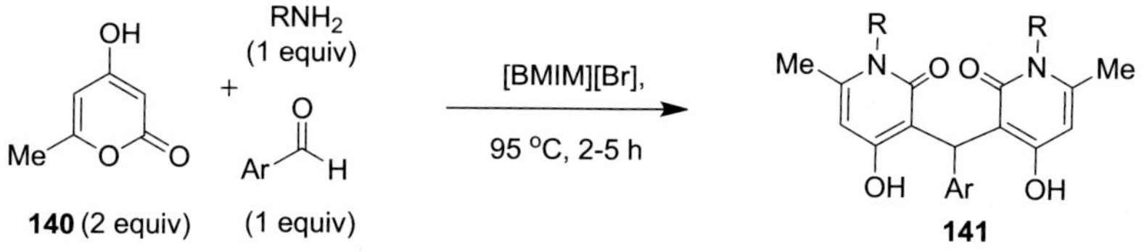

Scheme 31 Green synthesis of bislactam analogues in [BMIM][Br]

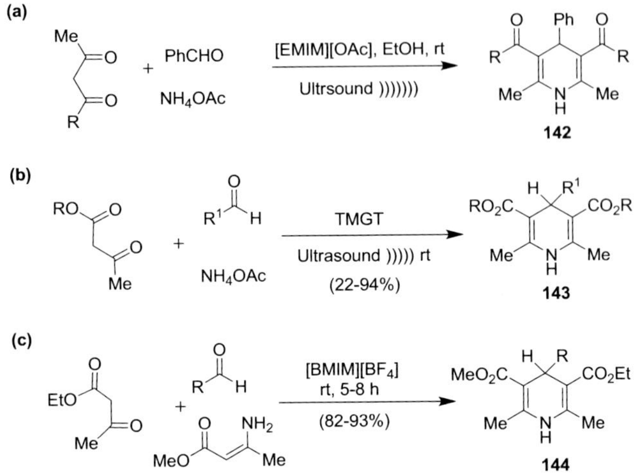

Scheme 32 RTILs-mediated expeditious synthesis of 1,4-dihydropyridine analogues

used as green media or environmental benign catalyst to overcome the unharmonious reaction circumstances for this important skeleton. Martins and coworkers explored the *MW*-induced reaction of 2-aminoacetophenone **145** with 4-alkoxy-3-alken-2-ones **146** in [HMIM][OTs] under acidic catalysis (*p*TSA) for an easy access of quinoline **147** (Scheme 33) [172]. Srinivasan and coworkers employed the reaction of aminoacetophenone **145** with ketones or ketoesters in presence of [HBIM][BF$_4$] and achieved the desired functionalized quinoline analogues **148** in good-to-excellent yields [173]. Likewise, the same quinoline system **148** was developed by Perumal in almost quantitative yield simply by reacting 2-aminoketones with ketones or ketoesters in [BMIM][Cl]/ZnCl$_2$ melt (in molar ratio 1:2) [174].

Wu and coworkers reported an expeditious 4-CR of cyclohexanone, aldehydes, malononitrile, and amines in the presence of basic IL [BMIM][OH] at 80 °C to afford 2-amino-4-aryl-5,6,7,8-tetrahydroquinoline-3-carbonitriles **149** (Scheme 34) [175].

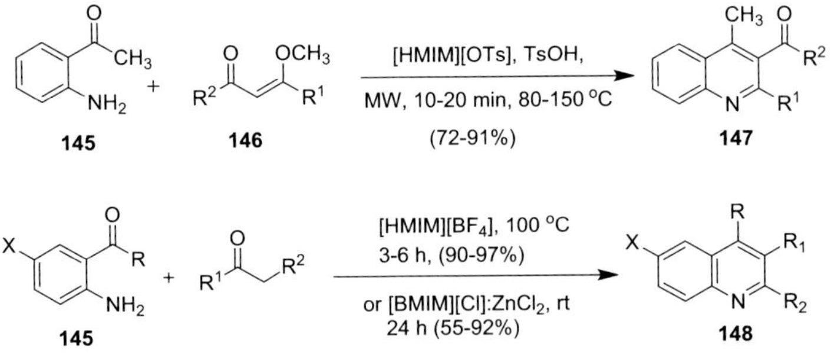

Scheme 33 Ionic liquid-mediated synthesis of quinoline analogues (**147, 148**)

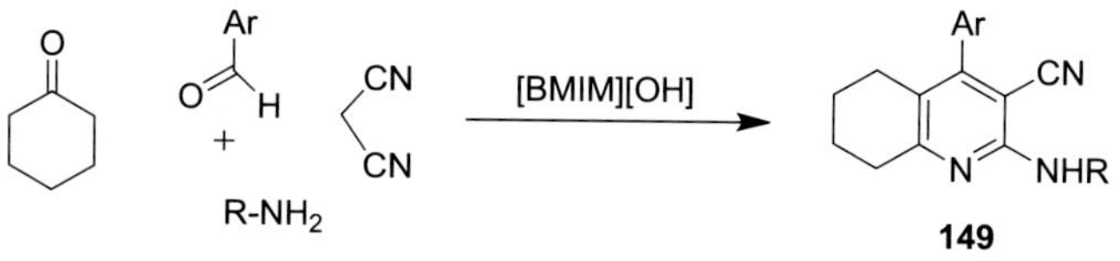

5 Impact of RTILs in the Synthesis of Biologically Relevant … 149

Scheme 34 [BMIM][OH]-mediated synthesis of tetrahydroquinolines **149**

Furthermore, the three-component reaction of cyclic β-enaminones **150**, aldehydes, and active methylene compounds in presence of [BMIM][BF$_4$] as environmental benign green media at 90 °C afforded good-to-excellent yield of respective tetrahydroquinolines (**151**, **152**) (Scheme 35) [176].

Zare and coworkers used [DSIM][HSO$_4$] for a facile one-pot MCR of β-ketoesters, ammonium acetate, aldehyde, and dimedone under solvent-free conditions to furnish high-to-excellent yield of target hexahydroquinolines **153** (Scheme 36) [177]. Another ionic liquid, e.g., [pyridine-SO$_3$H]Cl has also been employed in such synthesis.

Interestingly, fused pyrimido-[4,5-*b*]quinoline analogues **155** has been obtained in good-to-excellent yield via three-component reaction of 6-aminopyrimidine-2,4-dione **154**, 5,5-dimethyl-1,3-cyclohex anedione, and aldehydes in the presence of [BMIM][Br] at 95 °C (Scheme 37) [178].

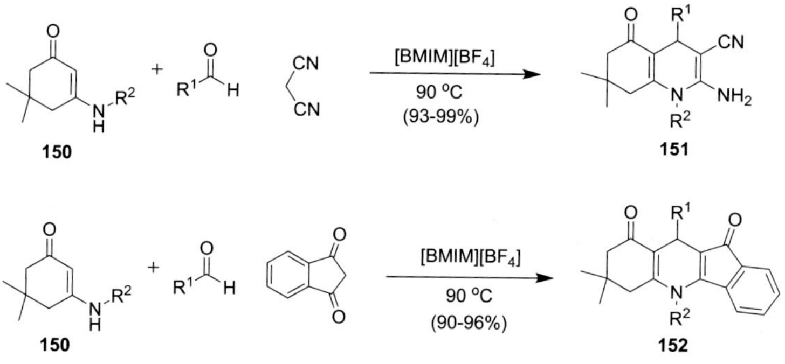

Scheme 35 Synthesis of tetrahydroquinolines (**150**, **151**) using [BMIM][BF$_4$]

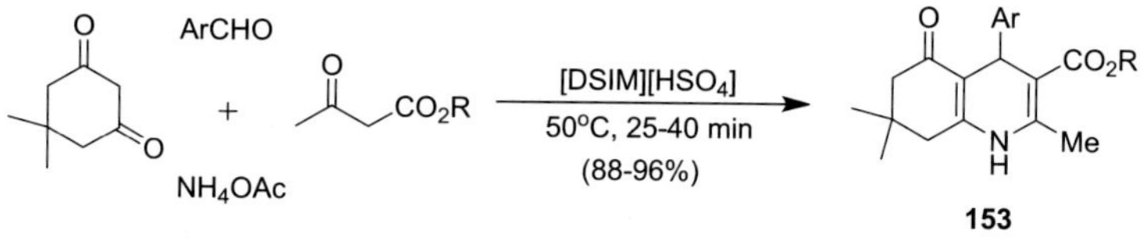

Scheme 36 [DSIM][HSO$_4$]-mediated synthesis of hexahydroquinolines **152**

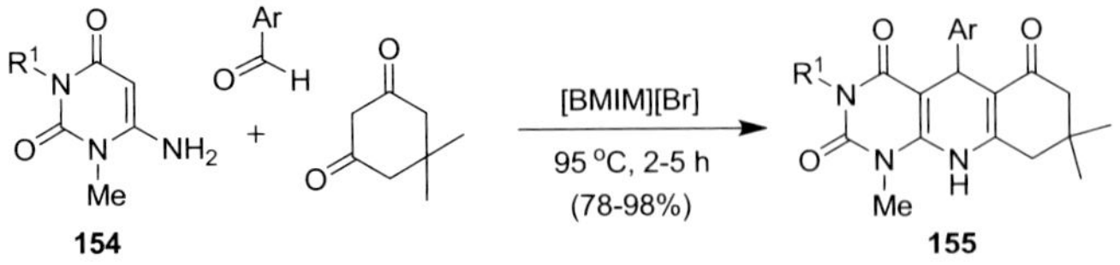

Scheme 37 [BMIM][Br]-mediated synthesis of pyrimido-[4,5-*b*]quinolines **155**

Acridine, structurally resemble to anthracene except '10-CH' being replaced by "*N*" atom, display a range of potent bioactivities. Due to their hydrophobicity, it is susceptible to diffuse in the cell membrane and can intercalate with nucleic acids. Zong and coworkers reported a convenient synthesis of 9-arylpolyhydroacridines **156** by reacting 5,5-dimethyl-1,3-cyclohexanedione with aromatic aldehydes and ammonium acetate in the presence of [BMIM][BF$_4$] at 80 °C in 3–5 h (Scheme 38) [179].

Furthermore, Yu and coworkers explored a similar cyclocondensation chemistry with dimedone, aromatic aldehydes, and anilines using Bronsted acid imidazolium salt containing perfluoroalkyl tails monomeric ionic liquid **157** for an easy access of functionalized acridines **158** (Scheme 39) [180].

Coumarin is an important class of benzopyrones and this heterocycle is well-known for their imperative pharmacological activities such as Wedelolactone, Novobiocin, etc. Phenols, e.g., phloroglucinol, resorcinol, etc. on treatment with β-ketoester using 1-butanesulfonic acid-3-methylimidazolium tosylate [BSMIm][Ts] (10 mol%) under solvent-free condition at 85 °C afforded high yield of title coumarin

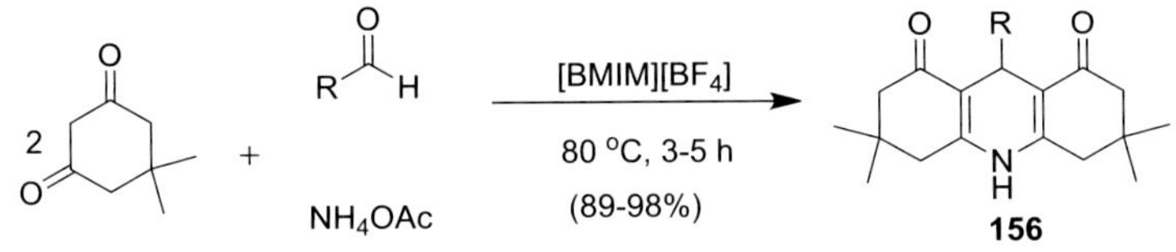

Scheme 38 [BMIM][BF$_4$]-mediated synthesis of 9-arylpolyhydroacridines **156**

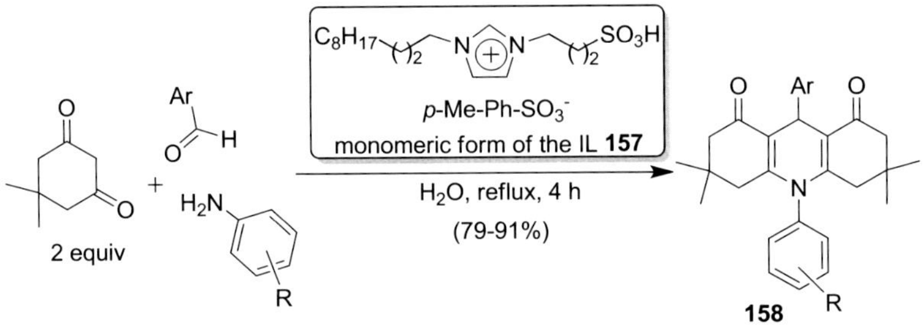

Scheme 39 Perfluoroalkyl appended IL-mediated synthesis of acridines

5 Impact of RTILs in the Synthesis of Biologically Relevant ...

analogues **159** (Scheme 40) [181]. Likewise, a [MSIM][HSO$_4$]-mediated Pechmann reaction was employed for the high-yielding synthesis of functionalized coumarin analogues **160** [182].

The **flavone**, a benzo-γ-pyrone derivative, has similar look with coumarin. This skeleton has active bioceutical ingredients in plants known for their potential antioxidant and metal chelator activities. A classical Baker–Venkatraman rearrangement includes the reaction of benzoyl ester of 2-hydroxyacetophenone with pyridine/KOH and the 1,3-diketone thus obtained was then subjected to further cyclization in strongly acidic condition to afford high yield of desired flavone. Similar approach was employed by Pawar and coworkers [183] for an easy access of diverse flavones **162** by heating diketones **161** under *MW* irradiation in presence [EtNH$_3$][NO$_3$] (Scheme 41a) [184]. Likewise, substituted flavone analogues **164** have been developed by Wang and coworkers via [BMIM][NTf$_2$]-induced Cu(I)-catalyzed cascade reaction (intramolecular *oxa*-Michael-oxidation) of chalcones **163** (Scheme 41b) [185]. In addition to the simplicity, high reaction yield, this cascade protocol demonstrated noteworthy tolerance toward varied functional groups.

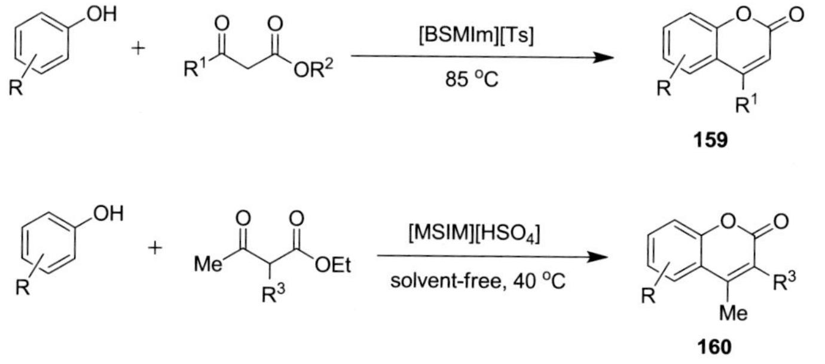

Scheme 40 IL-mediated expeditious synthesis of coumarin analogues (**159**, **160**)

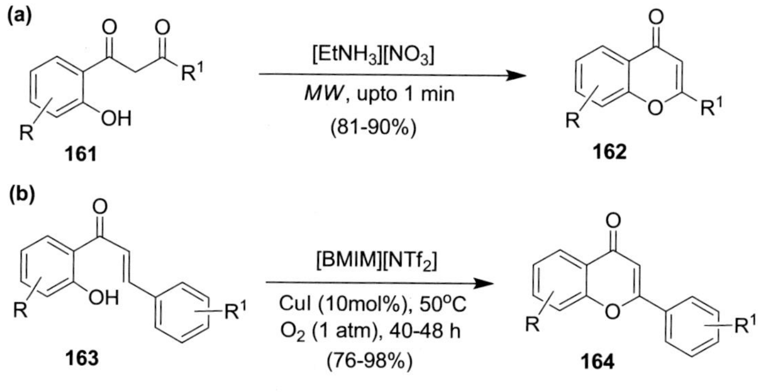

Scheme 41 IL-mediated expeditious synthesis of flavones analogues (**162**, **164**)

4 Growing Impact of Ionic Liquids in Heterocyclic Chemistry

Pyrimidinones and **pyrimidines** are other important biologically potent six-membered heterocyclic skeletons with two heteroatoms. The classical synthesis of dihydropyrimidones utilized an acid-catalyzed three-component reaction of β-dicarbonyl compounds with aldehydes and urea (or thiourea) in protic solvent (Biginelli protocol). The similar Biginelli pyrimidinone route was employed with an aid of RTILs, e.g., [BMImBF$_4$] or [BMImPF$_6$] for MCR of aldehydes, β-ketoesters, and urea under solvent-free condition at 100 °C (Scheme 42) [186]. Interestingly, a silica-supported sulfonic acid ionic liquid, e.g., (Si-[SbSipim][PF$_6$]) in refluxing ethanol was also utilized for the similar synthesis of 3,4-dihydropyrimidinones **165** [187]. Furthermore, Hajipour and coworkers used Brønsted acidic IL, e.g., 3-carboxypyridinium hydrogensulfate [HCPY][HSO$_4$] for the synthesis of 3,4-dihydropyrimidin-2(1H)-ones [188]. Similar reaction when performed in glycine nitrate [GlyNO$_3$] as green and biodegradable IL under sonication condition, it afforded high-to-excellent yield of title 3,4-dihydropyrimidin-2(1H)-ones [189].

Likewise, the similar condensation of aromatic aldehydes, β-ketoesters and urea (or thiourea) in presence of [Hbim]BF$_4$ on ultrasonication afforded high-to-excellent yield of 3,4-dihydropyrimidin-2-(1)-ones (**166** and **167**) (Scheme 43) [190]. Dadhania and coworkers used an carboxy functionalized ionic liquid [CMMIM][BF$_4$] as green catalyst under microwave irradiation for similar high-yielding synthesis of 3,4-dihydropyrimidin-2-(1H)-ones [191].

Furthermore, Zarei and coworkers reported MCR of diverse aldehydes with ethyl aceto acetate and urea (or thiourea) in the presence of [HMIM][HSO$_4$] under solvent-free refluxing condition to afford high-to-excellent yield of 3,4-dihydropyrimidin-2(1H)-ones **168** or 3,4-dihydropyrimidin-2(1H)-thiones **169** (Scheme 44) [192].

Next, Zlotin and coworkers reported an interesting synthesis of 3,4-dihydropyrimidin-2-(1H)-ones (**172** and **173**) by reacting 2-arylidenes **170** with O-methylisourea sulfate **171** in presence of [BMIM][BF$_4$] as green ionic media

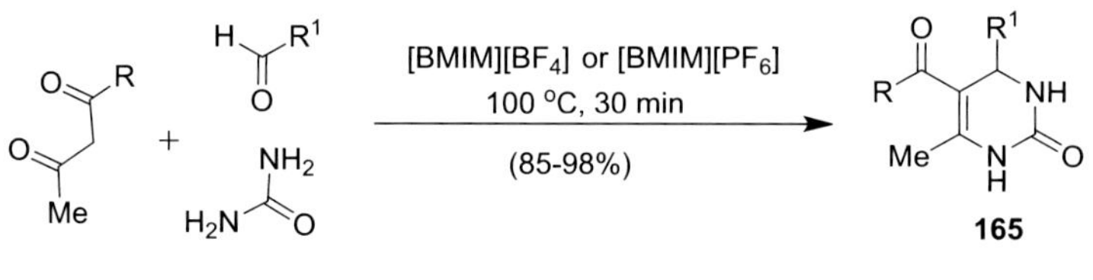

Scheme 42 Synthesis of pyrimidinones **165** using [BMIM][BF$_4$]

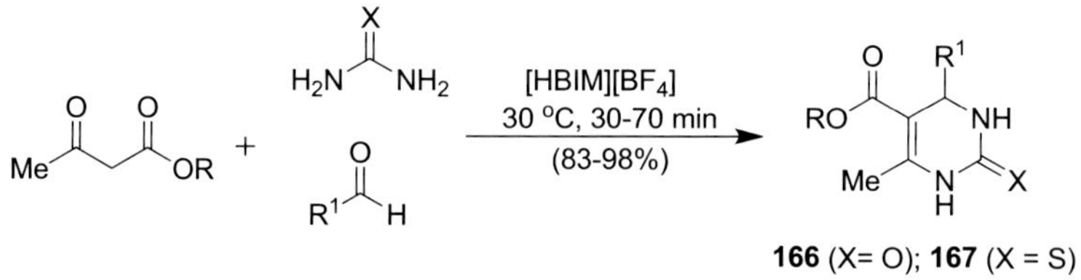

Scheme 43 IL-mediated synthesis of 3,4-dihydropyrimidin-2-(1H)-ones

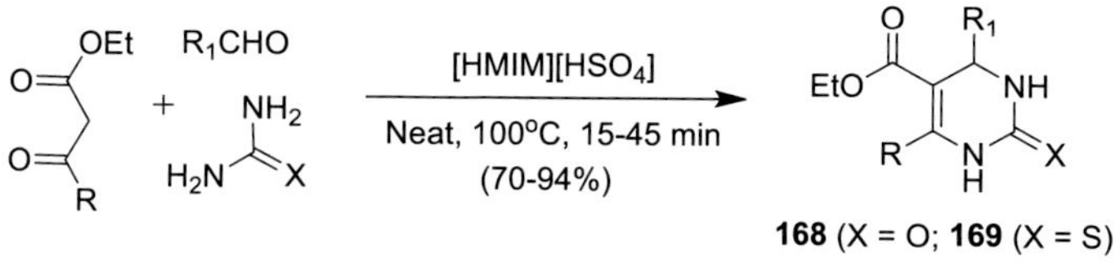

Scheme 44 [HMIM][HSO$_4$]-mediated synthesis of 3,4-dihydropyrimidin-2-(1H)-ones

(Scheme 45) [193]. The regioselective N-alkoxycarbonylation and O-demethylation of compound **172** may led via reacting in presence of NaHCO$_3$ and [BMIM][BF$_4$]. The reaction performed in ionic liquid required less reaction over the conventional method.

Singh and coworkers employed MCR of thiazolidine-2,4-dione **174**, aromatic aldehydes, and urea using [BMIM][Br] as green media for an expeditious synthesis of 7-phenyl-1,4,6,7-tetrahydrothiazolo[5,4-d]pyrimidine-2,5-diones **175** (Scheme 46) [194].

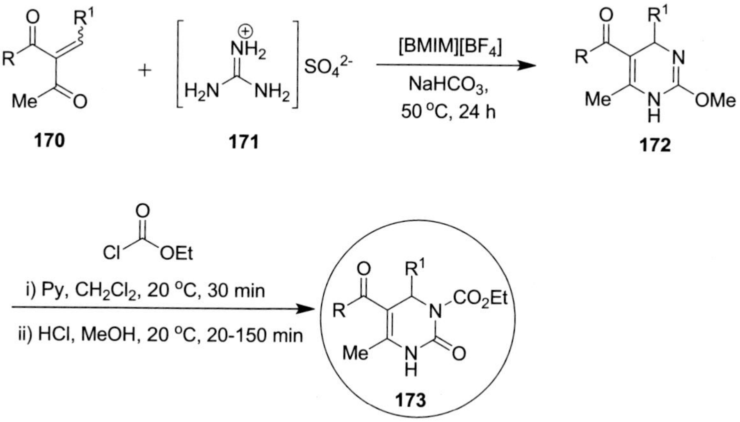

Scheme 45 Synthesis of 3,4-dihydropyrimidin-2-(1H)-ones **173** using [BMIM][BF$_4$]

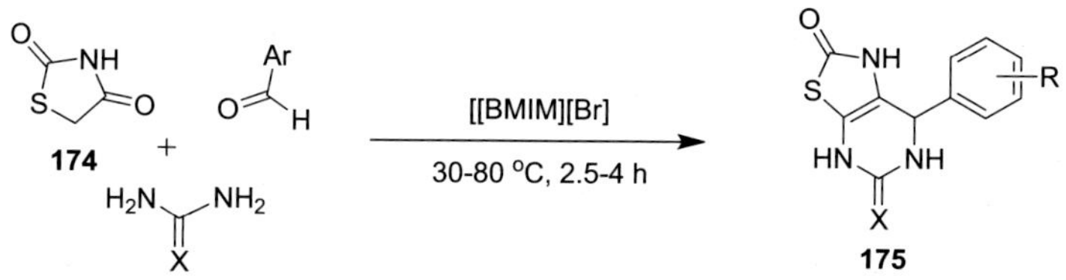

Scheme 46 IL-mediated synthesis of fuzed-pyrimidine-2,5-diones **175**

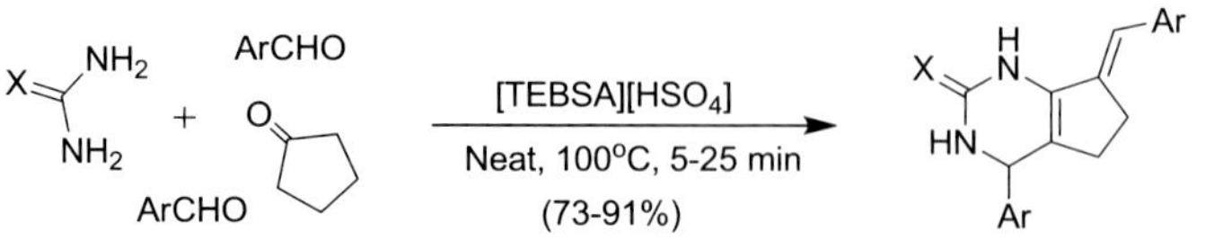

Scheme 47 Synthesis of pyrimidinone derivatives using [TEBSA][HSO$_4$]

Hajipour and coworkers exemplified a three-component condensation reaction of aromatic aldehydes, cyclopentanone and urea (or thiourea) in the presence of N-(4-sulfonic acid) butyl triethyl ammonium hydrogen sulfate under solvent-free heating condition to afford respective pyrimidinones (**176, 177**) in good-to-excellent yields (Scheme 47) [195].

Likewise, Basiri and coworkers reported an expeditious synthesis of biologically potent piperidone-grafted pyridopyrimidines **178** via ionic liquid-mediated multi-component reaction of aldehydes, thiourea, and cyclic ketone, e.g., piperidone system (Scheme 48) [196].

Karthikeyan and coworkers used L-amino acid decorated ionic liquid such as [L-AAIL]/AlCl$_3$ in Biginelli reaction of aldehyde, β-keto ester and thioureas under solvent-free conditions at 80 °C for 6 h and devised a cheap and straightforward synthesis of functionalized 3,4-dihydropyrimidine-2-(1H)-thione analogues **179** (Scheme 49) [197].

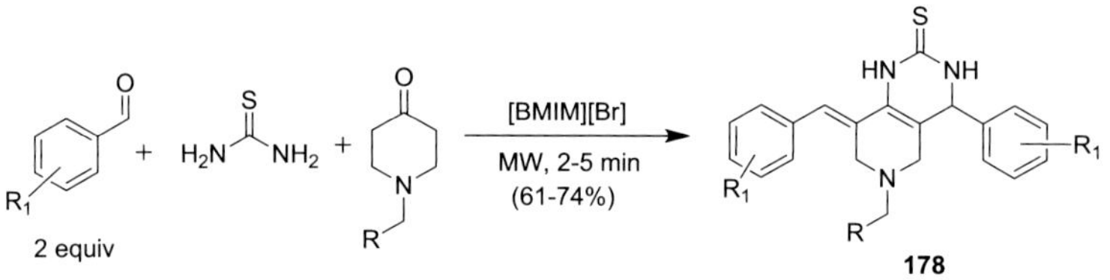

Scheme 48 Synthesis of piperidone-grafted pyridopyrimidines using [BMIM][Br] ILs

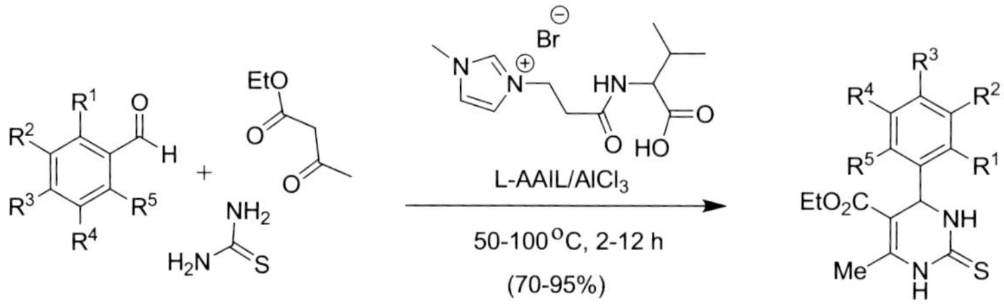

Scheme 49 [L-AAIL]/AlCl$_3$-mediated synthesis of 3,4-dihydropyrimidine-2-(1H)-thione analogues

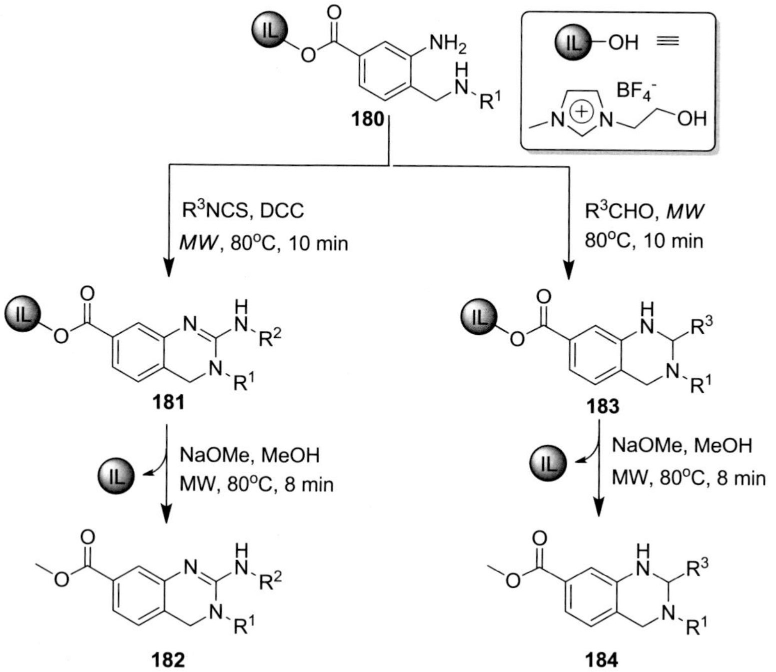

5 Impact of RTILs in the Synthesis of Biologically Relevant …

Scheme 50 Green synthesis of dihydroquinazolines **182** and tetrahydroquinazolines **184** in presence of a water-soluble ionic liquid

Quinazoline and related analogues are one of the widely explored heterocyclic systems in medicinal chemistry. Sun and coworkers used a water-soluble IL-supported tool and reported a practical synthesis of dihydroquinazolines **182** and tetrahydroquinazolines **184** (Scheme 50) [198]. IL-bound 4-((alkylamino)methyl)-3-nitrobenzoates **180**, obtained via S_N2 substitution of IL-bound 4-bromomethyl-3-nitrobenzoic acid with amines, when treated with isothiocyanates or aldehydes, it underwent cyclization to give IL-bound molecules **181** and **183**, respectively. This individually, at the end were subjected to methanolysis and leading to the respective dihydro- and tetrahydroquinazolines (**182** and **184**) in almost quantitative yields via cleavage of ionic liquid from the support.

Sekar and coworkers explored an environmental benign route for the facile synthesis of biologically relevant quinazoline scaffold **186** through cyclocondensation of 3-aminobenzene-1,2-dicarboxylate **185** with organic nitriles in presence of [NMP][HSO$_4$] at room temperature (Scheme 51) [199]. In another effort, the MCR of 2-aminobenzothiazole **187** with isatoic anhydride **188** and aldehydes in [BMIM][Br] as reaction media afforded high-to-excellent yield of desired 2,3-dihydroquinazolin-4(1H)-ones **189** (Scheme 51) [200]. Dabiri and coworkers reported a tandem synthesis of 2-styryl-4(3H)-quinazolinones **190** by reacting isatoic anhydride **188**, primary amine (or ammonium acetate), and triethylorthoacetate in presence of

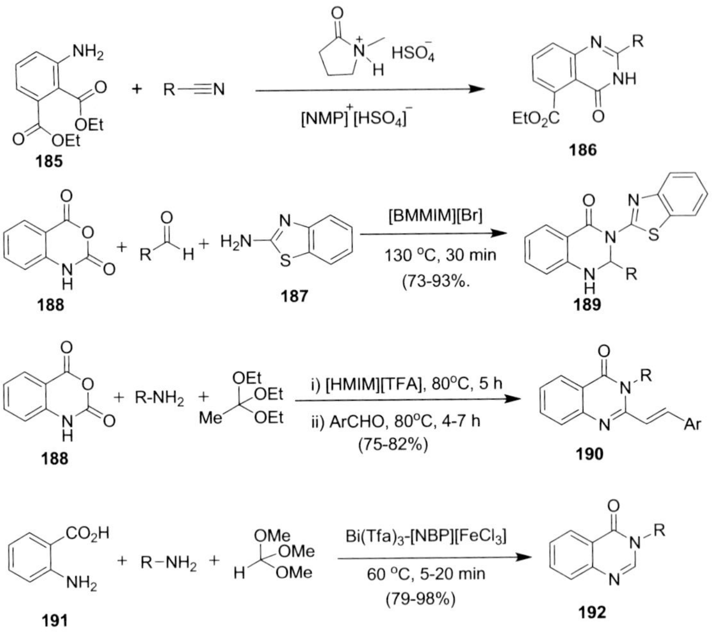

Scheme 51 IL-mediated facile synthesis of diverse quinazolinone analogues

[Hmim]TFA followed by subsequent reaction with aromatic aldehydes [201]. Likewise, Khosropour and coworkers exemplified a facile one-pot MRC of anthranilic acid **191**, trimethyl orthoformate, and primary amines in presence of Bi(Tfa)$_3$ (5 mol %) immobilized on [NBP][FeCl$_4$] at room-temperature synthesis to afford high yield of 4-(3*H*)-quinazolinones **192** (Scheme 51) [202]. Similar chemistry was explored for an easy access of quinazolinones and quinazolines [203].

Carboline is considered as one of the imperative pharmacophore in medicinal chemistry. Toward this end, Ghahremanzadeh and coworkers reported a MCR of indolin-2-one **193**, 3-oxo-3-phenylpropanenitrile **194**, aldehydes and hydrazines using catalytic amount of *p*-TSA in [BMIM][Br] as green media at 140 °C to afford high yield of desired pyrido[2,3-b]indoles **195** (Scheme 52) [204]. This multicompo-

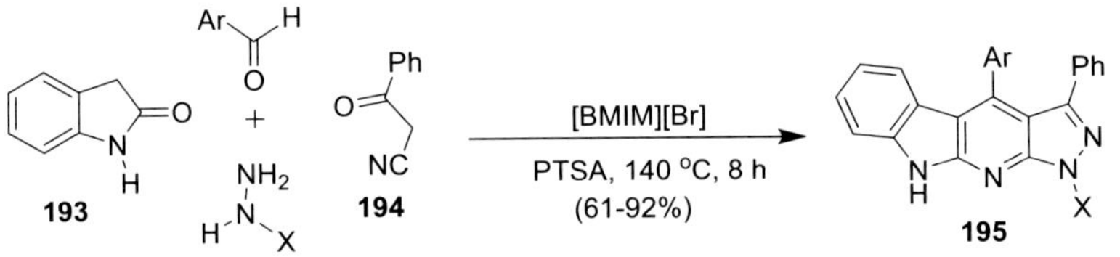

Scheme 52 [BMIM][Br]-mediated synthesis of pyrido[2,3-b]indoles

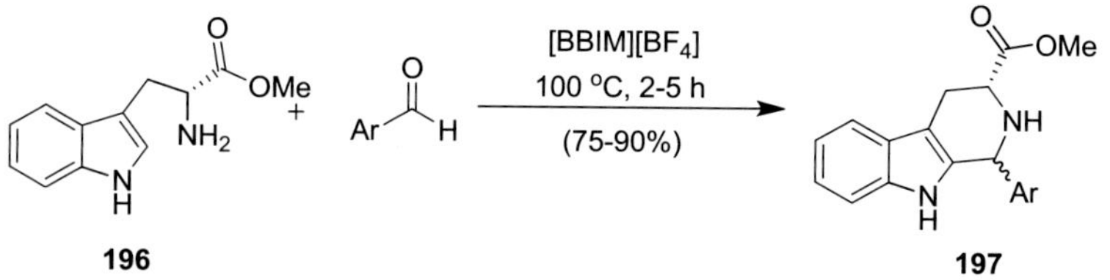

5 Impact of RTILs in the Synthesis of Biologically Relevant …

Scheme 53 [BBIM][BF$_4$]-mediated synthesis of 1,3-disubstituted 1,2,3,4-tetrahydro-β-carbolines **197**

nent condensation reaction is convenient with aromatic aldehydes, get complicated with aliphatic aldehydes, e.g., butanal or pentanal.

Pictet–Spengler reaction in strong acidic condition (ranging from catalytic to stoichiometric amounts) is frequently utilized to develop tetrahydro-β-carbolines as mixture of *cis*- and *trans*-diastereoisomers [205]. Interestingly, Joshi and coworkers explored the Pictet–Spengler condensation for an easy access 1,3-disubstituted 1,2,3,4-tetrahydro-β-carbolines **197** by treating D-tryptophan methyl ester **196** with aldehydes in presence of catalytic amount of TFA in [BBIM][BF$_4$] (Scheme 53) [206]. The Pictet–Spengler cyclization in [BBIM][BF$_4$] is advantageous as reaction yield was 82% in just 2–5 h, while similar conventional Pictet-Spengler cyclization using TFA as catalyst in DCM (in absence of IL) gave 62% of desired product **197** even after 4d [207].

A less reactive tetrahydro-β-carbolines **198** was subjected to coupling reaction with 2-nitrofluorobenzene in the ionic liquid [BDMIM][Tf$_2$N]/[BDMIM][PFBuSO$_3$] and resulted coupling product **199** thus obtained was subjected to reductive cyclization to afford the desired fused-tetrahydro-β-carbolinequinoxalinone **200** as sole product (Scheme 54) [208].

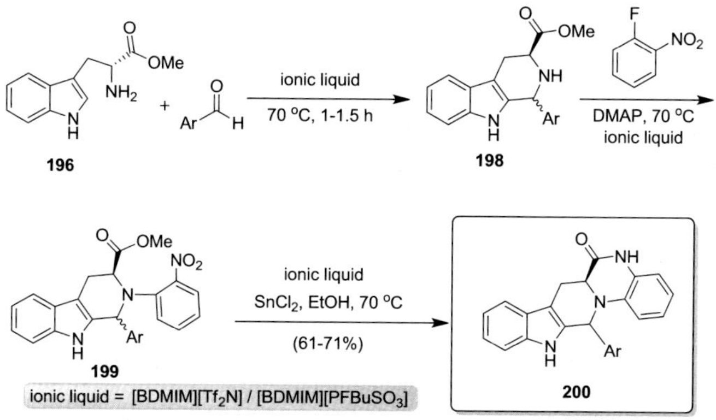

Scheme 54 IL-mediated synthesis of fused-tetrahydro-β-carbolinequinoxalinones **200**

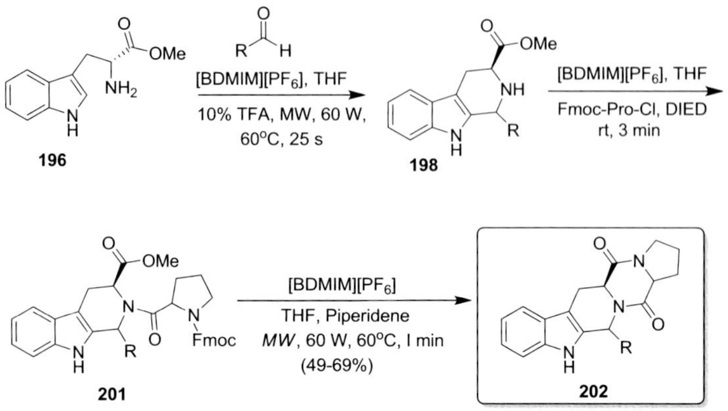

158   4 Growing Impact of Ionic Liquids in Heterocyclic Chemistry

Scheme 55 [BDMIM][PF$_6$]-mediated synthesis of tetrahydro-β-carbolinediketopyperazines **202**

Yen and Chu nicely utilized Pictet–Spengler cyclization followed by Schotten–Baumann acylation to afford respective amide **201** that at the end underwent an intramolecular ester amidation and furnish high yield of biologically potent tetrahydro-β-carbolinediketopyperazine analogues **202** (Scheme 55) [209]. All the three synthetic steps were performed in presence of ionic liquids [BDMIM][Tf$_2$N] and [BDMIM][PFBuSO$_3$] in a molar ratio of 1:1 (reactant/IL). In addition, the first and last steps of reactions carried out under microwave irradiation and thus make the protocol more advantageous as per concern of green chemistry.

The acid-catalyzed Prins reaction of olefins with aldehydes has been widely used for an easy access of diverse 1,3-diols, 1,3-dioxanes, and unsaturated alcohols [210]. Toward this end, Yadav and coworkers reported a proficient synthesis of **1,3-dioxanes (203)** in good-to-excellent yield by treating styrenes with paraformaldehyde using catalytic amount of InBr$_3$ in [BMIM][PF$_6$] (Scheme 56) [211]. Increased in reaction rates, improvement in reaction yields, and high selectivity are the notable advantage of protocol over the conventional one [212].

In addition to the interesting biological activities associated with **1,2-oxazines**, this heterocyclic skeleton also found of significant utilities as intermediates in organic synthesis particularly for the synthesis of unnatural amino acid. Hetero–Diels–Alder reactions either alkenes with ene-nitroso compounds or dienes with nitroso derivatives were greatly explored for the synthesis of this heterocycles [213]. However, Kitazume et al. established as practical synthesis of oxazines **205** by

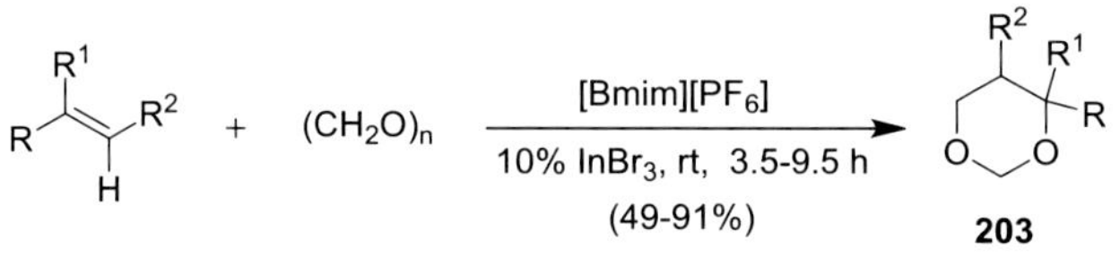

Scheme 56 [BMIM][PF$_6$]-mediated synthesis of 1,3-dioxanes **203** under Prins condition

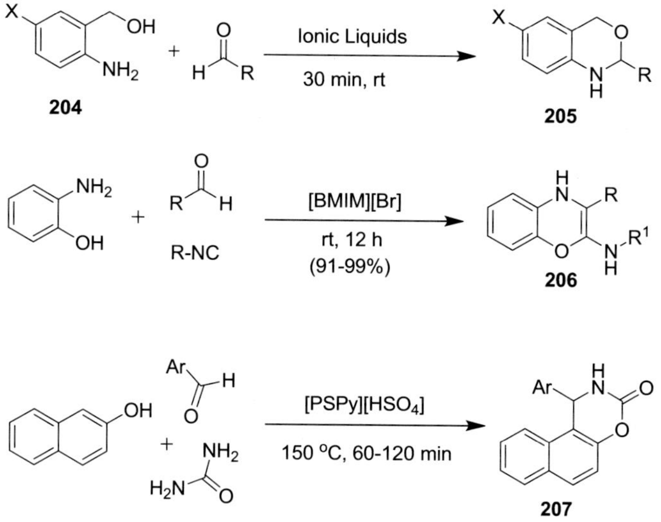

5 Impact of RTILs in the Synthesis of Biologically Relevant …

Scheme 57 IL-mediated convenient synthesis of oxazine analogues

treating aldehydes with 2-aminobenzyl alcohols **204** in presence of ionic liquids, e.g., [EMIM][CF3SO3] or [BMIM][BF4] or [BMIM][PF6] or [EtDBU][OTf] (Scheme 57a).

The author claimed that the separation of chiral product from racemic mixture was quite easy in ionic liquid than reaction carried out in conventional polar aprotic solvent [214]. Interestingly, 2-aminophenol on three-component condensation with aldehyde and isocyanide in presence of [BMIM][Br] at ambient temperature furnished respective benzo[b][1,4]oxazines **206** in high-to-excellent yield (Scheme 57b) [215]. Next, Dong and coworkers utilized pyridinium-based functionalized ionic liquid in MCR of β-naphthol, aromatic aldehydes, and urea heated at 150 °C for around 1–2 h to afford high-to-excellent yield of 1,2-dihydro-1-arylnaphtho[1,2-e][1,3]oxazine-3-ones **207** (Scheme 57c) [216].

1,4-Benzothiazines is the well-known anti-psychotic pharmacophore. Toward this end, Yadav and coworkers used a masked amino or mercapto acids and reported a [BMIM][Br]-mediated convenient and diastereoselective three-component reaction to furnish yield of respective thiazinone derivatives (**208** and **209**) (Scheme 58) [217].

A fused thiazinone e.g., 3-oxo-1,4-benzothiazines **210** could be obtained in good yield by reacting alkylsulfanylphenylamines with bromoacetyl bromide in presence of [BMIM][BF$_4$] (Scheme 59) [218]. The reaction when performed in absence of ILs (in refluxing DMSO) for 30–40 min gave comparatively low yield of product [219].

1,3,5-Triazine is frequently used as flame-retardant additives in many resins and found as an structural component in fire-resistant polymers [220]. A library of diaminotriazines **212** were obtained in high yield by treating dicyandiamide **211** with arylnitriles in presence of [BMIM] [PF$_6$] under microwave heating at 130 °C (Scheme 60) [221, 222].

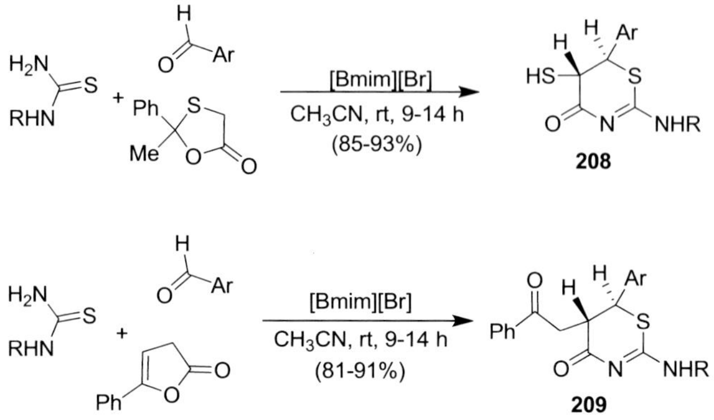

Scheme 58 [BMIM][Br]-mediated synthesis of thiazinone analogues (**208, 209**)

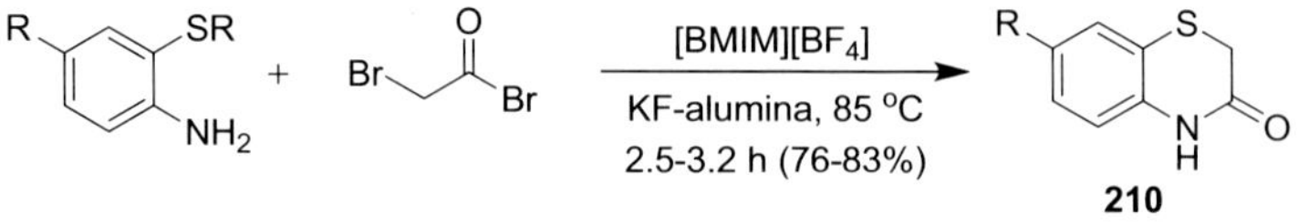

Scheme 59 [BMIM][BF$_4$]-mediated synthesis of benzothiazines **210**

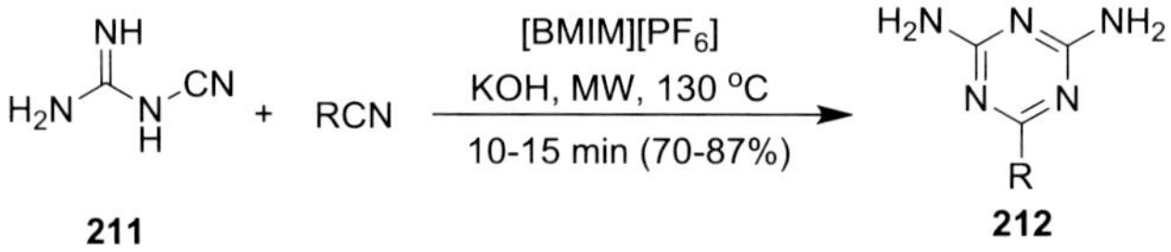

Scheme 60 [BMIM][PF$_6$]-mediated synthesis of diaminotriazine analogues

Benzothiazin-3-ones (**215**), a pharmacologically active heterocyclic system was developed in good yields by reacting 2-aminothiophenols **213** with 2-bromoalkanoates **214** in presence of [OMIM][NO$_3$] at room temperature (Scheme 61) [223].

Eligeti and coworkers reported a practical synthesis isoxazolyl-**1,3-benzoxazines** using [HMIM][BF$_4$], a Brønsted acidic IL (Scheme 62). Thus, treatment of O-hydroxy benzaldehydes **216** with amine **217** and formaldehyde in presence of

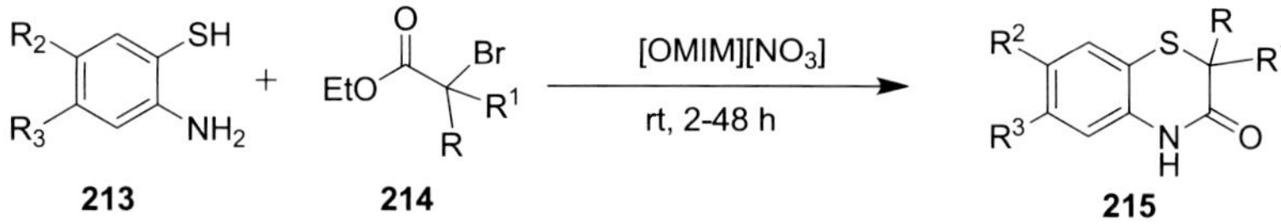

Scheme 61 [OMIM][NO$_3$]-mediated synthesis of benzothiazin-3-ones

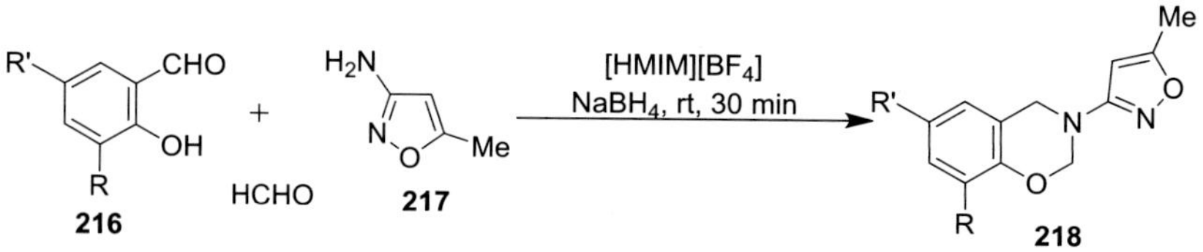

5 Impact of RTILs in the Synthesis of Biologically Relevant … 161

Scheme 62 [HMIM][BF$_4$]-mediated synthesis of isoxazolyl-1,3-benzoxazines

NaBH$_4$ and [HMIM][BF$_4$] under mild condition afforded high-to-excellent yield of 3-(5-methyl-3-isoxazolyl)-3,4-dihydro-2H-1,3-benzoxazine **218** [224].

5.3.1 IL-Mediated Synthesis of Some Selected Spiro Heterocycles

Ramin and coworkers utilized [BMIM][PF$_6$] as a green media and reported an expeditious synthesis of spiro[chromeno[2,3-d]pyrimidine-5,3′-indoline]-tetraone analogues **222** by treating isatin **219**, barbituric acid **220**, and cyclohexane-1,3-diones **221** in the presence of catalytic amount of alum (KAl(SO$_4$)$_2$ · 12H$_2$O) (Scheme 63) [225].

Phenylsulfonylacetonitrile on one-pot reaction with 1,3-dicarbonyl compounds and isatins or acenaphthenequinone in ethanol using catalytic amount of basic ionic liquid, e.g., 2-hydroxyethyl ammonium acetate [H$_3$N$^+$CH$_2$CH$_2$OH][AcO$^-$] afforded high-to-excellent yield of desired spiro-2-amino-3-phenylsulfonyl-4H-pyrans **223** (Scheme 64) [226].

1,3-Dipolar cycloaddition of azomethine ylide in equimolar mixture of the reactants was explored in [BMIM][Br] at 100 °C for 1–2 h to afford good-to-excellent yield of desired spiro analogues, including 1-methyl-4-arylpyrrolo-(spiro[2.2′]indan-1′,3′-dione)-spiro[3.3″]-1″-methyl/benzyl-5″-(arylmethylidene)piperidin-4″-ones **227** and 1-methyl-4-arylpyrrolo-(spiro[2.11′]-11H-indeno[1,2-b]quinoxaline)-spiro[3.3″]-1″-methyl/benzyl-5″-(arylmethylidene)piperidin-4″-ones **228** (Scheme 65) [227]. Furthermore, one-pot condensation of ninhydrin **224**, sarcosine

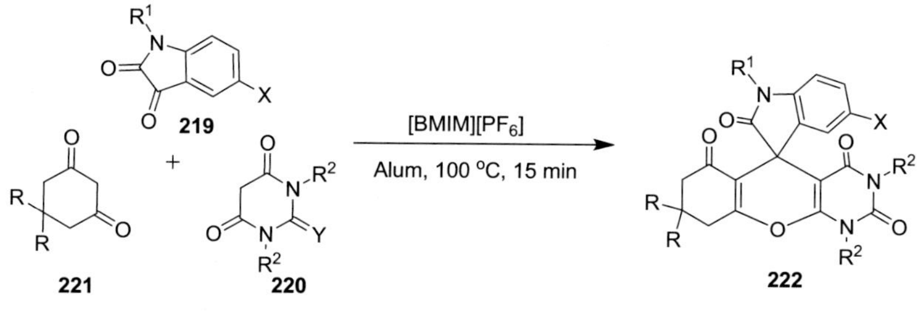

Scheme 63 [BMIM][PF$_6$]-mediated synthesis of spiro[chromeno[2,3-d]pyrimidine-5,3′-indoline]-tetraone analogues **222**

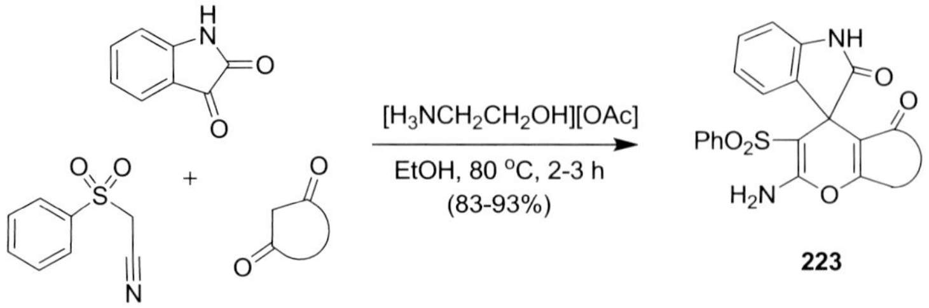

Scheme 64 Basic IL-mediated synthesis of spiro-2-amino-3-phenylsulfonyl-4*H*-pyrans **223**

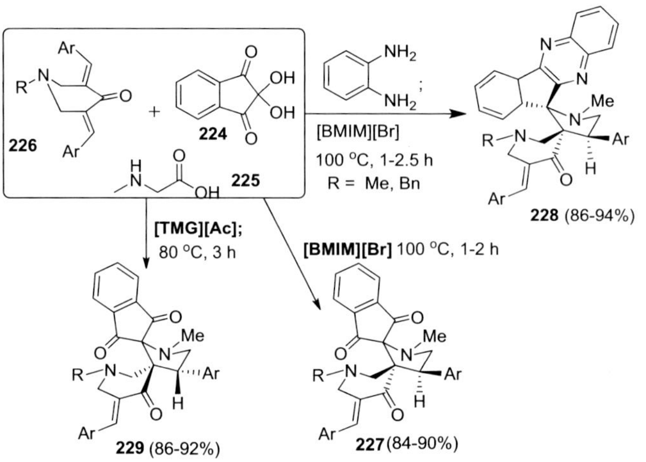

Scheme 65 [BMIM][Br]-mediated continued synthesis of spiro-heterocycles (**227–229**)

225, and 1-benzyl/methyl-3,5-*bis*[(*E*)-arylidene]-piperidin-4-one **226** in 1,1,3,3-tetramethylguanidine acetate ([TMG][Ac]) as green reaction media exemplify a convenient and regioselective synthesis of biologically relevant di-spiro heterocycles **229** (Scheme 65) [228].

Likewise, decarboxylative condensation of isatin with sarcosine resulted to in situ cycloaddition trapping of azomethine ylides that subsequently in presence of [BMIM][PF$_6$] afforded high yield of desired anti-cancer dispiropyrolidine-bisoxindoles **230** (Scheme 66a) [229]. In continuation, a dispirobisoxindole skeleton **231** was obtained by reacting dipolarophile, e.g., alkyl 2-cyano-2-(2-oxoindolin-3-ylidene)acetate with sarcosine azomethine ylide (generated in situ by decarboxylative condensation of isatin and sarcosine or proline) (Scheme 66b) [230].

Spirooxindole appended heterocyclic pharmacophores **232** has been developed in high yield through MCR of isatins, 1,3-dimethyl-2-amino uracil, and barbituric acid (or thiobarbituric acid) in presence of magnetic supported acidic ionic liquid (Scheme 67) [221]. Similar reaction of isatins, 1,3-dimethyl-2-amino uracil, and

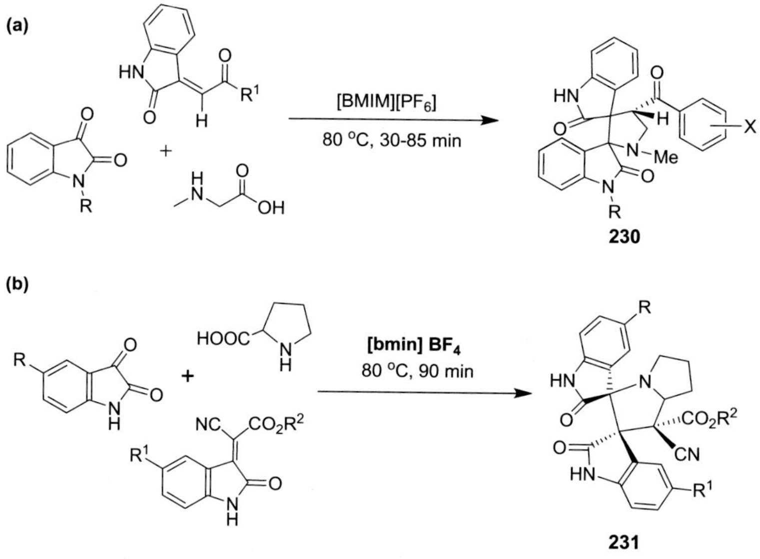

5 Impact of RTILs in the Synthesis of Biologically Relevant …

(a)

(b)

Scheme 66 [BMIM][PF$_6$]-mediated synthesis of dispiropyrrolidine-bisoxindole analogues (**230**, **231**)

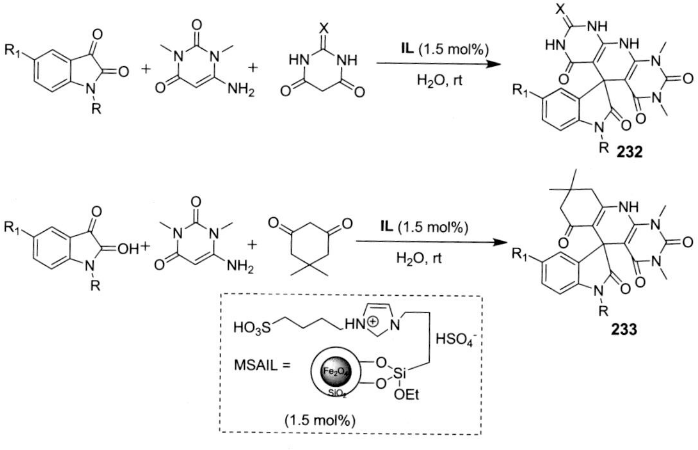

Scheme 67 Synthesis of spirooxindole derivatives (**232**, **233**) using [MIMPSA][HSO$_4$]

dimedon in presence of MSAIL afforded high yield of respective spirooxindole **233** [231].

Furthermore, a pharmacologically active spiro[3*H*-indole-3,2′-thiazolidine]-2,4′(1*H*)-diones **237** has been obtained in high yield by treating 1*H*-indole-2,3-diones

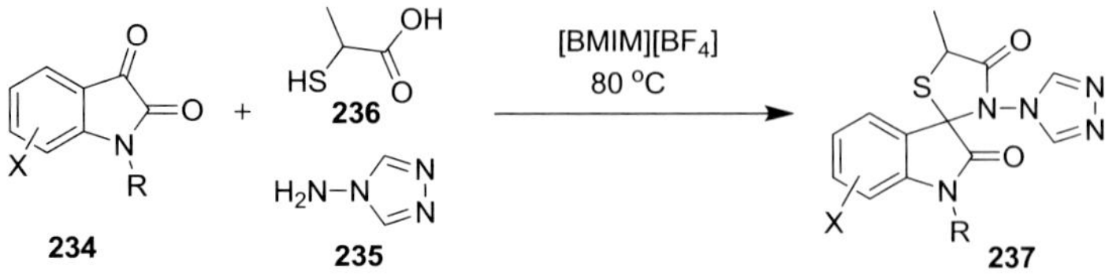

Scheme 68 [BMIM][PF$_6$]-mediated facile synthesis of spiro[3H-indole-3,2′-thiazolidine]-2,4′(1H)-dione analogues **237**

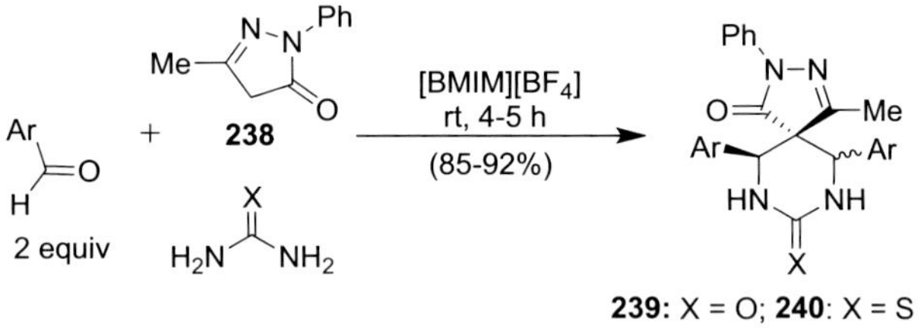

Scheme 69 [BMIM][BF$_4$]-mediated synthesis of aza-spiro[4.5]decene analogues (**239, 240**)

234, 4H-1,2,4-triazol-4-amine **235** and 2-sulfanylpropanoic acid **236** in presence of [BMIM][PF$_6$] (Scheme 68) [232].

A versatile pharmacophore, e.g., aza-spiro[4.5]decane analogues (**239** and **240**) with promising anti-microbial potency has been developed in good-to-excellent yield through MCR of urea/thiourea, aryl aldehydes, and 3-methyl-1-phenyl-2-pyrazolin-5-one **238** in presence of [BMIM][BF$_4$] (Scheme 69) [233].

6 Conclusions and Future Outlook

This chapter illustrated a brief overview about the properties, classification, and the growing impact of room-temperature ionic liquids (RTILs) in the synthesis of biologically relevant heterocyclic skeletons. The chapter also represents the structure, typical experimental preparation, and tunable physical properties of ionic liquids with their notable advantages as alternative green media, impact to enhanced reaction rates, reduces the reaction times and improvement in reaction yields, easy-workup, their recyclability, cost-effectiveness, etc. ILs justifies the title of "green solvents" or "designer solvents" by enabling the synthesis or modifications of bioactive molecules in a cost-effective and environmentally benign manner. Such notable features of ILs have productively secured their place as homogeneous catalysts in a number of common catalytic reactions of industrial importance. Although the versatile applications of RTILs as catalysts or solvents or both catalyst and solvent in various fields are increasing day by day, some important issues such as toxicity, non-biodegradability, viscosity, etc. still required to their reconsideration in order to complete avoid for any probable hazardous outcome. We hope that the chapter greatly helps the readers to consider the possible utilities of ionic liquid in the heterocyclic synthesis under environmental benign condition.

References

1. Lei, Z., Chen, B., Koo, Y.-M., MacFarlane, D.R.: Introduction: ionic liquids. Chem. Rev. **117**, 6633–6635 (2017)
2. Wilkes, J.S.: A short history of ionic liquids—from molten salts to neoteric solvents. Green Chem. **4**, 73–80 (2002)
3. Vekariya, R.L.: A review of ionic liquids: applications towards catalytic organic transformations. J. Mol. Liq. **227**, 44–60 (2017)
4. Wasserscheid, P., Schroer, W. (eds.): J. Mol. Liq. **192**, 1–208 (2014)
5. Hallett, J.P., Welton, T.: Room-temperature ionic liquids: solvents for synthesis and catalysis. Chem. Rev. **111**, 3508–3576 (2011)
6. Dong, K., Liu, X., Dong, H., Zhang, X., Zhang, S.: Multi-scale studies on ionic liquids. Chem. Rev. **117**, 6636–6695 (2017)
7. Mishra, B.B., Kumar, D., Singh, A.S., Tripathi, R.P., Tiwari, V.K.: Ionic liquids-prompted synthesis of biologically relevant five- and six-membered heterocyclic skeletons: an update. In: Brahmachari, G. (ed.) Green Synthetic Approaches for Biologically Relevant Heterocycles, pp. 437–493. Elsevier, The Netherlands (2015)
8. Bose, P., Agrahari, A.K., Singh, S.K., Singh, A.S., Yadav, M.S., Rajkhowa, S., Tiwari, V.K.: Recent developments on ionic liquids-mediated synthetic protocols of biologically relevant five and six-membered heterocyclic skeletons. In: Brahmachari, G. (ed.) Green Synthetic Approaches for Biologically Relevant Heterocycles, vol. 2, pp. 301–364. Elsevier, The Netherlands (2021)
9. Keglevich, G., Grun, A., Hermecz, I., Odinets, I.L.: Quaternary phosphonium salt and 1,3-dialkylimidazolium hexafluorophosphate ionic liquids as green chemical tools in organic syntheses. Curr. Org. Chem. **15**, 3824–3848 (2011)
10. Itoh, T., Takagi, Y.: Laccase-catalyzed reactions in ionic liquids for green sustainable chemistry. ACS Sustain. Chem. Eng. **9**, 1443–1458 (2021)
11. Prasad, V., Kale, R.R., Kumar, V., Tiwari, V.K.: Carbohydrate chemistry and room temperature ionic liquids (RTILs): recent trends, opportunities, challenges and future perspectives. Curr. Org. Syn. **7**, 506–531 (2010)
12. Rajkhowa, S., Kale, R.R., Sarma, J., Kumar, A., Mohapatra, P.P., Tiwari, V.K.: Room temperature ionic liquids in glycoscience: opportunities and challenges. Curr. Org. Chem. **25**, 2542–2578 (2021)
13. El Seoud, O.A., Koschella, A., Fidale, L.C., Dorn, S., Heinze, T.: Applications of ionic liquids in carbohydrate chemistry: a window of opportunities. Biomacromol **8**, 2629–2647 (2007)
14. Murugesan, S., Linhardt, R.J.: Ionic liquids in carbohydrate chemistry—current trends and future directions. Curr. Org. Syn. **2**, 437–451 (2005)
15. Farran, A., Cai, C., Sandoval, M., Xu, Y., Liu, J., Hernaiz, M.J., Linhardt, R.J.: Green solvents in carbohydrate chemistry: from raw materials to fine chemicals. Chem. Rev. **115**, 6811–6853 (2015)
16. Smiglak, M., Metlen, A., Rogers, R.D.: The second evolution of ionic liquids: from solvents and separations to advanced materials-energetic examples from the ionic liquid cookbook. Acc. Chem. Res. **40**, 1182–1192 (2007)
17. Qiao, Y., Ma, W., Theyssen, N., Chen, C., Hou, Z.: Temperature-responsive ionic liquids: fundamental behaviors and catalytic applications. Chem. Rev. **117**, 6881–6928 (2017)

18. Watanabe, M., Thomas, M.L., Zhang, S., Ueno, K., Yasuda, T., Dokko, K.: Application of ionic liquids to energy storage and conversion materials and devices. Chem. Rev. **117**, 7190–7239 (2017)
19. Ventura, S.P.M., Silva, F.A., Quental, M.V., Mondal, D., Freire, M.G., Coutinho, J.A.P.: Ionic-liquid-mediated extraction and separation processes for bioactive compounds: past, present, and future trends. Chem. Rev. **117**, 6984–7052 (2017)
20. Dai, C., Zhang, J., Huang, C., Lei, Z.: Ionic liquids in selective oxidation: catalysts and solvents. Chem. Rev. **117**, 6929–6983 (2017)
21. Egorova, K.S., Gordeev, E.G., Ananikov, V.P.: Biological activity of ionic liquids and their application in pharmaceutics and medicine. Chem. Rev. **117**, 7132–7189 (2017)
22. Majumdar, K.C., Chattopadhyay, S.K. (eds.): Heterocycles in Natural Product Synthesis. Wiley-VCH, Weinheim (2011)
23. Katritzky, A.R., Ramsden, C., Scriven, E., Taylor, R.J.K. (eds.): Comprehensive Heterocyclic Chemistry, vol. 1–15. Elsevier, Amsterdam (2010)
24. Mishra, B.B., Kumar, D., Mishra, A., Mohapatra, P.P., Tiwari, V.K.: Cyclorelease strategy in solid-phase combinatorial synthesis of heterocyclic skeletons. Adv. Heterocycl. Chem. **107**, 41–99 (2012)
25. Tiwari, V.K.: Development of biologically relevant glycohybrid molecules: twenty years of our journey. Chem. Rec. **11**, 3029–3048 (2021)
26. Walden, P.: Ueber die Molekulargroesse und elektrische Leitfähigkeiteinigergeschmolzenen Salze. Bull. Acad. Imp. Sci. St. Petersbourg **8**, 405–422 (1914)
27. Hurley, F.H.: US Patent, 4, 446, 331 (1948)
28. Wier Jr., T.P., Hurley, F.H.: US Patent, 4, 446, 349 (1948)
29. Wilkes, J.S., Levisky, J.A., Wilson, R.A., Hussey, C.L.: Dialkylimidazolium chloroaluminate melts: a new class of room-temperature ionic liquids for electrochemistry, spectroscopy, and synthesis. Inorg. Chem. **21**, 1263–1264 (1982)
30. Wilkes, J.S., Zaworotko, M.J.: Air and water stable 1-ethyl-3-methylimidazolium based ionic liquids. Chem. Commun. **13**, 965–967 (1992)
31. Davis, J.H.: Task-specific ionic liquids. Chem. Lett. **33**, 1072–1077 (2004)
32. Robinson, J., Osteryoung, R.A.: An electrochemical and spectroscopic study of some aromatic hydrocarbons in the room temperature molten salt system aluminum chloride-n-butylpyridinium chloride. J. Am. Chem. Soc. **101**, 323–327 (1979)
33. Wilkes, J.S., Levisky, J.A., Pflug, J.L., Hussey, C.L., Scheffler, T.B.: Composition determinations of liquid chloroaluminate molten salts by nuclear magnetic resonance spectrometry. Anal. Chem. **54**, 2378–2379 (1982)
34. Wilkes, J.S., Zaworotko, M.J.: Air and water stable 1-ethyl-3-methylimidazolium based ionic liquids. J. Chem. Soc. Chem. Commun. **13**, 965–967 (1992)
35. Naert, P., Rabaey, K., Stevens, C.V.: Ionic liquid ion exchange: exclusion from strong interactions condemns cations to the most weakly interacting anions and dictates reaction equilibrium. Green Chem. **20**, 4277–4286 (2018)
36. Wasserscheid, P., Welton, T. (eds.): Ionic Liquids in Synthesis, 2nd edn, vol. 1, p. 367. Wiley-VCH Verlags GmbH & Co. KGaA, Weinheim (2008)
37. Plechkovaa, N.V., Seddon, K.R.: Applications of ionic liquids in the chemical industry. Chem. Soc. Rev. **37**, 123–150 (2008)
38. Fukaya, Y., Iizuka, Y., Sekikawa, K., Ohno, H.: Bio ionic liquids: room temperature ionic liquids composed wholly of biomaterials. Green Chem. **9**, 1155–1157 (2007)
39. Abbott, A.P., Boothby, D., Capper, G., Davies, D.L., Rasheed, R.K.: Deep eutectic solvents formed between choline chloride and carboxylic acids: versatile alternatives to ionic liquids. J. Am. Chem. Soc. **126**, 9142–9147 (2004)
40. MacFarlane, D.R., Pringle, J.M., Johansson, K.M., Forsyth, S.A., Forsyth, M.: Lewis base ionic liquids. Chem. Commun. **18**, 1905–1917 (2006)
41. Winkel, A., Reddy, P.V.G., Wilhelm, R.: Recent advances in the synthesis and application of chiral ionic liquids. Synthesis **7**, 999–1016 (2008)
42. Ohno, H., Fukumoto, K.: Amino acid ionic liquids. Acc. Chem. Res. **40**, 1122–1129 (2007)

References 167

43. Fukumoto, K., Ohno, H.: Design and synthesis of hydrophobic and chiral anions from amino acids as precursor for functional ionic liquids. Chem. Commun. 3081–3083 (2006)
44. Gadenne, B., Hesemann, P., Polshettiwar, V., Moreau, J.J.E.: Highly ordered functional organosilicas by template-directed hydrolysis-polycondensation of chiral camphorsulfonamide precursors. Eur. J. Inorg. Chem. 18, 3697–3702 (2006)
45. Bourissou, D., Guerret, O., Gabbai, F.P., Bertrand, G.: Stable carbenes. Chem. Rev. 100, 39–92 (2000)
46. Siriwardana, A.I., Crossley, I.R., Torriero, A.A.J., Burgar, I.M., Dunlop, N.F., Bond, A.M., Deacon, G.B., MacFarlane, D.R.: Methimazole-based ionic liquids. J. Org. Chem. 73, 4676–4679 (2008)
47. Hough, W.L., Smiglak, M., Rodriguez, H., Swatloski, R.P., Spear, S.K., Daly, D.T., Pernak, J., Grisel, J.E., Carliss, R.D., Soutullo, M.D., Davis, J.H., Rogers, R.D.: The third evolution of ionic liquids: active pharmaceutical ingredients. New J. Chem. 31, 1429–1436 (2007)
48. Schneider, S., Hawkins, T., Rosander, M., Vaghjiani, G., Chambreau, S., Drake, G.: Ionic liquids as hypergolic fuels. Energy Fuels 22, 2871–2872 (2008)
49. Bates, E.D., Mayton, R.D., Ntai, I., Davis, J.H.: CO_2 capture by a task-specific ionic liquid. J. Am. Chem. Soc. 124, 926–927 (2002)
50. Zhang, J.M., Zhang, F.J., Dong, K., Zhang, Y.Q., Shen, Y.Q., Lv, X.M.: Supported absorption of CO_2 by tetrabutylphosphonium amino acid ionic liquids. Chem. Eur. J. 12, 4021–4026 (2006)
51. Yoshizawa-Fujita, M., Johansson, K., Newman, P., MacFarlane, D.R., Forsyth, M.: Novel Lewis-base ionic liquids replacing typical anions. Tetrahedron Lett. 47, 2755–2758 (2006)
52. Cai, Y., Peng, Y., Song, G.: Amino-functionalized Ionic liquid as an efficient and recyclable catalyst for Knoevenagel reactions in water. Catal. Lett. 109, 61–64 (2006)
53. Li, W., Zhang, Z., Han, B., Hu, S., Song, J., Xie, Y., Zhou, X.: Switching the basicity of ionic liquids by CO_2. Green Chem. 10, 1142–1145 (2008)
54. Wasserscheid, P., Bosmann, A., Bolm, C.: Synthesis and properties of ionic liquids derived from the chiral pool. Chem. Commun. 3, 200–201 (2002)
55. Ishida, Y., Miyauchi, H., Saigo, K.: Design and synthesis of a novel imidazolium-based ionic liquid with planar chirality. Chem. Commun. 19, 2240–2241 (2002)
56. SciFinder-Chemical Abstract Service. http://scifinder.cas.org/. Accessed on 2 Oct 2021
57. Anastas, P., Warner, J.C.: Green Chemistry: Theory and Practice, p. 30. Oxford University Press, New York (1998)
58. Horvath, I.T., Anastas, P.T.: Innovations and green chemistry. Chem. Rev. 107, 2169–2173 (2007)
59. Feng, J., Loussala, H.M., Han, S., Ji, X., Li, C., Sun, M.: Recent advances of ionic liquids in sample preparation. Trends Anal. Chem. 125, 115–833 (2020)
60. Benedetto, A., Ballone, P.: Room temperature ionic liquids meet biomolecules: a microscopic view of structure and dynamics. ACS Sustain. Chem. Eng. 4, 392–412 (2016)
61. Krossing, I., Slattery, J.M., Daguenet, C., Dyson, P.J., Oleinikova, A., Weingartner, H.: Why are ionic liquids liquid? A simple explanation based on lattice and solvation energies. J. Am. Chem. Soc. 128, 13427–13434 (2006)
62. Huddleston, J.G., Visser, A.E., Reichert, W.M., Willauer, H.D., Broker, G.A., Rogers, R.D.: Characterization and comparison of hydrophilic and hydrophobic room temperature ionic liquids incorporating the imidazolium cation. Green Chem. 3, 156–164 (2001)
63. Seddon, K.R., Stark, A., Torres, M.J.: Influence of chloride, water, and organic solvents on the physical properties of ionic liquids. Pure Appl. Chem. 72, 2275–2287 (2000)
64. Rooney, D., Jacquemin, J., Gardas, R.: Thermophysical properties of ionic liquids. Top Curr. Chem. 290, 185–212 (2009)
65. Handy, S.T.: Room temperature ionic liquids: different classes and physical properties. Curr. Org. Chem. 9, 959–988 (2005)
66. Ngo, H.L., LeCompte, K., Hargens, L., McEwen, A.B.: Thermal properties of imidazolium ionic liquids. Thermochim. Acta. 357, 97–102 (2000)

67. Fayer, M.D.: Dynamics and structure of room temperature ionic liquids. Chem. Phy. Lett. **616**, 259–274 (2014)
68. Smiglak, M., Reichert, W.M., Holbrey, J.D., Wilkes, J.S., Sun, L.Y., Thrasher, J.S., Kirichenko, K., Singh, S., Katritzky, A.R., Rogers, R.D.: Combustible ionic liquids by design: is laboratory safety another ionic liquid myth? Chem. Commun. **24**, 2554–2556 (2006)
69. Chen, Y., Mu, T.: Thermal stability of ionic liquids. In: Zhang, S. (ed.) Encyclopedia of Ionic Liquids. Springer, Singapore (2020)
70. Maton, C., De Vos, N., Stevens, C.V.: Ionic liquid thermal stabilities: decomposition mechanisms and analysis tools. Chem. Soc. Rev. **42**, 5963–5977 (2013)
71. Siedlecka, E.M., Czerwicka, M., Stolte, S., Stepnowski, P.: Stability of ionic liquids in application conditions. Curr. Org. Chem. **15**, 1974–1991 (2011)
72. Wang, B., Qin, L., Mu, T., Xue, Z., Gao, G.: Are ionic liquids chemically stable? Chem. Rev. **117**, 7113–7131 (2017)
73. Neves, C.M., Kurnia, K.A., Coutinho, J.A., Marrucho, I.M., Lopes, J.N.C., Freire, M.G., Rebelo, L.P.N.: Systematic study of the thermophysical properties of imidazolium-based ionic liquids with cyano-functionalized anions. J. Phys. Chem. B. **117**, 10271–10283 (2013)
74. Chernikova, E.A., Glukhov, L.M., Krasovskiy, V.G., Kustov, L.M., Vorobyeva, M.G., Koroteev, A.A.E.: Ionic liquids as heat transfer fluids: comparison with known systems, possible applications, advantages and disadvantages. Russ. Chem. Rev. **84**, 875–890 (2015)
75. Paul, A., Muthukumar, S., Prasad, S.: Review—room-temperature ionic liquids for electrochemical application with special focus on gas sensors. J. Electrochem. Soc. **167**, 037511 (2020)
76. Hapiot, P., Lagrost, C.: Electrochemical reactivity in room-temperature ionic liquids. Chem. Rev. **108**, 2238–2264 (2008)
77. Matsumoto, H., Matsuda, T., Tsuda, T., Hagiwara, R., Ito, Y., Miyazaki, Y.: The application of room temperature molten salt with low viscosity to the electrolyte for dye-sensitized solar cell. Chem. Lett. **30**, 26–27 (2001)
78. Noda, A., Hayamizu, K., Watanabe, M.: Pulsed-gradient spin-echo ^{1}H and ^{19}F NMR ionic diffusion coefficient, viscosity, and ionic conductivity of non-chloroaluminate room-temperature ionic liquids. J. Phys. Chem. B **105**, 4603–4610 (2001)
79. Soman, D.P., Kalaichelvi, P., Radhakrishnan, T.K.: Thermal conductivity enhancement of aqueous ionic liquid and nanoparticle suspension. Braz. J. Chem. Eng. **36**, 855–868 (2019)
80. Marsh, K.N., Boxall, J.A., Lichtenthaler, R.: Room temperature ionic liquids and their mixtures—a review. Fluid Phase Equilib. **219**, 93–98 (2004)
81. Chiappe, C., Malvaldi, M., Pomelli, C.S.: Ionic liquids: solvation ability and polarity. Pure Appl. Chem. **81**, 767–776 (2009)
82. Reichardt, C.: Polarity of ionic liquids determined empirically by means of solvatochromic pyridinium N-phenolate betaine dyes. Green Chem. **7**, 339–351 (2005)
83. Bright, F.V., Baker, G.: Comment on "how polar are ionic liquids? Determination of the static dielectric constant of an imidazolium-based ionic liquid by microwave dielectric spectroscopy." J. Phys. Chem. B **110**, 5822–5823 (2006)
84. Chiappe, C., Pomelli, C.S., Rajamani, S.: Influence of structural variations in cationic and anionic moieties on the polarity of ionic liquids. J. Phys. Chem. B **115**, 9653–9661 (2011)
85. Wang, X., Chen, K., Yao, J., Li, H.: Recent progress in studies on polarity of ionic liquids. Sci. China Chem. **59**, 517–525 (2016)
86. Guan, W., Chang, N., Yang, L., Bu, X., Wei, J., Liu, Q.: Determination and prediction for the polarity of ionic liquids. J. Chem. Eng. Data **62**, 2610–2616 (2017)
87. Marullo, S., D'Anna, F., Rizzo, C., Billeci, F.: Ionic liquids: "normal" solvents or nanostructured fluids? Org. Biomol. Chem. **19**, 2076–2095 (2021)
88. (a) Tiwari, V.K., Kumar, A., Rajkhowa, S.: Green chemistry: opportunity in drug discovery research (part 1). Curr. Org. Chem. **25**, 1455–1456 (2021). (b) Tiwari, V.K., Kumar, A., Rajkhowa, S.: Green chemistry: opportunity in drug discovery research (part 2). Curr. Org. Chem. **25**, 2257–2259 (2021)

References

89. Wang, X., Ohlin, C.A., Lu, Q., Fei, Z.F., Hu, J., Dyson, P.J.: Cytotoxicity of ionic liquids and precursor compounds towards human cell line HeLa. Green Chem. **9**, 1191–1197 (2007)
90. Gathergood, N., Scammels, P.J., Garcia, M.T.: Biodegradable ionic liquids. Part III: The first readily biodegradable ionic liquids. Green Chem. **8**, 156–160 (2006)
91. Fukumotu, K., Yoshizawa, M., Ohno, H.: Room temperature ionic liquids from 20 natural amino acids. J. Am. Chem. Soc. **127**, 2398–2399 (2005)
92. Tao, G., He, L., Liu, W., Xu, L., Xiong, W., Wang, T., Kou, Y.: Preparation, characterization and application of amino acid-based green ionic liquids. Green Chem. **8**, 639–646 (2006)
93. Mena, I.F., Diaz, E., Palomar, J., Rodriguez, J.J., Mohedano, A.F.: Cation and anion effect on the biodegradability and toxicity of imidazolium- and choline-based ionic liquids. Chemosphere **240**, 124947 (2020)
94. Mc Farlane, D.R., Sun, J., Golding, J., Meakin, P., Forsyth, M.: High conductivity molten salts based on the imide ion. Electrochim. Acta **45**, 1271–1278 (2000)
95. Ganesh, I.: BMIM-BF$_4$ RTIL: synthesis, characterization and performance evaluation for electrochemical CO_2 reduction to CO over Sn and MoSi$_2$ cathodes. J. Carbon Res. **6**, 47 (2020). https://doi.org/10.3390/c6030047
96. Himmler, S., Horma, S., van Hal, R., Schulz, P.S., Wasserscheid, P.: Transesterification of methylsulfate and ethyl sulfate ionic liquids-an environmentally benign way to synthesize long-chain and functionalized alkyl sulfate ionic liquids. Green Chem. **8**, 887–894 (2006)
97. Kuhlmann, E., Himmler, S., Giebelhaus, H., Wasserscheid, P.: Imidazolium dialkyl phosphates—a class of versatile, halogen-free and hydrolytically stable ionic liquids. Green Chem. **9**, 233–242 (2007)
98. Glorius, F., Altenhoff, G., Goddard, R., Lehmann, C.: Oxazolines as chiral building blocks for imidazolium salts and N-heterocyclic carbene ligands. Chem. Commun. **22**, 2704–2705 (2002)
99. Handy, S.T., Okella, M., Dickenson, G.: Solvents from biorenewable sources: ionic liquids based on fructose. Org. Lett. **5**, 2513–2515 (2003)
100. Ebner, G., Schiehser, S., Potthast, A., Rosenau, T.: Side reaction of cellulose with common 1-alkyl-3-methylimidazolium-based ionic liquids. Tetrahedron Lett. **49**, 7322–7324 (2008)
101. Toh, H., Naka, K., Chujo, Y.: Synthesis of gold nanoparticles modified with ionic liquid based on the imidazolium cation. J. Am. Chem. Soc. **126**, 3026–3027 (2004)
102. Carmichael, A.J., Earle, M.J., Holbrey, J.D., McCormac, P.B., Seddon, K.R.: The heck reaction in ionic liquids: a multiphasic catalyst system. Org. Lett. **1**, 997–1000 (1999)
103. Earle, M.J., McCormac, P.B., Seddon, K.R.: Diels-alder reaction in ionic liquids. A safe recyclable alternative to lithium perchlorate-diethyl ether mixtures. Green Chem. **1**, 23–25 (1999)
104. (a) Liu, K.K.C., Sakya, S.M., O'Donnell, C.J., Flick, A.C., Li, J.: Synthetic approaches to the 2009 new drugs. Bioorg. Med. Chem. **19**, 1136–1154 (2011). (b) Liu, K.K.-C., Sakya, S.M., O'Donnell, C.J., Flick, A.C., Ding, H.X.: Synthetic approaches to the 2010 new drugs. Bioorg. Med. Chem. 20, 1155–1174 (2012). (c) Ding, H.X., Liu, K.K.-C., Sakya, S.M., Flick, A.C., O'Donnell, C.J.: Synthetic approaches to the 2011 new drugs. Bioorg. Med. Chem. 21, 2795–2825 (2013). (d) Ding, H.X., Leverett, C.A., Kyne, R.E., Liu, K.K.-C., Sakya, S.M., Flick, A.C., O'Donnell, C.J.: Synthetic approaches to the 2012 new drugs. Bioorg. Med. Chem. **22**, 2005–2032 (2014). (e) Ding, H.X., Leverett, C.A., Kyne, R.E., Liu, K.K.-C., Fink, S.J., Flick, A.C., O'Donnell, C.J.: Synthetic approaches to the 2013 new drugs. Bioorg. Med. Chem. **23**, 1895–1922 (2015). (f) Flick, A.C., Ding, H.X., Leverett, C.A., Kyne, R.E., Liu, K.K.-C., Fink, S.J., O'Donnell, C.J.: Synthetic approaches to the 2014 new drugs. Bioorg. Med. Chem. **24**, 1937–1980 (2016)
105. (a) Flick, A.C., Ding, H.X., Leverett, C.A., Kyne, R.E., Liu, K.K.-C., Fink, S.J., O'Donnell, C.J.: Synthetic approaches to the new drugs approved during 2015. J. Med. Chem. **60**, 6480−6515 (2017). (b) Flick, A.C., Ding, H.X., Leverett, C.A., Fink, S.J., O'Donnell, C.J.: Synthetic approaches to new drugs approved during 2016. J. Med. Chem. **61**, 7004−7031 (2018)

106. (a) Flick, A.C., Leverett, C.A., Ding, H.X., McInturff, E.L., Fink, S.J., Helal, C.J., O'Donnell, C.J.: Synthetic approaches to the new drugs approved during 2017. J. Med. Chem. **62**, 7340–7382 (2019). (b) Flick, A.C., Leverett, C.A., Ding, H.X., McInturff, E., Fink, S.J., Helal, C.J., DeForest, J.C., Morse, P.D., Mahapatra, S., O'Donnell, C.J.: Synthetic approaches to new drugs approved during 2018. Med. Chem. **63**, 10652–10704 (2020). (c) Flick, A.C., Leverett, C.A., Ding, H.X., Fink, S.J., Mahapatra, S., Carney, D.W., McInturff, E., Lindsey, E.A., Helal, C.J., DeForest, J.C., France, S.P., Berritt, S., Bigi-Botterill, S.V., Gibson, T.S., Liu, Y., O'Donnell, C.J.: Synthetic approaches to new drugs approved during 2019. Med. Chem. 64, 3604–3657 (2021)
107. (a) Knorr, L.: Einwirkung des diacetbernsteinsäureesters auf ammoniak und primäre aminbasen. Chem. Ber. **18**, 299–311 (1885). (b) Paal, C.: Synthese von thiophen- und pyrrolderivaten. Chem. Ber. **18**, 367–371 (1885)
108. Yadav, J.S., Reddy, B.V.S., Eeshwaraiah, B., Gupta, M.K.: Bi(OTf)$_3$/[bmim]BF$_4$ as novel and reusable catalytic system for the synthesis of furan, pyrrole and thiophene derivatives. Tetrahedron Lett. **45**, 5873–5876 (2004)
109. Meshram, H.M., Prasad, B.R.V., Kumar, D.A.: A green approach for efficient synthesis of N-substituted pyrroles in ionic liquid under microwave irradiation. Tetrahedron Lett. **51**, 3477–3480 (2010)
110. Shamekhi, S.S.: Microwave-assisted synthesis of diketopyrrolopyrrole pigments in ionic liquid. Pigm. Resin Technol. **42**, 215–222 (2013)
111. Kathiravan, S., Raghunathan, R.: Synthesis of pyrrolo[2,3-a]pyrrolizidino derivatives through intramolecular 1,3-dipolar cycloaddition in ionic liquid medium. Synth. Commun. **43**, 147–155 (2013)
112. Yadav, J.S., Reddy, B.V.S., Shubashree, S., Sadashiv, K., Naidu, J.: Ionic liquids-promoted multi-component reaction: green approach for highly substituted 2-aminofuran derivatives. Synthesis **14**, 2376–2380 (2004)
113. Shaabani, A., Soleimani, E., Darvishi, M.: Ionic liquid promoted one-pot synthesis of furo[2,3-d]pyrimidine-2,4(1H,3H)-diones. Monatsh. Chem. **138**, 43–46 (2007)
114. (a) Villemin, D., Liao, L.: Application of microwaves in organic synthesis: a rapid and efficient synthesis of new 3-aryl-2-imino-4-methyl-2,5-dihydrofurans and 3-aryl-3-2-(5H)-furanones. Synth. Commun. **33**, 1575–1586 (2003). (b) Villemin, D., Mostefa-Kara, B., Bar, N., Choukchou-Braham, N., Cheikh, N., Benmeddah, A., Hazimeh, H., Ziani-Cherif, C.: Base-promoted reactions in ionic liquid solvent: synthesis of butenolides. Lett. Org. Chem. **3**, 558–559 (2006)
115. Kim, B., Jeong, J., Lee, D., Kim, S., Yoon, H.-J., Lee, Y.-S., Cho, J.K.: Direct transformation of cellulose into 5-hydroxymethyl-2-furfural using a combination of metal chlorides in imidazolium ionic liquid. Green Chem. **13**, 1503–1506 (2011)
116. Hu, L., Sun, Y., Lin, L.: Efficient conversion of glucose into 5-hydroxymethylfurfural by chromium(III) chloride in inexpensive ionic liquid. Ind. Eng. Chem. Res. **51**, 1099–1104 (2012)
117. Qu, Y., Huang, C., Song, Y., Zhang, J., Chen, B.: Efficient dehydration of glucose to 5-hydroxymethylfurfural catalyzed by the ionic liquid, 1-hydroxyethyl-3-methylimidazolium tetrafluoroborate. Bioresour. Tech. **121**, 462–466 (2012)
118. Ma, H., Zhou, B., Li, Y., Argyropoulos, D.S.: Conversion of fructose to 5-hydroxymethyl-furfural with a functionalized ionic liquid. BioResources **7**, 533–544 (2012)
119. Guo, X., Cao, Q., Jiang, Y., Guan, J., Wang, X., Mu, X.: Selective dehydration of fructose to 5-hydroxymethylfurfural catalyzed by mesoporous SBA-15-SO$_3$H in ionic liquid [bmim]Cl. Carbohydr. Res. **351**, 35–41 (2012)
120. Hu, Y., Wei, P., Huang, H., Han, S.-Q., Ouyang, P.-K.: Synthesis of 2-aminothiophenes on ionic liquid phase support using the Gewald reaction. Synth. Commun. **36**, 1543–1548 (2006)
121. Kumar, A., Gupta, G., Srivastava, S.: Functional ionic liquid mediated synthesis (FILMS) of dihydrothiophenes and tacrine derivatives. Green Chem. **13**, 2459–2463 (2011)
122. Schmidt, A., Dreger, A.: Recent advances in the chemistry of pyrazoles. Properties, biological activities, and syntheses. Curr. Org. Chem. **15**, 1423–1463 (2011)

References 171

123. Isambert, N., Duque, M.M.S., Plaquevent, J.-C., Genisson, Y., Rodriguez, J., Constantieux, T.: Multicomponent reactions and ionic liquids: a perfect synergy for eco-compatible heterocyclic synthesis. Chem. Soc. Rev. **40**, 1347–1357 (2011)

124. Ebrahimi, J., Mohammadi, A., Pakjoo, V., Bahramzade, E., Habibi, A.: Highly efficient solvent-free synthesis of pyranopyrazoles by a Bronsted-acidic ionic liquid as a green and reusable catalyst. J. Chem. Sci. **124**, 1013–1017 (2012)

125. Shekouhy, M., Hasaninejad, A.: Ultrasound-promoted catalyst-free one-pot four component synthesis of 2*H*-indazolo[2,1-*b*]phthalazine-triones in neutral ionic liquid 1-butyl-3-methylimidazolium bromide. Ultrason. Sonochem. **19**, 307–313 (2012)

126. Safaei, S., Mohammadpoor-Baltork, I., Khosropour, A.R., Moghadam, M., Tangestaninejad, S., Mirkhani, V., Ki, R.: Application of a multi-SO$_3$H Brønsted acidic ionic liquid in water: a highly efficient and reusable catalyst for the regioselective and scaled-up synthesis of pyrazoles under mild conditions. RSC Adv. **2**, 5610–5616 (2012)

127. Siddiqui, S.A., Narkhede, U.C., Palimkar, S.S., Daniel, T., Lahoti, R.J., Srinivasan, K.V.: Room temperature ionic liquid promoted improved and rapid synthesis of 2,4,5-triaryl imidazoles from aryl aldehydes and 1,2-diketones or α-hydroxyketone. Tetrahedron **61**, 3539–3546 (2005)

128. Fang, D., Yang, J., Jiao, C.: Thermal-regulated PEG1000-based ionic liquid/PM for one-pot three-component synthesis of 2,4,5-trisubstituted imidazoles. Catal. Sci. Technol. **1**, 243–245 (2011)

129. Spitzer, W.A., Victor, F., Pollock, G.D., Hayes, J.S.: Imidazo[1,2-*a*]pyrimidines and imidazo[1,2-*a*]pyrazines: the role of nitrogen position in inotropic activity. J. Med. Chem. **31**, 1590–1595 (1988)

130. Parchinsky, V.Z., Shuvalova, O., Ushakova, O., Kravchenko, D.V., Krasavin, M.: Multicomponent reactions between 2-aminopyrimidine, aldehydes and isonitriles: the use of a nonpolar solvent suppresses formation of multiple products. Tetrahedron Lett. **47**, 947–951 (2006)

131. Xie, Y.-Y.: Organic reactions in ionic liquids: ionic liquid-accelerated one-pot synthesis of 2-arylimidazo[1,2-*a*]pyrimidine. Synth. Commun. **35**, 1741–1746 (2005)

132. Artyomov, V.A., Shestopalov, A.M., Litvinov, V.P.: Synthesis of imidazo[1,2-*a*]pyridines from pyridines and *p*-bromophenacyl bromide o-methyloxime. Synthesis 927–929 (1996)

133. Hou, R.-S., Wang, H.-M., Huang, H.-Y., Chen, L.-C.: Synthesis of imidazo[2,1-*a*]isoquinolines from α-tosyloxyketones and 1-aminoisoquinoline in ionic liquid solvent. J. Chin. Chem. Soc. **51**, 1417–1420 (2004)

134. Yadav, A.K., Kumar, M., Yadav, T., Jain, R.: An ionic liquid mediated one-pot synthesis of substituted thiazolidinones and benzimidazoles. Tetrahedron Lett. **50**, 5031–5034 (2009)

135. Mishra, A., Mishra, B.B., Tiwari, V.K.: Regioselective synthesis of novel isoxazole-linked glycoconjugates. RSC Adv. **5**, 41520–41535 (2015)

136. Martins, M.A.P., Frizzo, C.P., Moreira, D.N., Zanatta, N., Bonacorso, H.G.: Ionic liquids in heterocyclic synthesis. Chem. Rev. **108**, 2015–2050 (2008)

137. Wu, B., Wen, J., Zhang, J., Li, J., Xiang, Y.-Z., Yu, X.-Q.: One-pot Van Leusen synthesis of 4,5-disubstituted oxazoles in ionic liquids. Synlett 500–504 (2009)

138. Singh, A.S., Kumar, D., Tiwari, V.K.: Lewis acid-mediated benzotriazole ring cleavage (BtRC) strategy for the facile synthesis of 2-aryl benzoxazoles from *N*-acylbenzotriazoles. ACS Omega **2**, 5044–5051 (2017)

139. Fan, X., He, X., Wang, Y., Zhang, Y., Wang, X., Chin, J.: An efficient synthesis of 2-substituted benzoxazoles *via* RuCl$_3$·3H$_2$O catalyzed tandem reactions in ionic liquid. J. Chem. **29**, 773–777 (2011)

140. Nadaf, R.N., Siddiqui, S.A., Daniel, T., Lahoti, R.J., Srinivasan, K.V.: Room temperature ionic liquid promoted regioselective synthesis of 2-aryl benzimidazoles, benzoxazoles and benzthiazoles under ambient conditions. J. Mol. Catal. A: Chem. **214**, 155–160 (2004)

141. Saravanan, P., Corey, E.J., Short, A.: Stereocontrolled and practical synthesis of α-methylomuralide, a potent inhibitor of proteasome function. J. Org. Chem. **68**, 2760–2764 (2003)

142. Kamakshi, R., Reddy, B.S.R.: An efficient, eco-friendly, one-pot protocol for the synthesis of 2-oxazolines promoted by ionic liquid/indium chloride. Aust. J. Chem. **59**, 463–467 (2006)
143. Hodgetts, K.J., Kershaw, M.T.: Regiocontrolled synthesis of substituted thiazoles. Org. Lett. **4**, 1363–1365 (2002)
144. Hou, R.-S., Wang, H.-M., Tsai, H.-H., Chen, L.-C.: Synthesis of 2-phenylthiazoles from α-tosyloxyketones and thiobenzamide in [Bmim][PF$_6$] ionic liquid at ambient temperature. J. Chin. Chem. Soc. **53**, 863–866 (2006)
145. Kumar, D., Mishra, A., Mishra, B.B., Bhattacharya, S., Tiwari, V.K.: Synthesis of glycoconjugate benzothiazoles *via* cleavage of benzotriazole ring. J. Org. Chem. **78**, 899–909 (2013)
146. Kumar, D., Mishra, B.B., Tiwari, V.K.: Synthesis of 2-*N*/*S*/*C*-substituted benzothiazoles *via* intramolecular cyclative cleavage of benzotriazole ring. J. Org. Chem. **79**, 251–265 (2014)
147. Le, Z.-G., Xu, J.-P., Rao, H.-Y., Ying, M.: One-pot synthesis of 2-aminobenzothiazoles using a new reagent of [BMIM]Br 3 in [BMIM]B$_F$4. J. Heterocycl. Chem. **43**, 1123–1124 (2006)
148. Hassan, Z.-B., Arash, S., Abbas, Z., Kamal, G.: Highly efficient synthesis of 3,5-disubstituted 1,2,4-thiadiazoles using pentylpyridinium tribromide as a solvent/reagent ionic liquid. J. Sulfur Chem. **33**, 165–170 (2012)
149. Rostovtsev, V.V., Green, L.G., Fokin, V.V., Sharpless, K.B.: A stepwise Huisgen cycloaddition process: copper(I)-catalyzed regioselective ligation of azides and terminal alkynes. Angew. Chem. Int. Ed. **41**, 2596–2599 (2002)
150. Tornoe, C.W., Christensen, C., Meldal, M.: Peptidotriazoles on solid phase: 1,2,3-triazoles by regiospecific copper(I)-catalyzed 1,3-dipolar cycloadditions of terminal alkynes to azides. J. Org. Chem. **67**, 3057–3062 (2002)
151. Tiwari, V.K., Mishra, B.B., Mishra, K.B., Mishra, N., Singh, A.S., Chen, X.: Cu(I)-catalyzed click reaction in carbohydrate chemistry. Chem. Rev. **116**, 3086–3240 (2016)
152. Agrahari, A.K., Bose, P., Singh, A.S., Rajkhowa, S., Jaiswal, M.K., Hotha, S., Mishra, N., Tiwari, V.K.: Cu(I)-catalyzed click chemistry in glycoscience and their applications. Chem. Rev. **121**, 7638–7956 (2021)
153. Marra, A., Vecchi, A., Chiappe, C., Melai, B., Dondoni, A.: Validation of the copper(I)-catalyzed azide-alkyne coupling in ionic liquids: synthesis of a triazole-linked *C*-disaccharide as a case study. J. Org. Chem. **73**, 2458–2461 (2008)
154. Keshavarz, M., Karami, B., Ahmady, A.Z., Ghaedi, A., Vafaei, H.: [bmim]BF$_4$/[Cu(Im^{12})$_2$]CuCl$_2$ as a novel catalytic reaction medium for click cyclization. C. R. Chim. **17**, 570–576 (2014)
155. Hassan, V., Hamid, G., Manzar, M.: Facile synthesis of benzotriazole derivatives using nanoparticles of organosilane-based nitrite ionic liquid immobilized on silica and two room-temperature nitrite ionic liquids. Synth. Commun. **43**, 2801–2808 (2013)
156. Margarita, A.E., Alexander, S.K., Nikolai, V.I., Schulte, M., Makhova, N.N.: Ionic liquid-assisted synthesis of 5-monoand 1,5-disubstituted tetrazoles. Mendeleev Commun. **21**, 334–336 (2011)
157. Li, T.-J., Yao, C.-S., Yu, C.-X., Wang, X.-S., Tu, S.-J.: Ionic liquid-mediated one-pot synthesis of 5-(trifluoromethyl)-4,7-dihydrotetrazolo[1,5-*a*]pyrimidine derivatives. Synth. Commun. **42**, 2728–2738 (2012)
158. Khosropour, A.R., Iraj, M.-B., Forough, K.: Green, new and efficient tandem oxidation and conversion of aryl alcohols to 2,4,6-triarylpyridines promoted by [HMIM]NO$_3$-[BMIM]BF$_4$ as a binary ionic liquid. C. R. Chim. **14**, 441–445 (2011)
159. Karthikeyan, G., Perumal, P.T.: Ionic liquid promoted simple and efficient synthesis of β-enamino esters and *β*-enaminones from 1,3-dicarbonyl compounds. One-pot, three-component reaction for the synthesis of substituted pyridines. Can. J. Chem. **83**, 1746–1751 (2005)
160. Mohammad, R., Poor, H., Farnazalsadat, F.: Ultrasound-promoted synthesis of 2-amino-6-(arylthio)-4-arylpyridine-3,5-dicarbonitriles using ZrOCl$_2$·8H$_2$O/NaNH$_2$ as the catalyst in the ionic liquid [bmim]BF$_4$ at room temperature. Tetrahedron Lett. **52**, 6779–6782 (2011)

References

161. Sobhani, S., Honarmand, M.: 2-Hydroxyethylammonium acetate: a reusable task-specific ionic liquid promoting one-pot, three-component synthesis of 2-amino-3,5-dicarbonitrile-6-thio-pyridines. C. R. Chim. **16**, 279–286 (2013)
162. Ranu, B.C., Banerjee, S.: Ionic liquid as catalyst and reaction medium. The dramatic influence of a task-specific ionic liquid, [bmim]OH, in Michael addition of active methylene compounds to conjugated ketones, carboxylic esters, and nitriles. Org. Lett. **7**, 3049–3052 (2005)
163. Huang, Z., Hu, Y., Zhou, Y., Shi, D.: Efficient one-pot three-component synthesis of fused pyridine derivatives in ionic liquid. ACS Comb. Sci. **13**, 45–49 (2011)
164. Chavan, S.S., Degani, M.S.: Ionic liquid catalyzed 4,6-disubstituted-3-cyano-2-pyridone synthesis under solvent-free conditions. Catal. Lett. **141**, 1693–1697 (2011)
165. Moreira, D.N., Frizzo, C.P., Longhi, K., Soares, A.B., Marzari, M.R.B., Buriol, L., Brondani, S., Zanatta, N., Bonacorso, H.G., Martins, M.A.P.: Ionic liquid and Lewis acid combination in the synthesis of novel (E)-1-(benzylideneamino)-3-cyano-6-(trifluoromethyl)-1*H*-2-pyridones. Monatsh. Chem. **142**, 1265–1270 (2011)
166. Shi, D.Q., Ni, S.N., Yang, F., Ji, S.J.: An efficient and green synthesis of 3,3′-benzylidenebis(4-hydroxy-6-methylpyridin-2(1*H*)-one) derivatives through multi-component reaction in ionic liquid. J. Heterocycl. Chem. **45**, 1275–1280 (2008)
167. de Graaff, C., Ruijter, E., Orru, R.V.A.: Recent developments in asymmetric multicomponent reactions. Chem. Soc. Rev. **41**, 3969–4009 (2012)
168. Reddy, B.P., Rajesh, K., Vijayakumar, V.: Ionic liquid [EMIM]OAc under ultrasonic irradiation towards synthesis of 1,4-DHP's. J. Chin. Chem. Soc. **58**, 384–388 (2011)
169. Shaabani, A., Rezayan, A.H., Rahmati, A., Sharifi, M.: Ultrasound-accelerated synthesis of 1,4-dihydropyridines in an ionic liquid. Monatsh. Chem. **137**, 77–81 (2006)
170. Yadav, J.S., Reddy, B.V.S., Basak, A.K., Narsaiah, A.V.: Three-component coupling reactions in ionic liquids: an improved protocol for the synthesis of 1,4-dihydropyridines. Green Chem. **5**, 60–63 (2003)
171. Sridhar, R., Perumal, P.T.: A new protocol to synthesize 1,4-dihydropyridines by using 3,4,5-trifluorobenzeneboronic acid as a catalyst in ionic liquid: synthesis of novel 4-(3-carboxyl-1H-pyrazol-4-yl)-1,4-dihydropyridines. Tetrahedron **61**, 2465–2470 (2005)
172. Prola, L.D.T., Buriol, L., Frizzo, C.P., Caleffi, G.S., Marzari, M.R.B., Moreira, D.N., Bonacorso, H.G., Zanatta, N., Martins, M.A.P.: Synthesis of novel quinolines using TsOH/ionic liquid under microwave. J. Braz. Chem. Soc. **23**, 1663–1668 (2012)
173. Palimkar, S.S., Siddiqui, S.A., Daniel, T., Lahoti, R.J., Srinivasan, K.V.: Ionic liquid-promoted regiospecific Friedlander annulation: novel synthesis of quinolines and fused polycyclic quinolines. J. Org. Chem. **68**, 9371–9378 (2003)
174. Karthikeyan, G., Perumal, P.T.: A mild, efficient and improved protocol for the Friedländer synthesis of quinolines using lewis acidic ionic liquid. J. Heterocycl. Chem. **41**, 1039–1041 (2004)
175. Wan, Y., Yuan, R., Zhang, F.-R., Pang, L.-L., Ma, R., Yue, C.-H., Lin, W., Yin, W., Bo, R.-C., Wu, H.: One-pot synthesis of N2-substituted 2-amino-4-aryl-5,6,7,8-tetrahydroquinoline-3-carbonitrile in basic ionic liquid [bmim]OH. Synth. Commun. **41**, 2997–3015 (2011)
176. Wang, X., Zhang, M., Jiang, H., Yao, C., Tu, S.: Three-component green synthesis of N-arylquinoline derivatives in ionic liquid [Bmim$^+$][BF$_4$$^-$]: reactions of arylaldehyde, 3-arylamino-5,5-dimethylcyclohex-2-enone, and active methylene compounds. Tetrahedron **63**, 4439–4449 (2007)
177. Zare, A.K., Abi, F., Moosavi-Zare, A.R., Beyzavi, M.H., Zolfigol, M.A.: Synthesis, characterization and application of ionic liquid 1,3-disulfonic acid imidazolium hydrogen sulfate as an efficient catalyst for the preparation of hexahydroquinolines. J. Mol. Liq. **178**, 113–121 (2013)
178. Shi, D.-Q., Ni, S.-N., Yang, F., Shi, J.-W., Dou, G.-L., Li, X.-Y., Wang, X.-S., Ji, S.-J.: An efficient synthesis of pyrimido[4,5-b]quinoline and indeno[2′,1′:5,6]pyrido[2,3-*d*]pyrimidine derivatives *via* multicomponent reactions in ionic liquid. J. Heterocycl. Chem. **45**, 693–702 (2008)

179. Li, Y.-L., Zhang, M.-M., Wang, X.-S., Shi, D.-Q., Tu, S.-J., Wei, X.-Y., Zong, Z.-M.: One pot three component synthesis of 9-arylpolyhydroacridine derivatives in an ionic liquid medium. J. Chem. Res. 600–604 (2005)
180. Shen, W., Wang, L.-M., Tian, H., Tang, J., Yu, J.-J.: Brønsted acidic imidazolium salts containing perfluoroalkyl tails catalyzed one-pot synthesis of 1,8-dioxo-decahydroacridines in water. J. Fluorine Chem. **130**, 522–527 (2009)
181. Das, S., Majee, A., Hajra, A.: A convenient synthesis of coumarins using reusable ionic liquid as catalyst. Green Chem. Lett. Rev. **4**, 349–353 (2011)
182. Khaligh, N.G.: Synthesis of coumarins *via* Pechmann reaction catalyzed by 3-methyl-1-sulfonic acid imidazolium hydrogen sulfate as an efficient, halogen-free and reusable acidic ionic liquid. Catal. Sci. Technol. **2**, 1633–1636 (2012)
183. Sarda, S.R., Pathan, M.Y., Paike, V.V., Pachmase, P.R., Jadhav, W.N., Pawar, R.P.: A facile synthesis of flavones using recyclable ionic liquid under microwave irradiation. ARKIVOC **xvi**, 43–48 (2006)
184. Ganguly, A.K., Mahata, P.K., Biswas, D.: Synthesis of oxygen heterocycles. Tetrahedron Lett. **47**, 1347–1349 (2006)
185. Du, Z., Ng, H., Zhang, K., Zeng, H., Wang, J.: Ionic liquid mediated Cu-catalyzed cascade oxa-Michael-oxidation: efficient synthesis of flavones under mild reaction conditions. Org. Biomol. Chem. **9**, 6930–6933 (2011)
186. Peng, J., Deng, Y.: Ionic liquids catalyzed Biginelli reaction under solvent-free conditions. Tetrahedron Lett. **42**, 5917–5919 (2001)
187. Kang, L.-Q., Jin, D.-Y., Cai, Y.-Q.: Silica-supported ionic liquid Si-[SbSipim][PF$_6$]: an efficient catalyst for the synthesis of 3,4-dihydropyrimidine-2-(1H)-ones. Synth. Commun. **43**, 1896–1901 (2013)
188. Hajipour, A.R., Seddighi, M.: Pyridinium-based Brønsted acidic ionic liquid as a highly efficient catalyst for one-pot synthesis of dihydropyrimidinones. Synth. Commun. **42**, 227–235 (2012)
189. Sharma, N., Sharma, U.K., Kumar, R., Richa, Sinha, A.K.: Green and recyclable glycine nitrate (GlyNO$_3$) ionic liquid triggered multicomponent Biginelli reaction for the efficient synthesis of dihydropyrimidinones. RSC Adv. **2**, 10648–10651 (2012)
190. Gholap, A.R., Venkatesan, K., Daniel, T., Lahott, R.J., Srinivasan, K.V.: Ionic liquid promoted novel and efficient one pot synthesis of 3,4-dihydropyrimidin-2-(1H)-ones at ambient temperature under ultrasound irradiation. Green Chem. **6**, 147–150 (2004)
191. Dadhania, A.N., Patel, V.K., Raval, D.K.: A facile approach for the synthesis of 3,4-dihydropyrimidin-2-(1H)-ones using a microwave promoted Biginelli protocol in ionic liquid. J. Chem. Sci. **124**, 921–926 (2012)
192. Hajipour, A.R., Khazdooz, L., Zarei, A.: Brønsted acidic ionic liquid–catalyzed one-pot synthesis of 3,4-dihydropyrimidin-2(1H)-ones and thiones under solvent-free conditions. Synth. Commun. **41**, 2200–2208 (2011)
193. Putilova, E.S., Troitskii, N.A., Zlotin, S.G.: Reaction of aromatic aldehydes with β-dicarbonyl compounds in a catalytic system: piperidinium acetate-1-butyl-3-methylimidazolium tetrafluoroborate ionic liquid. Russ. Chem. Bull., Int. Ed. 54, 1233–1238 (2005)
194. Singh, P., Kumari, K., Dubey, M., Vishvakarma, V.K., Mehrotra, G.K., Pandey, N.D., Chandra, R.: Ionic liquid catalyzed synthesis of 7-phenyl-1,4,6,7-tetrahydro-thiazolo[5,4-d]pyrimidine-2,5-diones. C. R. Chim. **15**, 504–510 (2012)
195. Hajipour, A.R., Ghayeb, Y., Sheikhan, N., Ruoho, A.E.: Brønsted acidic ionic liquid as an efficient and reusable catalyst for one-pot three-component synthesis of pyrimidinone derivatives *via* Biginelli-type reaction under solvent-free conditions. Synth. Commun. **41**, 2226–2233 (2011)
196. Basiri, A., Murugaiyah, V., Osman, H., Kumar, R.S., Kia, Y., Awang, K.B., Ali, M.A.: An expedient, ionic liquid mediated multi-component synthesis of novel piperidone grafted cholinesterase enzymes inhibitors and their molecular modeling study. Eur. J. Med. Chem. **67**, 221–229 (2013)

References

197. Karthikeyan, P., Kumar, S.S., Arunrao, A.S., Narayan, M.P., Bhagat, P.R.: A novel amino acid functionalized ionic liquid promoted one-pot solvent-free synthesis of 3,4-dihydropyrimidin-2-(1H)-thiones. Res. Chem. Int. **39**, 1335–1342 (2013)
198. Hsu, H.-Y., Tseng, C.-C., Matii, B., Sun, C.-M.: Ionic liquid-supported synthesis of dihydroquinazolines and tetrahydroquinazolines under microwave irradiation. Mol. Diversity **16**, 241–249 (2012)
199. Patil, V.S., Padalkar, V.S., Chaudhari, A.S., Sekar, N.: Intrinsic catalytic activity of an acidic ionic liquid as a solvent for quinazoline synthesis. Catal. Sci. Technol. **2**, 1681–1684 (2012)
200. Shaabani, A., Rahmati, A., Rad, J.M.: Ionic liquid promoted synthesis of 3-(2'-benzothiazolo)-2,3-dihydroquinazolin-4(1H)-ones. C. R. Chim. **11**, 759–764 (2008)
201. Dabiri, M., Baghbanzadeh, M., Delbari, A.S.: Novel and efficient one-pot tandem synthesis of 2-styryl-substituted 4(3H)-quinazolinones. J. Comb. Chem. **10**, 700–703 (2008)
202. Khosropour, A.R., Mohammadpoor-Baltork, I., Gohrbankhani, H.: Bi(TFA)$_3$–[nbp]FeCl$_4$: a new, efficient and reusable promoter system for the synthesis of 4(3H)-quinazolinone derivatives. Tetrahedron Lett. **47**, 3561–3564 (2006)
203. Connolly, D.J., Cusack, D., O'Sullivan, T.P., Guiry, P.J.: Synthesis of quinazolinones and quinazolines. Tetrahedron **61**, 10153–10202 (2005)
204. Ghahremanzadeh, R., Ahadi, S., Bazgir, A.: A one-pot, four-component synthesis of α-carboline derivatives. Tetrahedron Lett. **50**, 7379–7381 (2009)
205. Cox, E.D., Cook, J.M.: The Pictet-Spengler condensation: a new direction for an old reaction. Chem. Rev. **95**, 1797–1842 (1995)
206. Joshi, R.A., Muthukrishnan, M., More, S.V., Garud, D.R., Ramana, C.V., Joshi, R.R., Joshi, R.A.: Pictet-Spengler cyclization in room temperature ionic liquid: a convenient access to tetrahydro β-carbolines. J. Heterocycl. Chem. **43**, 767–772 (2006)
207. Daugan, A., Grondin, P., Ruault, C., de Gouville, A.C.L., Coste, H., Kirilovsky, J., Hyafil, F., Labaudiniere, R.: The discovery of Tadalafil: a novel and highly selective PDE5 inhibitor. 1: 5,6,11,11a-tetrahydro-1H-imidazo[1′,5′:1,6]pyrido[3,4-b]indole-1,3(2H)-dione analogues. J. Med. Chem. **46**, 4525–4532 (2003)
208. Tseng, M.-C., Liang, Y.-M., Chu, Y.-H.: Synthesis of fused tetrahydro-β-carbolinequinoxalinones in 1-n-butyl-2,3-dimethylimidazolium bis(trifluoromethylsulfonyl)imide ([bdmim][Tf$_2$N]) and 1-n-butyl-2,3-dimethylimidazolium perfluorobutylsulfonate ([bdmim][PFBuSO$_3$]) ionic liquids. Tetrahedron Lett. **46**, 6131–6135 (2005)
209. Yen, Y.H., Chu, Y.H.: Synthesis of tetrahydro-β-carbolinediketopiperazines in [bdmim][PF$_6$] ionic liquid accelerated by controlled microwave heating. Tetrahedron Lett. **45**, 8137–8140 (2004)
210. Prins, H.J.: Condensation of formaldehyde with some unsaturated compounds. Chem. Weekblad **16**, 1072–1073 (1919)
211. Yadav, J.S., Reddy, B.V.S., Bhaishya, G.: InBr 3–[bmim]P$_F$6: a novel and recyclable catalytic system for the synthesis of 1,3-dioxane derivatives. Green Chem. **5**, 264–266 (2003)
212. Tateiwa, J., Kimura, A., Takasuka, M., Uemura, S.: Metal cation-exchanged montmorillonite (M^{n+}-Mont)-catalysed carbonyl-ene reactions. J. Chem. Soc. Perkin Trans. 1, 2169–2174 (1997)
213. Miyashita, M., Awen, B.Z.E., Yoshikoshi, A.: A new synthetic aspect of acetic nitronic anhydrides. Tetrahedron **46**, 7569–7586 (1990)
214. Kitazume, T., Zulfiqar, F., Tanaka, G.: Molten salts as a reusable medium for the preparation of heterocyclic compounds. Green Chem. **2**, 133–136 (2000)
215. Soleimani, E., Khodaei, M.M., Koshvandi, A.T.K.: Three-component, one-pot synthesis of benzo[b][1,4]oxazines in ionic liquid 1-butyl-3-methylimidazolium bromide. Synth. Commun. **42**, 1367–1371 (2012)
216. Fang, D., Yang, L.-F., Yang, J.-M.: Synthesis of 1,2-dihydro-1-arylnaphtho(1,2-e)(1,3)oxazine-3-one catalyzed by pyridinium based-ionic liquid. Res. Chem. Intermed. **39**, 2505–2512 (2013)

217. Yadav, L.D.S., Rai, V.K., Yadav, B.S.: The first ionic liquid-promoted one-pot diastereoselective synthesis of 2,5-diamino-/2-amino-5-mercapto-1,3-thiazin-4-ones using masked amino/mercapto acids. Tetrahedron **65**, 1306–1315 (2009)
218. Tandon, V., Mishra, A.K., Chhikara, B.S.: KF-alumina immobilized in ionic liquids: a novel heterogeneous base for heterocyclization of alkylsulfanylphenylamines into 1,4-benzothiazine. Heterocycles **63**, 1057–1065 (2004)
219. Miyano, S., Abe, N., Sumoto, K., Teramoto, K.: Reactions of enamino-ketones. Part II. Synthesis of 4H-1,4-benzothiazines. J. Chem. Soc. Perkin Trans. 1, 1146–1149 (1976)
220. Fukue, Y., Ishikawa, M., Mizusawa, K., Tanaka, N.: Methods for modifying 1,3,5-triazine derivatives. World Patent WO 1997024338
221. Piesch, S.D., Wolf, A., Sinsel, S.: Modifiziertes melaminharz, seine herstellung und seine verwendung sowie das modifizierungsmittel. German Patent DE3512446 (1986)
222. Peng, Y., Song, G.: Microwave-assisted clean synthesis of 6-aryl-2,4-diamino-1,3,5-triazines in [bmim][PF$_6$]. Tetrahedron Lett. **45**, 5313–5316 (2004)
223. Sharifi, A., Abaee, M.S., Rouzgard, M., Mirzaei, M.: Ionic liquid [omim][NO$_3$], a green medium for room-temperature synthesis of benzothiazinone derivatives in one pot. Green Chem. Lett. Rev. **5**, 649–698 (2012)
224. Eligeti, R., Kundur, G.R., Atthunuri, S.R., Modugu, N.R.: Brønsted acidic ionic liquid [HMIm]BF$_4$ promoted simple and efficient one-pot green synthesis of isoxazolyl-1,3-benzoxazines at ambient temperature. Green Chem. Lett. Rev. **5**, 699–705 (2012)
225. Moghaddam, M., Bazgir, M., Mehdi, A.M., Ghahremanzadeh, A., Ramin: Alum (KAl(SO$_4$)$_2$·12H$_2$O) catalyzed multicomponent transformation: simple, efficient, and green route to synthesis of functionalized spiro[chromeno[2,3-d]pyrimidine-5,3′-indoline]-tetraones in ionic liquid media. Chin. J. Chem. **30**, 709–714 (2012)
226. Jin, S.-S., Wang, H., Guo, H.-Y.: Ionic liquid catalyzed one-pot synthesis of novel spiro-2-amino-3-phenylsulfonyl-4H-pyran derivatives. Tetrahedron Lett. **54**, 2353–2356 (2013)
227. Rajesh, S.M., Bala, B.D., Perumal, S.: Multi-component, 1,3-dipolar cycloaddition reactions for the chemo-, regio- and stereoselective synthesis of novel hybrid spiroheterocycles in ionic liquid. Tetrahedron Lett. **53**, 5367–5371 (2012)
228. Dandia, A., Jain, A.K., Sharma, S.: An efficient and highly selective approach for the construction of novel dispiro heterocycles in guanidine-based task-specific [TMG][Ac] ionic liquid. Tetrahedron Lett. **53**, 5859–5863 (2012)
229. Jain, R., Sharma, K., Kumar, D.: Ionic liquid mediated 1,3-dipolar cycloaddition of azomethine ylides: a facile and green synthesis of novel dispiro heterocycles. Tetrahedron Lett. **53**, 1993–1997 (2012)
230. Dandia, A., Jain, A.K., Laxkar, A.K., Bhati, D.S.: Synthesis and stereochemical investigation of highly functionalized novel dispirobisoxindole derivatives via [3+2] cycloaddition reaction in ionic liquid. Tetrahedron **69**, 2062–2069 (2013)
231. Nezhad, A.K., Mohammadi, S.: Magnetic, acidic, ionic liquid-catalyzed one-pot synthesis of spirooxindoles. ACS Comb. Sci. **15**, 512–518 (2013)
232. Jain, R., Sharma, K., Kumar, D.: One-pot, three-component synthesis of novel spiro[3H-indole-3,2′-thiazolidine]-2,4′(1H)-diones in an ionic liquid as a reusable reaction media. Helv. Chim. Acta **96**, 414–418 (2013)
233. Dandia, A., Jain, A.K.: Ionic liquid-mediated facile synthesis of novel spiroheterobicyclic rings as potential antifungal and antibacterial drugs. J. Heterocycl. Chem. **50**, 104–113 (2013)

Chapter 5
Growing Impact of Ionic Liquids in Carbohydrate Chemistry

1 Introduction

This is well evidenced through the examples dealing with the wide application of ionic liquids in the development of diverse range of biologically relevant heterocyclic compounds [1–3]. A number of widely explored RTILs are itself contains heterocyclic skeletons such as imidazoles, morpholines, pyridines, triazoles, etc. and their impact in the high-yielding synthesis of five-membered heterocycles, six-membered heterocycles, fused-heterocycles, and also spiro heterocycles under environmentally benign condition is exponentially growing in recent years [1–4]. Furthermore, ILs have great applications in the wide range of catalysis [5–8] and other emerging areas like enzyme-induced synthesis [9], energy storage [10], extraction and separation of bioactive molecules [11, 12], API-IL concept in drug development [13]. In continuation of previous chapter on growing impact of ILs in the synthesis of pharmacologically active heterocyclic skeletons, we now extended their widespread applications in carbohydrate chemistry. In addition to heterocycles, carbohydrates have also been identified as promising scaffold for their great impact in drug discovery and development [14, 15].

Carbohydrates play a pivotal role in biological systems and also involve in energy conversion and storage in cells. Interestingly, their application as foods, pharmaceuticals and drugs, personal care products, surfactants, and chemical feed stocks makes them widely studied biomolecules with numerous future aspects [14–21]. This most abundant biomolecule bears pathological and physiological importance due to the ability of adhesion, cellular recognition, invasion, migration, communication, etc. [16–19] Furthermore, carbohydrates possess fascinating structural features including chiral pool [22–25] and thus can be considered to be an excellent scaffold for the practical synthesis of carbohydrate-based chiral ionic liquids (CILs), a cheap and biocompatible candidate of diverse applications [26].

Owing to the clean and environment benign (green) nature, RTILs have attracted the attention of synthetic chemists for last two decades, and they serve as polar green media in the synthesis of diverse range of carbohydrate-containing molecules for

© The Author(s), under exclusive license to Springer Nature Singapore Pte Ltd. 2022
V. K. Tiwari et al., *Green Chemistry*,
https://doi.org/10.1007/978-981-19-2734-8_5

various purposes [27–31]. However, a plethora of research has been carried out on mixtures of RTIL-biomolecule, their interactions, complex phases, behaviors, applications in biomedicine and nanotechnology; a systematic study on the importance of RTILs will provide insights at their role in dissolution, modulation, glycosylation, functionalization, and modification of carbohydrates. This chapter discusses all these aspects of RTILs, their challenges, and future perspectives in emerging issues in science and technology in special reference to their impact in carbohydrate chemistry.

2 Representative Example for the Synthesis of Carbohydrate-Based Chiral ILs

A few basic ILs derived from naturally occurring amino acids display interesting properties such as biodegradability [32]. In recent years, a number of research groups successfully explored readily available carbohydrates particularly monosaccharides as chiral scaffold to develop scale-up synthesis of chiral and biorenewable ionic liquids [26, 33–35]. A sugar-based functional IL **5** could be obtained in high-yield starting from phenyl 2,3,4-tribenzyl-1-thio-β-glucopyranoside **1**. [33] Thus, β-thioglycoside **1** was first subjected to acylation reaction with bromoacetic acid using DCC/DMAP and ester **2** thus obtained on further reaction with *N*-methyl imidazole **3** followed by treatment with sodium tetrafluoroborate in acetone at rt gave [mim] [BH$_4$] anchored glycoside **4** in 98% yield. Compound **4** on oxidation using *m*-chloroperbenzoic acid in CH$_2$Cl$_2$ at $-78\,^{\circ}$C afforded respective IL-anchored sulfoxide **5** as diastereomeric mixture (due to sulfoxide chirality) in 97% yield (Scheme 1) [33]. The resulted [mim] [BH$_4$]-tagged sulfoxide is fruitfully used as donor in glycosylation reaction for the expeditious synthesis of oligosaccharides.

Few chiral IL helps to form diastereomeric salts which were successfully used as a suitable chiral medium in asymmetric synthesis and also in the resolution of enantiomers. Few representative chiral ILs displayed promising catalytic activities for the asymmetric Aldol reaction of acetone with 4-nitrobenzaldehyde under optimized condition [34, 35].

Handy and coworkers developed a biorenewable RTIL **9** from D-Fructose. Monosaccharide **6** was first reacted with NH$_3$/CH$_2$O using catalytic amount of CuCO$_3$ followed by alkylation under basic condition to give imidazole analogues (**7** and **8**) as an inseparable mixture of regioisomers in 9:1 ratio [36]. These analogues were then further alkylated with methyl iodide in DCM to give the respective iodide salt **9a**, which at the end could be converted to the target ILs **9b–f** in good-to-excellent yield (Scheme 2).

Ebner and coworkers exemplified 1-^{13}C-D-glucopyranose **10** which on treatment with butylmethylimidazolium acetate **11** under basic condition at room temperature for 7 days gave the desired D-glucopyranose-based butylmethylimidazolium ion IL **12** and IL **13** (Scheme 3) [37].

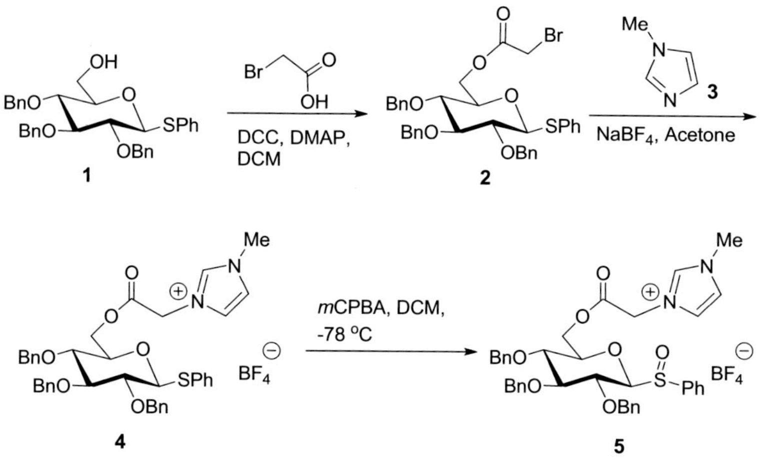

3 Application of RTILs in Carbohydrate Chemistry

Scheme 1 Synthesis of IL-anchored chiral sulfoxide **5** from β-thioglycoside **1**

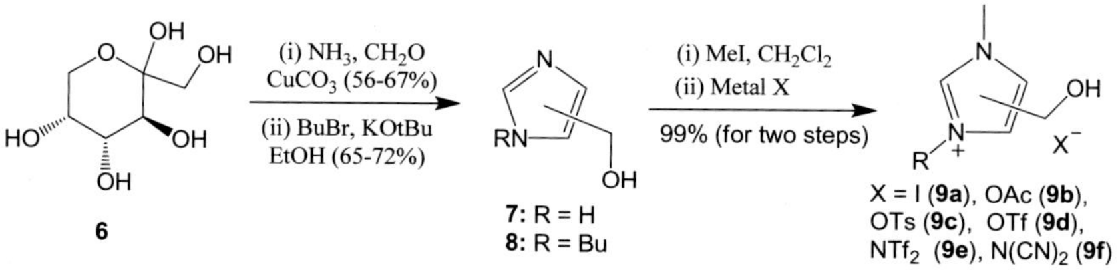

Scheme 2 Functional ILs (**9a-f**) derived from common monosaccharide, D-fructose

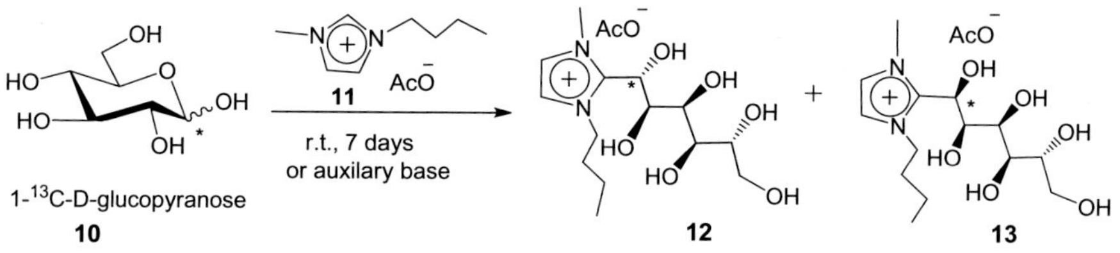

Scheme 3 Synthesis of 1-^{13}C-D-glucopyranose-linked butylmethylimidazolium-based ILs (**12**, **13**)

3 Application of RTILs in Carbohydrate Chemistry

Carbohydrate, an important components of biomolecules, plays imperative roles in a number of biological events notably, cell–cell communication, cell adhesion, cell proliferation, protein folding, and many more in metabolic systems [14–19]. In addition, they are involved in energy storage (e.g., starch and glycogen). Also, they may contain an aldehyde (an aldose) or a ketone (ketoses) and mainly due to the presence of hydroxyl groups, they simply participate in hydrogen-bonding phenomena in the aqueous environment as large number. Furthermore, they may contain also amino, carboxyl, phospho, and sulfo functional groups that eventually can increases their hydrophilicity and consequently enable interaction with the ILs. In general, ILs are nonflammable and also nonvolatile nature, and thus, their utilities in carbohydrate chemistry such as promising solvents for its dissolution, structural chemistry in modern spectroscopic and analytical methods, functionalization, further modifications, glycosylation, and imperative reactions in glycoscience are ever increasing in recent years [27–31]. RTILs have successfully been used as environmental benign solvents and/or catalysts to convert carbohydrates (both chemically and enzymatically) into new versatile chemicals and materials with a number of potential applications such as building materials, textiles, as well paper.

3.1 Dissolution and Gelation of Carbohydrates in ILs

ILs is known to improve the solubility of carbohydrates and thus may increase the potency of sugar-derived analogues to be useful as drug candidate. Solubility of different carbohydrates in various ILs is depicted in Table 1 [27]. The solubility of various monosaccharides (e.g., xylose, fructose, and glucose) and disaccharides (e.g., sucrose) were investigated in [hmIm] Cl, ranging from 0.005 g/mL (for sucrose) to 0.062 g/mL (for fructose) [38].

In 1934, the dissolution of cellulose in molten N-ethylpyridinium chloride in the presence of base was known for the first time [39]. Later on, its high m.p. (118–120 °C) was reduced to 77 °C by mixing it with 50% DMSO in order to get a novel cellulose solution [40]. Despite of the solubility of saccharides in DMSO and DMF, environmental toxicity, non-compatibility with sugars and moreover the deactivation of enzyme used in enzymatic esterification are few serious limitations that hurdle for their wide utilities [30].

Molecular weight, size of carbohydrates, and other factors such as different types of inter and intermolecular interactions remarkably affect the solubility of carbohydrates in ILs. Fascinatingly, a sugar which is insoluble in an IL having the same imidazolium cation may be soluble in other anion (e.g., hexafluorophosphate), and this way RTILs may showed tunable properties. Swatloski and coworkers demonstrated the dissolution of crystalline cellulose **14** in ILs (Fig. 1) [41].

The solubility of cellulose in BmimCl was 10 wt% that could be increased to 25 wt% under microwave-assisted condition [42]. A number of such ILs with [HCOO], [CH$_3$(CH$_2$)$_{0-3}$COO], [PhCOO], [HSCH$_2$COO], and [(MeO)(R)PO$_2$] anions were effective when dissolving cellulose at room temperature. The solubility of cellulose in ILs with [BF$_4$], [PF$_6$], [N(CN)$_2$], [MeSO$_4$], [HSO$_4$], and [Tf$_2$N] counter ions was less than the ILs having counter anions of superior hydrogen-bonding capacity [43, 44]. While the side chain attached with methylimidazolium cation in ILs can also influence the dissolution of cellulose [45, 46]. In general, enhanced solubility of cellulose was noticed with ILs having symmetrical cation and also with long-chain

3 Application of RTILs in Carbohydrate Chemistry

Table 1 Solubility of different carbohydrates in various ILs [27]

Carbohydrate	Solvent	Solubility (g/L)	T (°C)
Glucose	$[C_1OCH_2mIm][N(TFMS)_2]$	0.5	25
	$[C_1OCH_2mIm][BF_4]$	4.4	25
	$[C_1OC_2H_4mIm][BF4]$	5	55
	$[C_1OCH_2mIm][N(CN)_2]$	66	25
	$[C_1OC_2H_4mIm][N(CN)_2]$	91	25
	$[C_2OC_2H_4mIm][N(CN)_2]$	70	25
	$[C_4mIm][N(CN)_2]$	145	25
	$[C_4mIm][Cl]$	50	70
Fructose	$[C_4mIm][Cl]$	50	70
	$[C_4mIm][Cl]$	560	110
	$[C_4m_2Im][Cl]$	400	120
Lactose	$[C_4mIm][N(CN)_2]$	51/225	25/75
Sucrose	$[C_1OC_2H_4mIm][TFMS]$	2.1	25
	$[C_1OCH_2mIm][N(CN)_2]$	249/352	25/60
	$[C_1OC_2H_4mIm][N(CN)_2]$	220	25
	$[C_2OC_2H_4mIm][N(CN)_2]$	50/240	25/60
	$[C_4mIm][N(CN)_2]$	195/282	25/60
	$[C_4mIm][Cl]$	50	70
	$[C_4mIm][Cl]$	180	110
	$[C_4m_2Im][Cl]$	140	120
α-CD	$[C_1OCH_2mIm][Br]$	350	–
β-CD	$[C_4mIm][N(CN)_2]$	450	75
	Water	18.5	25
Agarose	$[C_1OCH_2mIm][Br]$	20	–
Amylose	$[C_4mIm][N(CN)_2]$	4	25
	$[C_1OCH_2mIm][Br]$	30	–
	Water	<0.5	25
Amylopectin	$[C_4mIm][Cl]$	50	70

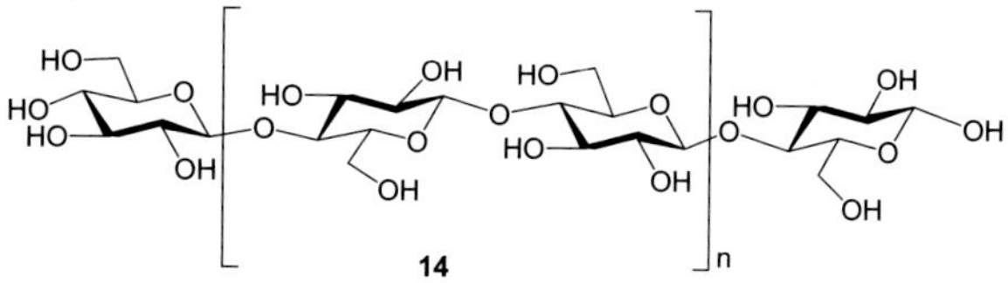

Fig. 1 Structure of typical cellulose (polymer chain with $n = 400$–1000) [41]

alkyl group [47]. [Emim], [Amim], and [Bmim] cations are even more efficient in ILs for dissolving cellulose. The hydroxyl group when present with C4 side chain in IL displayed the maximum dissolution ability [48]. Thus it is obvious to conclude that both anions and cations certainly participate in the dissolution process (Fig. 2).

A **gelation** phenomenon is noticed during the dissolution of carbohydrates. Thus, a solution of agarose in ether-based RTILs when cooled to room temperature then it resulted to the formation of 'Ionogel'. [49] Such materials displayed the properties of hydrogels (ionic character) and also organogels [50]. They have potential utilities

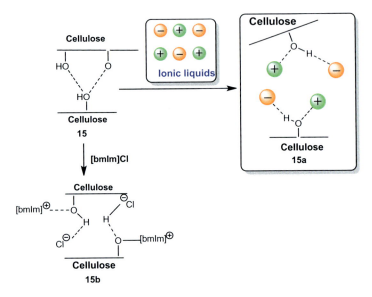

Fig. 2 Possible dissolution of cellulose in [bmim] based ILs

in electrolytic membranes, actuators, biosensors, separation membranes, lithium-ion batteries, and fuel cells [51]. ILs was commonly used as media to develop any suitable gels. Homogeneous gels with improved thermal stabilities and mechanical properties could be easily obtained and displayed strong interactions between the ions of ILs and polysaccharide. A common route for the development of cellulose-based ion gels is depicted in Fig. 3. [Bmim] Cl is identified as an effective IL media for breaking down the strong intra- and intermolecular interactions occurs in cellulose.

Ether-based RTILs (commonly known as sugar-philic ILs) are the potential solvents for the effective dissolution of glycolipids [49]. Glycolipids **17** and **18** dissolve in ether-based ILs on heating glycolipid **17** is insoluble in aqueous media at low conc (1 mM), while compound **18** having three amide and flexible ether linkage is a water-soluble amphiphile. Gelation was noticed when glycolipid **18** (Fig. 4) was

Solid cellulose-based ion gel **16**

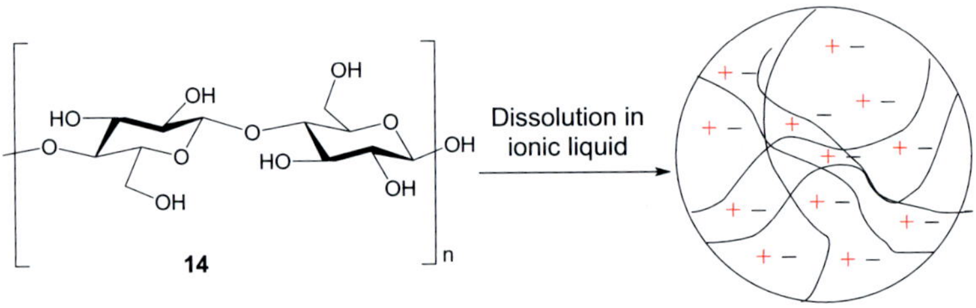

Fig. 3 Common route for the development of cellulose containing gel **16**

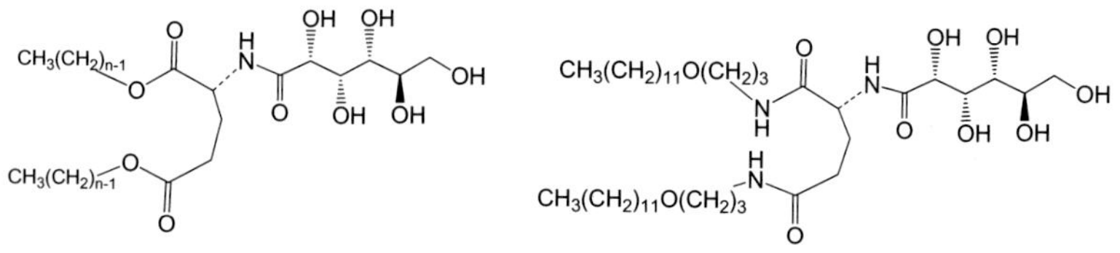

3 Application of RTILs in Carbohydrate Chemistry 183

Fig. 4 Chemical structure of glycolipids 17 and 18

dissolved in both ether-linked ILs at conc above 10 mM, mainly due to the formation of fibrous nanostructures. Other lipids when evaluated in ILs, it could not result in gelation that confirms the impact of carbohydrate in gelation in ILs [52].

3.2 IL-Mediated Some Common Reactions in Carbohydrate Chemistry

In addition, the remarkable solubility issue of carbohydrates in ILs also had known for their notable catalytic ability in a broad range of important chemical and enzymatic reactions in carbohydrate chemistry including common protection–deprotection method and glycosylation reactions [27–31].

ILs is widely used for a rapid peracetylation of simple and protected monosaccharides under mild reaction condition [53, 54]. Forsyth and coworkers reported high-yielding O-acetylation of sugars in ILs such as [emIm] dicyanamide and [bmIm] dicyanamide [53]. Monosaccharides, e.g., α-D-glucose, methyl-β-glucopyranoside, N-acetylneuraminic acid, disaccharide, e.g., sucrose, and the trisaccharide, e.g., raffinose on peracetylation afforded the acetylated products in 98%, 92%, 72%, 93%, and 90% yield, respectively. The sulfated sugars, due to their solubility issue in organic solvents, are difficult to peracetylate. However, sulfated sugars are soluble in [emIm][ba]. Thus, Murugesan and coworkers used benzoate-based ILs and devised a practical and high-yielding method for the peracetylation of sulfated monosaccharides [55].

Likewise, perbenzoylation of a number of carbohydrates including α-D-Glu, β-D-Glu, and α- & β- combination of D-Man can be done by treating them with benzoate-based RTILs [56]. In addition, ILs is also used for the per-O-acetylation and also for the regioselective benzylidene ring-opening of some common carbohydrate analogues [57]. Benzylidenation is another very important protection of carbohydrates particularly, hexopyranosides. Zhang and coworkers used imidazolium-based ILs (e.g., [bmIm]BF$_4$, [emIm]BF$_4$, [bmIm]PO$_4$, and [bmIm]MeSO$_4$) in the benzylidenation of methyl glycosides [54].

Cyclic 1,2-orthoesters of carbohydrates are used as suitable protecting groups for C_1-OH and C_2-OH group. They are used as glycosyl donors in a number of glycosylation reactions and considered as valuable building blocks for an easy access of biologically potent carbohydrate-containing molecules. *n*-Pentenyl orthoesters on treatment with a Lewis acid resulted to a stable dioxolenium ions. The classical synthesis of 1,2-orthoesters **21** include the reaction of peracetylated or perbenzoylated glycosyl bromides **19** with alcohol **20** in presence of silver triflate or quaternary ammonium salts. The moderate yield and reflux condition for 24 h limit the reaction, while the similar reaction performed in [C_4mIm][PF_6] using 2,6-lutidine required 8 h (Scheme 4) [58].

Tiwari and coworkers reported a stereoselective synthesis of bicyclic iminosugar, such as 1-deoxy-norcastanospermine starting from readily available D-glucose. The key step of the reaction was *N*-methylmorpholinium hydrogen sulfate [NMM][HSO$_4$][-] catalyzed Michael addition of benzyl amine to glycosyl olefinic ester **22** for 15 min at room temperature to give respective glycosylated amino esters **23a** and **23b** (in 80:20 ratio) in 95% yield (Scheme 5) [59]. The similar reaction in the absence of IL required longer reaction time (18–24 h) [60–62]. This sugar amino ester serves as promising scaffold and well explored in drug discovery and development [63, 64].

Furthermore, the same IL namely [NMM][HSO$_4$][-] is used for one-pot high-yielding synthesis of glycosyl carboxamides **26**. Thus, aldehydes **24** on reaction with primary and secondary amine **25** in presence of diacetoxyiodobenzene using

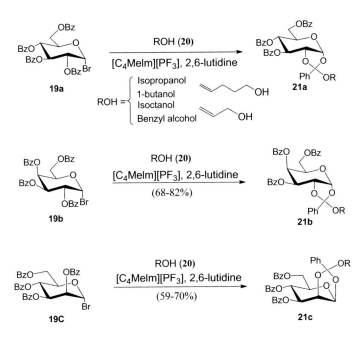

Scheme 4 ILs mediated synthesis of cyclic 1,2-orthoesters [58]

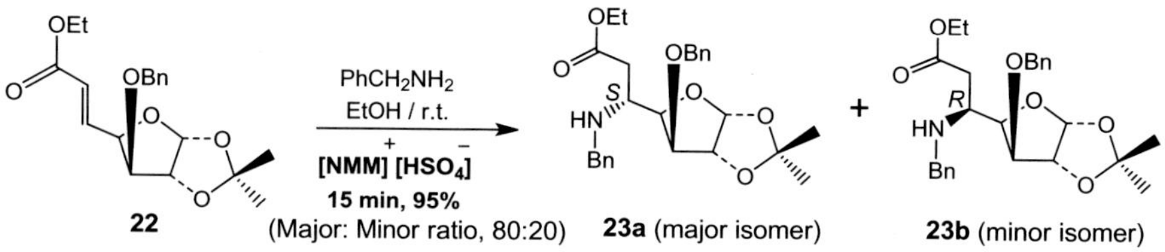

3 Application of RTILs in Carbohydrate Chemistry

Scheme 5 [NMM]$^+$[HSO$_4$]$^-$ mediated Michael addition of amine to glycosyl olefinic ester [58]

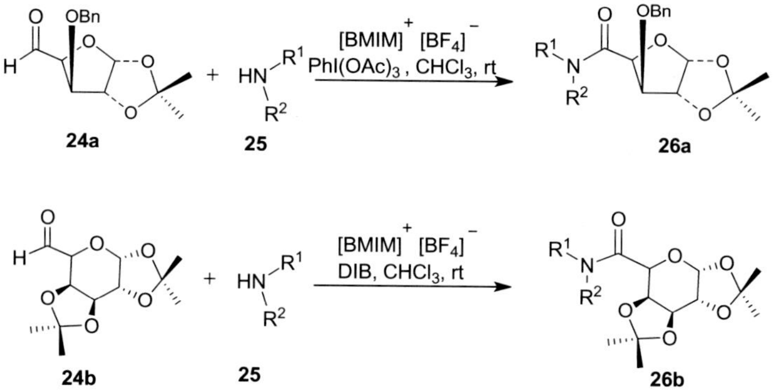

Scheme 6 [NMM]$^+$[HSO$_4$]$^-$ mediated synthesis of glycosyl carboxamide **26a,b** from glycosylated aldehydes **24a,b** [65]

catalytic amount of IL in anhydrous chloroform at room temperature afforded high-to-excellent yield of desired sugar-containing carboxamides **26** (Scheme 6) [65]. The method is considered as an environmentally benign protocol as it avoids the use of toxic metals.

Carbohydrates, because of their fascinating structural features including chiral pool, were considered to be a good candidate for the synthesis preparation of carbohydrate-based ILs [26]. Ferlin and coworkers used regioselective CuAAC tool and developed a glycosyl triazolium-based ionic liquid **30** from D-xylose [66]. β-Propargyl xyloside **27** was first subjected to CuAAC conjugation with hexyl azide **28** using CuSO$_4$/NaAsc in THF/water at r.t. for 18 h and glycohybrid **29** thus obtained on acetate removal followed by reaction with trimethyloxonium tetrafluoroborate afforded high yield of biocompatible D-xylose-based IL **30** (Scheme 7) [66].

Likewise, CuAAC click reaction of 1,2,5,6-di-*O*-isopropylidine-3-*O*-propargyl-α-D-glucofuranose **31** with 1-azido-2,3,4,6-tetra-*O*-acetyl-β-D-glucose **32** using CuI/DIPEA catalytic system afforded good yield of glycohybrid triazole **33**. The triazole **33** was then reacted with CAN followed by reaction with iodomethane to give required triazolium-based chiral IL **34**. This glycosyl 1,2,3-triazolium IL **34** is reported to catalyze the amination of aryl halides using aqueous ammonia solution at 90 °C in THF (Scheme 8) [67].

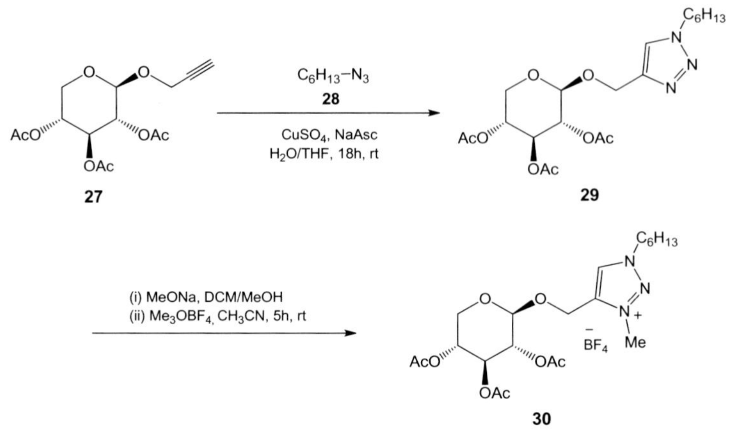

Scheme 7 CuAAC 'Click Chemisty' mediated D-xylose derived biocompatible IL **30** [66]

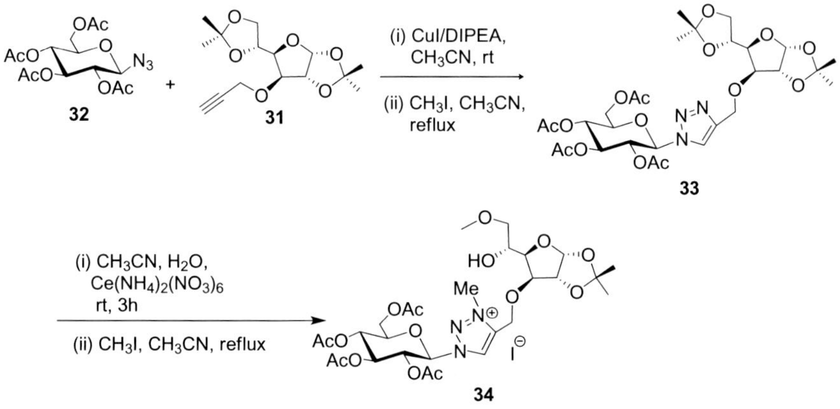

Scheme 8 CuAAC-inspired chiral triazolium IL **34**, suitable for amination reaction [67]

ILs is also used for the development of value-aided product starting from widely available carbohydrates. Toward this end, Lansalot-Matras and coworkers reported a useful method for dehydration of fructose **6** using hydrophobic [bmIm][PF$_6$] and hydrophilic [bmIm][BF$_4$] RTILs in DMSO to give 5-hydroxymethylfurfural (HMF, **35**) (Scheme 9) [68]. Similar dehydration of Fructose using polyethylene glycol-400-functionalized dicationic acidic ILs (PEG400-DAILs) afforded high yield of HMF [69]. This product is useful starting material for the development of nonpetroleum-derived polymers.

A number of interesting biopolymer composite was developed through simple functionalization of cellulose **14** in ILs, for example **36**. Such biopolymers consist of polar (hydroxyl) groups and displayed hydrophilic behavior (Scheme 10) [70]. IL [bmim][Cl] **37a** operate as the cellulose solvent, and IL **37b** having an acrylate

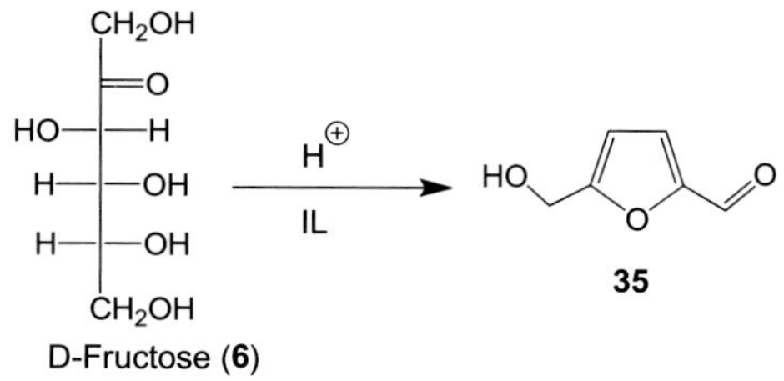

3 Application of RTILs in Carbohydrate Chemistry

Scheme 9 Dehydration of fructose to HMF in presence of ILs [69]

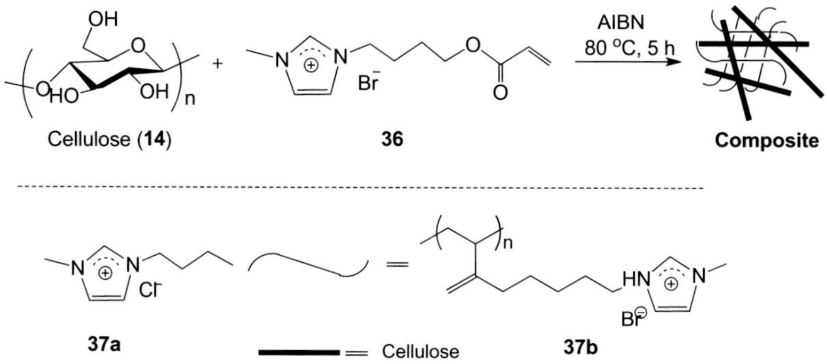

D-Fructose (**6**)

35

Cellulose (**14**) + **36** → Composite

AIBN
80 °C, 5 h

37a

━━━ = Cellulose

37b

Scheme 10 Cellulose/IL polymer composite [70]

group in the side chain exemplified as a suitable monomer for the polymerization. The resulted composite consists of cellulose with IL-linked biopolymer strands [70]. In another investigation, one side chain of imidazolium IL contains a vinyl group, while other chain contains a styrene residue, and this way it facilitates to act as both cellulose solvent and required monomer [71].

Cationic functionalization of cellulose product is widely used for the successful removal of anionic dyes from textile effluents. A eutectic mixture of choline chloride and urea (1:2 ratio) is used to dissolve cellulose (Scheme 11) [72]. Thus, cellulose **14** was treated with NaOH in eutectic medium for 15 h and 90 °C to afford biopolymer **38** having half of the nitrogen bound as a quaternary ammonium group. The product formed (was found to be efficient in removing a typical water-soluble dye, orange II [72]. The (bmIm–OAc)-mediated chemistry with cellulose was further extended

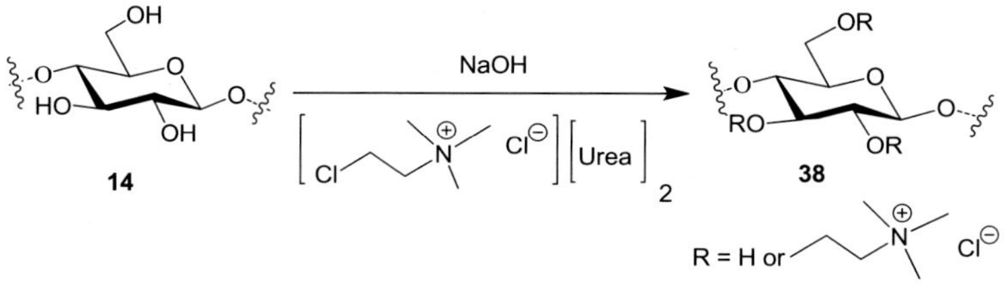

14

NaOH

38

R = H or

Scheme 11 Cationic modification of cellulose in choline-urea eutectic mixture by Ebner and coworkers [37]. Prasad and coworkers reported a cellulose-based ionic porous material using templating technique and oil/IL emulsion in the presence of sorbitan monooleate [73].

3.3 Ionic Liquids in Enzyme-Induced Carbohydrate Modifications

To achieve the desired selectivity in products under environmentally benign condition, a number of enzymatic reactions are well explored in presence various ionic liquids as green media or catalyst or both. In addition to other, oxido-reductases and hydrolases preserve their activities in the suspensions of suitable ILs [74]. Since last two decades; RTILs received great attention in chemistry, biology and material sciences. They serve significant features such as non-toxic, biocompatibility, environmental benign polar media, and compatibility with functional group in carbohydrate analogues and thus widely explored in glycoscience particularly for dissolution, modulation, glycosylation, functionalization, and modification of carbohydrates [27–31]. Their impact is rather growing in various plethora of research on RTIL-biomolecule, their interactions, behaviors, complex phases, and notable applications in biomedical science [7, 13]. Some selected anions in a particular ILs including PF_6^-, BF_4^-, and NTf_2^- and few more are more suitable in enzyme-induced organic reactions over the other anions, e.g. Cl^-, NO_3^-, Tf^-, ACO^-, and $CF_3CO_2^-$, etc. [75]. The dissimilarity about the activities of such anions in a particular ILs in enzymatic reaction is mainly due to their difference in the ability to participate in hydrogen bonding. Because of the less obstruction in internal hydrogen bonding with a particular enzyme, their compatibility in catalysis is determined and noticed. On the other hand, the hydrogen-bonding ability of ILs is further known to enhance the dissolution of glycopolymers, e.g., cellulose anions. Consequently, the selection about the suitability of IL is an extremely important feature in any enzyme-induced modification of carbohydrate or their derived analogues.

As a representative example dealing with enzyme catalysis with carbohydrates in presence of ILs, a fundamental lipase-mediated enantio- and regioselective acylation of hydroxyl groups in a particular sugar can be considered. Poor or many times extremely low solubility of carbohydrates in common non-aqueous organic solvents frequently hinders the lipase-mediated enzymatic esterification reaction. This problem was solved in great extent once the lipase-mediated acylation is performed in presence of dialkyl imidazolium-based ILs (Scheme 12) [76]. Lipases are in general inactivated in polar organic solvents such as N-methylformamide and methanol. The selective 6-acylation of glucose **39** under lipase-mediated acylation reaction is rather difficult when performed in conventional organic solvents. As a result of second acylation of 6-acyl product **40**, a 3,6-diacyl analogue **41** is also obtained in this reaction. The poor reaction yield and also poor regioselectivity in this acylation reaction is possibly due to the poor solubility of D-glucose. However,

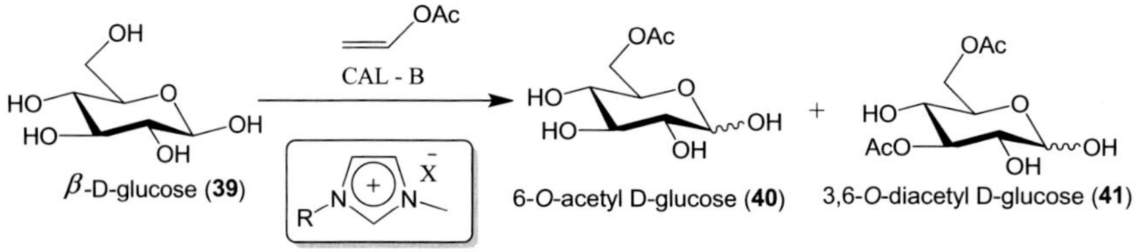

3 Application of RTILs in Carbohydrate Chemistry

Scheme 12 Lipase-catalyzed acylation of β-D-Glucose in dialkyl imidazolium ILs

when this reaction is carried out in 1-methoxy ethyl-3-methyl imidazolium [BF$_4$]$^-$, the regioselectivity of 6-O-acetyl glucopyranoside **40** is improved to be optimum and the reaction yield was almost quantitative (99%). The authors claim that the solubility of D-glucose is nearly 100 times more in ILs in comparison to THF or acetone [76]. AcOH was obtained as a byproduct in acylation reaction (in trace amount) that is accountable for the anomerization of sugar at higher reaction temperature. Likewise, regioselective acylation of maltose monohydrate afforded product in 50% yield. In fact, the improvement in regioselectivity of the lipase-mediated reaction in IL is mainly due to an increase in the solubility of reactant in concerned ILs.

Murugesan and coworkers reported the peracetylation of carbohydrates in IL where the yield was quantitative even when the reactant was insoluble. IL-mediated biphasic reaction was proceed which was quite useful to afford excellent yields of products [55]. The detailed investigation of Murugesan and Park Group might imitate the fundamental divergence between chemical and enzymatic reactions [55, 76].

Candida rugosa lipase was further explored as an enzyme catalyst in ILs for the selective acylation of monoprotected glycosides including α- and β-methyl-6-O-trityl glucoside **42** and galactoside **44** to give the respective products **43** and **45** (Scheme 13) [77].

Lipase-mediated acylation works superior when performed in ILs, e.g., [bmim][pf$_6$] and [moemim][PF$_6$]) than conventional organic solvents. The acylation of α-glycosides proceeded smoothly (3–11 h) to afford high regioselectivity (>98:1, 2-OAc: 3-OAc) in comparison to the β-glycosides (50–120 h and 9:1 to

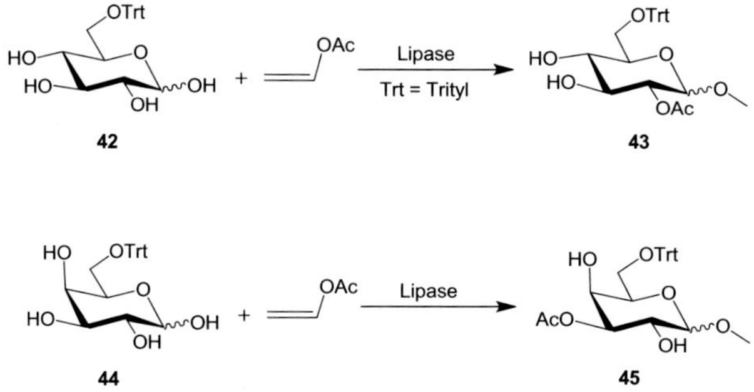

Scheme 13 Lipase-catalyzed acylation of monoprotected glycosides

190   5 Growing Impact of Ionic Liquids in Carbohydrate Chemistry

95:5 selectivity). ILs was recycled and reused in subsequent reactions without any notable drop in reaction yields (~92%). Park and coworkers attributed the improved reactivity in IL to higher solubility of substrates in more polar RTIL solvents [76].

Carbohydrate fatty acid esters are non-ionic surfactants and were widely explored in various emerging fields including pharmaceutical, cosmetic, and food industry [78]. Again, the poor solubility of fatty acid in organic solvents, it is rather difficult to synthesize them from respective carbohydrate and fatty acid. Ganske and coworkers reported a practical synthesis of glucose fatty acid esters (**48**) by reacting glucose with fatty acids (**46**) or their vinyl esters (**47**) in ILs as the reaction media (Scheme 14) [79]. Thus, poly(ethylene glycol)-modified *C. antarctica* lipase B (CAL-B)-mediated acylation of monosaccharide when carried out in [BMIM][BF$_4$] and [BMIM] [PF$_6$], the desired glucose fatty acid esters were obtained respectively in 30% and 35% yields. Furthermore, fatty acid as the acyl donors and commercial CAL-B-mediated route when carried out in an IL with 40% *t*-BuOH, fatty acid vinyl esters were obtained in 89% isolated yield. By using this lipase-mediated protocol in ILs/*t*-BuOH, glucose fatty acid ester **48** was obtained with quantitative conversion.

Oxidase–peroxidase-mediated sulfoxidation is quite convenient reaction once performed in suitable ILs. Peroxidases could be deactivated if hydrogen peroxide used in little excess and for that reason, hydrogen peroxide is suggested to be added slowly just to maintain the catalytic activity required to attain the stereoselectivity in reaction. Okrasa and coworkers reported a glucose oxidase-mediated oxidation of glucose (**39**) with O$_2$ to gluconic acid (**49**). Interestingly, the peroxidase needed to convert H$_2$O$_2$ (steadily generated in situ during the reaction) again utilized for the oxidation of sulfides **50** to respective chiral sulfoxides **51** (Scheme 15) [80]. 1-Butyl,3-methyl imidazolium hexafluorophosphate increases the solubility for both organic substrates and oxygen in comparison to the similar reaction in aqueous media and thus found more advantageous in peroxidase chemistry [81].

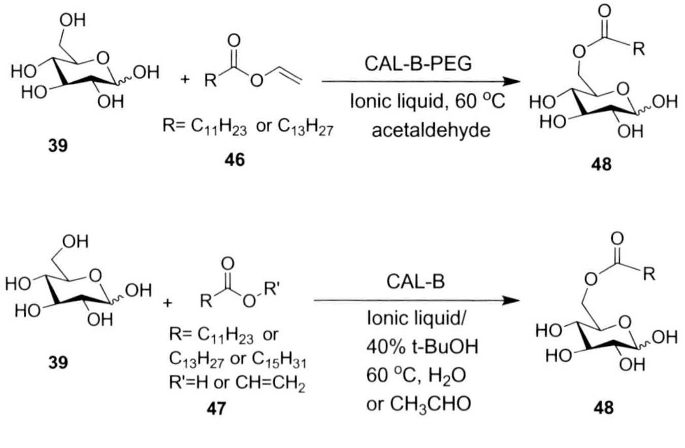

Scheme 14 IL-CAL-B-PEG mediated synthesis of glucose fatty acid ester (**48**) in ILs

3 Application of RTILs in Carbohydrate Chemistry

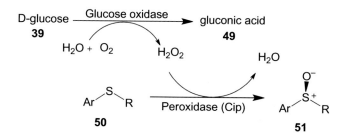

Scheme 15 Oxidase–peroxidase mediated asymmetric sulfoxidation reaction

In general, the performance of an enzyme-induced reaction in non-aqueous media together with ILs usually depends on the water activity of reaction media [82, 83]. Different percentages of water in IL can be optimized for the conversion of glucose to gluconic acid, where 10% v/v water in IL gave 66% yield. Reaction temperature when increased from r.t. to 40 °C, release of H_2O_2 increases, and yield was dropped to 46%, and thus, the excess of H_2O_2 inhibits the enzyme (peroxidase) activity.

3.4 ILs in Glycosidic Bond Formation Methodology

In carbohydrate chemistry, the most important area of investigation is the "Glycosidic Bond Forming Methodology" [84]. Since last two decades, a number of such glycosylation reactions were performed in presence of RTILs to attain desired selectivities in glycosylation products [55, 85–89].

Park and coworkers investigated the glycosylation of unprotected and unactivated monosaccharides (including D-Man, D-Glu, D-GalNAc, D-Gal) with acceptors **52** in 1-ethyl-3-methylimidazolium benzoate [emIm][ba] and Amberlite IR-120 (H⁺) resin as promoter and reported a one-step method for an easy access of *O*-glycosides **53** including oligosaccharides with α-selectivity (Scheme 16) [90]. Similar method using [BMIM][OTf] as a recyclable solvent and $Sc(OTf)_3$ at 80 °C was explored to develop α-(1 → 6)-linked galactose oligosaccharides [91, 92].

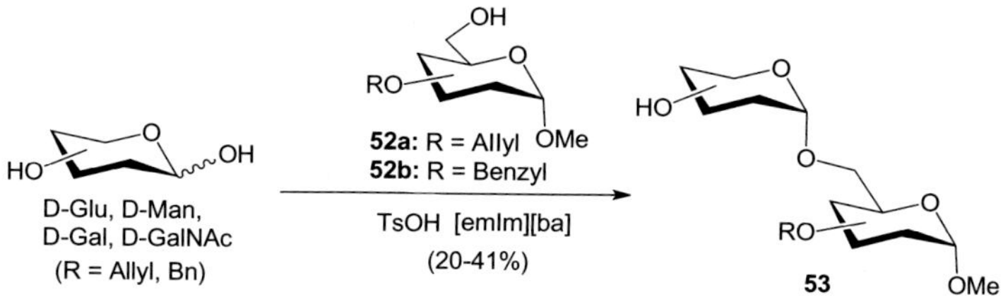

Scheme 16 α-Glycosylation of unprotected monosaccharides using [emIm][ba] [90]

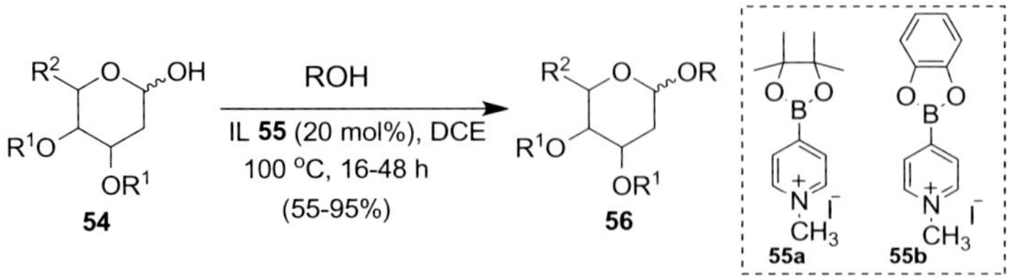

Scheme 17 Boronic acid mediated dehydrative glycosylation [93]

Taylor and coworkers used N-methylpyridinium-4-boronic esters **55** as the proficient catalyst and devised a dehydrative glycosylation method of 2-deoxy sugar-derived hemiacetal **54** (Scheme 17) [93]. The protocol tolerance a wide range of 2-deoxy- and 2,6-dideoxyglycosides and demonstrated as the modest level of selectivity to get the respective α-glycosides **56**.

Wang and coworkers used diphenylammonium triflate (DPAT) for the direct dehydrogenative glycosylation of hemiacetal **57** (as donor) with various alcohols **58** (as acceptor) under *MW* condition and developed desired glycosides **59** in high-to-excellent yields (Scheme 18) [94]. The method was further extended with an acceptor immobilized on solid support under standard solid-supported condition.

Kancharla and coworkers explored 2-deoxy-glycosides **60** as suitable donor in glycosylation reaction with alcohols as acceptors and reported amine salt-mediated controlled glycosylation protocol for an easy access of 2-deoxy-ribofuranosides **61** as desired 2-deoxy-glycosides (Scheme 19) [95].

Sasaki and coworkers exemplified the glycosyl phosphite, e.g., glucopyranosyl diether phosphate **62** as suitable glycosyl donor in glycosidation reaction with various acceptors using protic acid (e.g., HOTf or HNTf$_2$ and HBF) as promoter in suitable

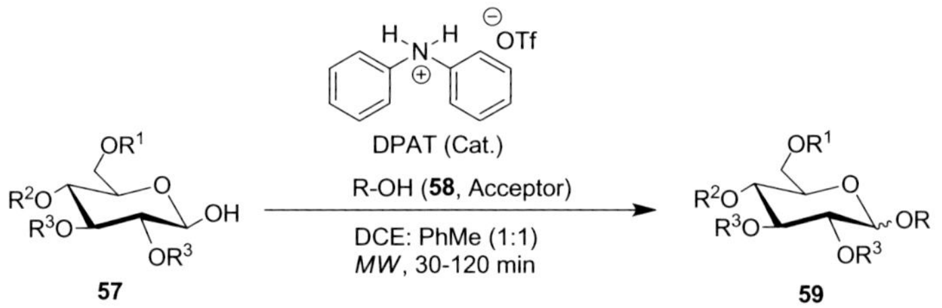

Scheme 18 DPAT-catalyzed direct dehydrogenative glycosylation reaction [94]

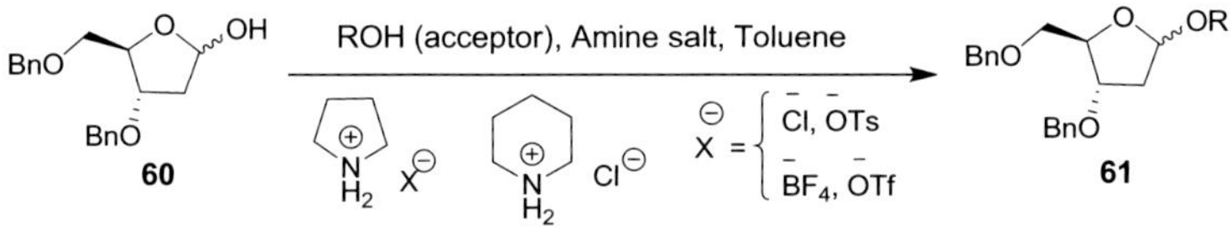

Scheme 19 Pyrrolidine salts mediated synthesis of 2-deoxy-glycosides [95]

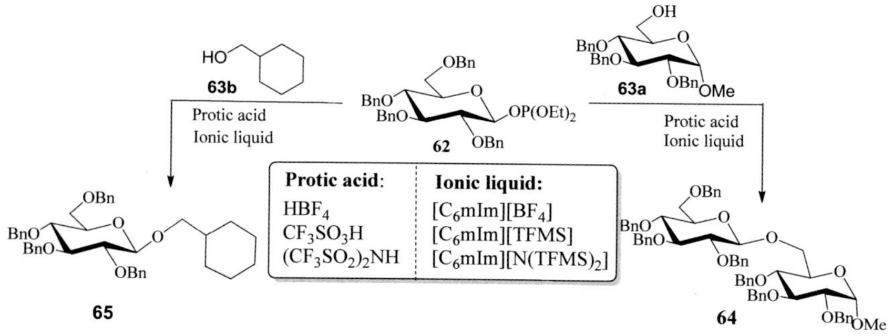

3 Application of RTILs in Carbohydrate Chemistry

Scheme 20 Glycosyl phosphite as donor in glycosidation using acid-ILs catalytic system [96]

ILs (e.g., 1-*n*-hexyl imidazolium-based RTILs having distinctive anions BF_4, trifluoromethanesulfonate, and trifluoromethanesulfonimidide (Scheme 20) [96]. Use of [hmIm][NTf_2] and $HNTf_2$ in glycosylation of donor **62** and acceptor **63a** exclusively afforded product **64** with β-selectivity. Halides were found to be the major impurities in RTILs and thus analysis of its purity needed. The basicity of Cl^- decreases the reaction yield even when performed in the presence of 5 mol% of [hmIm][Cl] [97].

In another investigation, Yadav and coworkers reported a high-yielding synthesis of 2,3-unsaturated *O*-glycopyranosides **68** with α-selectivity by reacting 3,4,6-tri-*O*-acetyl-D-glucal **66** separately with alcohols, phenols, and hydroxyl α-amino acids (**67**) using 5 mol% $Dy(OTf)_3$ in [bmIm][PF_6] as recyclable catalytic system (Scheme 21) [98].

Interestingly, RTIL is used in glycosylation reaction for the total synthesis of carbohydrate-containing natural products, e.g., Vineomycin-B_2. Sasaki and coworkers used 1-hexyl-3-methylimidazolium trifluromethanesulfonate (C_6mim [OTf]) in a key step of glycosylation reaction. The glycosylation of rhodinose **69** with β-oxo-*tert*-alcohol **70** in presence of [C_6mim] [OTf] gave a high yield of desired glycosylation product **71** with high stereoselectivity (Scheme 22) [99]. The volatile byproduct AcOH was easily removed under reduced pressure and thus avoid the formation of retro-Ferrier-type rearrangement product once this nonvolatile C_6mim [OTf] was used in glycosylation and thus the method displayed a significant advantage.

Thioglycosides is one the frequently explored donors in a number of glycosylation reactions particularly for an easy access of glycosyl fluorides [100], sulfoxides, and

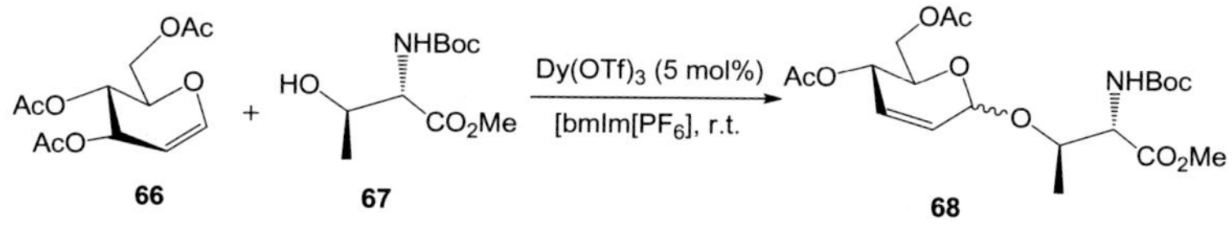

Scheme 21 Chemical glycosylation of D-glucals using $Dy(OTf)_3$ in [bmIm][PF_6] [98]

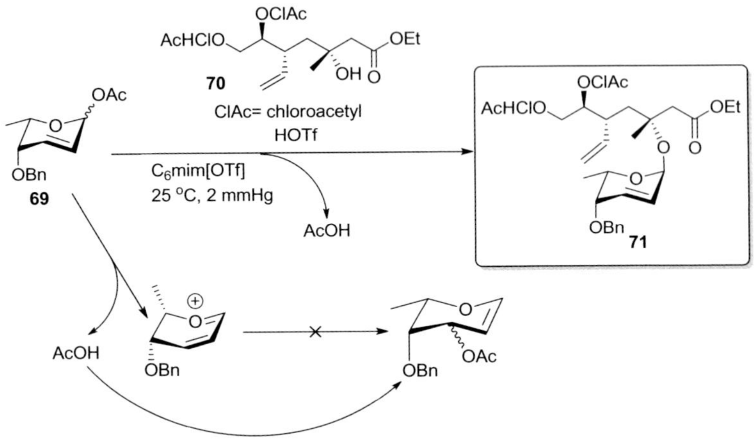

Scheme 22 ILs mediated glycosylation of D-glucal using *N*-Boc protected amino acids [99]

sulfones [101, 102]. Due to the notable stability of thioglycosides under a wide range of reaction condition, they are widely used both as effective glycosyl donors as well glycosyl acceptors. Santra and coworkers used [Bmim][BF$_4$] as green media and reported an environmentally benign protocol for the high-yielding synthesis thioglycosides [103]. Thus, the glycosylation reaction of β-D-Glu pentaacetate **72** with thiols (1.2 equiv) in [Bmim][BF4] (1.0 mL) using BF$_3$.OEt$_2$ (1.2 equiv) as promoter at 20 °C gave good-to-excellent yield of thioglycosides **73** (Scheme 23) [104].

Cui and coworkers used combination of [C$_6$mim] [OTf], Ag$_2$O and TBAI for synergistic catalytic system for the high-yielding synthesis of β-glycosyl 1-ester **75** (Scheme 24) [105]. [C$_6$mim] [OTf] exhibited synergistic effect in this glycosylation reaction with notable increase in yield to 93%, while 39% yield was noticed when reaction performed in presence of Ag$_2$O/TBAI in absence of IL [105].

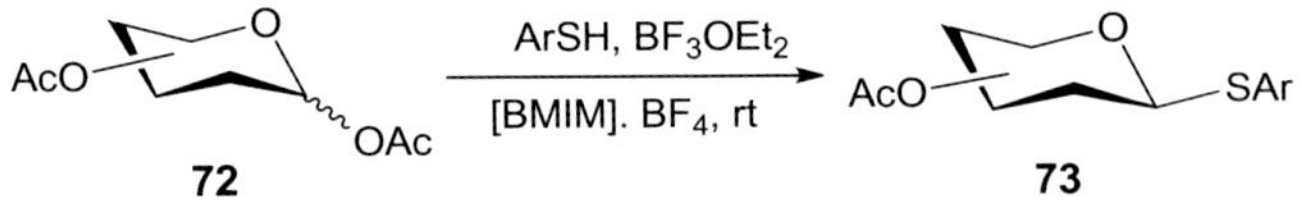

Scheme 23 [BMIM] [BF$_4$] mediated synthesis of thioglycosides [104]

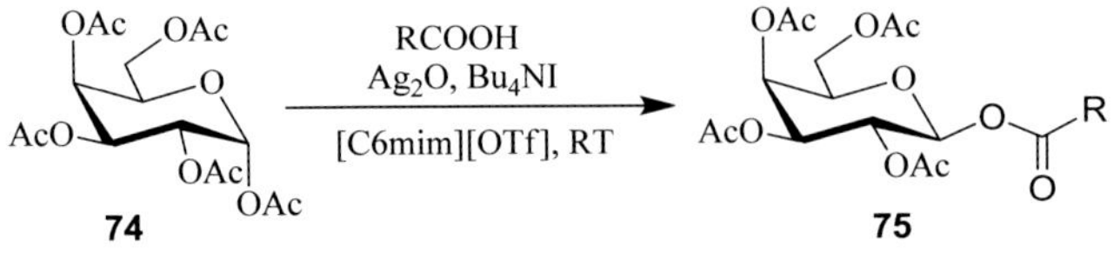

Scheme 24 RTIL- promoted synthesis of glycosyl-1-ester [105]

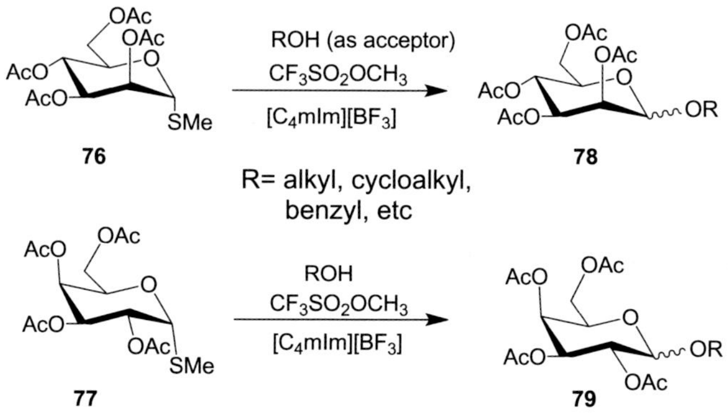

3 Application of RTILs in Carbohydrate Chemistry

Scheme 25 Chemical glycosylation of thioglycoside in [C$_4$mIm][BF$_4$]

Likewise, glycosylation of methyl 2,3,4,6-tetra-*O*-acetyl-α-D-thiomannopyranoside **76** or methyl 2,3,4,6-tetra-*O*-acetyl-β-D-thiogalactopyranoside **78** with alcohols (acceptor) in ILs (such as [C$_4$mIm][X], X = BF$_4^-$, PF$_6^-$, and CH$_3$SO$_4^-$) using methyl trifluoromethane sulfonate as promoter is performed to give high yield of respective *O*-glycosides **77** and **79** (Scheme 25) [85, 86].

Furthermore, Galan and coworkers nicely explored some surfactant ILs (**82a–d**) in combination with NIS for the required activation of thioglycosides **80** in glycosylation methodology with acceptor **81** to provide good-to-excellent yields of respective *O*-glycosides **83** with desired anomeric selectivity [88]. The condition is mild, amenable to NH$_2$-masking, e.g., trichloroethylcarbamate (Troc) and also compatible with a several standard groups of hydroxyl protection. The surfactant ILs (**82a-d**) proceeds through the slow activation of NIS to release I$^+$ in situ which then activate the thioglycosides (Scheme 26) [88].

Furthermore, trichloroacetimidate donors were also exemplified in glycosylation with various acceptors (e.g., alcohols) in [bmIm][PF$_6$] or [emIm][OTf] as green

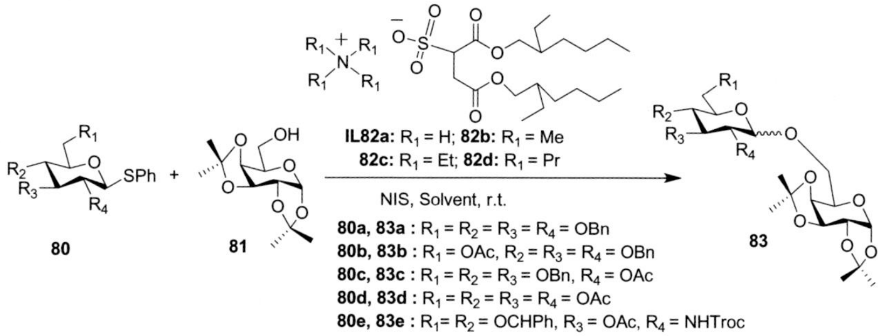

Scheme 26 Chemical glycosylation reaction in surfactant sulphonate anion [88]

media using TMSOTf as standard promoter [55, 106]. Anomeric selectivity is greatly influenced by solvent used and also catalyst. Fascinatingly, the glycosylation when carried out in [bmIm][PF$_6$] favored α-anomer (α:β = 76:24), while similar reaction in [emIm][OTf] favored β-anomer (α:β = 20:80) [106].

In addition to the selection of glycosyl donors and glycosyl receptors, efficiency of catalytic system greatly explored for the successful synthesis of desired saponins.[295] The glycosylation of glycosyl donor **84** separately with oleanolic acid **85a** and ursolic acid **85b** when carried out in combination of [C$_4$mim][OTf] and TMSOTf, it resulted to the desired glycosylation product, for example, **86** with β-selectivity as the sole product (Scheme 27) [107].

Song and coworkers explored 1-ethyl-3-methylimidazolium trifluoromethanesulfonate ([emim] [OTf]) as an efficient cosolvent and copromoter and reported a practical one-pot tandem glycosylation strategy using thioglycosides and trichloroacetimidate (or N-phenyltrifluoroacetimidate) donors at room temperature for an easy access of triterpenoid saponins **87** (Scheme 28) [108].

Galan and coworkers used [bmim][OTf] in regio- and chemoselective glycosylation for an easy synthesis of complex oligosaccharide [109]. The reactivity of various orthogonally protected thioglycosides was examined by utilizing [bmim][OTf] and NIS in glycosylation of glucose-derived thioglycoside donors **88a-h** separately with 1,2:3,4-di-O-isopropylidene-α-D-galactopyranose **89** required for regioselective and chemoselective delivery of glycosylation products **90a-h** (Scheme 29). The IL/NIS-mediated activation of donors **88d** and **88e** proceed well to give oligosaccharides **90d** (60%, β) and **90e** (83%, 0.95/1 (α/β)).

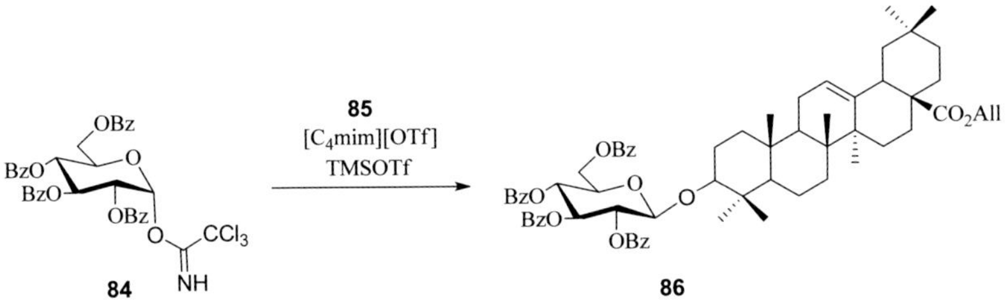

Scheme 27 IL promoted synthesis of triterpenoid sapogenins [107]

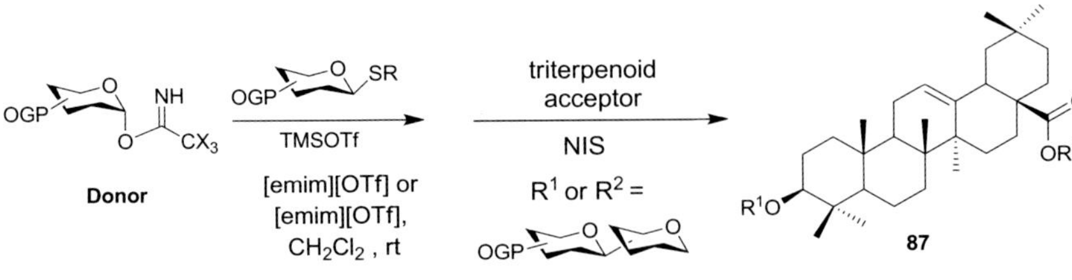

Scheme 28 [emim][OTf] mediated tandem glycosylations at r.t. [108]

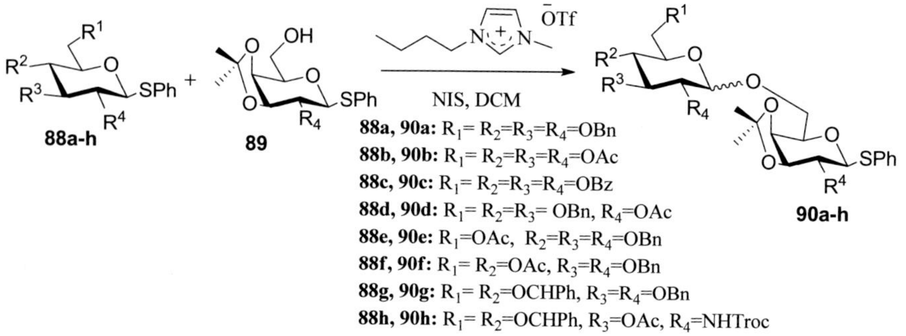

3 Application of RTILs in Carbohydrate Chemistry

Scheme 29 Regio- and chemoselective synthesis of oligosaccharides using protected thioglycosides in [bmim][OTf] [109]

IL-supported α-(1–4)-glycosylation methodology for an expeditious synthesis of complex oligosaccharide through activation of imidate is reported [110]. The glycosylation of trichloroacetimidate donor **91** with IL-linked acceptor **92** afforded product **93** with α-selectivity (Scheme 30) [110]. Use of an easy liquid/liquid extraction method for the purification is the notable merit.

A number of automated oligosaccharide synthesis [111] and solid-supported synthesis [112–114] have been explored for an easy access of complex oligosaccharides. In the IL-supported protocol, the IL is covalently attached to the donor/acceptor system, which at the end of synthetic steps can be removed (Scheme 31) [33]. Here, the desired trisaccharide, e.g., phenyl-2,3,4-tri-O-benzyl-β-D-glucopyranosyl-$(1 \rightarrow 6)$-2,3,4-tri-O-benzyl-β-D-glucopyranosyl-$(1 \rightarrow 6)$-2,3,4-tri-O-benzyl-1-thio-β-D-glucopyranoside **96**, was developed in good yield from phenyl 2,3,4,6-tetra-O-benzyl-1-thio-β-D-glucopyranoside **1** via sulfoxide glycosylation as key-step (Scheme 31) [33].

Furthermore, Pathak and coworkers exemplified similar IL-supported glycosylation reaction and developed a tetrasaccharide [115]. Alternative to the solid-supported synthesis, this method also excludes the tedious purification steps. The authors used

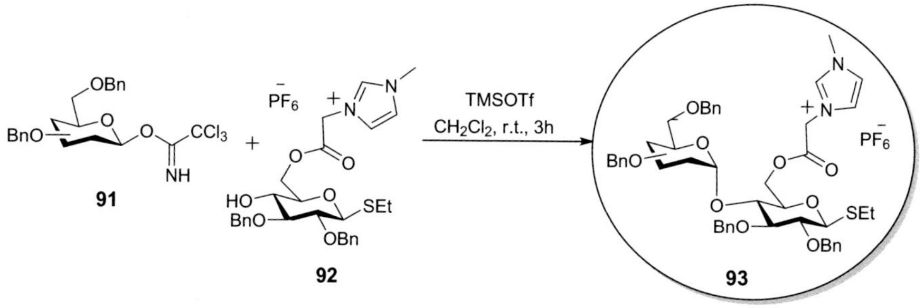

Scheme 30 α-(1–4)-Glycosylation of trichloroacetimidate donors **91** with IL-appended acceptor **92** [110]

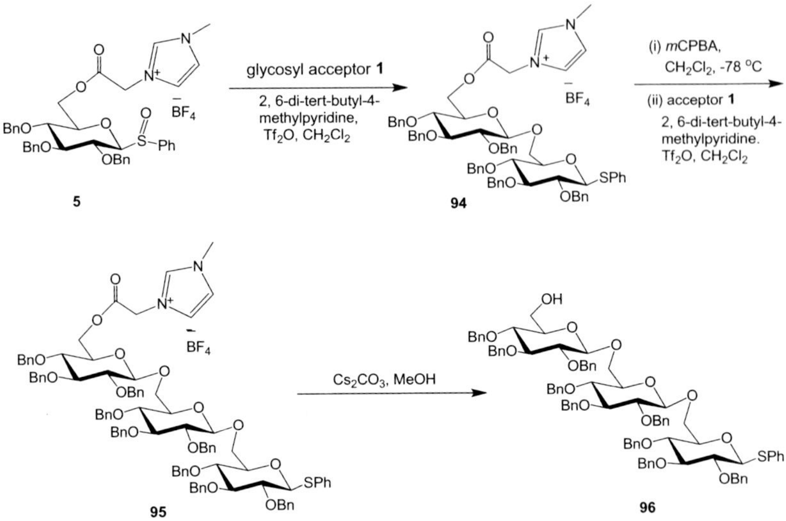

Scheme 31 IL-supported synthesis of oligosaccharide **96** [33]

IL-tagged fluoride glycoside as the donor (which was linked to the IL support via an ester linkage) and p-thiotolyl glycoside as acceptor to develop high yield of a linear $\alpha(1–6)$ tetramannopyranose thioglycoside (Scheme 32) [115]. As per demand, the IL support was removed simply by treating it with aq. saturated $NaHCO_3$ in H_2O/ether (1:1) and TBAI. The fluoride glycoside **97** was first heated with N-methylimidazole **3** in anhydrous acetonitrile followed by anion exchange reaction gave IL-tagged mannopyranosyl fluoride **98**. Next, the donor **98** was subjected to glycosylation with thioglycoside **99** using promoter $AgClO/SnCl_2$ and disaccharide **100** thus obtained was further treated with aq. saturated $NaHCO_3$ in H_2O/ether (1:1)/TBAI for the removal of IL tag. The resulted disaccharide **101** on glycosylation with glycoside **98** followed by removal of IL-tagged trisaccharide **102** gave desired trisaccharide **103**. At the end, glycosylation of trisaccharide acceptor **103** with IL-tagged fluoride donor **98** followed by removal of IL support from the resulted IL-tagged tetrasaccharide **104** afforded high yield of required homolinear $\alpha(1–6)$ tetramanno-pyranose thioglycoside **105** (Scheme 32) [115].

4 Conclusions and Future Outlook

The chapter illustrated a brief overview on growing impact of RTIL in some fascinating aspects of carbohydrate chemistry. This provides an insight at their role in dissolution, modulation, glycosylation, functionalization, and modification of carbohydrates. Through the exemplified discussion, this is obvious to conclude that a number of biocompatible ILs can be used for the expeditious and high-yielding

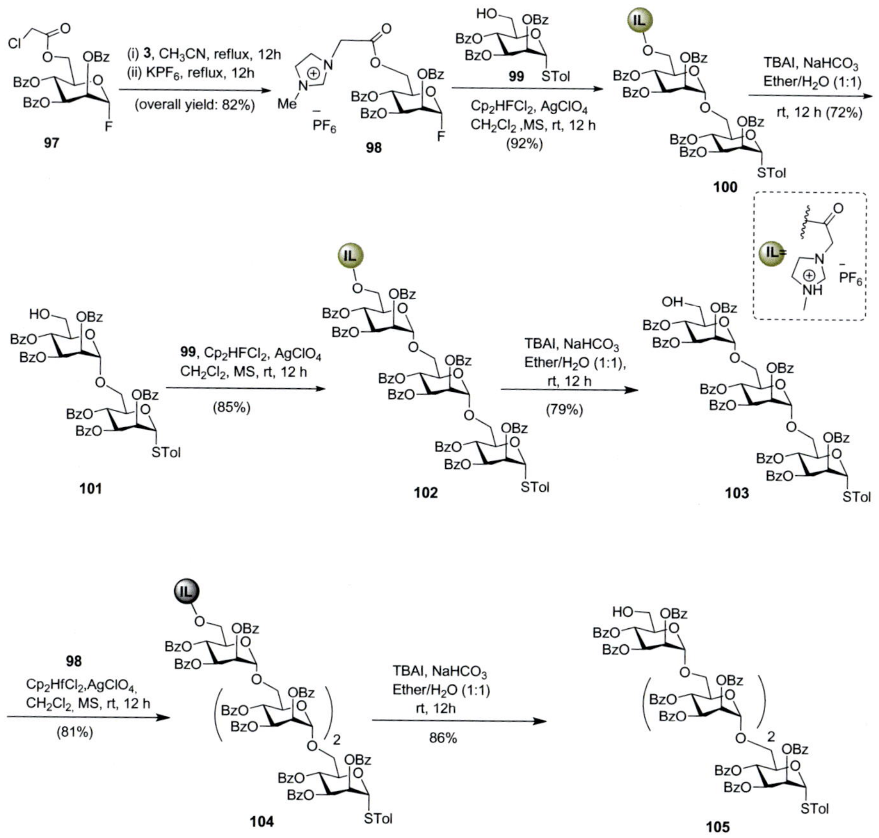

4 Conclusions and Future Outlook

Scheme 32 IL-Supported synthesis of homo-linear α(1–6) tetramannopyranose thioglycoside **105** [115]

synthesis of biologically relevant carbohydrate-containing molecules. Glycosylation methodology often performed in presence of suitable ILs to get the product with desired anomeric selectivity. Sugar-derived chiral ionic liquids are relatively easy to scale-up in the laboratory. The tunable properties of ILs can make them an even advantageous over the traditional solvents. Such reaction in once performed in green media, they usually impact to enhanced the reaction rate, reduces the reaction time in addition to the improvement in reaction yields. The easy work-up, recyclability, and cost effectiveness are the additional features of ILs. Despite of their growing impact, toxicity, non-biodegradability, etc. still need to consider to make the protocol non-hazardous.

References

1. Martins, M.A.P., Frizzo, C.P., Moreira, D.N., Zanatta, N., Bonacorso, H.G.: Ionic liquids in heterocyclic synthesis. Chem. Rev. **108**, 2015–2050 (2008)
2. Mishra, B.B., Kumar, D., Singh, A.S., Tripathi, R.P., Tiwari, V.K.: Ionic liquids-prompted synthesis of biologically relevant five- and six-membered heterocyclic skeletons: an update. In: Brahmachari, G. (ed.) Green Synthetic Approaches for Biologically Relevant Heterocycles, pp 437–493. Elsevier, The Netherlands (2015)
3. Bose, P., Agrahari, A.K., Singh, S.K., Singh, A.S., Yadav, M.S., Rajkhowa, S., Tiwari, V.K.: Recent developments on ionic liquids-mediated synthetic protocols of biologically relevant five and six-membered heterocyclic skeletons. In: Brahmachari, G. (ed.) Green Synthetic Approaches for Biologically Relevant Heterocycles (Second Edition) Vol 2: Green Catalytic Systems and Solvents. Advances in Green and Sustainable Chemistry, pp. 301–364. Elsevier, The Netherlands (2021)
4. Plechkovaa, N.V., Seddon, K.R.: Applications of ionic liquids in the chemical industry. Chem. Soc. Rev. **37**, 123–150 (2008)
5. Vekariya, R.L.: A review of ionic liquids: applications towards catalytic organic transformations. J. Mol. Liq. **227**, 44–60 (2017)
6. Hallett, J.P., Welton, T.: Room-temperature ionic liquids: solvents for synthesis and catalysis. Chem. Rev. **111**, 3508–3576 (2011)
7. Qiao, Y., Ma, W., Theyssen, N., Chen, C., Hou, Z.: Temperature-responsive ionic liquids: fundamental behaviors and catalytic applications. Chem. Rev. **117**, 6881–6928 (2017)
8. Dai, C., Zhang, J., Huang, C., Lei, Z.: Ionic liquids in selective oxidation: catalysts and solvents. Chem. Rev. **117**, 6929–6983 (2017)
9. Itoh, T., Takagi, Y.: Laccase-catalyzed reactions in ionic liquids for green sustainable chemistry. ACS Sustain. Chem. Eng. **9**, 1443–1458 (2021)
10. Watanabe, M., Thomas, M.L., Zhang, S., Ueno, K., Yasuda, T., Dokko, K.: Application of ionic liquids to energy storage and conversion materials and devices. Chem. Rev. **117**, 7190–7239 (2017)
11. Ventura, S.P.M., Silva, F.A., Quental, M.V., Mondal, D., Freire, M.G., Coutinho, J.A.P.: Ionic-liquid-mediated extraction and separation processes for bioactive compounds: past, present, and future trends. Chem. Rev. **117**, 6984–7052 (2017)
12. Smiglak, M., Metlen, A., Rogers, R.D.: The second evolution of ionic liquids: from solvents and separations to advanced materials-energetic examples from the ionic liquid cookbook. Acc. Chem. Res. **40**, 1182–1192 (2007)
13. Egorova, K.S., Gordeev, E.G., Ananikov, V.P.: Biological activity of ionic liquids and their application in pharmaceutics and medicine. Chem. Rev. **117**, 7132–7189 (2017)
14. Tiwari, V.K. (ed.): Carbohydrates in drug discovery and development. Elsevier Inc., The Netherlands (2020)
15. Tiwari, V.K.: Development of biologically relevant glycohybrid molecules: twenty years of our journey. Chem. Rec. **11**, 3029–3048 (2021)
16. Varki, A.: Biological roles of oligosaccharides: all of the theories are correct. Glycobiology **3**, 97–130 (1993)
17. Dwek, R.A.: Glycobiology: towards understanding the function of sugars. Chem. Rev. **96**, 683–720 (1996)
18. Angata, T., Varki, A.: Chemical dversity in the sialic acds and related α-keto acids. Chem. Rev. **102**, 439–469 (2002)
19. Bertozzi, C.R., Kiessling, L.L.: Chemical glycobiology. Science **291**, 2357–2364 (2001)
20. Tiwari, V.K., Mishra, R.C., Sharma, A., Tripathi, R.P.: Carbohydrate-based potential chemotherapeutic agents: recent developments and their scope in future drug discovery. Mini-Rev. Med. Chem. **12**, 1497–1519 (2012)
21. Mishra, S., Upadhayay, K., Mishra, K.B., Tripathi, R.P., Tiwari, V.K.: Carbohydrate-based therapeutics: a Frontier in drug discovery and development. Stud. Nat. Prod. Chem. **50**, 307–361 (2016)

References

22. Agrahari, A.K., Bose, P., Singh, A.S., Rajkhowa, S., Jaiswal, M.K., Hotha, S., Mishra, N., Tiwari, V.K.: Chem. Rev. **121**, 7638–7956 (2021)
23. Tiwari, V.K., Mishra, B.B., Mishra, K.B., Mishra, N., Singh, A.S., Chen, X.: Cu(I)-catalyzed click reaction in carbohydrate chemistry. Chem. Rev. **116**, 3086–3240 (2016)
24. Zhu, X., Schmidt, R.R.: New principles for glycoside-bond formation. Angew. Chem. Int. Ed. **48**, 1900–1934 (2009)
25. Mishra, A., Mishra, N., Tiwari, V.K.: Carbohydrate-based organocatalysts: recent developments and future perspectives. Curr. Org. Synth. **13**, 176–219 (2016)
26. Reib, M., Brietzke, A., Eickner, T., Stein, F., Villinger, A., Vogel, C., Kragl, U., Jopp, S.: Synthesis of novel carbohydrate based pyridinium ionic liquids and cytotoxicity of ionic liquids for mammalian cells. RSC Adv. **10**, 14299–14304 (2020)
27. Murugesan, S., Linhardt, R.J.: Ionic liquids in carbohydrate chemistry—current trends and future directions. Curr. Org. Syn. **2**, 437–451 (2005)
28. El Seoud, O.A., Koschella, A., Fidale, L.C., Dorn, S., Heinze, T.: Applications of ionic liquids in carbohydrate chemistry: a window of opportunities. Biomacromol **8**, 2629–2647 (2007)
29. Prasad, V., Kale, R.R., Kumar, V., Tiwari, V.K.: Carbohydrate chemistry and room temperature ionic liquids (RTILs): recent trends, opportunities, challenges and future perspectives. Curr. Org. Syn. **7**, 506–531 (2010)
30. Farran, A., Cai, C., Sandoval, M., Xu, Y., Liu, J., Hernaiz, M.J., Linhardt, R.J.: Green solvents in carbohydrate chemistry: from raw materials to fine chemicals. Chem. Rev. **115**, 6811–6853 (2015)
31. Rajkhowa, S., Kale, R.R., Sarma, J., Kumar, A., Mohapatra, P.P., Tiwari, V.K.: Room temperature ionic liquids in glycoscience: opportunities and challenges. Curr. Org. Chem. **25**, 2542–2578 (2021)
32. Li, W., Zhang, Z., Han, B., Hu, S., Song, J., Xie, Y., Zhou, X.: Switching the basicity of ionic liquids by CO_2. Green Chem. **10**, 1142–1145 (2008)
33. He, X., Chan, T.H.: Ionic-tag-assisted oligosaccharide synthesis. Synthesis 1645–1651 (2006)
34. Gauchot, V., Schmitzer, A.R.: Asymmetric aldol reaction catalyzed by the anion of an ionic liquid. J. Org. Chem. **77**, 4917–4923 (2012)
35. Gonzalez, L., Escorihuela, J., Altava, B., Burguete, M.I., Luis, S.V.: Chiral room temperature ionic liquids as enantioselective promoters for the asymmetric Aldol reaction. Eur. J. Org. Chem. 5356–5363 (2014)
36. Handy, S.T., Okella, M., Dickenson, G.: Solvents from Biorenewable sources: ionic liquids based on Fructose. Org. Lett. **5**, 2513–2515 (2003)
37. Ebner, G., Schiehser, S., Potthast, A., Rosenau, T.: Side reaction of cellulose with common 1-alkyl-3-methylimidazolium-based ionic liquids. Tetrahedron Lett. **49**, 7322–7324 (2008)
38. Spear, S.K., Visser, A.E., Rogers, R.D.: In: SPRI Conference on Sugar Processing Research (2002)
39. Graenacher C. U.S. Patent, **1**, 946,176 (1934)
40. Husemann, V.E., Siefert, E.N.: N-Ethyl-N-pyridinium chloride as a solvent and reaction medium for cellulose. Macromol. Chem. Phys. **128**, 288–291 (1969)
41. Swatloski, R.P., Spear, S.K., Holbrey, J.D., Rogers, R.D.: Dissolution of cellulose with ionic liquids. J. Am. Chem. Soc. **124**, 4974–4975 (2002)
42. Zhang, Y.J., Xu, A.R., Lu, B.L., Li, Z.Y., Wang, J.J.: Dissolution of cellulose in 1-allyl-3-methylimizodalium carboxylates at room temperature: a structure property relationship study. Carbohydr. Polym. **117**, 666–672 (2015)
43. Wu, J., Zhang, J., Zhang, H., He, J., Ren, Q., Guo, M.: Homogeneous acetylation of cellulose in a new ionic liquid. Biomacromol **5**, 266–268 (2004)
44. Brandt, A., Ray, M.J., To, T.Q., Leak, D.J., Murphy, R.J., Welton, T.: Ionic liquid pretreatment of lignocellulosic biomass with ionic liquid–water mixtures. Green Chem. **13**, 2489–2499 (2011)
45. Barthel, S., Heinze, T.: Acylation and carbanilation of cellulose in ionic liquids. Green Chem. **8**, 301–306 (2006)

46. Xiao, W.J., Chen, Q., Wu, Y., Wu, T.H., Dai, L.Z.: Dissolution and blending of chitosan using 1,3-dimethylimidazolium chloride and 1-H-3-methylimidazolium chloride binary ionic liquid solvent. Carbohydr. Polym. **83**, 233–238 (2011)
47. Liu, Y., Jing, S., Carvalho, D., Fu, J., Martins, M., Cavaco-Paulo, A.: Cellulose dissolved in ionic liquids for modification of the shape of keratin fibres. ACS Sustain. Chem. Eng. **9**, 4102–4110 (2021)
48. Feng, L., Chen, Z.-L.: Erratum to molecular dynamics of molten Li_2CO_3–K_2CO_3. J. Mol. Liq. **142**, 161 (2008)
49. Gathergood, N., Scammels, P.J., Garcia, M.T.: Biodegradable ionic liquids Part III. The first readily biodegradable ionic liquids. Green Chem. **8**, 156–160 (2006)
50. Kawasaki, M., Iwasa, Y.: "Cut and stick" ion gels. Nature **489**, 510–511 (2012)
51. Neouze, M.-A., Bideau, J.L., Gaveau, P., Bellayer, S., Vioux, A.: Ionogels, new materials arising from the confinement of ionic liquids within silica derived networks. Chem. Mater. **18**, 3931–3936 (2006)
52. Kimizuka, N., Nakashima, T.: Spontaneous self-assembly of glycolipid bilayer membranes in sugar-philic ionic liquids and formation of ionogels. Langmuir **17**, 6759–6761 (2001)
53. Forsyth, S.A., MacFarlane, D.R., Thomson, R.J., Itzstein, M.V.: Rapid, clean, and mild O-acetylation of alcohols and carbohydrates in an ionic liquid. Chem. Commun. **7**, 714–715 (2002)
54. Zhang, J.G., Ragauskas, A.J.: Synthesis of benzylidenated hexopyranosides in ionic liquids. Carbohydr. Res. **340**, 2812–2815 (2005)
55. Murugesan, S., Karst, N., Islam, T., Wiencek, J.M., Linhardt, R.J.: Dialkyl imidazolium benzoates-room temperature ionic liquids useful in the peracetylation and perbenzoylation of simple and sulfated saccharides. Synlett **9**, 1283–1286 (2003)
56. Baker, D.C.: Preparative carbohydrate chemistry; Stephen Hanessian. Marcel Dekker: New York. 1997. J. Am. Chem. Soc. **119**, 12028–12029 (1997)
57. Thul, M., Wu, Y.P., Lin, Y.J., Du, S.L., Wu, H.R., Ho, W.Y., Luo, S.Y.: Ionic liquid catalyzed per-O-acetylation and benzylidene ring-openng reaction. Catalysis **10**(6), 642 (2020)
58. Radhakrishnan, K.V., Sajisha, V.S., Chacko, J.M.: A facile and eco-friendly method for the synthesis of 1,2-orthoesters of carbohydrates in ionic liquid, [bmim]PF6. Synlett **6**, 997–999 (2005)
59. Prasad, V., Kumar, D., Tiwari, V.K.: A highly expeditious synthesis of bicyclic iminosugar using novel key step of [NMM]+[HSO4]−promoted conjugate addition and Mitsunobu reaction. RSC Adv. **3**, 5794–5797 (2013)
60. Singh, A., Mishra, B.B., Kale, R.R., Tiwari, V.K.: Facile synthesis of novel glycosyl azetidine under mild reaction condition. Synth. Commun. **42**, 3598–3613 (2012)
61. Tiwari, V.K., Tripathi, R.P.: Unexpected isomerisation of double bond with DBU: a convenient synthesis of tetrasubstituted α-alkylidene tetrahydrofuranoses. Ind. J. Chem. **41B**, 1681–1685 (2002)
62. Tripathi, R.P., Tripathi, R., Tiwari, V.K., Bala, L., Sinha, S., Srivastava, A., Srivastava, R., Srivastava, B.S.: Synthesis of glycosylated β-amino acids as new class of antitubercular agents. Eur. J. Med. Chem. **37**, 773–781 (2002)
63. Mishra, A., Tiwari, V.K.: One-pot synthesis of glycosyl-β-azidoester via diazotransfer reaction towards an easy access of glycosyl-β-triazolyl ester. J. Org. Chem. **80**, 4869–4881 (2015)
64. Pandey, J., Sharma, A., Tiwari, V.K., Dube, D., Ramachandran, R., Chaturvedi, V., Sinha, S., Mishra, N., Shulka, P.K., Tripathi, R.P.: Synthesis, molecular modeling and antitubercular activities of glycopeptide analogs with both furanose and pyranose ring structures. J. Comb. Chem. **11**, 422–427 (2009)
65. Prasad, V., Kale, R.R., Mishra, B.B., Kumar, D., Tiwari, V.K.: Diacetoxyiodobenzene mediated one-pot synthesis of diverse carboxamides from aldehydes. Org. Lett. **14**, 2936–2939 (2012)
66. Ferlin, N., Gatard, S., Nguyen, A., Nhien, V., Courty, M., Bouquillon, S.: Click reactions as a key step for an efficient and selective synthesis of D-xylose-based ILs. Molecules **18**, 11512–11525 (2013)

References

67. Jha, A.K., Jain, N.: Synthesis of glucose-tagged triazolium ionic liquids and their application as solvent and ligand for copper(I) catalyzed amination. Tetrahedron Lett. **54**, 4738–4741 (2013)
68. Lansalot-Matras, C., Moreau, C.: Dehydration of fructose into 5-hydroxymethylfurfural in the presence of ionic liquids. Catal. Commun. **4**, 517–520 (2003)
69. Liu, W., Wang, Y., Li, W., Yang, Y., Wang, N., Song, Z., Xia, X.-F., Wang, H.: Polyethylene glycol-400-functionalized dicationic acidic ionic liquids for highly efficient conversion of fructose into 5-hydroxymethylfurfural. Catal Lett. **145**, 1080–1088 (2015)
70. Murakami, M.-A., Kaneko, Y., Kadokawa, J.-I.: Preparation of cellulose–starch composite gel and fibrous material from a mixture of the polysaccharides in ionic liquid. Carbohydr. Polym. **69**, 378–381 (2007)
71. Kadokawa, J.-I., Murakami, M.-A., Kaneko, Y.: A facile method for preparation of composites composed of cellulose and a polystyrene-type polymeric ionic liquid using a polymerizable ionic liquid. Compos. Sci. Technol. **68**, 493–498 (2008)
72. Abbott, A.P., Bell, T.J., Handa, S., Stoddart, B.: Cationic functionalisation of cellulose using a choline based ionic liquid analogue. Green Chem. **8**, 784–786 (2006)
73. Prasad, K., Mine, S., Kaneko, Y., Kadokawa, J.: Preparation of cellulose-based ionic porous material compatibilized with polymeric ionic liquid. Polym. Bull. **64**, 341–349 (2010)
74. Park, S., Kazlauskas, R.J.: Biocatalysis in ionic liquids: advantageous beyond green technology. Curr. Opin. Biotech. **14**, 432–437 (2003)
75. Kaar, J.L., Jesionowski, A.M., Berberich, J.A., Moulton, R., Russell, A.J.: impact of ionic liquid physical properties on lipase activity and stability. J. Am. Chem. Soc. **125**, 4125–4131 (2003)
76. Park, S., Kazlauskas, R.J.: Improved preparation and use of room-temperature ionic liquids in lipase-catalyzed enantio- and regioselective acylations. J. Org. Chem. **66**, 8395–8401 (2001)
77. Kim, M.J., Choi, M.Y., Lee, J.K., Ahn, Y.: Enzymatic selective acylation of glycosides in ionic liquids: significantly enhanced reactivity and regioselectivity. J. Mol. Catal. B: Enzym. **26**, 115–118 (2003)
78. Kale, R.R., Jadhav, N.K., Rajkhowa, S., Muntode, B.B., Gaikwad, V.B., Kajale, D.D., Sarma, J.: Carbohydrate-based surfactant: promises, challenges and future prospective. Trends Carbohyd. Res. **13**(2), 50–77 (2021)
79. Ganske, F., Bornscheuer, U.T.: Lipase-catalyzed glucose fatty acid ester synthesis in ionic liquids. Org. Lett. **7**, 3097–3098 (2005)
80. Okrasa, K., Guibe-Jampel, E., Therisod, M: Tandem peroxidase-glucose oxidase catalysed enantioselective sulfoxidation of thioanisoles. J. Chem Soc. Perkin Trans-1 **7**, 1077–1079 (2000)
81. Okrasa, K., Guibe-Jampel, E., Therisod, M: Ionic liquids as a new reaction medium for oxidase–peroxidase- catalyzedsulfoxidation. Tetrahedron: Asymmetry **14**, 2487–2490 (2003)
82. Bell, G., Halling, P.J., Moore, B.D., Partridge, J., Rees, D.G.: Biocatalyst behaviour in low-water systems. Trends Biotechnol. **13**, 468–473 (1995)
83. Echstein, M., Sesing, M., Kragle, U., Adlercreutz, P.: At low water activity-chymotrypsin is more active in an ionic liquid than in non-ionic organic solvents. Biotechnol. Lett. **24**, 867–872 (2002)
84. Schmidt, R.R.: New methods for the synthesis of glycosides and oligosaccharides-are there alternatives to the Koenigs-Knorr method. Angew Chem. Int. Edn. **25**, 212–235 (1986)
85. Sasaki, K., Nagai, H., Matsumura, S., Toshima, K.: A novel greener glycosidation using an acid ionic liquid containing a protic acid. Tetrahedron Lett. **44**, 5605–5608 (2003)
86. Sasaki, K., Matsumura, S., Toshima, K.: A novel glycosidation of glycosyl fluoride using a designed ionic liquid and its effect on the stereoselectivity. Tetrahedron Lett. **45**, 7043–7047 (2004)
87. Zhang, J.G., Ragauskas, A.: Study of thioglycosylation in ionic liquids. Beil. J. Org. Chem. **2**, 2–12 (2006)
88. Galan, C.M., Tran, A.T., Boisson, J., Benito, D., Butts, C., Eastoe, J., Brown, P.: [R$_4$N] [AOT]: a surfactant ionic liquid as a mild glycosylation promoter. J. Carbohyd. Chem. **30**, 486–497 (2011)

89. Paulski, Z.: Glycosylation in ionic liquids. Synthesis **13**, 2074–2078 (2003)
90. Park, T.-J., Weiwer, M., Yuan, X., Sultan, N.B., Eva, M.M., Saravanababu, M., Robert, J.L.: Glycosylation in room temperature ionic liquid using unprotected and unactivated donors. Carbohyd. Res. **342**, 614–620 (2007)
91. Monasson, O., Sizun-Thomé, G., Lubin-Germain, N., Uziel, J., Auge, J.: Straightforward glycosylation of alcohols and amino acids mediated by ionic liquid. Carbohyd. Res. **352**, 202–205 (2012)
92. Auge, J., Sizun, G.: Ionic liquid promoted atom economic glycosylation under Lewis acid catalysis. Green Chem. **11**, 1179–1183 (2009)
93. Manhas, S., Taylor, M.S.: Dehydrative glycosidations of 2-deoxysugar derivatives catalyzed by an arylboronic ester. Carbohyd. Res. **470**, 42–49 (2018)
94. Hsu, M.Y., Lam, S., Wu C.H., Lin, M.H., Lin, S.C., Wang C.C.: Direct Dehydrative Glycosylation Catalyzed by Diphenylammonium Triflate. Molecules, **25**, 1103–1111 (2020).
95. Ghosh, T., Mukherji, A., Kancharla, P.K.: Open-close strategy toward the organocatalytic generation of 2-Deoxyribosyl Oxocarbenium Ions: Pyrrolidine Salt-Catalyzed Synthesis of 2-Deoxyribofuranosides. Eur. J. Org. Chem. **45**, 7488–7498 (2019)
96. Sasaki, K., Nagai, H., Matsumura, S., Toshima, K.: A novel greener glycosidation using an acid–ionic liquid containing a protic acid. Tetrahedron Lett. **44**, 5605–5608 (2003)
97. Aggarwal, A., Lancaster, N.L., Sethi, A.R., Welton, T.: The role of hydrogen bonding in controlling the selectivity of Diels-Alder reactions in room-temperature ionic liquids. Green Chem. **4**, 517–520 (2002)
98. Yadav, J.S., Reddy, B.V.S., Reddy, J.S.S.: Dy(OTf)$_3$-immobilized in ionic liquids: a novel and recyclable reaction media for the synthesis of 2,3-unsaturated glycopyranosides. J. Chem. Soc. Perkin Trans-1. **21**, 2390–2394 (2002)
99. Sasaki, K., Matsumura, S., Toshima, K.: Toward the total synthesis of vineomycin B$_2$: application of efficient glycosylation methodology using 2, 3-unsaturated sugars. Tetrahedron Lett. **48**, 6982–6986 (2007)
100. Nicolaou, K.C., Dolle, R.E., Papahatjis, D.P., Randall, L.J.: Reactions of glycosyl fluorides. Synthesis of O-, S-, and N-glycosides. J. Chem. Soc. Chem. Commun. **17**, 1155–1156 (1984)
101. Crich, D., Sun, S.: Direct synthesis of β-mannopyranosides by the sulfoxide method. J. Org. Chem. **62**, 1198–1199 (1997)
102. Vlahov, I.R., Vlahova, P.I., Linhardt, R.J.: Diastereocontrolled synthesis of carbon glycosides of N-acetylneuraminic acid via glycosyl samarium(III) intermediates. J. Am. Chem. Soc. **119**, 1480–1481 (1997)
103. Santra, A., Sau, A., Misra, A.K.: Synthesis of Thioglycosides in room temperature ionic liquid. J. Carbohyd. Chem. **30**, 85–93 (2011)
104. Rencurosi, A., Lay, I., Russo, G., Caneva, E., Poletti, I.: Glycosylation with trichloroacetimidates in ionic liquids: influence of the reaction medium on the stereochemical outcome. J. Org Chem. **70**, 7765–7768 (2005)
105. Cui, Y., Xu, M., Yao, W., Mao, J.: Room-temperature ionic liquids enhanced green synthesis of β-glycosyl 1-ester. Carbohyd. Res. **407**, 51–54 (2015)
106. Poletti, L., Rencurosi, A., Lay, L., Russo, G.: Trichloroacetimidates as glycosyl donors in recyclable ionic liquids. Synlett **15**, 2297–2300 (2003)
107. Zhang, T., Li, X., Song, H., Yao, S.: Ionic liquid-assisted catalysis for glycosidation of two triterpenoid sapogenins. New J. Chem. **43**, 16881–16888 (2019)
108. Song, Y., Guo, T., Liu, Q., Song, W., Li, F.: A sequential one-pot strategy for the synthesis of triterpenoid, saponins in ionic liquid [emim][OTf]. J. Chem. Res. **43**, 20–25 (2019)
109. Galan, C.M., Tran, A.T., Whitaker, S.: [bmim][OTf] as co-solvent/promoter in room temperature reactivity-based one-pot glycosylation reactions. Chem. Comm. **46**, 2106–2108 (2010)
110. Matthieu, P., Hubert-Roux, M., Martin, C., Guillen, F., Lange, C., Gouhier, G.: First examples of α-(1–4)-Glycosylation reactions on ionic liquid supports. Eur. J. Org. Chem. **33**, 6366–6371 (2010)

References

111. Hsu, C.H., Hung, S.C., Wu, C.Y., Wong, C.H.: Toward automated Oligosaccharide synthesis. Angew. Chem. Int. Ed. **50**, 11872–11923 (2011)
112. Seeberger, P.H., Haase, W.C.: Solid-phase Oligosaccharide synthesis and combinatorial carbohydrate libraries. Chem. Rev. **100**, 4349–4394 (2000)
113. Majumdar, D., Zhu, T., Boons, G.J.: Synthesis of Oligosaccharides on soluble high-molecular-weight branched polymers in combination with purification by nanofiltration. Org. Lett. **5**, 3591–3594 (2003)
114. Palmacci, E.R., Hewitt, M.C., Seeberger, P.H.: 'Cap-Tag'—Novel methods for the rapid purification of oligosaccharides prepared by automated solid-phase synthesis. Angew. Chem. Int. Ed. **40**, 4433–4437 (2001)
115. Pathak, A.K., Yerneni, C.K., Young, Z., Pathak, V.: Oligomannan synthesis using ionic liquid supported Glycosylation. Org. Lett. **10**, 145–148 (2008)

Chapter 6
Catalysis: Application and Scope in Organic Synthesis

1 Introduction

Catalysis is one of the most important aspects of green chemistry which deals with the design of chemical products and processes which limits the use and production of hazardous chemicals. Today catalysis is the backbone of organic synthesis in laboratory, pharmaceutical industries, polymer industry, and fine chemical production. According to estimates, about 90% of the chemicals produced at large scale meets a catalyst at least at one stage in the journey of their production. The economic benefit achieved via employment of catalysts in huge number is due to their incredible activity. A catalyst may transform the substrate tens of million times to the product in comparison with their own weight.

A catalyst is known for lowering down the activation energy without itself being consumed in the overall reaction. In other words, they interact with the starting material and deviate from the path having lower activation energy barrier (Fig. 1). However, catalyst does not change the equilibrium position of the reaction. Hence, by increasing the rate of attainment of equilibrium via proceeding through lowered activation energy pathway, the consumption of energy is reduced. This makes the catalytic processes inherently green. For example, nitrogen and hydrogen do not react to form ammonia in ambient condition according to the equation.

$$N_2 + 3H_2 \rightleftharpoons 2NH_3 \quad \Delta H° = -92\,kJ/mol$$

Since moles of product are lesser than reactants and the reaction is exothermic, according to the Le Chatelier's principle, high pressure and low temperature will push the equilibrium to right. In 1909, Haber demonstrated that at 200 °C and 300 atmospheric pressure the yield of ammonia was about 90%. Unfortunately, to overcome the activation energy of the reaction, more than 1000 °C is required. This drastic condition was not viable for industrial use. Later, Bosch employed iron-based catalyst to overcome the requirement of high temperature.

© The Author(s), under exclusive license to Springer Nature Singapore Pte Ltd. 2022
V. K. Tiwari et al., *Green Chemistry*,
https://doi.org/10.1007/978-981-19-2734-8_6

Fig. 1 Activation energy with and without catalyst in a reaction

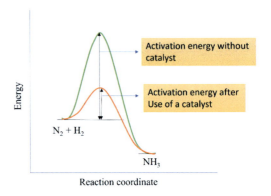

An ideal catalyst not only alters the reaction condition via changing the activation energy barrier but should have higher selectivity, turnover frequency, and turnover number. Selectivity is referred as the percentage of conversion of substrate to desired product over the by-product. Selectivity is one of the most desirable properties of a catalyst in the designing of new catalytic processes. More selective catalysis is greener and cheaper as it produces more desirable product, involves less complicated separation procedures, and produces less by-products. Turnover frequency is defined as amount of product formed per unit time by the help of catalyst. Lesser turnover frequency requires more amount of catalyst and hence produces more waste products. Turnover number is the moles of product formed per mole of the catalyst used. Higher turnover number requires less amount of catalyst, and so the process will be both economically and environmentally beneficial.

1.1 Type of Catalysts

Basically, there are two types of catalysis depending upon their physical state relative to the reaction medium.

(a) Heterogenous and
(b) Homogenous catalysts

Heterogenous catalyst, also known as surface catalysts, acts upon the substrate by virtue of their surface area. Heterogenous catalysts have different phases relative to the reaction medium. In most of the cases, these catalysts are in solid phase and the reactants are in gaseous phase. The reaction takes place on the surface or inside the pores present on the surface of a solid catalyst. Homogenous catalysts, on the other hand, have the same phase as the substrate have and are evenly distributed in whole reaction mixture. In general, the reaction rate of homogenous catalysis is greater than the heterogenous catalysis. Considering the separation process, heterogenous catalysts are easily separable compared to the homogenous catalysts.

Table 1 A comparison between homogenous and heterogenous catalysis [1]

Features	Homogenous catalysis	Heterogenous catalysis
Phase	Same phase as reaction medium	Different phase than the reaction medium
Separation	Difficult and expensive	Easy
Recyclability	Often difficult to recycle	Easily regenerated and recycled
Kinetics	Fast	Often slow
Sensitivity to poison	Sensitive to poison	Usually robust to poison
Selectivity	Usually high selectivity	Lower selectivity
Service life	Short service life	Longer service life
Energy consumption	Often takes place in mild condition	High energy process
Mechanism	Often mechanism well understood	Poor mechanistic understanding

In context to the greenness of the heterogenous and homogenous catalysis, the comparisons are often difficult to make and there are always some exceptions to any generalization. There are obvious differences between the two catalytic processes which have significant effect on the environment (Table 1).

As mentioned in Table 1, some characteristic features of homogenous catalysis are better than heterogenous catalysis and vice-versa. The ultimate goal of researchers in this field is to choose the best qualities from both the methods. For instance, one can choose fast rate and high selectivity of homogenous catalysis and easy separation and recyclability of heterogenous catalysis depending upon the required reaction conditions. In a nutshell, the approach should be heterogenization of homogenous catalysts.

2 Catalytic Oxidation Process

Oxidation reaction is one of the key chemical transformations usually needed in industries and laboratories around the globe. Most of the oxidations are still largely carried out with stoichiometric inorganic (or organic) oxidants such as Cr(VI) reagents, permanganate, manganese dioxide, and periodate which make the utter need of development of green catalytic alternatives for oxidation process. The stoichiometric oxidants, in general, lead to poor atom economy and hence add hazardous by-products in the environment.

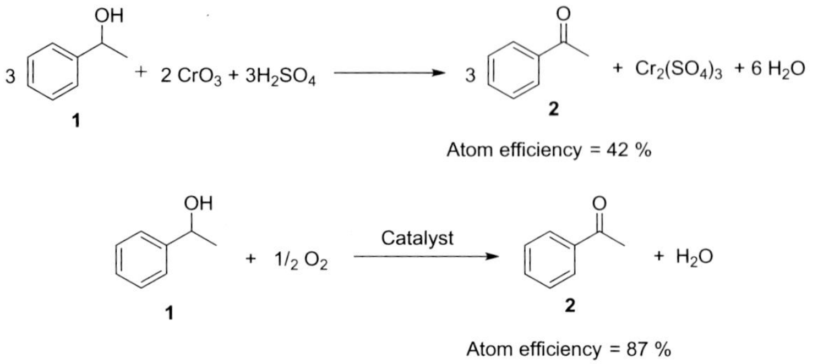

Scheme 1 showing two reactions comparing atom efficiencies:

Reaction 1:
$$3 \text{ (1-phenylethanol, OH)} + 2\,CrO_3 + 3H_2SO_4 \longrightarrow 3 \text{ (acetophenone, 2)} + Cr_2(SO_4)_3 + 6\,H_2O$$

Atom efficiency = 42 %

Reaction 2:
$$\text{(1-phenylethanol, 1)} + 1/2\,O_2 \xrightarrow{\text{Catalyst}} \text{(acetophenone, 2)} + H_2O$$

Atom efficiency = 87 %

Scheme 1 Comparison of atom efficiencies in stoichiometric and catalytic oxidation of secondary alcohol

The parameters like E-factor and atom efficiency or atom economy are extremely helpful in rapid evaluation of the amounts of waste that will be generated in different processes. *E-factor* can be evaluated as mass ratio of waste to the desired product. Higher the E-factor greater is the waste production, hence the negative impact on the environment is elevated. On the other hand, *Atom efficiency* is defined as molecular weight of desired product divided by sum total of the molecular weight of all the reactants appeared in stoichiometric equation. A catalytic process is better option to obtain the desired products in environmentally benign way. For instance, a comparison of atom efficiencies between stoichiometric and catalytic oxidation of a secondary alcohol **1** is shown in Scheme 1.

The partial catalytic oxidation of hydrocarbons (alkanes, alkenes, and aromatics) obtained from oil and natural gas feedstocks is the most important techniques to obtain useful organic compounds to the chemical industries. Table 2 illustrates few useful products as the result of catalytic oxidations. Traditionally, the fine chemical industries are more dependent upon the stoichiometric way to oxidize the hydrocarbons involving dichromates and permanganates as oxidizing agents. These oxidants result in concomitant generation of inorganic salt-containing effluents which are toxic and poor in terms of atom economy. The increasing pressure of stringent environmental regulations encourages the use of catalytic oxidation alternatives in manufacturing of fine chemicals. Most of these oxidations are carried out in the presence of transition metal catalysts and dioxygen as the prime oxidants. These routes are more economical, mild, and selective. The combination of H_2O_2 and O_2 is another greener alternative that can be used stoichiometrically in the reaction which produces water as by-product.

2 Catalytic Oxidation Process

Table 2 Industrial products as a result of catalytic oxidation [2]

Product	Feedstock	Oxidant
Terephthalic acid	*p*-xylene	O_2 (l)
Formaldehyde	Methanol	O_2 (g)
Ethene oxide	Ethene	O_2 (g)
Phenol	Benzene, toluene	O_2 (l)
Acetic acid	*n*-Butane, ethene	O_2 (l)
Propene oxide	Propene	RO_2H (l)
Acrylonitrile	Propene	O_2 (g)
Vinyl acetate	Ethene	O_2 (l, g)
Benzoic acid	Toluene	O_2 (l)
Adipic acid	Benzene	O_2 (l)
ε-Caprolactam	Benzene	O_2 (l)
Phthalic anhydride	*o*-Xylene	O_2 (g)
Acrylic acid	Propene	O_2 (g)
Methyl methacrylate	Isobutene	O_2 (g)
Maleic anhydride	*n*-Butane	O_2 (g)

In parenthesis, l = liquid; g = gas

2.1 Oxidation of Alkenes

Olefins are the most important substrates for oxidation to produce a variety of industrial products involving different chemical transformations such as epoxidation, oxidative cleavage, dihydroxylation, ketonization, and allylic oxidation.

2.1.1 Epoxidation

The epoxidation of propene **3** with peroxide like tert-butylhydroperoxide (TBHP) **4** or ethylbenzene hydroperoxide (EBHP) **5** remains the most favorite route for fine chemical industries which produces more than a million ton of propene oxide per year. The reaction is catalyzed by early transition metals in higher oxidation states such as Mo(VI), W(VI), V(V), and Ti(IV). TBHP or EBHP plays role of oxidant in this reaction (Scheme 2).

Scheme 2 Epoxidation of propene

4, R = $(CH_3)_3$C- or
5, PhCH(CH_3)-

Catalyst: Mo (VI) or Ti (IV)/SiO_2

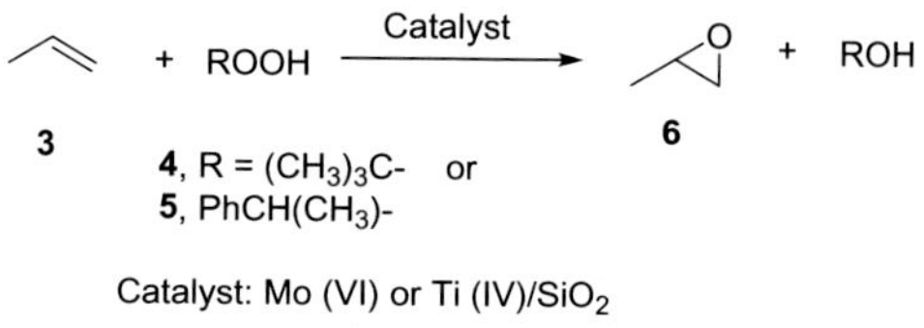

Scheme 3 Peroxometal mechanism

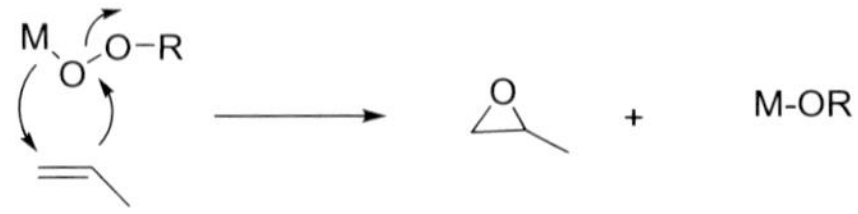

Scheme 4 Epoxidation by TS-1 and H_2O_2 as oxidant

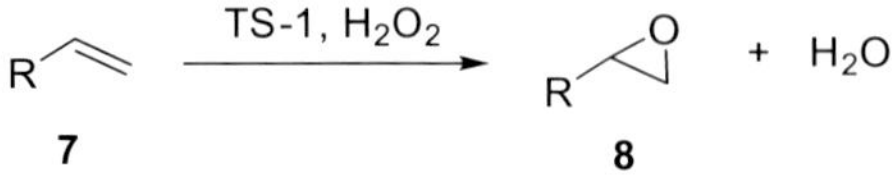

The epoxidation reaction employing alkyl peroxides and homogenous (Mo, W, V, Ti) or heterogenous catalysts (Ti(IV)/SiO$_2$) undergoes peroxometal mechanism in which the rate determining step is the transfer of single oxygen atom from electrophilic peroxometal species to nucleophilic olefinic center (Scheme 3). The metal does not undergo any change in oxidation state. They act like a Lewis acid and pull electron density from the peroxy group making them more electrophilic. The active metal centers are strong Lewis acids.

Clerici and Ingallina found that Titanium (IV) silicalites (TS-1) catalysts are extremely useful in epoxidation of unhindered alkenes **7** with 30% aqueous H_2O_2 as oxidant to produce **8** in good yield (Scheme 4). Yield and rate of the reaction decrease as the chain length increases. The electron-withdrawing groups (1-butene > allyl chloride > allyl alcohol) and the solvent system (MeOH > EtOH > t-BuOH) also have impact on the kinetics of the reaction. The rate of reaction also depends upon position and steric configuration of double bond (trans-2-butene < isobutene < 1-butene < cis-2-butene, 2-methyl-1-butene < 1-pentene). Catalytic activity was found to be increased by acids and inhibited by bases [3].

Venturello et al. developed an effective catalytic system for epoxidation in phase transfer conditions. They employed two component associations (tungstate and phosphate) with dilute H_2O_2 in acidic condition for epoxidation of olefins **9** which affords **10** (Scheme 5) [4]. Noyori et al. modified Venturello's approach and developed new greener catalytic system consisting of Na_2WO_4 dihydrate, (aminomethyl)phosphonic acid, and methyltrin-octylammonium hydrogen sulfate in a 2:1:1 molar ratio. They avoid any use of halogenated organic/inorganic solvents. Terminal olefins **11** which are in general not very active for oxidation produced **12** with an excellent yield of yield 94–99% with 2 mol% use of catalyst (Scheme 6) [5].

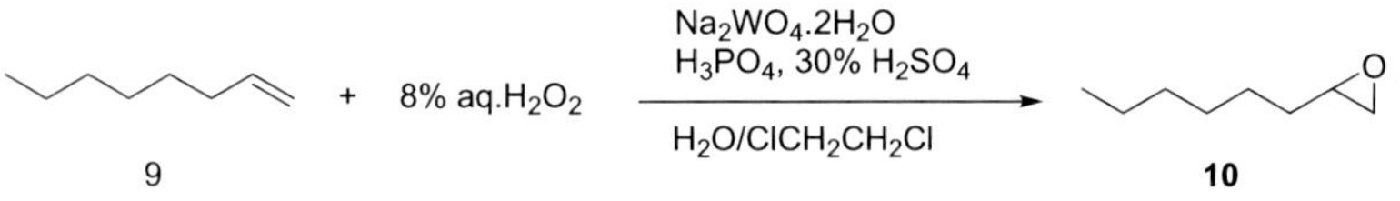

Scheme 5 Venturello's approach of terminal epoxidation

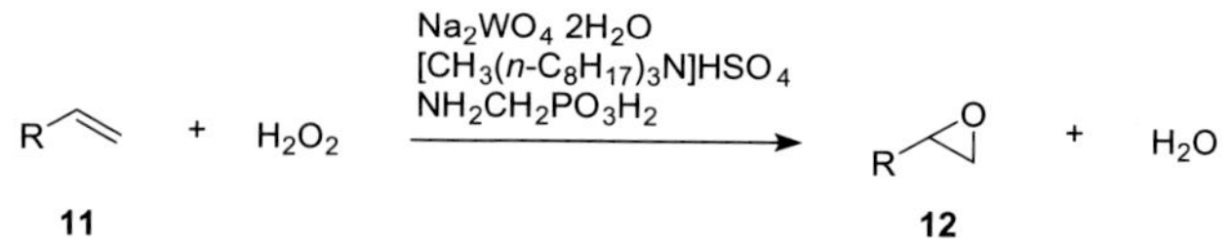

2 Catalytic Oxidation Process

$$R\diagdown + H_2O_2 \xrightarrow[\text{NH}_2\text{CH}_2\text{PO}_3\text{H}_2]{\overset{\text{Na}_2\text{WO}_4\ 2\text{H}_2\text{O}}{[\text{CH}_3(n\text{-}C_8\text{H}_{17})_3\text{N}]\text{HSO}_4}} R\diagdown\!\!\overset{O}{\triangle} + H_2O$$

11   **12**

Scheme 6 Noyori's approach of terminal epoxidation

Subsequently, several approaches have been developed to date using different catalytic and oxidant system. Transition metals like Re [6–8], Ru [9–12], Mn [13–15], Fe [16] have eminent role in epoxidation of least active olefins. Some of them are shown in Scheme 7.

2.1.2 Vicinal Dihydroxylation

Vicinal dihydroxylation of alkenes is a well-known reaction catalyzed by Osmium tetroxide and single electron donors like N-methylmorpholine-N-oxide or TBHP which produce *cis* diols exclusively. Sharpless and coworkers came with a better catalytic system addressing the limitations of previous methods. Notably, the previous methods were based on the stoichiometric use of OsO_4 in pyridine followed by reductive hydrolysis [17]. Sharpless employed the alkaline condition and 0.2% of OsO_4 in tert-butyl alcohol to yield vicinal diol **24** exclusively (Scheme 8) [18]. The reaction follows oxometal mechanism in which $Os^{(VIII)}O_4$ undergoes [2 + 2] cycloaddition reaction with alkene to produce an oxametallocycle as intermediate which subsequently rearrange to give an Os(VI)-diol complex (Scheme 8). Further, Os(VI)-diol complex reacts with an oxidant to afford diol product and regenerates the original Osmium tetroxide catalyst. Sharpless and coworkers further modified the process for Osmium-catalyzed reaction in acidic medium. They found improved yield when citric acid is used as additive and 4-methylmorpholine-N-oxide (NMO) played the role of oxidant in comparison with the reaction in standard condition (Table 3). This reaction efficiently transforms the electron-deficient substrates like α,β-unsaturated esters and amides to corresponding diols with excellent yield [19].

Recently, Beller and coworkers used more greener alternative, the molecular oxygen as oxidant in Osmium-catalyzed dihydroxylation reaction. The reaction has several advantages as it is chemoselective, enantioselective, and greener in terms of atom economy. The reaction is carried out in basic medium (pH = 10.4) with potassium osmate as catalyst in tert-butanol affords the vicinal diols **29** in high yield. The Osmium re-oxidized in basic medium in the presence of molecular oxygen. The chemoselectivity of the catalyst is found to be improved as 1.5 mol% 1,4-diazabicyclo[2.2.2]octane (DABCO) or 1.5 mol% quinuclidine used as ligands (Scheme 9) [20, 21].

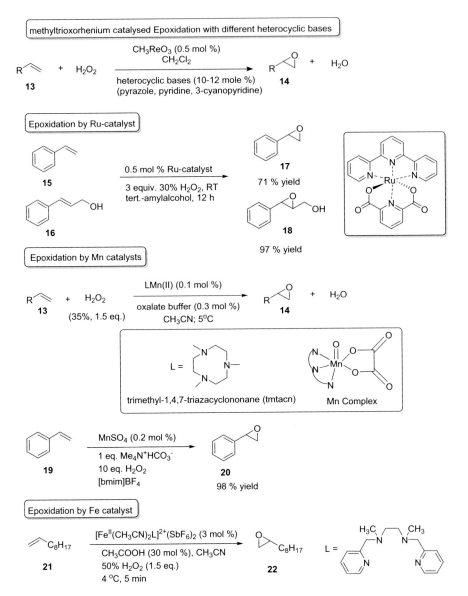

Scheme 7 Epoxidation of olefins by different catalyst systems

2.1.3 Oxidative Cleavage of Alkenes

The oxidative cleavage of olefinic double bond is an important synthetic tool for degrading the large molecule or introducing oxygen as functionalities such as ketone, carboxylic acid, and aldehyde into the molecule. Ruthenium-catalyzed oxidative cleavage is well-documented in the literature with oxone, NaIO$_4$ and NaOCl as

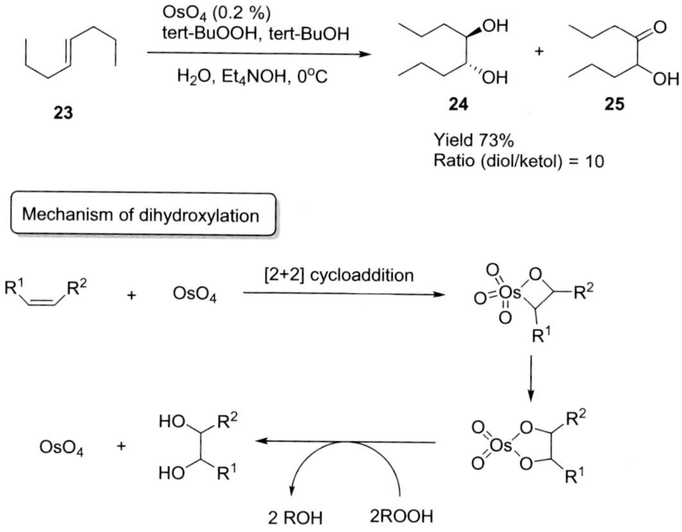

2 Catalytic Oxidation Process

Scheme 8 Vicinal dihydroxylation and its mechanism

primary oxidants. Ruthenium tetroxide was first used as oxidant in 1953 [22]. Since then, a number of different oxidation routes have been developed to transform several organic substrates using $NaIO_4$ and perchlorates as stoichiometric oxidants [23, 24]. Sharpless et al. used catalytic route consisting of $RuCl_3$ and $NaIO_4$ in CCl_4-CH_3CN-H_2O (2:2:3) solvent system. This method afforded carboxylic acid **31** in excellent yield as the oxidized product (Scheme 10a) [25]. Further, two popular methods were developed to obtain aldehyde as the oxidized products from olefins. One is ozonization followed by reductive workup and another one is the use of Osmium tetroxide-periodate system [26, 27]. Yang and Zhang developed three new protocols for oxidative cleavage of olefins (**32, 34, 36**) to aldehydes (**33, 35, 37**, respectively) rather than carboxylic acid (Scheme 10b–d). They exploited ruthenium trichloride as the catalyst (3.5 mol %), Oxone as the primary oxidant, $NaHCO_3$ as buffer medium and acetonitrile–water (1.5:1) as solvent system for aryl olefins. Aliphatic olefins gave excellent results in $RuCl_3$-$NaIO_4$ in 1,2-dichloroethane-H_2O (1:1) system. Further, the terminal aliphatic olefins were cleaved to the corresponding aldehydes in excellent yields by using $RuCl_3$-$NaIO_4$ in CH_3CN-H_2O (6:1) [28].

Various attempts have been made to introduce some greener approach for oxidative cleavage of olefins. Heterogenous recyclable Ru nanoparticles [29] and solvents like hydrogen peroxides were used as primary oxidants. Ogawa and coworkers reported a facile conversion of olefin **39** to carboxylic acid **40** with 35% H_2O_2 and tungstic acid in t-BuOH (Scheme 11a) [30]. Similarly, Venturello and colleagues transformed alkenes **42** to corresponding carboxylic acids **43** and **44** with 40% aq. H_2O_2 and methyltrioctylammonium tetrakis(oxodiperoxotungsto)phosphate(3-) **45** as catalyst with high yield and selectivity in a biphasic solvent system (Scheme 11b) [31]. Klein Gebbink et al. developed iron catalyst **49** and reported one-pot synthetic method

Table 3 Dihydroxylation of electron-deficient olefins under (a) "standard" versus (b) new, "acidic" conditions

$$R^1 \diagdown R^2 \quad \mathbf{26} \xrightarrow[\substack{\text{(a) water/acetone/tert-BuOH, 5:2:1} \\ \text{(b) water/tert-BuOH, 1:1} \\ \text{citric acid, 25 mol\%}}]{\substack{\text{NMO, 1.1 equiv.} \\ \text{OsO}_4,\ 0.2\ \text{mol\%}}} \quad \substack{HO\ R^2 \\ R^1\ OH} \quad \mathbf{27}$$

Entry	Product	Yield (%)	
		(a) Standard conditions	(b) New conditions
1	**27a**	50	96
2	**27b**	<10	76
3	**27c**	45	67

(continued)

Table 3 (continued)

Entry	Product	Yield (%)	
		(a) Standard conditions	(b) New conditions
4	**27d**	<40	78
5	**27e**	30	77

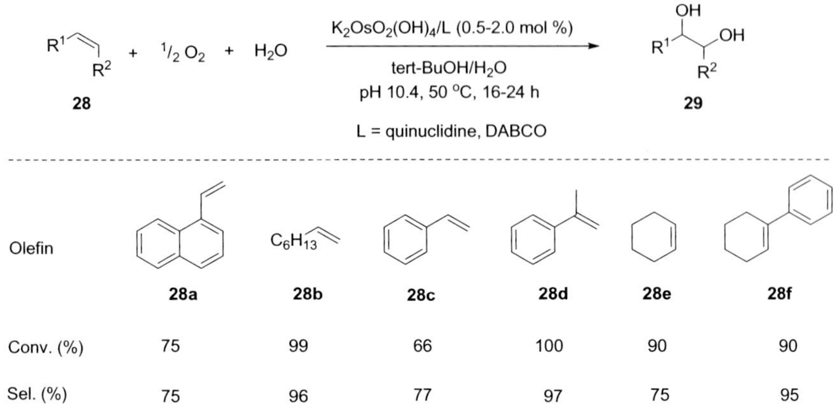

218   6 Catalysis: Application and Scope in Organic Synthesis

Scheme 9 Osmium-catalyzed hydroxylation using molecular oxygen as oxidant

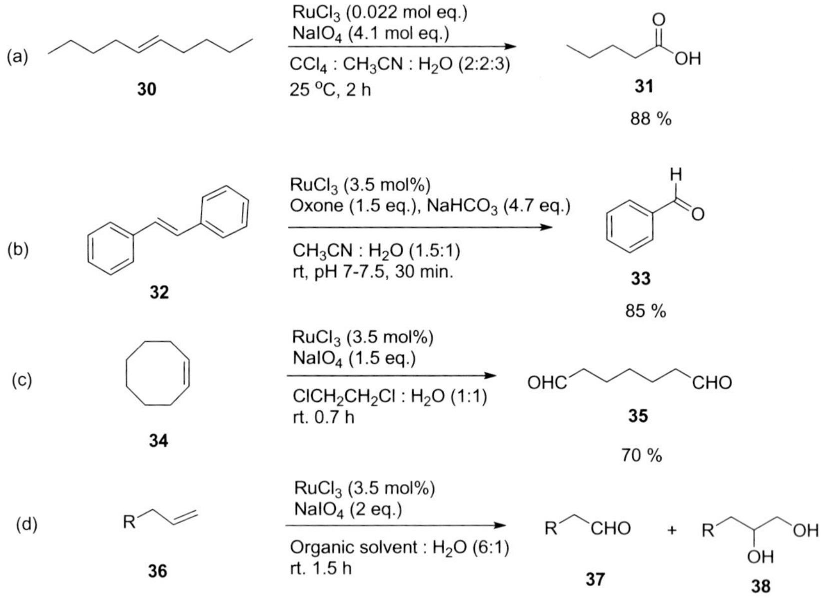

Olefin	28a	28b	28c	28d	28e	28f
Conv. (%)	75	99	66	100	90	90
Sel. (%)	75	96	77	97	75	95

Scheme 10 Oxidative cleavage of olefins

for oxidative cleavage of internal olefins **46** into aldehydes **48** (Scheme 11c) [32]. Recently, Liu developed 1,4-dioxane promoted oxidative cleavage of olefins in the presence of O_2 without any use of catalysts and additives [33].

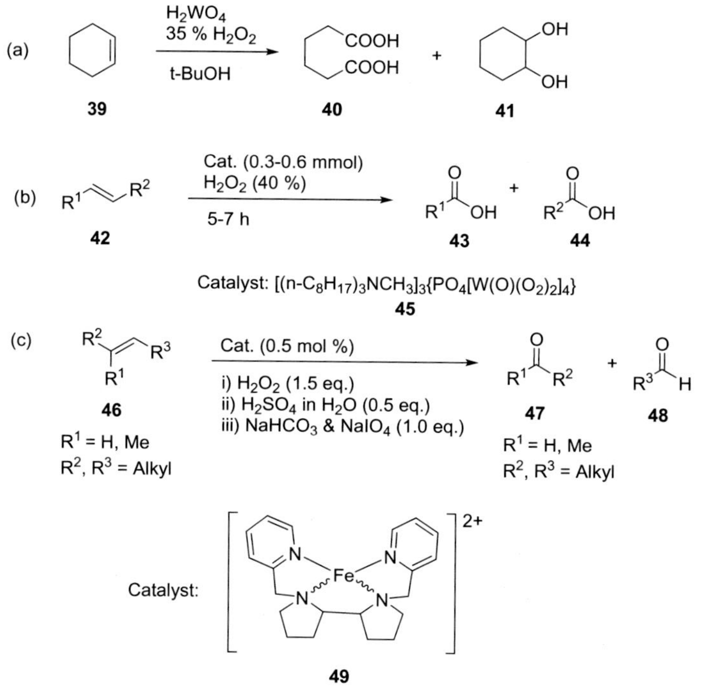

2 Catalytic Oxidation Process

219

Scheme 11 Oxidative cleavage of olefins with tungsten and iron catalyst in the presence of H_2O_2

2.1.4 Oxidative Ketonization of Alkenes: Wacker Oxidation

Aerobic oxidation of ethene to acetaldehyde and methyl ketonization of terminal alkenes in the presence of $PdCl_2/CuCl_2$ catalyst is referred as Wacker oxidation and is one of the well-studied oxidation processes of alkenes. Wacker oxidation is a multistep reaction in which one step involves the hydroxypalladation of alkene through nucleophilic attack by hydroxide ion on a palladium–olefin coordinated complex **50** followed by a β-hydride shift which results in acetaldehyde/methylketone and Pd(0) formation. Further, the Pd(0) re-oxidized by $CuCl_2$ in acidic medium (Scheme 12).

The major disadvantage of this process is the emission of huge amount of environmentally hazardous chlorinated by-products and its highly corrosive nature. This necessitates the use of greener approach towards oxidation of olefins. Several attempts have been made to make it acid-free $PdCl_2/CuCl_2$-catalyzed oxidations. For instance, Ansari and coworkers used ionic liquids like [bmim][BF_4] and water as solvent [34], Hou et al. used supercritical CO_2/[bmim][PF_6] [35] and Kulkarni et al. explored Pd(0)/C-$KBrO_3$ catalyst system as alternate approach [36].

Sheldon and coworkers reported copper and chlorine-free, recyclable, water-soluble Pd (II)-diamine catalytic system **51** which selectively oxidize the terminal olefins **52** to corresponding ketones **53** in a biphasic system (Scheme 13) [37]. Simi-

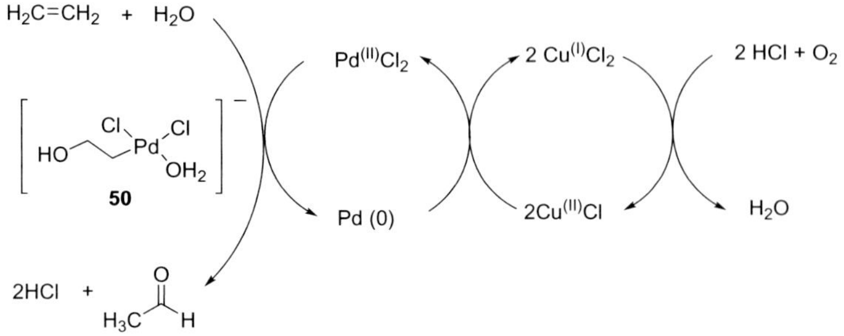

Scheme 12 Mechanism of Wacker oxidation of ethene

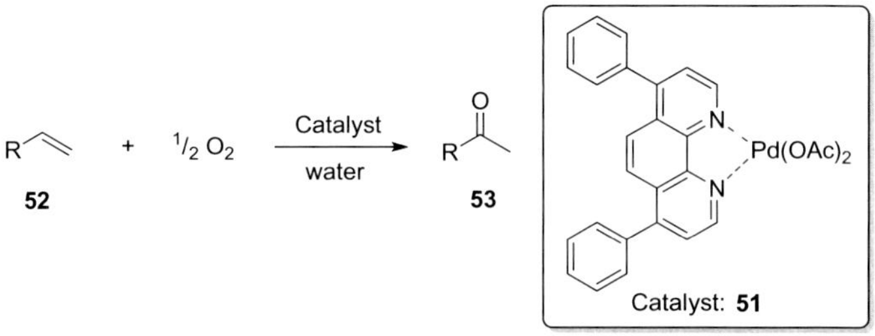

Scheme 13 Wacker oxidation with soluble Pd catalyst

Scheme 14 Wacker oxidation with PdCl$_2$/N,N-dimethylacetamide system

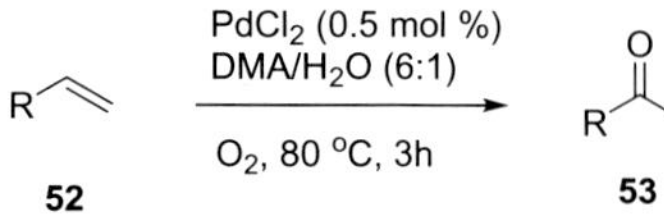

larly, PdCl$_2$ with N,N-dimethylacetamide (DMA) solvent was found to be simple, efficient, and Cu/HCl-free oxidizing catalyst system for alkenes **52** transformation to ketones **53** was explored by Kaneda and coworkers (Scheme 14) [38].

2.2 Oxidation of Alkanes

Selective oxidation of poorly reactive C–H bond is one of the challenging tasks for synthetic chemists [39]. The problem lies in transformation and stabilization of more reactive products from less reactive hydrocarbons. To observe the oxidation process in such hydrocarbons, generally drastic conditions are employed. Several enzymes like Cytochrome-450 and methane oxygenase employ highly reactive Fe–Oxo species to oxidize unreactive alkanes [40]. Influenced by this, several high-valent transition metal-oxo catalyst systems were developed such as Ru, Os, Fe, Cr, and

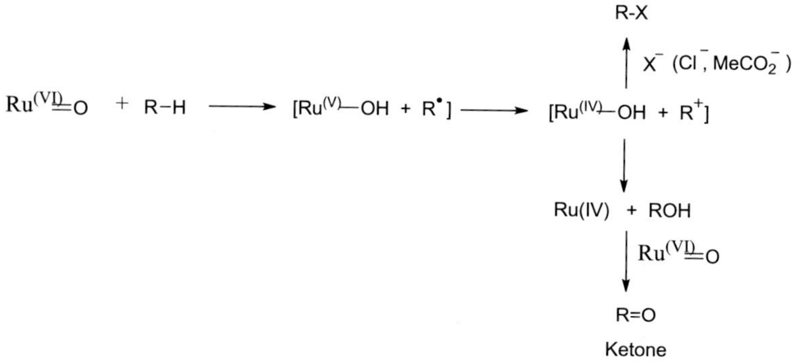

2 Catalytic Oxidation Process

$$Ru^{(VI)} = O + R-H \longrightarrow [Ru^{(V)}-OH + R^{\bullet}] \longrightarrow [Ru^{(IV)}-OH + R^{+}]$$

R-X

$\uparrow$ X⁻ (Cl⁻, MeCO₂⁻)

Ru(IV) + ROH

$\downarrow$ Ru$^{(VI)}$=O

R=O

Ketone

Scheme 15 Ruthenium-oxo-catalyzed reaction mechanism

Mn along with stoichiometric amounts of peroxides and peracids for the oxidation of alkanes. Among these transition metal-oxo complexes, Ru–oxo catalytic system was widely explored [41]. For example, Lau and Mak reported the oxidation of alkane to corresponding alcohols and ketones using barium ruthenate(VI) in acetic acid [42]. The proposed mechanism showed the involvement of free radical (Scheme 15). Same group developed a protocol for oxidation of alkanes using different peroxides (tert-BuOOH, H_2O_2, $PhCH_2C(CH_3)_2OOH$) and $[Os^{VI}(N)Cl_4]^-$/Lewis acid ($FeCl_3$ or $Sc(OTf)_3$) as catalytic system in acetic acid and dichloromethane. They also proposed that the mechanism involves the transfer of oxygen from peroxide to $[Os^{VI}(N)Cl_4]^-$/Lewis acid to generate $[Os^{VIII}(N)(O)Cl_4]^-$/Lewis acid followed by oxidation of alkane by hydrogen abstraction. [43]. Drago et al. used *cis*-dioxoruthenium complex **54** as catalyst and H_2O_2 as primary oxidant for hydroxylation of methane (Scheme 16) [44]. Periana et al. reported platinum-bipyrimidine complex in concentrated sulfuric acid as catalytic system for hydroxylation of methane.

Scheme 16
Ruthenium-oxo-catalyzed
hydroxylation of methane

$$CH_4 \xrightarrow[\text{Catalyst}]{H_2O_2/H_2O,\ 75\ ^\circ C} CH_3OH + CH_2O$$

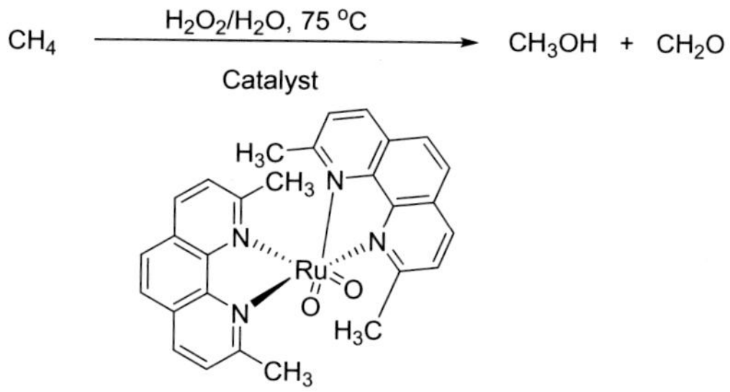

54

2.3 Oxidation of Aromatic Hydrocarbon

Oxidative transformation of aromatic hydrocarbons offers a number of aromatic feedstocks to petrochemical industries such as carboxylic acids, ketone, aldehydes, alcohols, and epoxides. The oxidation of aromatic ring is quite challenging task as the radical intermediates abstract hydrogen from alkyl side chain rather than from the nucleus and the product formed, i.e., the phenol, is more reactive than the precursor compound. The oxidation of alkyl side chain is comparatively easier than the oxidation of benzene ring. Oxidation of aromatic hydrocarbons may undergo one of the three pathways. (1) Formation of oxygen-containing compounds without cleavage of aromatic ring, i.e., formation of phenolic **56** or benzoquinone compound **57** (Scheme 17a). (2) Formation of oxygen-containing compounds due to the cleavage of the aromatic rings, e.g., formation of maleic anhydride **58** from benzene **55** (Scheme 17b). (3) Oxidation of side chain of alkyl aromatics, e.g., formation of benzaldehyde or benzoic acid **60** from toluene **59** (Scheme 17c).

2.3.1 Benzene Ring Oxidation

The vapor phase oxidation of benzene to maleic anhydride is one of the oldest catalytic processes adopted by industries around 1930s. Recently, Bianchi et al. reported a novel route for direct oxidation of benzene to phenol. Notably, this reaction is important because selective oxidation of benzene to phenol is a challenging task as over-oxidation of benzene **55** occurs to produce by-products like catechol, hydroquinone, and benzoquinones. They employed $FeSO_4.7H_2O$ as catalyst, 5-carboxy-2-methylpyrazine-N-oxide **61** as ligand and trifluoroacetic acid as cocatalyst. The reaction is carried out in biphasic medium in the presence of H_2O_2 (Scheme 18) [45].

Scheme 17 Formation of various oxygenates from aromatic hydrocarbons

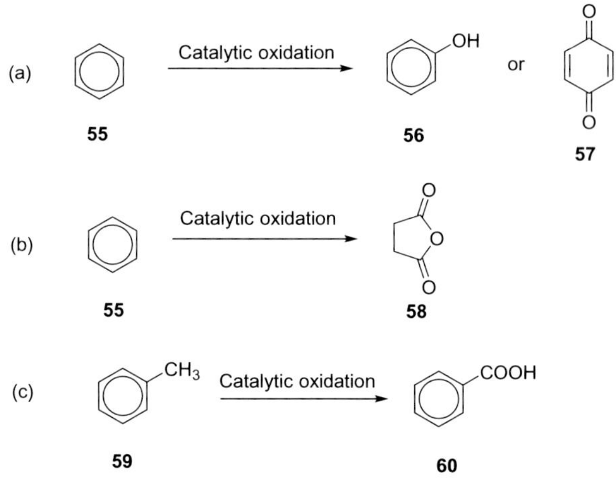

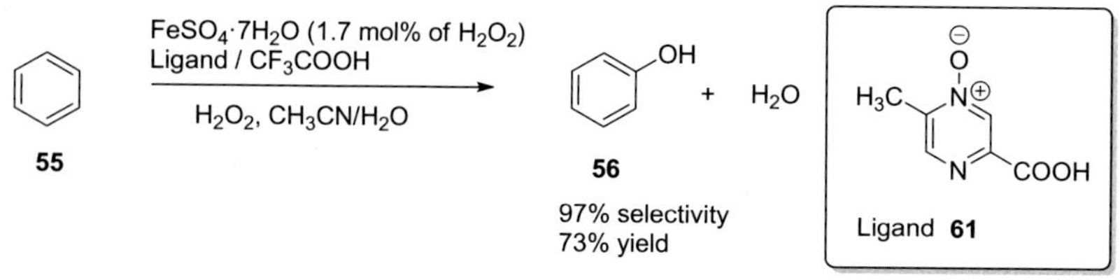

3 Catalytic Reduction

Scheme 18 Direct oxidation of benzene to phenol

2.3.2 Aromatic Side Chain Oxidation

The classical oxidation processes for the oxidation of side chain of aromatic system employed hazardous substrates, like chlorine or nitric acids in stoichiometric amount which in turn not only produces notorious by-products but the atom economy is also substantially low (Scheme 19). On the other hand, catalytic aerobic oxidation results in green by-product like water, and the atom utilization is higher as well.

3 Catalytic Reduction

Hydrogenation of organic compounds using molecular dihydrogen and heterogenous catalysts is one of the important methods used by organic chemists both at the laboratory and industrial levels. Hydrogen gas is one of the simple and clean reducing agents. However, alcohols, ethers, and amines are also used as hydrogen donors via transfer hydrogenation reactions. While hydrogenation directly employs hydrogen gas and transition metals for catalytic reduction, transfer hydrogenation reaction uses hydrogen donors like amines, alcohols, and ethers (other than dihydrogen molecule) to transfer H to acceptor molecules [46]. Hydrogenation is the first catalytic process used in fine chemical manufacturing industries. The hydrogenation of unsaturated

1) Chlorination, Atom economy: 36%

$$ArCH_3 \ + \ 3Cl_2 \longrightarrow ArCCl_3 \ + \ 3\,HCl$$

$$ArCCl_3 \ + \ 2H_2O \longrightarrow ArCOOH \ + \ 3\,HCl$$

2) Nitric acid oxidation, Atom economy: 56%

$$ArCH_3 \ + \ 2\,HNO_3 \longrightarrow ArCOOH \ + \ 2\,NO \ + \ 2H_2O$$

3) Catalytic aerobic oxidation, Atom economy: 87%

$$ArCH_3 \ + \ {}^1/_2\,O_2 \longrightarrow ArCOOH \ + \ H_2O$$

Scheme 19 Classical versus catalytic oxidation of aromatic side chain hydrocarbons, ketones, and reductive amination, with the aid of heterogenous catalysts, has been widely used more than 30 years [47]. Heterogenous catalysis is usually faster and easy to separate the catalyst from reaction mixture but enantioselectivity is generally lower than homogenous and biocatalytic processes. The major breakthrough in the field of homogenous asymmetric catalysis was provided by the pioneer works of Noyori, Knowles and Kagan by the discovery of ligands like BINAP, DIPAMP, and DIOP [48, 49]. Asymmetric catalysis furnish maximum chiral efficiency through the combination of these chiral ligands and transition metals like Rh, Ru, and Ir in the suitable reaction conditions. Besides heterogenous and homogenous catalysis, biocatalytic reduction has been emerged as striking field in the synthesis of chiral organic molecules for more than two decades not only for academic purpose but also at production scale [50].

3.1 Reaction Conditions in Heterogenous Catalysis

For hydrogenation, the choice of catalyst is made based on trial-and-error method. This is the reason why the optimal conditions like choice of catalyst, support system, additives/modifiers are not found in the literature. Various factors which influence the reactivity are the following:

A Type of catalyst: There may be two types of heterogenous catalysis-supported and unsupported. Furthermore, supported catalysts go through slurry process or fixed bed operation.

B Type of metal: Noble metals like palladium, platinum, chromium, rhodium, nickel is generally used. They are usually supported system. Bimetallic systems are also popular.

C Type of support: Charcoal, Al_2O_3, or silicas.

D Metal loading: Slurry-type noble metal batch usually employed 1–10% of metal. Fixed bed type is usually employed with low metal loading, often 0.1–1.0%.

E Solvent: Catalytic hydrogenation is very often carried out in different solvents. It leads to easy handling, easy recovery of catalyst, enhanced reactivity and selectivity and moderates exothermic reactions. Generally, used solvents are methanol, ethanol, ethyl acetate, water, acetic acid, and methoxyethanol.

F Modifiers and promoters: Catalyst poisons are generally used to slow down the activity of the metal catalyst. Common poison includes divalent sulfur compounds, carbon monoxide, nitrogenous compounds, and phosphines.

3.2 Hydrogenation of Alkenes

The unsaturated carbon–carbon double bond among organic compounds is most commonly hydrogenated unless they are highly sterically hindered. The most common catalyst systems for alkene hydrogenation in liquid phase are palladium,

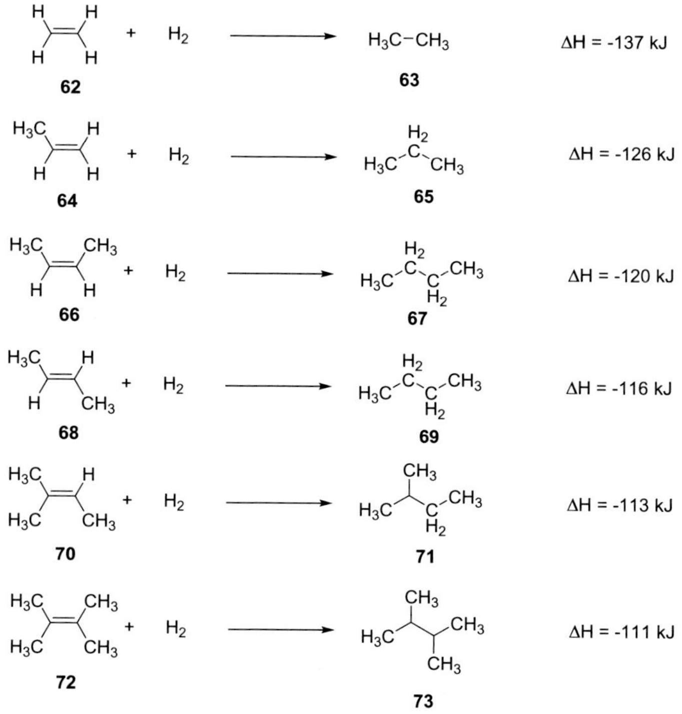

3 Catalytic Reduction   225

Scheme 20 Heat of hydrogenation of differently substituted alkenes

platinum and activated metals like Raney Nickel under mild conditions. However, elevated temperature and pressure result in rapid reaction with reduced metal loading and at larger-scale production. Mono- and disubstituted alkenes react rapidly with higher heat of hydrogenation as compared to tri- and tetra-substituted alkenes (Scheme 20).

3.2.1 Mechanism of Hydrogenation of Alkenes

The classical hydrogenation of ethylene is carried out in the presence of activated Raney Ni or supported metals like Ni and Cu. Since the reaction is operated on the surface of metal, larger surface area is required for rapid reaction. The whole process of hydrogenation is supposed to proceed in stepwise manner rather than in a concerted way (Fig. 2). The activated surface of metal adsorbs the dihydrogen molecule on its surface which results in cleavage of H–H bond and formation of M–H bond. This step is called dissociative adsorption and the newly formed M–H bonds supplies the energy required to break the stronger H–H bond. Meanwhile, the ethylene molecule is also adsorbed on the surface of metal through π bonds and gets activated. The metal brings the two substrates (hydrogen atom and activated

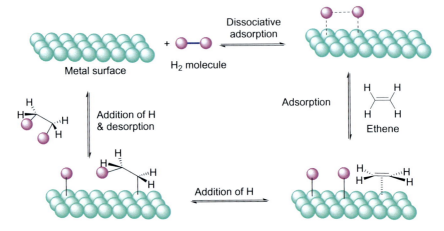

Fig. 2 Mechanism of ethene hydrogenation

ethylene) close enough so that the addition of hydrogen to ethylene would take place successfully. The final step is addition of second hydrogen atom in *cis* fashion to the ethylene, transforming them to ethane [51, 52].

The ease of hydrogenation of isolated double bond depends on the degree of substitution around the C=C. Lesser substituted ethylenes hydrogenate faster than those which are highly substituted. Moreover, the size and degree of branching of substituents also contribute to variation in reactivity of ethylenes. The order of hydrogenation is following:

3.2.2 Selective Hydrogenation of Isolated Alkenes

Depending upon the degree substitution in the compounds containing two or more isolated double bonds, selective hydrogenation can be achieved easily with specified catalyst system. For example, (R)-(+)-Limonene **74** gets hydrogenated to (R)-(+)-carvomenthene **75** in excellent quantity in the presence of 5% Pt–C [53] or in the presence of Raney Nickel [54] (Scheme 21). This selectivity is not found when palladium catalyst was used.

Similarly, 4-Vinylcyclohexene **76** selectively hydrogenated to produce 4-ethylcyclohexene **77** in almost quantitative yield by P-2 Ni (Nickel boride) and Nic (finely divided Ni reduced with NaH-*t*-AmOH) in ethanol at 25 °C and 1 atm hydrogen (Scheme 22a). Furthermore, 5-methylenenorbornene **78** selectively hydro-

3 Catalytic Reduction

Scheme 21 Selective hydrogenation in the presence of Pt–C and Raney Ni

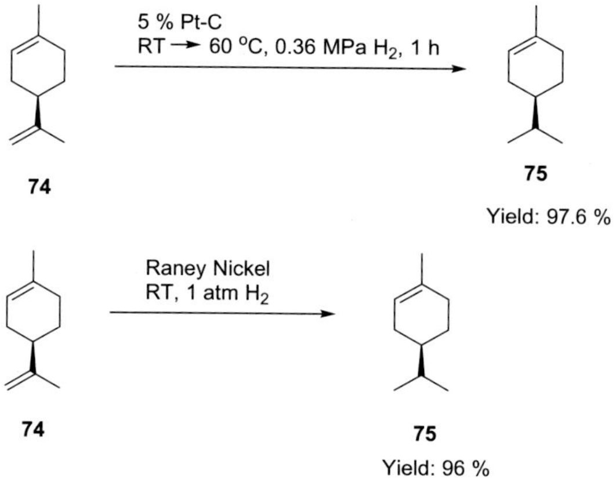

5 % Pt-C
RT → 60 °C, 0.36 MPa H$_2$, 1 h

74 **75**
Yield: 97.6 %

Raney Nickel
RT, 1 atm H$_2$

74 **75**
Yield: 96 %

Scheme 22 Examples of selective hydrogenation

(a)
P-2 Ni
25 °C, 1 atm H$_2$
EtOH
97% yield
76 **77**

(b)
P-2 Ni
EtOH
96 % yield
78 **79**

(c)
P-2 Ni
95% EtOH
RT, 1 atm H$_2$
90 % yield
80 **81**

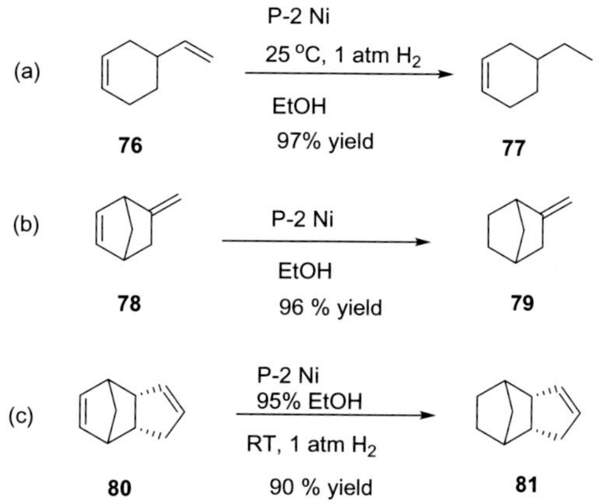

genated to produce 2-methylenenorbornane **79** with 96% yield in the presence of P-2 Ni catalyst (Scheme 22b). *Endo*-dicyclopentadiene **80** readily hydrogenated over P-2 Ni to afford 5,6-dihydro-endo-dicyclopentadiene **81** in excellent yield (Scheme 22c).

3.2.3 Hydrogenation of Aryl-Substituted Ethylenes

Zartman and Adkins studied the hydrogenation of phenyl-substituted ethylene in the presence of Ni-kieselguhr and Cu–Cr oxide as catalyst [55]. The rate of hydrogenation markedly depends upon the temperature, pressure of hydrogen gas, and the structure of ethylenic linkage (Scheme 23).

228   6 Catalysis: Application and Scope in Organic Synthesis

Scheme 23 Hydrogenation of aryl-substituted ethylenes

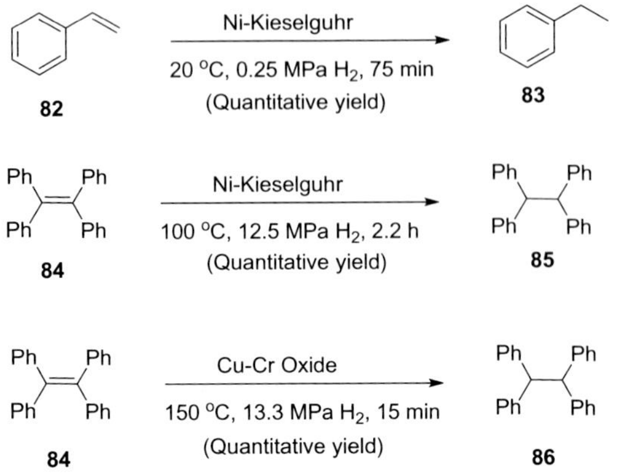

3.2.4 Hydrogenation of α,β-Unsaturated Acids and Esters

The carbon–carbon double bond conjugated to carbonyl functionality gets hydrogenated faster than usual olefinic double bond in the presence of Ni and Pd catalyst (Scheme 24) [56, 57].

3.3 Hydrogenation of Alkynes

Hydrogenation of acetylenes to saturated hydrocarbons is carried out on a variety of catalytic systems under mild condition with uptake of two moles of dihydrogen. Broadly speaking, the hydrogenation of alkyne is two-step process; the second hydrogenation of resulted olefin is usually faster than the first hydrogenation of alkyne.

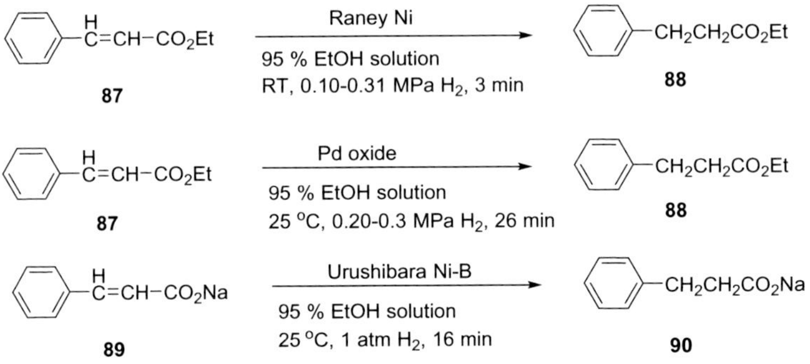

Scheme 24 Hydrogenation of α,β-unsaturated acids and esters

3 Catalytic Reduction

Selection of proper catalytic system, proper modifier and the suitable reaction condition markedly influence the reactivity and selectivity of hydrogenation. As the hydrogenation of acetylenes is exothermic process, therefore a careful handling of reaction regarding rise in temperature and amount of catalyst should be taken for high scale production. Bond and Wells studied the selectivity of hydrogenation process of various catalytic systems on acetylene, methylacetylene, and dimethylacetylene. They found that selectivity of hydrogenation decreases in the order: Pd $\gg$ Rh $\geq$ Pt > Ru $\gg$ Ir > Os [58, 59].

Semi-hydrogenation of carbon–carbon triple bond to olefin is of immense utility in fine chemical industries as acetylenic function is one of the most usable tools for C–C bond formation. Further, the semi-hydrogenation technique is utilized in the synthesis of vitamins and insect sex pheromones [60]. Palladium catalyst, among the transition metals, in combination with modifiers (supporter/inhibitor) have been most widely used for selective hydrogenation. One of the most popular selective catalysts was developed by Lindlar who poisoned $CaCO_3$-supported Pd catalyst by lead acetate in the presence of organic base quinoline. In general, the solvent used in partial hydrogenation with Lindlar's catalyst should be aprotic, and it is noted that protic solvents decrease the poisoning effect of modifiers. Some selected examples of partial hydrogenation of alkynes are depicted in Scheme 25.

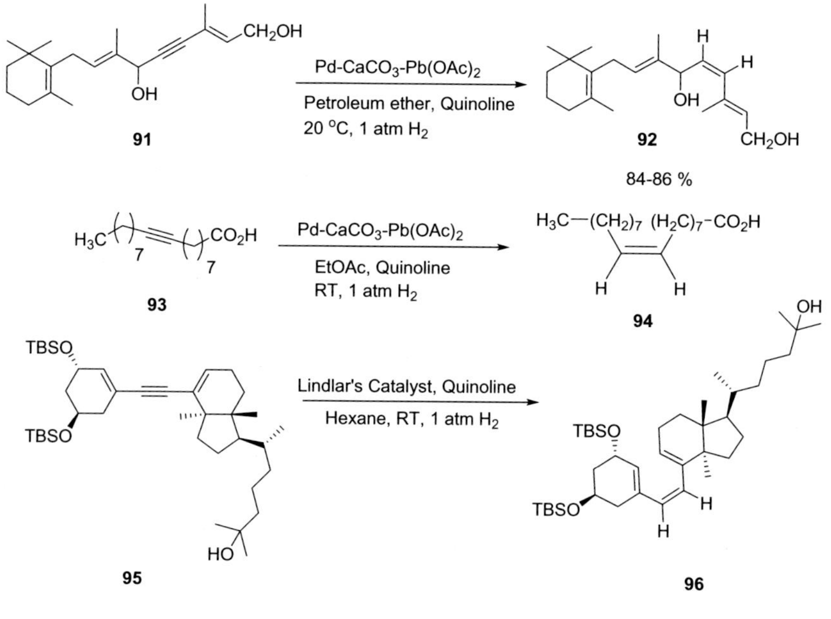

Scheme 25 Selected examples of partial hydrogenation of alkenes

3.4 Hydrogenation of Aldehydes and Ketones

Hydrogenation of aldehydes and ketones to corresponding alcohols is comparatively easier and could be carried out in the presence of several transition metals. The rate of hydrogenation and selectivity may vary depending upon the nature of catalyst, structure of the substrate, reaction medium, and reaction conditions. Aldehydes readily hydrogenated over Nickel and Cu–Cr Oxide catalyst to yield corresponding alcohols (Scheme 26). Pd catalyst is least active for the reduction of aliphatic aldehydes, but aromatic aldehydes are easily hydrogenated in mild conditions.

The unsaturated aldehydes, which are not conjugated, are prone to the reduction of C–C double bond selectively unless it is highly hindered. For example, 3-cyclohexenecarboxaldehyde **109** selectively reduced to corresponding alcohol **110** over Cu–Cr Oxide catalyst. Citronellal **111** an aldehyde also gets selectively hydrogenated over Ru catalyst poisoned by lead acetate (Scheme 27).

Similarly, aliphatic and alicyclic ketones are easily hydrogenated over a number of transition metals until they are highly hindered. Palladium catalysts are seldom active for hydrogenation of aliphatic ketone **115** but are highly active for aromatic ketones **113**. Raney Nickel and copper–chromium oxide, on the other hand, are highly active for both aliphatic and aromatic ketones (Scheme 28).

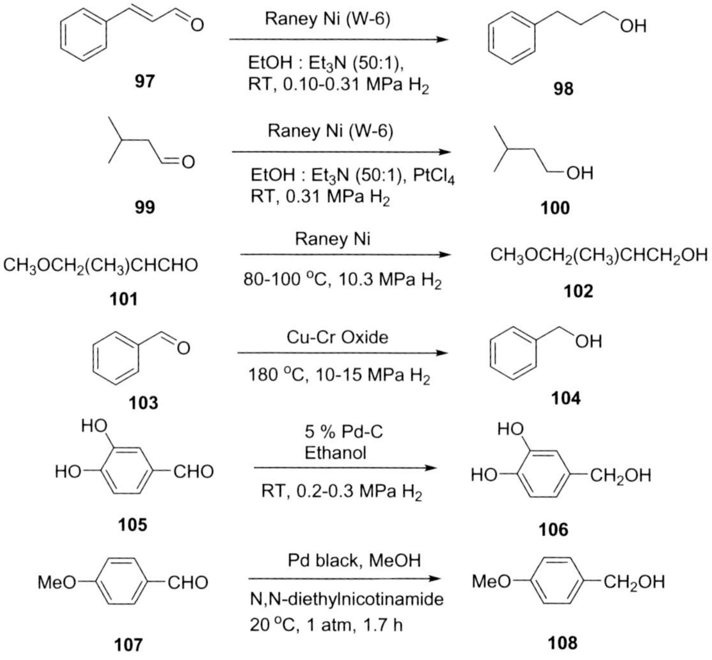

Scheme 26 Hydrogenation of aliphatic and aromatic aldehydes

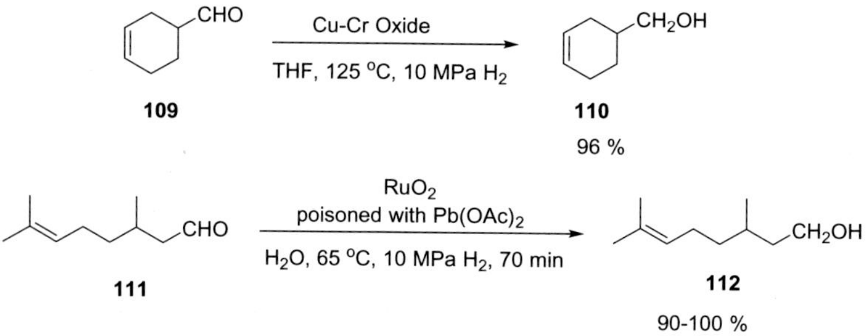

3 Catalytic Reduction

231

Scheme 27 Selective hydrogenation of aldehyde

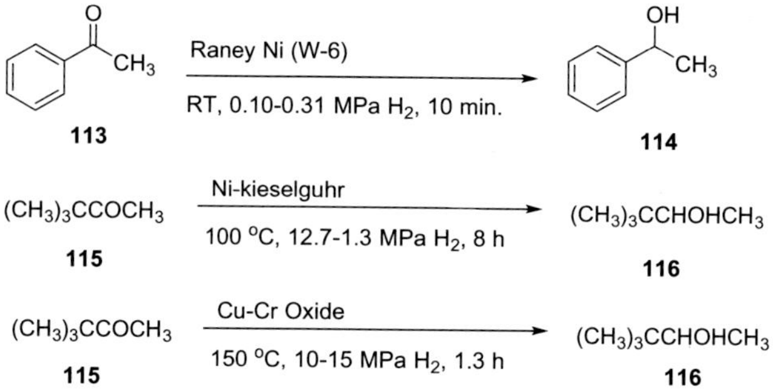

Scheme 28 Hydrogenation of aliphatic and aromatic ketones to corresponding alcohols

3.5 Catalytic Reductive Amination

Reductive amination is one of the powerful strategies for the synthesis of amines which could be further transformed into complex natural products, agrochemicals, pharmaceuticals, etc. The compounds with amine functionalities have a very special place in organic chemistry due to its presence in biological systems like amino acids, proteins, nucleic acids, and alkaloids. Many commercially used drugs have amines as key structural elements, and several organocatalysts, which are important for asymmetric synthesis, are composed of primary and secondary amines [61–63] (for more details see Chap. 7). Furthermore, amines act as bases and ionic liquids [64] in several organic transformations. Reduction of nitrogen-containing functional groups such as nitro, cyano, azides, carboxamide derivatives is an important strategy for amine preparation. Another generally used methods are the alkylation of ammonia, primary, and secondary amines.

Reductive amination is one of the widely used process in which aldehydes or ketones react with ammonia and amines (primary or secondary) in the presence reducing agents to give primary, secondary and tertiary amines, respectively (Scheme 29). The reaction initiates with the formation of adduct called aminol inter-

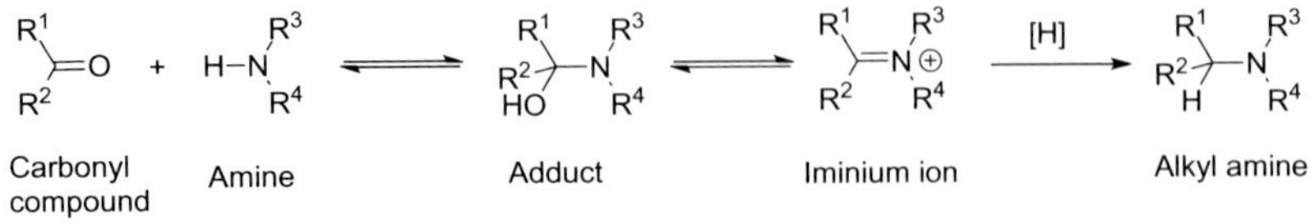

Scheme 29 A generalized reductive amination reaction

mediate which in suitable conditions dehydrate to yield iminium ions. Iminium ions, in the presence of suitable reducing agent get reduced to suitable alkyl amines.

Reductive amination is categorized on the basis of reducing agent used in a particular reaction as homogenous and heterogenous catalytic reductive amination. Various transition metals like Fe, Ru, Pd, Pt, and Ir along with symmetric and asymmetric ligands played the crucial role in catalytic reductive amination reaction. For instance, Boyle and Keating hydrogenated the imine bonds of folic acid **117** in the presence of optically active Rh catalyst which leads to chiral tetrahydrofolic acid **118** in good yield (Scheme 30a) [65]. Tetrahydrofolate is used in biosynthesis of nucleic acids. Similarly, Scorrano et al. asymmetrically hydrogenated the imine **119** using the chiral Rh catalyst, [Rh(nbd)(diop)]ClO₄ (Scheme 30b) [66].

Ruthenium metals with chiral ligands are used as an effective catalytic system for asymmetric hydrogenation of imines. A pertinent work was reported by Oppolzer and coworkers in which they asymmetrically hydrogenated the cyclic *N*-arylsulfonyl imines **122** to afford R-**124** and S-Sultam **123** using chiral Ru-BINAP complex (Scheme 31a) [67]. Similarly, Burk et al. came up with highly enantioselective

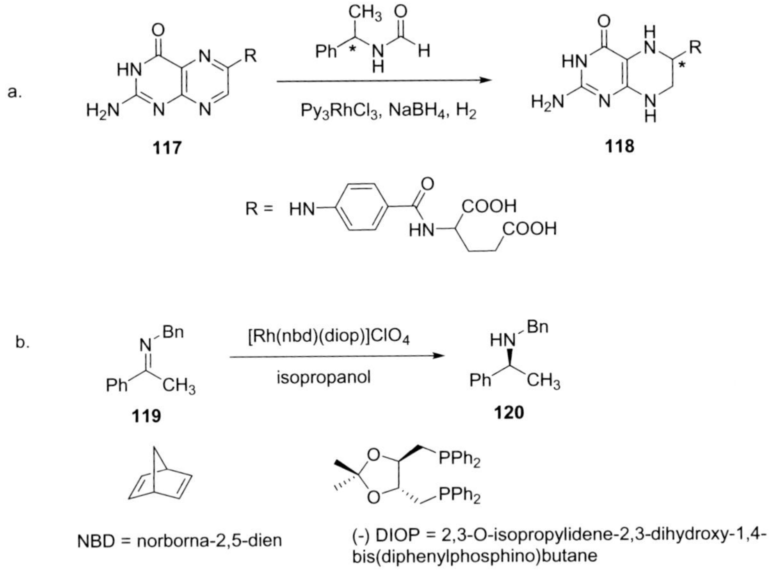

Scheme 30 Catalytic reductive hydrogenation of imines

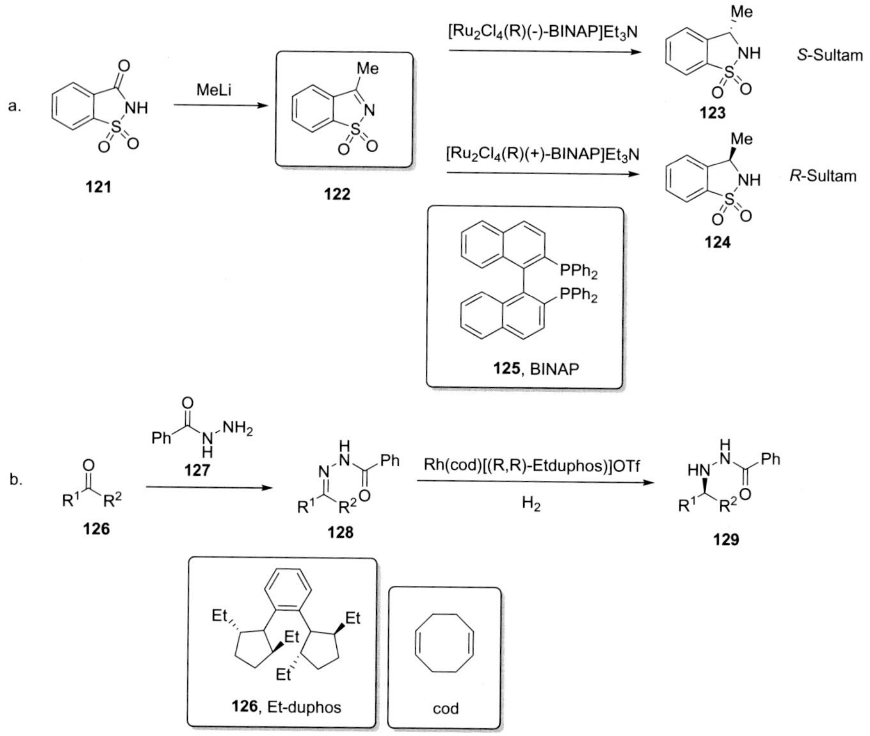

3 Catalytic Reduction

Scheme 31 Catalytic reductive amination using chiral ligands

reduction of *N*-acylhydrazones **128** with 2 mol% chiral Rh(I)-DuPHOS complex (Scheme 31b) [68].

Heterogenous catalysts such as Pt, Pd, Ni, or Ru are being used for cost-effective reductive amination of imines with hydrogen in industries for a long time. Hosseini and coworkers reported a two-step reductive amination using PtO_2 for a library of pyrrolidinone-containing dipeptide derivatives **131** in high yield and excellent diastereoselectivity (Scheme 32) [69].

Reduction of carbonyl compounds possessing other reducible functional groups like NO_2, cyano, C–C multiple bonds have serious limitations with heterogenous catalysis [70].

Scheme 32 Reductive amination using heterogenous PtO_2 catalyst

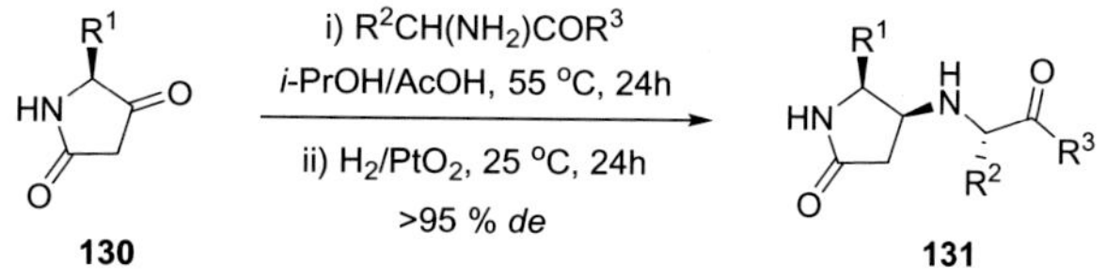

4 Catalytic C–C Bond Formation

The C–C bond formation is one of the most important key transformations which entirely changed the synthetic approach in organic chemistry. It is the heart of organic synthesis. The C–C bond formation could be achieved by carbonylation reaction in a convenient way. For example, Rhodium-catalyzed carbonylation of methanol is achieved to produce acetic acid in a significant yield [71]. These reactions are exemplified for 100% atom efficiency, and therefore, they have wide applications in fine chemical manufacturing industries [72]. The production of ibuprofen, an analgesic, with Hoechst-Celanese process is an elegant example of carbonylation with 100% atom efficiency. In this process, p-isobutylacetophenone **132** is first hydrogenated followed by carbonylation (Scheme 33). This reaction replaced the older route which had more numbers of steps and high E-factor.

Another elegant example of C–C bond formation is the production of lazabemide **136**, which is used as anti-Parkinsonian drug developed by Hoffmann-La Roche. In this process, 2,5-dichloropyridine **137** undergoes palladium-catalyzed amidocarbonylation in single step with 100% atom efficiency. This new route is highly efficient and replaced the classical synthetic route which had low yield and higher number of steps (Scheme 34) [73]. Similarly, palladium-catalyzed amidocarbonylation of aldehydes **139** in the presence of CO and amide **140** was developed to afford α-amino

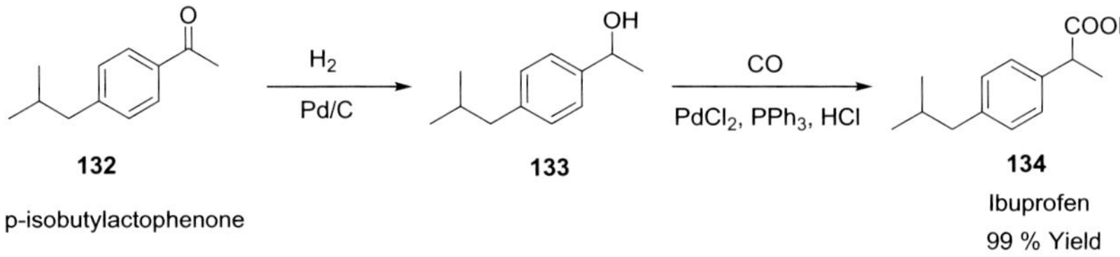

Scheme 33 Production of ibuprofen with Hoechst-Celanese process

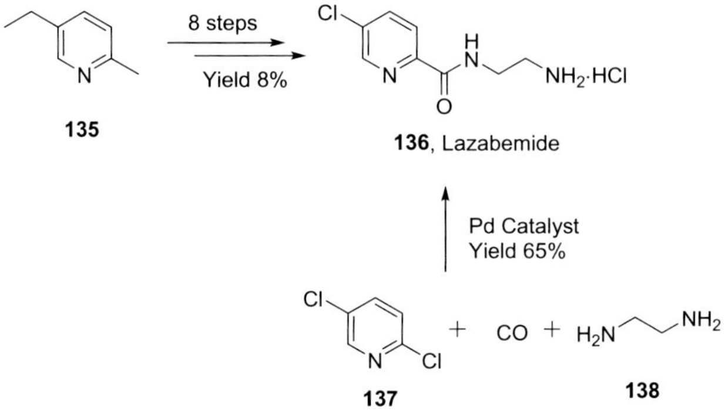

Scheme 34 Synthesis of lazabemide with classical and new routes

4 Catalytic C–C Bond Formation

acid **141** in excellent yield with 100% atom efficiency which further replaced the classical methods (Scheme 35) [74, 75].

Another widely used catalytic methodology for C–C bond formation is Heck and related coupling reactions. The Heck reaction is one of the most studied C–C coupling reactions which have a wide application in industries including agrochemicals, pharmaceuticals, fine chemicals, etc. [76–79]. This reaction drawn attention due to its simplicity, efficiency, chemoselectivity, and mild reaction conditions. The Heck reaction is known for palladium-catalyzed vinylation or arylation of alkenes where a large variety of olefins can be used (Scheme 36). The catalysts have a crucial role which comprise different transition metals along with a large variety of ligands [80].

Heck reaction is an alternative of Friedel–Crafts alkylation where an aromatic ring gets coupled with alkyl moiety. The Heck reaction has a great scope in syntheses of fine chemicals where classically longer and wasteful route were replaced with simple and shorter route which are cheap and environmentally benign. For example, the very first reported industrial use of Heck reaction is Matsuda–Heck coupling of a diazonium salt **142**, a step in the synthesis of herbicide Prosulfuron **146** (Scheme 37) [81, 82].

Similarly, Heck reaction is also found broad application in pharmaceutical industries. For instance, Naproxen, an anti-inflammatory drug sold under the brand name Aleve, was produced via new approach in which Heck coupling played the key role (Scheme 38) [83].

Another similar type of reaction which found wide application in pharmaceutical and fine chemical industries is the Suzuki coupling in which aryl boronic acid undergoes coupling with aryl halides to form C–C bond. The reaction affords high yield and more specifically, a clean linkage between two differently substituted aryl rings. For example, Valsartan **153**, an angiotensin II receptor antagonist synthesized by Hoechst process in which Suzuki reaction is the key step [82, 83]. In this reaction, p-tolylboronic acid **150** reacts with 2-chlorobenzonitrile **151** in the presence of palladium acetate and sulphonated triphenylphosphine **154** (acts as ligand) as catalytic system to yield 2-cyano-4'-methylbiphenyl **152** which further undergo multistep reactions to afford Valsartan (Scheme 39).

Furthermore, Merck synthesized Losartan **157**, an angiotensin receptor blocker used to treat high blood pressure and diabetic neuropathy, with Suzuki reaction at the

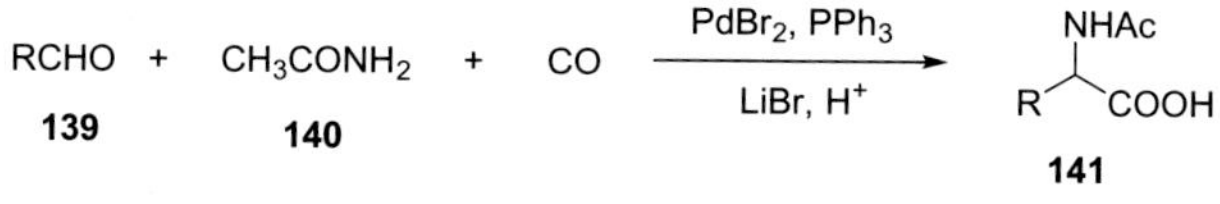

$$RCHO \; + \; CH_3CONH_2 \; + \; CO \quad \xrightarrow[\text{LiBr, } H^+]{PdBr_2, \; PPh_3} \quad R\overset{NHAc}{\underset{COOH}{\wedge}}$$

$$\mathbf{139} \qquad\qquad \mathbf{140} \qquad\qquad\qquad\qquad\qquad\qquad \mathbf{141}$$

Scheme 35 Amidocarbonylation of aldehyde

Scheme 36 A generalized Heck reaction

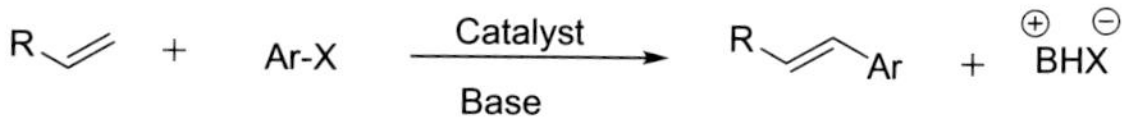

$$R\diagup \; + \; Ar\text{-}X \quad \xrightarrow[\text{Base}]{\text{Catalyst}} \quad R\diagdown\diagup Ar \; + \; \overset{\oplus}{B}\overset{\ominus}{H}X$$

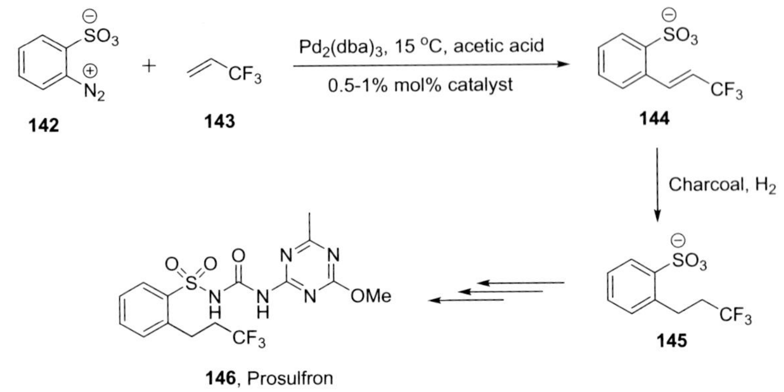

Scheme 37 First industrial application of Heck reaction

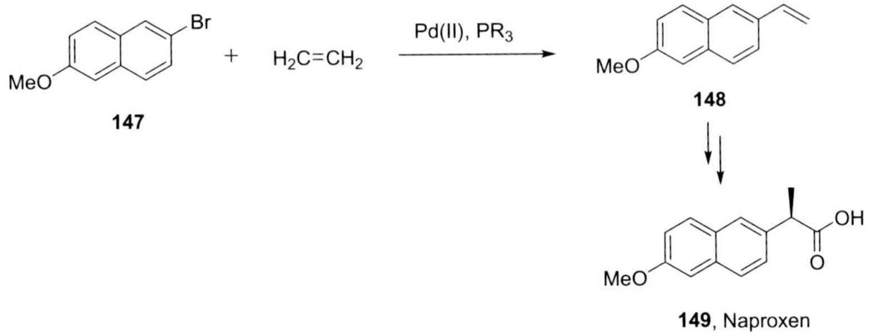

Scheme 38 Synthesis of Naproxen via Heck reaction

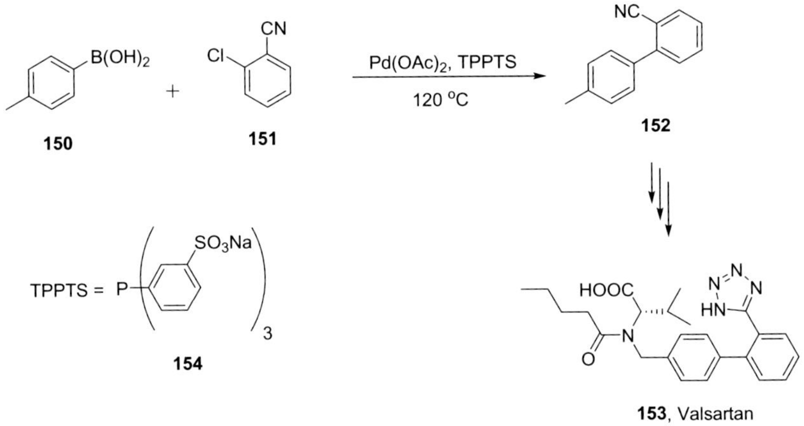

Scheme 39 Synthesis of Valsartan via Suzuki coupling reaction

4 Catalytic C–C Bond Formation

later stage of synthesis linking two highly substituted aryl rings with high selectivity (Scheme 40) [84].

Sonogashira reaction enables C–C cross-coupling reaction between terminal alkyne and an aryl or vinyl halide by employing palladium-phosphane ligand complex as catalyst and copper (I) salt as cocatalyst to afford alkynylated product which has wide applications in agrochemicals, natural products, fine chemicals, and pharmaceuticals (Scheme 41) [85]. For example, in the total synthesis of a natural product bulgaramine **164**, a benzindenoazepine alkaloid, the Sonogashira reaction is employed to couple aryl iodide **161** and the tris(isopropyl)silylacetylene **162** to synthesize a crucial intermediate **163** which further undergoes multistep reaction to afford the final product **164** (Scheme 42) [86].

In the same fashion, 2-Iodoindole **165** reacts with alkyne 166 in typical Sonogashira reaction conditions to form an intermediate which further employed in the total synthesis of $(-)$ aspidophytine **168** (Scheme 43) [87, 88].

The Negishi cross-coupling reaction of organozinc reagents with aryl halides in the presence of Pd or Ni catalysts is another valuable tool for C–C bond formation (Scheme 44) [89, 90]. It is reaction of choice for involvement of $C(sp^3)$ due to high rate of transmetalation of organozinc reagents. In contrast, the other organometals like organoboron and organotin show sluggish reactivity and react only when strong nucleophilic activators are used to facilitate transmetalation.

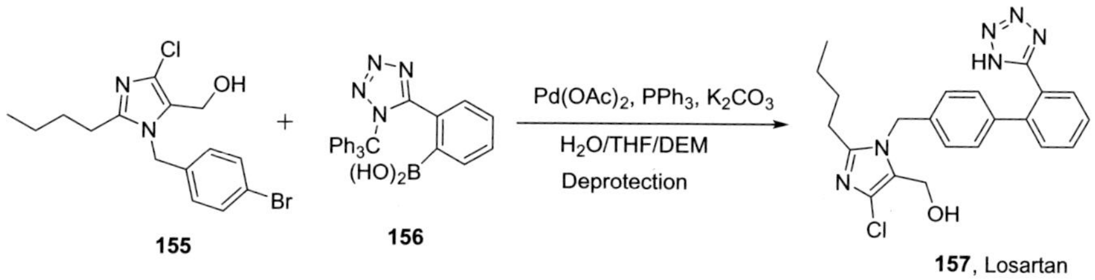

Scheme 40 Synthesis of Losartan via Suzuki coupling reaction

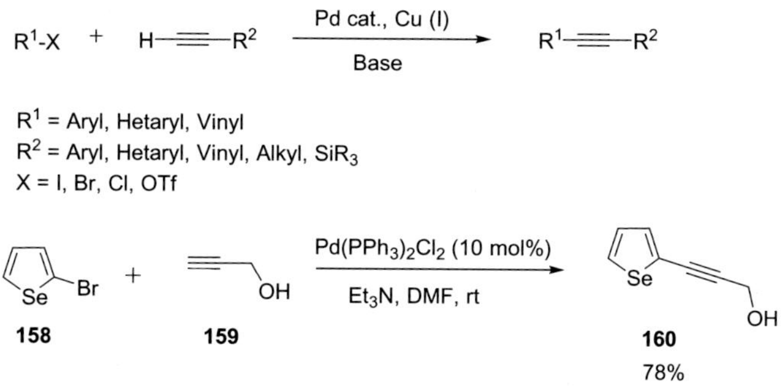

Scheme 41 An example of Sonogashira reaction

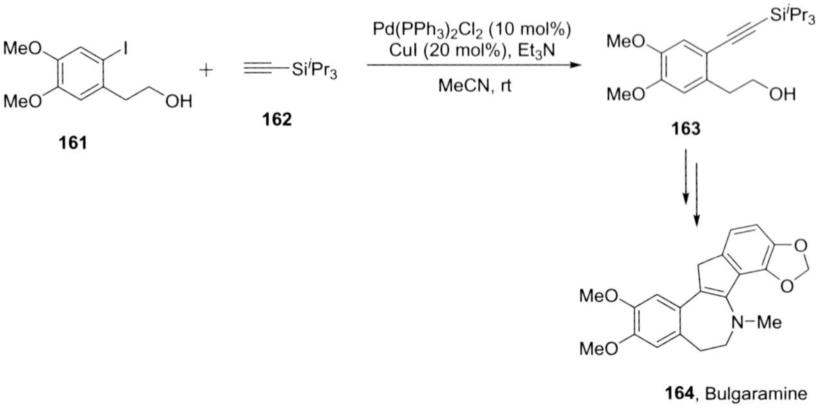

238   6 Catalysis: Application and Scope in Organic Synthesis

Scheme 42 Role of Sonogashira reaction in total synthesis of Bulgaramine

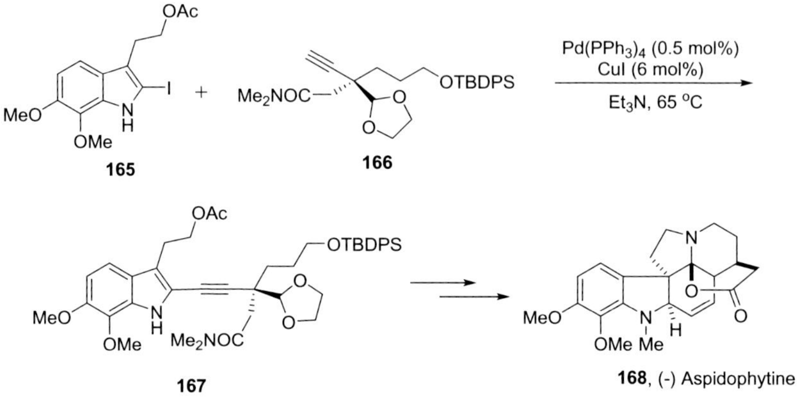

Scheme 43 Total synthesis of (−) Aspidophytine

Scheme 44 A generalized Negishi cross-coupling reaction

$$Ar^1\text{-}X \quad + \quad Ar^2\text{-}ZnX \quad \xrightarrow[\text{Solvent}]{\text{Pd/Ni Cat.}} \quad Ar^1\text{-}Ar^2 \quad + \quad ZnX_2$$

The Negishi reaction has found a wide application in the syntheses of pharmaceutical and natural products. For instance, Manley and colleagues scaled up 5-[2-Methoxy-5-(4-pyridinyl)phenyl]-2,1,3-benzoxadiazole (PDE472) **172**, an inhibitor of phosphodiesterase PDE4D isoenzyme which is a known drug target in the treatment of asthma, from laboratory to pilot-plant scale. They employed Negishi cross-coupling conditions at later stage of synthesis (Scheme 45) [91]. Similarly, Nolasco and coworkers synthesized cyclic tripeptides such as aminopeptidase inhibitor OF4949-III **176** and the ACE inhibitor K-13 **179** via intermolecular and intramolecular cyclization, respectively, employing Negishi coupling as a key step (Scheme 46)

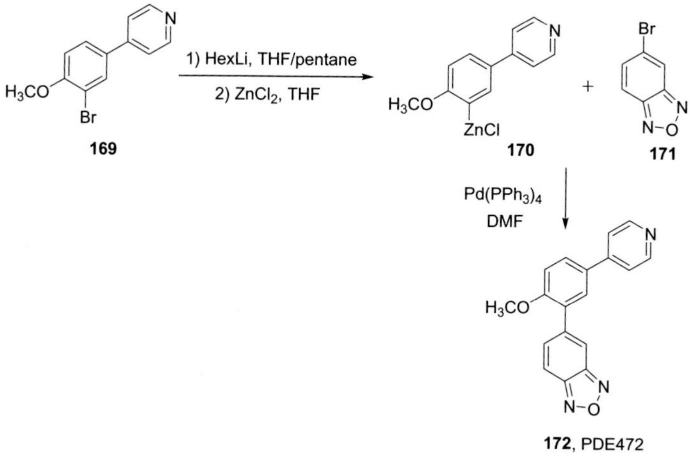

4 Catalytic C–C Bond Formation

Scheme 45 Synthesis of PDE472 via Negishi coupling

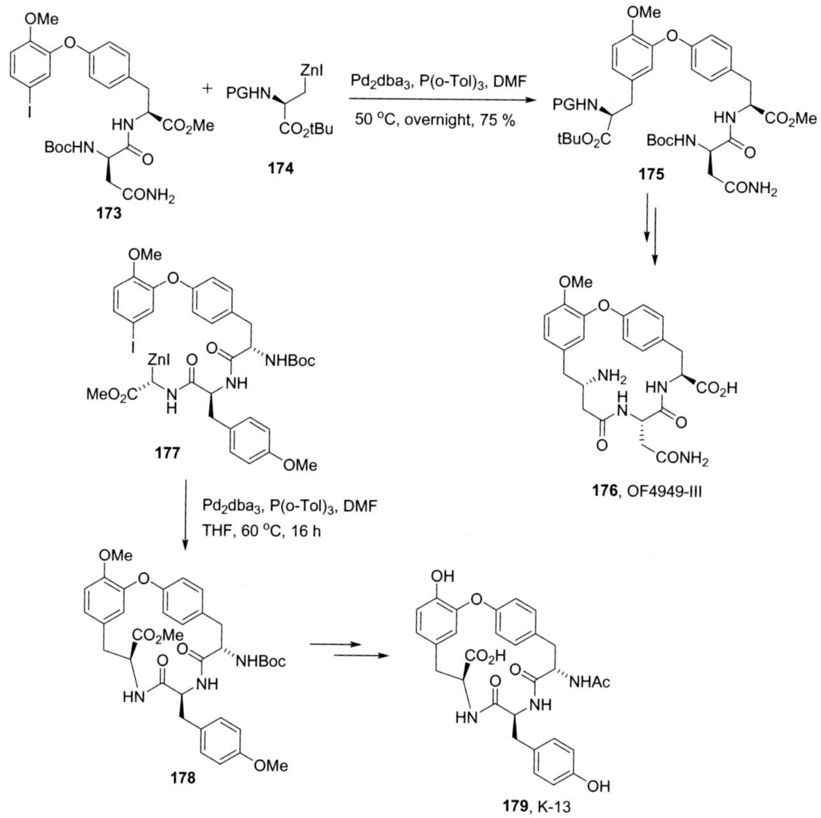

Scheme 46 Synthesis of OF4949-III and K-13 employing Negishi cross-coupling reaction

5 Catalytic C–N Bond Formation: Click Chemistry

Copper(I)-catalyzed 1,3-dipolar cycloaddition of organic azides with terminal alkynes, commonly known as "CuAAC Click Chemistry," was independently developed by Sharpless [93] and Meldal [94] which just after invention received great recognition and now identified as one of the widely explored protocols in various growing fields particularly glycoscience during the 2 decades [95, 96]. In contrast to Huisgen's catalyst-free thermal reaction of organic azides **180** with terminal alkynes **181**, similar reaction under CuAAC condition resulted selectively to the 1,4-disubstituted 1,2,3-triazole **182** while RuAAC-mediated reaction afforded exclusively 1,5-disubstituted 1,2,3-triazole **183** (Scheme 47) [93, 97]. The CuAAC tool is regioselective, modular, reliable, versatile, and most successful one for the rapid synthesis of diverse 1,4-disubstituted 1,2,3-triazoles **182** [93]. This modular CuAAC and cost-effective protocol has other notable advantages including high reaction rate (up to 107 times) over the uncatalyzed reaction and excellent reaction yield, easy to

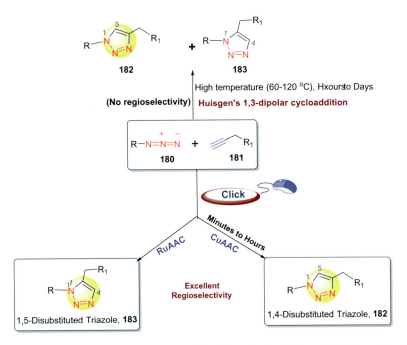

Scheme 47 Regioselectivity of 1,3-Dipolar Azide Alkyne Cycloaddition under Different Catalytic Conditions (CuAAC and RuAAC) [93, 97]

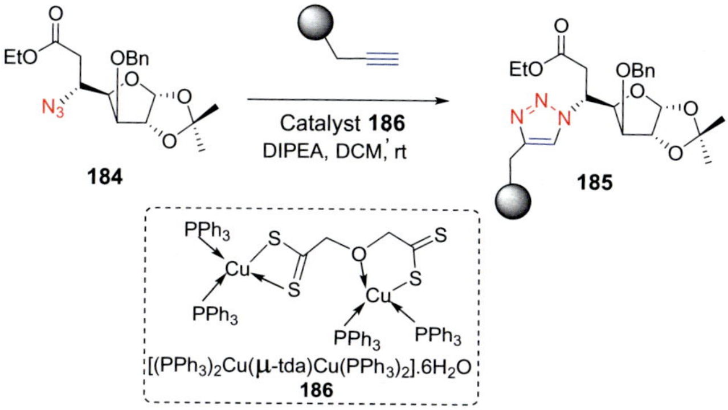

6 Catalysis by Acidic Clays and Zeolites

Scheme 48 Synthesis of triazolyl glycoconjugates exploiting Cu(I)-catalyst [98]

perform and scale-up, which all together make it one of the widely investigated tools considering the environmentally benign or green chemistry issues.

The widely used copper catalyst in CuAAC click conjugation includes a copper(I) source (e.g. CuI/base) or achieving in situ copper (I) via reduction reaction (e.g., $CuSO_4$/NaAsc) or also oxidizing elemental Cu to the oxidized copper source (e.g., Cu(I) in the reaction medium). A number of Cu-complexes have been further used to furnish high-to-excellent yield of desired 1,4-disubstituted triazoles. For a notable example, $[(PPh_3)_2Cu(\mu\text{-tda})Cu(PPh_3)_2]\cdot6H_2O$ (tda = thiodiacetate anion), a highly stable dinuclear Cu(I) catalyst displayed a significant catalytic efficiency for CuAAC reaction (Scheme 48) [98]. This catalyst utilized for the development of regioselective triazolyl glycoconjugates, noscapine glycohybrids, and also glycodendrimer. Easy synthesis of catalyst **186**, its high stability, excellent recyclability, wide substrate scope, excellent reaction yields, solvent-free, ligand-free, base-free condition and moreover easy workup, and short reaction time (up to 10 min) are the notable advantages. Notable advantages include the catalyst was recycled two times and worked well under base-free and solvent-free conditions, and thus, the method can be considered as environmentally benign protocol.

6 Catalysis by Acidic Clays and Zeolites

Reactions catalyzed by acids and bases are the backbone of manufacturing of fine chemicals, pharmaceuticals, agrochemicals, fragrances, etc. Many key reactions including catalytic cracking, esterification, alkylation, acylation, hydrolysis, oligomerization, isomerization, and condensation involve plethora of conventional Bronsted and Lewis acids like H_2SO_4, HF, HCl, BF_3, and $AlCl_3$ in liquid phase synthesis. Similarly, bases like NaOH, NaOMe, and KOH are used in industry and academia. In view of green synthesis, these acids and bases create undesirable consequences. Firstly, they produce hazardous inorganic salts after subsequent workup and hence end up with pollution of water bodies. Secondly, in general, they are used in stoichiometric quantity in several reactions, for example, Friedel–Crafts acylation and Aldol condensation. Though they might be used in catalytic amount in oil refining and petrochemical industries, the amount of effluents are considerable owing to the high volume production. Moreover, due to corrosive nature of these acids proper handling is an issue of concern.

Recyclable solid acids and bases came out as greener alternatives obviating the problem of enormous salt generation by use of traditional Bronsted and Lewis acids [99–102]. This will avoid the cost of hydrolytic workup and neutralization problem which in turn reduces the environmental burden which is associated with the use of traditional acids and bases. Solid acids and bases have several advantages over traditional acids and bases as summarized below.

1. Solid acids and bases are easily recyclable which lowers the manufacturing costs.
2. They can be easily separated from the reaction mixture which simplifies the overall process.
3. Milder reaction conditions are facilitated. They are easy and safe to handle as their liquid counterparts are corrosive and need safer material containers for storage and transportation.
4. Contamination would generally low in solid acids and bases as compared to their liquid counterparts.

Solid catalysis is a broad area of application of heterogenous catalysis. It constitutes mixed oxides like silica-alumina and sulfated zirconia, zeolites and zeotypes, acidic clay, supported heteropoly acids, hybrid organic–inorganic materials.

6.1 Acidic Clays

Clays are naturally occurring, amorphous, layered aluminosilicates which find a wide application in variety of catalytic processes. Clays are acidic catalysts which can function as both Bronsted and Lewis acids. The basic building blocks of clays are SiO_4 (tetrahedral) and MO_6 (Octahedral, M = Fe^{2+}, Fe^{3+}, Al^{3+}, Mg^{2+}) which polymerize to display a two-dimensional sheet structure. One of the most used clays is montmorillonite which is composed of sheet of octahedrally coordinated Al^{3+} sandwiched between sheets of tetrahedrally coordinated silicates. These three-layered structure repeat itself, and the void between the layers holds the key for physical and chemical properties.

6.1.1 Acidic Clay-Catalyzed Reactions

Many organic transformations which take place in the presence of various Bronsted acids like H_2SO_4, HF, HNO_3, and Lewis acids like $AlCl_3$, $FeCl_3$ could be performed

6 Catalysis by Acidic Clays and Zeolites

243

in the presence of montmorillonite with high efficiency, high selectivity, and low reaction time. For instance, primary alcohols react with olefins to yield mixed or unsymmetrical ethers in the presence of Al^{3+} montmorillonite (Scheme 49). This process is better than the conventional Williamson ether synthesis. However, secondary alcohol affords low yield and tertiary alcohol do not give the product. Few examples of ether synthesis with the aid of clay as a catalyst are shown in Scheme 49 [103, 104].

The conventional esterification reaction involves the following sequences.

1. Conversion of carboxylic acids to corresponding acid halides with the help of sulphuryl chloride or thionyl chloride followed by reaction with alcohol in the presence of base.
2. Heating the mixture of alcohol and carboxylic acid in the presence of dehydrating agents such as H_2SO_4 and H_3PO_4.

In both the above cases, the reactants like $SOCl_2$, SO_2Cl_2, PCl_5, and the ejected by-products are highly corrosive which imposes unwanted burden on the environment. Replacing these chemicals with acidic clay may provide green and cost-effective ester synthesis. There are some known clay-catalyzed reactions for the formation of esters (Scheme 50) [105, 106].

Similarly, acid anhydride can be formed by acidic clay-catalyzed reactions. Dicarboxylic acids (5 and 6 membered) undergo dehydration reaction to afford cyclic anhydrides (Scheme 51) [107].

Villemin and coworkers reported microwave-assisted synthesis of acid anhydrides catalyzed by montmorillonite KSF in the presence of isopropenyl acetate (Scheme 52) [108].

Furthermore, protection of aldehydes and ketones is also reported using natural kaolinitic clay. Protection of carbonyl group is considered as one of the essential steps in the sequence of multistep synthesis. It protects carbonyl groups from moderately strong nucleophilic attacks by various acidic, basic, oxidizing, or catalytic reagents. Kaolinites are clays known for material with which Chinawares are made. They have 1:1 stacking structure of tetrahedral and octahedral layers [109]. Ponde and coworkers reported the natural kaolinitic catalyzed chemoselective protection of aldehydes and

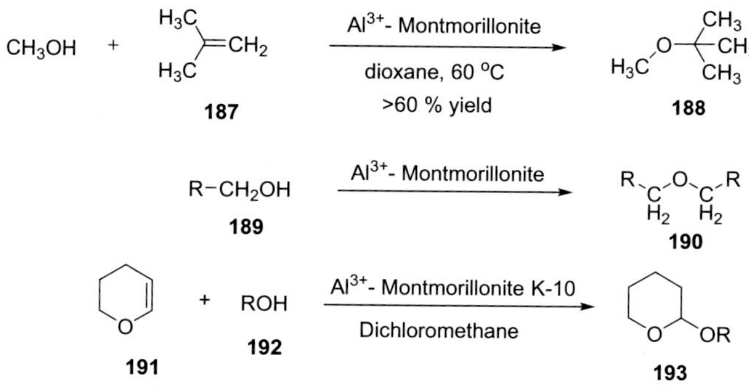

Scheme 49 Synthesis of ethers in the presence of acidic clay montmorillonite

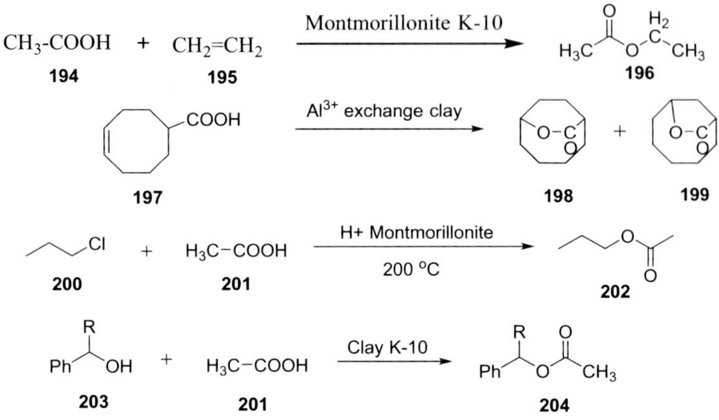

6 Catalysis: Application and Scope in Organic Synthesis

Scheme 50 Esterification reaction in the presence of clay

Scheme 51 Acid clay-catalyzed formation of cyclic anhydrides

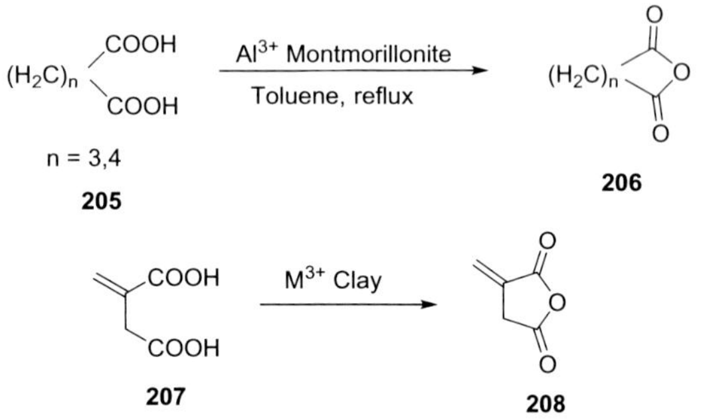

Scheme 52 Montmorillonite KSF-catalyzed formation of anhydrides

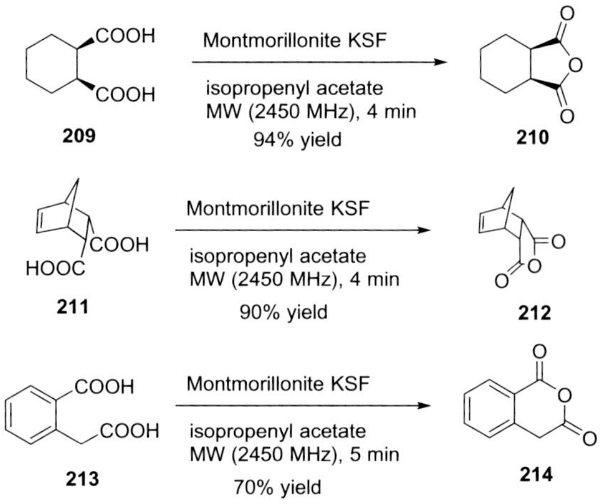

6 Catalysis by Acidic Clays and Zeolites

Scheme 53 Protection of carbonyl compounds using kaolinites clay

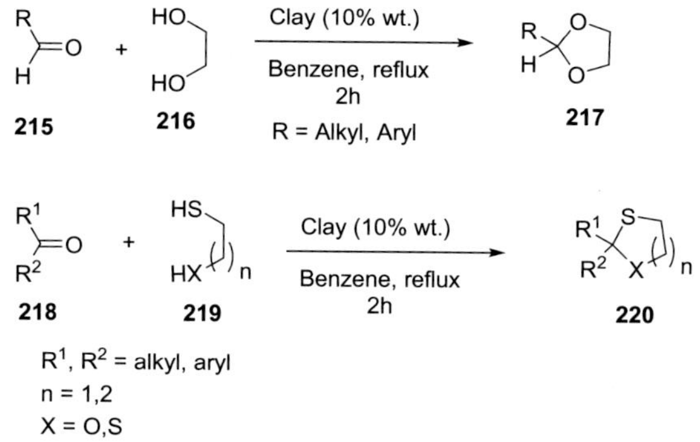

215 + **216** → **217** (R = Alkyl, Aryl)

Clay (10% wt.), Benzene, reflux 2h

218 + **219** → **220**

Clay (10% wt.), Benzene, reflux 2h

R^1, R^2 = alkyl, aryl
n = 1,2
X = O,S

ketone with variety of diols and dithiols like ethane-1,2-diol, ethane-1,2-dithiol, 1-hydroxyethane-2-thiol and propane-1,3-dithiol producing different dioxolanes, oxothiolanes, and dithiolanes (Scheme 53) [110].

6.2 Zeolites

Zeolites are crystalline, hydrated aluminosilicates which comprise a linked framework of AlO_4^- and SiO_4 tetrahedra having a regular system of pores and cavities with radius in nanometer range [111, 112]. A large number of zeolites are known till date. Some of them are naturally occurring but most of them have been synthesized in the laboratories. To maintain the electroneutrality with respect to AlO_4^- moiety, amorphous zeolites have an extraframework cation like Na^+. These extra framework cations can be readily replaced by other cations by ion exchange process. When these zeolites undergo ion exchange with proton donors/acids like ammonium ion followed by thermal dissociation, there is exchange of proton which affords acidic form of zeolites and ammonia. The Bronsted acid strength of acidic zeolite is comparable to the sulfuric acid and depends upon Si/Al ratio. A stable zeolite must not have two AlO_4^- units seated side by side. There must be a SiO_4 unit in between two AlO_4^-. So, for a stable zeolite, the Si/Al ratio must be equal to or greater than 1. The acidic strength of zeolites depends upon the proton donor hydroxyl group present on Si atom, i.e., popularly known as silanol. The acidic strength increases as the Si/Al ratio increases or in other words the reduction in AlO_4^- units (it should be noted that AlO_4^- could not be replaced completely as it will lead to decrease in acidic strength) (Scheme 54).

The most salient feature of zeolites, in the context of catalysis, is not their range of acid–base properties, but the presence of a regular structure possessing molecular-sized cavities and channels that attributes shape-selective catalysts for a wide variety of organic transformations. The structure of zeolites which actually matters for the organic chemists. The building block of zeolite is the ring-like structures in which each corner is occupied either by SiO_4 or AlO_4^- tetrahedron units (Fig. 3). These

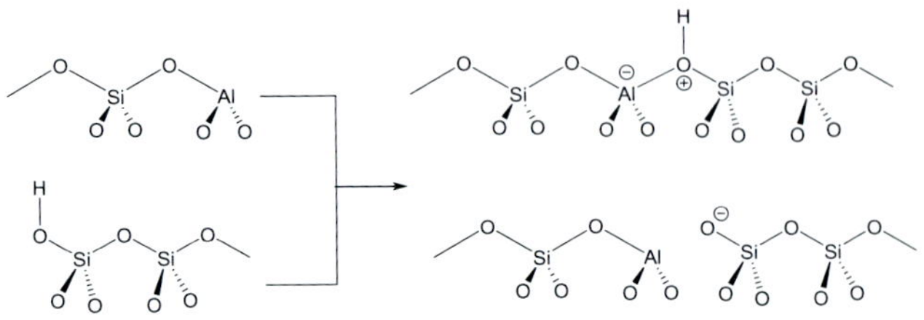

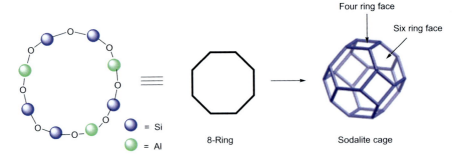

Scheme 54 Acid form of zeolites

Fig. 3 A building block of zeolite

tetrahedra units are linked to each other by O atom. This basic ring structure combined to form three-dimensional arrangements and ultimately form the structure of zeolites. For instance, sodalite cage in one of the building blocks of several zeolites in which the truncated octahedron is formed by six-membered rings as part of original octahedron and four-membered rings after truncation (Fig. 3).

Depending upon the pattern of assemblance of these sodalite cages, different types of zeolite structures are formed. For example, Zeolite A is the extension of four ring faces and Faujasites, a naturally occurring zeolite is the extension of six ring faces (Fig. 4). These sodalite cages assemble in a ring shape and create a super cage in the center with different cavity sizes. These cavity sizes vary depending upon type of assemblance of the sodalite cages. For instance, the cavity of Zeolite A has a diameter of 1.14 nm surrounded by eight sodalite cages. Similarly, Faujasites have a cavity with a diameter of 1.3 nm. Zeolite X and Y are synthetic zeolites having similar crystal structure as Faujasite but differ in Si/Al ratios. While Faujasite has Si/Al ratio of 2.2, Zeolite X and Y has Si/Al ratio of 1–1.5 and 1.5–3.0, respectively.

A naturally occurring zeolite is mordenite have Si/Al ratio about 10, and the structure is composed of 8 ring and 12 ring tunnels with diameter approx. 0.7 and 0.39, respectively. ZSM-5 is synthetic zeolite composed of highly siliceous material and has Si/Al ratio ranging from 25 to 2000. The entire framework if extended by 10 ring tunnels with a diameter of 0.55–0.6 nm (Fig. 5).

6 Catalysis by Acidic Clays and Zeolites

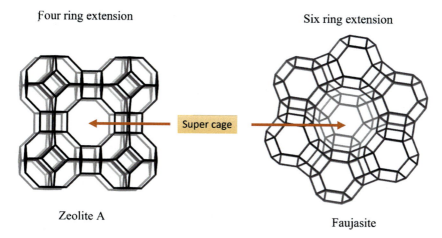

Fig. 4 Structures of zeolite A and Faujasite

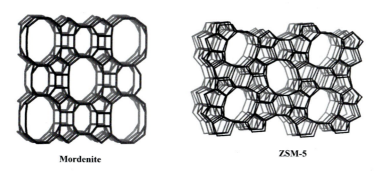

Fig. 5 Structure of mordenite and ZSM-5

It is now well-established that zeolites display a wide range of catalytic applications in organic transformation. Advancement in this field came with some very excellent examples of catalysis like electrophilic aromatic substitution (Friedel–Crafts alkylation and acylation), additions and eliminations, condensation, and cyclization.

6.2.1 Zeolite-Catalyzed Electrophilic Aromatic Substitution

Friedel–Crafts reaction is well-known reaction used in bulk and fine chemical industries. For example, alkylbenzene was produced with direct alkylation of benzene. The original synthesis was reported in 1940s where $AlCl_3$ was used as homogenous catalyst. Further, the process was modified and heterogenous catalyst, immobilized H_3PO_4 or BF_3 was used. These systems are highly corrosive and generate substantial amount of acidic wastes due to high volume production in industries. A major breakthrough in the FC technology was achieved in 1980s where H-ZSM-5 was used

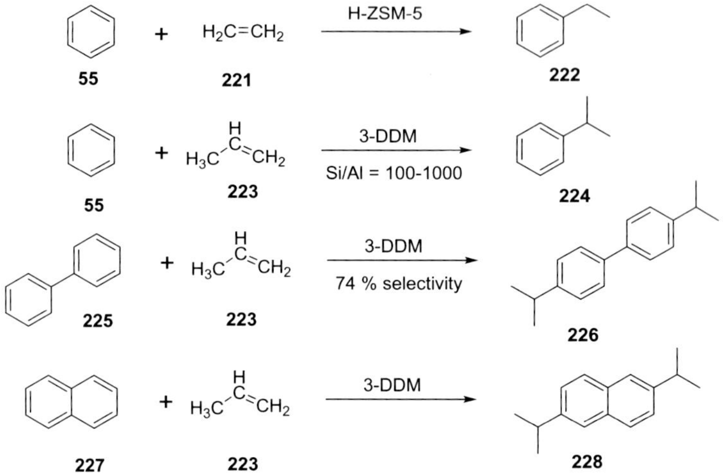

Scheme 55 Friedel–Crafts alkylation in the presence of zeolites

as stable and recyclable catalyst for mass production of ethylbenzene. This milestone was the beginning of intensive research in the area of zeolites and its application in petrochemical and fine chemical industries. Zeolite-based catalysis gradually replaced the use of immobilized H_3PO_4 and $AlCl_3$ as catalyst in the manufacturing of cumene, an important raw material for phenol synthesis. A three-dimensional dealuminated mordenite (3-DDM) catalyst was used for cumene **224** synthesis instead (Scheme 55). Recently, the use of nanomorphic zeolites was reported as catalyst in the synthesis of cumene. These nanomorphic zeolites displayed high efficiency and long catalytic lifetime [113].

Zeolites like H-MOR-type catalyst successfully transformed guaiacol to *p*-hydroxymethylguaiacol which is a precursor of vanillin through hydroxyalkylation process [114]. Aromatic hydroxyalkylation is important for the production of raw materials for fine chemicals. For instance, 2-phenylethanol is used in fragrance industries, synthesized by reaction of benzene and epoxide. The replacement of alkylating agent like $AlCl_3$ with zeolite gave little success. It is due to the large polarity difference between aromatic substrate and epoxide alkylating agent. The polarity difference results in poor adsorption ratio of aromatic substrate and alkylating agent which leads to low yield. This situation was avoided by intramolecular hydroxyalkylation in the presence of zeolite. For example, H-MOR and H-Beta result in intramolecular cyclialkylation of 4-phenyl-1-butene oxide [2]. Similarly, Pechmann coumarin synthesis also proceed via intramolecular hydroxyalkylation. H-Beta successfully replaced sulfuric acid, conventionally used as catalyst, in the synthesis of coumarin via condensation of phenols **234** with β-keto ester **235** (Scheme 56) [115].

Friedel–Crafts acylation, an important transformation in organic syntheses, classically requires stoichiometric amount of Lewis acid like $AlCl_3$. Replacement of these Lewis acids would be greener alternate for the production of various raw materials

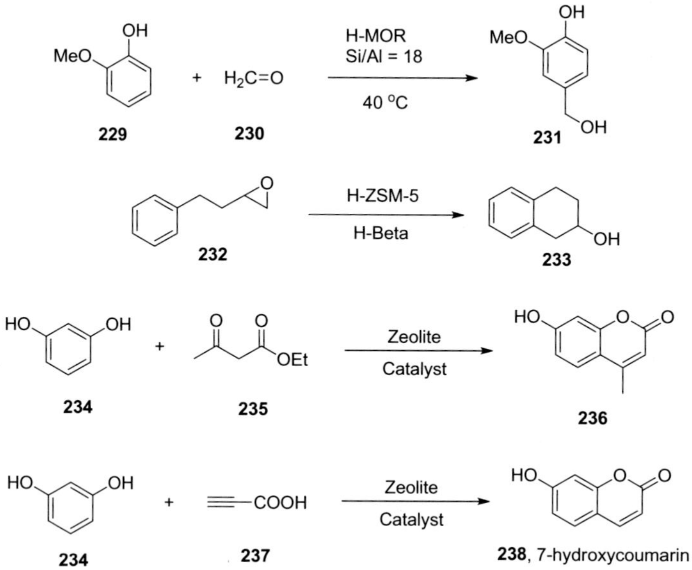

6 Catalysis by Acidic Clays and Zeolites

Scheme 56 Zeolite-catalyzed inter- and intramolecular hydroxyalkylation

for fine chemicals, agrochemicals, and pharmaceuticals. Like hydroxyalkylation of benzene with epoxides, acylation of aromatic compounds faces same difficulty of adsorption. Due to the polarity difference in the substrate and acylating agents like carboxylic acid and acid anhydrides, it is difficult to achieve proper adsorption ratio. Hence, heterogenous catalysis using zeolites in Friedel–Crafts acylation is more difficult than alkylation and poses tough challenges.

Intramolecular Friedel–Crafts acylation of 4-phenylbutyric acid **239** to afford α-tetralone **240** over a H-Beta catalyst combat the difficulty of adsorption ratio. Similarly, synthesis of p-acetylisobutylbenzene **245** was achieved via intramolecular acylation of isobutylbenzene **244** with acetic anhydride **242** over H-Beta catalyst. Corma et al. reported acylation of anisole with phenacetyl and phenylpropanoyl chlorides with the aid of H-ZSM-5 and H-Beta catalyst [116]. Geneste and coworkers reported the catalytic acylation of toluene and p-xylene with a variety of carboxylic acids over Y-faujasite-type zeolite exchanged with Ce^{3+} (Scheme 57) [117]. Ryoo et al. reported MFI zeolite nanosponge-catalyzed FC reaction of bulky aromatic compounds. MFI nanosponge is mesoporous and synthesized by seed-assisted hydrothermal method. They exhibited excellent catalytic activity due to a large number of Bronsted acid site were present in the mesoporous wall of the catalyst [118].

Other electrophilic aromatic substitutions, like nitration [119] and halogenation [120], are well-documented using zeolite as heterogenous catalysts. Conventionally, aromatic nitration requires mixture of sulfuric and nitric acid as nitrating agents

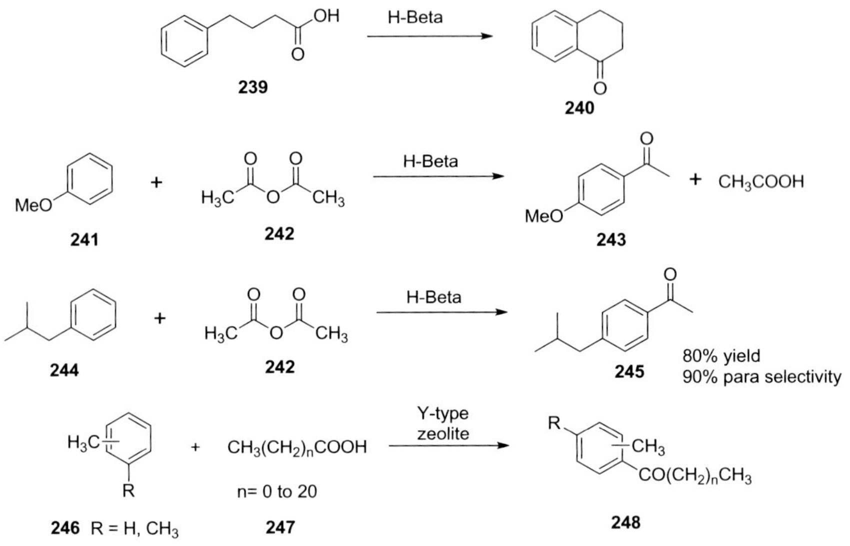

250   6 Catalysis: Application and Scope in Organic Synthesis

Scheme 57 Zeolite-catalyzed Friedel–Crafts acylation

which lead to generation of copious amount of acid wastes. Dealuminated mordenite is a robust alternate for vapor phase nitration using 65% aqueous nitric acid [121].

Aromatic halogenation process traditionally employs stoichiometric amounts of reagents like Cl_2 and $POCl_3$ which leads to hazardous salts as waste and atom economy of overall reaction being very low. Louis and coworkers developed a new process in which H-BEA zeolites were employed for halogenation reaction. This process is robust and greener in comparison with established conventional halogenation reactions. BEA catalyst exhibited high conversion rate and selectivity towards monochlorinated products up to 98% [122].

6.2.2 Zeolite-Catalyzed Addition Reactions

Zeolites have been widely used in hydration and dehydration reactions. One of the pertinent examples is the hydration of cyclohexene to cyclohexanol over H-ZSM-5-type (Si/Al > 20) reported by Asahi. Hydration of cyclohexene is an important step in the synthesis of cyclohexanone **251** from benzene **55** (Scheme 58) [123]. Conventional route involves hydrogenation of benzene followed by auto-oxidation leads to cyclohexanol **250** and cyclohexanone **251** with good selectivity but very low yield (~5%). Cyclohexanol **250** further dehydrogenated to yield cyclohexanone **251**. The disadvantage of this process is the recycling cyclohexane in enormous amount due to very low rate of conversion. In Asahi process, benzene is partially hydrogenated to afford cyclohexene in the presence of Ru catalyst. This cyclohexene oxidized to give

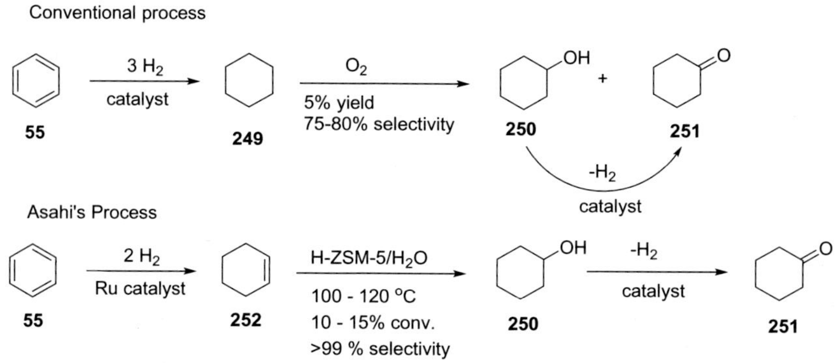

6 Catalysis by Acidic Clays and Zeolites 251

Conventional process

Scheme 58 A comparison between conventional and Asahi's process of cyclohexanone synthesis

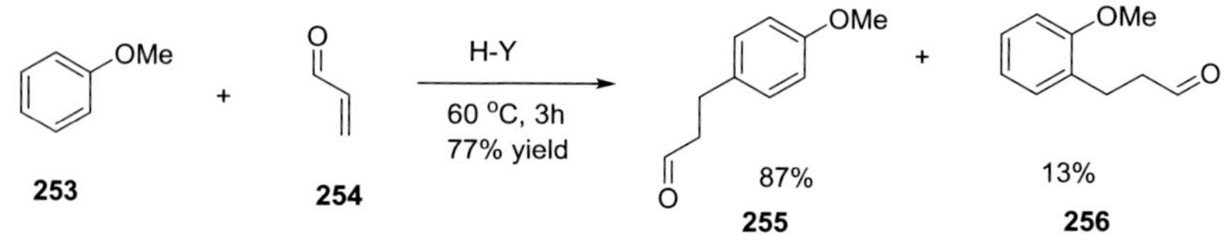

Scheme 59 1,4-addition of methoxy benzene to enal

cyclohexanol with comparatively higher yield and excellent selectivity. The resulted cyclohexanol dehydrogenated to give cyclohexanone.

Onaka and coworkers reported a versatile process for acrolein synthesis **255** and **256** via 1,4-addition of α,β-unsaturated enals **254** to benzene derivatives using H-Beta and H-Y catalyst (Scheme 59) [124]. They also disclosed that the content of monomeric acrolein depends upon the solvent polarity. Electron-rich benzene derivatives like phenol and N,N-dimethyl aniline showed similar reaction products in good yield and selectivity.

6.2.3 Zeolite-Catalyzed Cyclization Reactions

Zeolites are known for various acid-catalyzed cyclization reactions including the formation of terpenols, [125] heterocycles like benzofurans and [126, 127] isochromans[128].

Liu and coworkers reported the metal-doped Y-zeolites catalyze cyclization of geraniols **257** to afford α- and γ-cyclogeraniol **258** and **259,** respectively, with high yields (Scheme 60). They disclosed that selective cyclization in the presence of NaY and FeY zeolites depends upon the metal-doped and the activation temperatures. When the reaction was carried out 10 °C with NaY activated at 500 °C, the γ-cyclogeraniol **259** was the major product while at 60 °C, α-cyclogeraniol **258** was

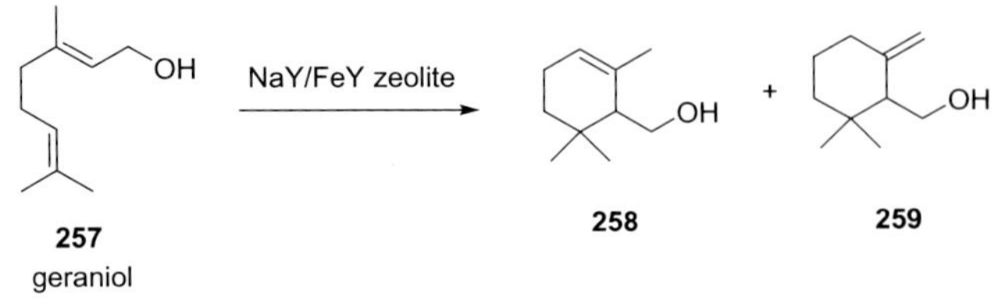

6 Catalysis: Application and Scope in Organic Synthesis

Scheme 60 Zeolite-catalyzed cyclization of geraniol

formed exclusively. When reaction temperature is elevated, the γ-product isomerizes to α-product showing that α-cyclogeraniol **258** is the thermodynamic product [125].

Hu et al. reported the tin-exchanged H-β zeolite (Sn-β)-catalyzed synthesis of 2,3-unsubstituted benzo[*b*]furans **261** via intramolecular cyclization of 2-aryloxyacetaldehyde acetals **260** in moderate to excellent yield (Scheme 61a) [127]. Kim and coworkers reported HY zeolite-catalyzed one-pot Claisen rearrangement cyclization reaction of the aryl methallyl ethers **262** to afford 2,3-dihydrobenzofuran **263** derivatives in good to excellent yield (Scheme 61b) [126]. The substituents on methallyl aryl ethers have effect on the conversion rate of the reaction. Electron-donating groups showed high conversion rate and excellent yield while electron-donating groups displayed lower conversion rate with low yield.

Hell and Hegedüs reported the synthesis of chromans **267** via the oxa-Pictet–Spengler reaction using modified version of Ersorb-4 (E4), i.e., E4a zeolite (Scheme 62) [128]. E4a zeolites have more acidic surface (pH = 3) than E4 (pH = 5.5). Previously they had published the utility of E4 zeolite catalysis in a wide variety of organic transformations including the synthesis of oxazoline derivatives [129], 2-arylimidazolines, arylbenzoxazoles [130], arylbenzimidazoles, and E4a in the synthesis of 3,4-dihydropyrimidin-2(1H)-ones [131] and benzodiazepine derivatives [132] via Pictet–Spengler cyclization reaction. Inspired by their own previous work, they investigated the reaction of 2-phenylethanol derivatives with aromatic/aliphatic aldehydes and ketone in the presence of E4a catalyst. The reaction gave good to

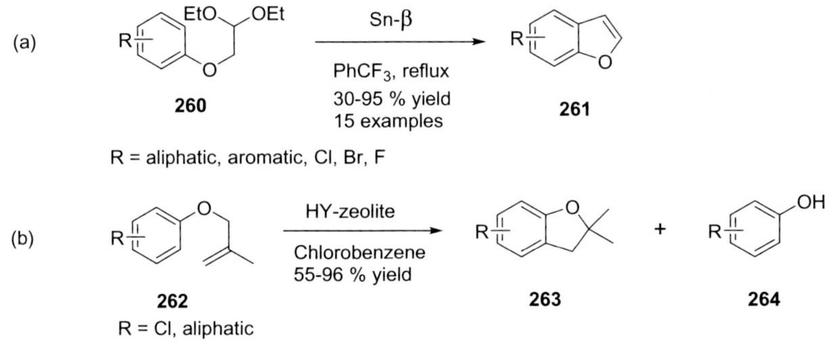

Scheme 61 Zeolite-catalyzed synthesis of **a** benzo[*b*]furan and **b** 2,3-dihydrobenzofuran derivatives

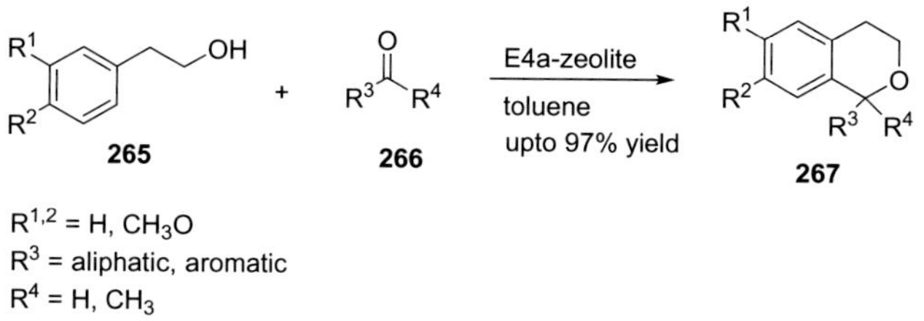

7 Organocatalysis: General Consideration

$R^{1,2}$ = H, CH$_3$O
R^3 = aliphatic, aromatic
R^4 = H, CH$_3$

Scheme 62 E4a-catalyzed intermolecular cyclization reaction

excellent results and interestingly, neither the low acidic condition (E4) nor the high acidic condition gave good results.

7 Organocatalysis: General Consideration

Asymmetric catalysis has an important place in the field of organic syntheses by virtue of the methodologies for the construction of novel chiral molecules or chiral building blocks which might be used in synthesis of several complex molecules such as natural products and commercial drug products. This field has been dominated by transition metal catalysis using chiral ligands and enzymes for most of its history. However, past two decades have witnessed organocatalysis as new strategy for addressing modern-day challenges in organic synthesis. Organocatalysts are small organic molecules used in sub-stoichiometric amount to catalyze the reaction. After the independent works of List and Macmillan at the beginning of this century, the development of asymmetric catalysis using organocatalysts took place with breathtaking pace. For the contribution in the field of organocatalysis, List and Macmillan shared the Nobel Prize of Chemistry in 2021. An elegant example which advocates the efficiency and power of organocatalysis in the synthesis of complex natural products is the synthesis of astoundingly complex strychnine molecule (Scheme 63) [133]. When first synthesized in 1952, it involved 29 steps and yield was exceptionally low, i.e., 0.0009% of the starting material. In 2011, by virtue of organocatalysis, the step was reduced to 12 and the production process was 7000 times more efficient. In view of the importance, a separate (next) chapter is dedicated to organocatalysis.

Finally, it is worth mentioning that role of catalysis in organic synthesis is very crucial as it replaces the stoichiometric reagents with cleaner and catalytic alternatives. There is no doubt that, in present time, organic synthesis has attained a high level of sophistication and precision with almost any kind of complexity can be synthesized, in context of chemo-, regio-, and stereoselectivity. The advent of biocatalysts, organocatalysts, chiral ligands, etc., became powerful tools for achieving sustainable and environmentally benign organic synthesis.

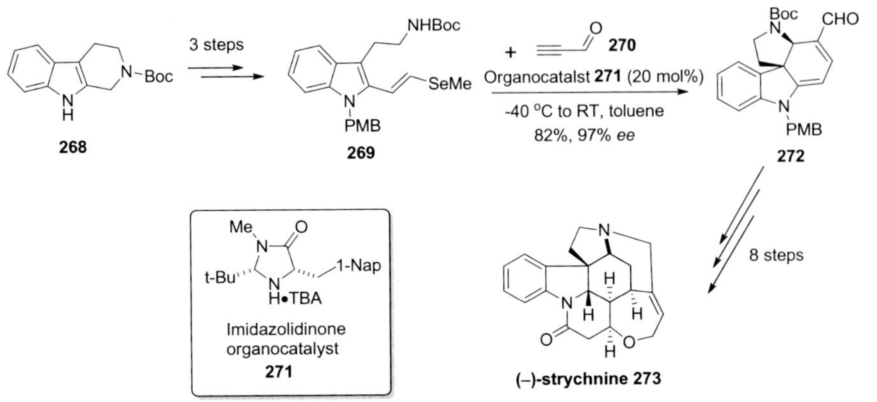

254   6 Catalysis: Application and Scope in Organic Synthesis

Scheme 63 Synthesis of (−)-strychnine using imidazolidinone as an organocatalyst

References

1. Lancaster, M.: Green Chemistry: An Introductory Text. Royal Society of Chemistry (2020)
2. Sheldon, R.A., Arends, I., Hanefeld, U.: Green Chemistry and Catalysis. Wiley (2007)
3. Clerici, M.G., Ingallina, P.: Epoxidation of lower olefins with hydrogen peroxide and titanium silicalite. J. Catal. **140**(1), 71–83 (1993)
4. Venturello, C., Alneri, E., Ricci, M.: A new, effective catalytic system for epoxidation of olefins by hydrogen peroxide under phase-transfer conditions. J. Org. Chem. **48**(21), 3831–3833 (1983)
5. Sato, K., Aoki, M., Ogawa, M., Hashimoto, T., Noyori, R.: A practical method for epoxidation of terminal olefins with 30% hydrogen peroxide under halide-free conditions. J. Org. Chem. **61**(23), 8310–8311 (1996)
6. Herrmann, W.A., Fischer, R.W., Marz, D.W.: Methyltrioxorhenium as catalyst for olefin oxidation. Angew. Chemie Int. Ed. Engl. **30**(12), 1638–1641 (1991)
7. Herrmann, W.A., Kratzer, R.M., Ding, H., Thiel, W.R., Glas, H.: Methyltrioxorhenium/pyrazole—a highly efficient catalyst for the epoxidation of olefins. J. Organomet. Chem. **555**(2), 293–295 (1998)
8. Rudolph, J., Reddy, K.L., Chiang, J.P., Sharpless, K.B.: Highly efficient epoxidation of olefins using aqueous H_2O_2 and catalytic methyltrioxorhenium/pyridine: pyridine-mediated ligand acceleration. J. Am. Chem. Soc. **119**(26), 6189–6190 (1997)
9. Nishiyama, H., Motoyama, Y.: Novel ruthenium-pyridinedicarboxylate complexes of terpyridine and chiral bis (oxazolinyl) pyridine: a new catalytic system for alkene epoxidation with [bis (acetoxy) iodo] benzene as an oxygen donor. Chem. Commun. **19**, 1863–1864 (1997)
10. Tse, M.K., Döbler, C., Bhor, S., Klawonn, M., Mägerlein, W., Hugl, H., Beller, M.: Development of a ruthenium-catalyzed asymmetric epoxidation procedure with hydrogen peroxide as the oxidant. Angew. Chemie Int. Ed. **43**(39), 5255–5260 (2004)
11. Tse, M.K., Klawonn, M., Bhor, S., Döbler, C., Anilkumar, G., Hugl, H., Mägerlein, W., Beller, M.: Convenient method for epoxidation of alkenes using aqueous hydrogen peroxide. Org. Lett. **7**(6), 987–990 (2005)
12. Stoop, R.: Asymmetric epoxidation of olefins. The first enantioselective epoxidation of unfunctionalised olefins catalysed by a chiral ruthenium complex with H_2O_2 as ixidant. Green Chem. **1**(1), 39–41 (1999)
13. Anelli, P.L., Banfi, S., Legramandi, F., Montanari, F., Pozzi, G., Quici, S.: Tailed Mn III-tetraarylporphyrins bearing an axial ligand and/or a carboxylic group: self-consistent catalysts for H_2O_2 or NaOCl alkene epoxidation. J. Chem. Soc. Perkin Trans. **1**(12), 1345–1357 (1993)

References

14. Hage, R., Iburg, J.E., Kerschner, J., Koek, J.H., Lempers, E.L.M., Martens, R.J., Racherla, U.S., Russell, S.W., Swarthoff, T., van Vliet, M.R.P.: Efficient manganese catalysts for low-temperature bleaching. Nature **369**(6482), 637–639 (1994)
15. Lane, B.S., Burgess, K.: A cheap, catalytic, scalable, and environmentally benign method for alkene epoxidations. J. Am. Chem. Soc. **123**(12), 2933–2934 (2001)
16. Chen, K., Que, L., Jr.: Evidence for the participation of a high-valent iron-oxo species in stereospecific alkane hydroxylation by a non-heme iron catalyst. Chem. Commun. **15**, 1375–1376 (1999)
17. Criegee, R.: Osmiumsäure-ester als Zwischenprodukte bei Oxydationen. Justus Liebigs Ann. Chem. **522**(1), 75–96 (1936)
18. Sharpless, K.B., Akashi, K.: Osmium catalyzed vicinal hydroxylation of olefins by tert-butyl hydroperoxide under alkaline conditions. J. Am. Chem. Soc. **98**(7), 1986–1987 (1976)
19. Dupau, P., Epple, R., Thomas, A.A., Fokin, V.V., Sharpless, K.B.: Osmium-catalyzed dihydroxylation of olefins in acidic media: old process new tricks. Adv. Synth. Catal. **344**(3/4), 421–433 (2002)
20. Döbler, C., Mehltretter, G., Beller, M.: Atom-efficient oxidation of alkenes with molecular oxygen: synthesis of diols. Angew. Chem. Int. Ed. **38**(20), 3026–3028 (1999)
21. Döbler, C., Mehltretter, G.M., Sundermeier, U., Beller, M.: Osmium-catalyzed dihydroxylation of olefins using dioxygen or air as the terminal oxidant. J. Am. Chem. Soc. **122**(42), 10289–10297 (2000)
22. Djerassi, C., Engle, R.R.: Oxidations with ruthenium tetroxide. J. Am. Chem. Soc. **75**(15), 3838–3840 (1953)
23. Berkowitz, L.M., Rylander, P.N.: Use of ruthenium tetroxide as a multi-purpose oxidant. J. Am. Chem. Soc. **80**(24), 6682–6684 (1958)
24. Lee, D.G., Van Den Engh, M.: Oxidation in Organic Chemistry. In: Trahanovsky, W.S. (ed.) No. Part B. Academic Press, New York (1973)
25. Carlsen, P.H.J., Katsuki, T., Martin, V.S., Sharpless, K.B.: A greatly improved procedure for ruthenium tetroxide catalyzed oxidations of organic compounds. J. Org. Chem. **46**(19), 3936–3938 (1981)
26. Pappas, J.J., Keaveney, W.P., Gancher, E., Berger, M.: A new and convenient method for converting olefins to aldehydes. Tetrahedron Lett. **7**(36), 4273–4278 (1966)
27. Pappo, R., Allen, D.J., Lemieux, R., Johnson, W.S.: Osmium tetroxide-catalyzed periodate oxidation of olefinic bonds. J. Org. Chem. **21**(4), 478–479 (1956)
28. Yang, D., Zhang, C.: Ruthenium-catalyzed oxidative cleavage of olefins to aldehydes. J. Org. Chem. **66**(14), 4814–4818 (2001)
29. Ho, C., Yu, W., Che, C.: Ruthenium nanoparticles supported on hydroxyapatite as an efficient and recyclable catalyst for *cis*-dihydroxylation and oxidative cleavage of alkenes. Angew. Chem. Int. Ed. **43**(25), 3303–3307 (2004)
30. Oguchi, T., Ura, T., Ishii, Y., Ogawa, M.: Oxidative cleavage of olefins into carboxylic acids with hydrogen peroxide by tungstic acid. Chem. Lett. **18**(5), 857–860 (1989)
31. Antonelli, E., D'Aloisio, R., Gambaro, M., Fiorani, T., Venturello, C.: Efficient oxidative cleavage of olefins to carboxylic acids with hydrogen peroxide catalyzed by methyltrioctylammonium tetrakis (oxodiperoxotungsto) phosphate (3−) under two-phase conditions. synthetic aspects and investigation of the reaction course. J. Org. Chem. **63**(21), 7190–7206 (1998)
32. Spannring, P., Yazerski, V., Bruijnincx, P.C.A., Weckhuysen, B.M., Klein Gebbink, R.J.M.: Fe-catalyzed one-pot oxidative cleavage of unsaturated fatty acids into aldehydes with hydrogen peroxide and sodium periodate. Chem. Eur. J. **19**(44), 15012–15018 (2013)
33. Ou, J., He, S., Wang, W., Tan, H., Liu, K.: Highly efficient oxidative cleavage of olefins with O_2 under catalyst-initiator-and additive-free conditions. Org. Chem. Front. **8**(12), 3102–3109 (2021)
34. Ansari, I.A., Joyasawal, S., Gupta, M.K., Yadav, J.S., Gree, R.: Wacker oxidation of terminal olefins in a mixture of [Bmim][BF4] and water. Tetrahedron Lett. **46**(44), 7507–7510 (2005)
35. Hou, Z., Han, B., Gao, L., Jiang, T., Liu, Z., Chang, Y., Zhang, X., He, J.: Wacker oxidation of 1-hexene in 1-n-butyl-3-methylimidazolium hexafluorophosphate ([bmim][PF6]), supercritical (SC) CO_2, and SC CO_2/[bmim][PF6] mixed solvent. New J. Chem. **26**(9), 1246–1248 (2002)

36. Kulkarni, M.G., Shaikh, Y.B., Borhade, A.S., Chavhan, S.W., Dhondge, A.P., Gaikwad, D.D., Desai, M.P., Birhade, D.R., Dhatrak, N.R.: Greening the Wacker process. Tetrahedron Lett. **54**(19), 2293–2295 (2013)

37. ten Brink, G.-J., Arends, I.W.C.E., Papadogianakis, G., Sheldon, R.A.: Catalytic conversions in water. Part 10.† aerobic oxidation of terminal olefins to methyl ketones catalysed by water soluble palladium complexes. Chem. Commun. 21, 2359–2360 (1998)

38. Mitsudome, T., Umetani, T., Nosaka, N., Mori, K., Mizugaki, T., Ebitani, K., Kaneda, K.: Convenient and efficient Pd-catalyzed regioselective oxyfunctionalization of terminal olefins by using molecular oxygen as sole reoxidant. Angew. Chem. **118**(3), 495–499 (2006)

39. Shilov, A.E., Shulpin, G. B.: Activation and Catalytic Reactions of Saturated Hydrocarbons in the Presence of Metal Complexes, vol. 21. Springer Science & Business Media, (2001)

40. Baik, M.-H., Newcomb, M., Friesner, R.A., Lippard, S.J.: Mechanistic studies on the hydroxylation of methane by methane monooxygenase. Chem. Rev. **103**(6), 2385–2420 (2003)

41. Griffith, W.P.: Transition Metal Chem. **15**, 251 (1990). (b) Griffith, W.P.: Chem. Soc. Rev. **21**, 179 (1992)

42. Lau, T.-C., Mak, C.-K.: Oxidation of alkanes by barium ruthenate in acetic acid: catalysis by Lewis acids. J. Chem. Soc. Chem. Commun. **9**, 766–767 (1993)

43. Yiu, S.-M., Man, W.-L., Lau, T.-C.: Efficient catalytic oxidation of alkanes by Lewis acid/[OsVI (N) Cl4]—using peroxides as terminal oxidants. Evidence for a metal-based active intermediate. J. Am. Chem. Soc. **130**(32), 10821–10827 (2008)

44. Goldstein, A.S., Drago, R.S.: Hydroxylation of methane by a sterically hindered ruthenium complex. J. Chem. Soc. Chem. Commun. **1**, 21–22 (1991)

45. Bianchi, D., Bortolo, R., Tassinari, R., Ricci, M., Vignola, R., Werts, H.V., Hofstraat, J.W., Geurts, F.A.J., Verhoeven, J.W., Slooff, H., Polman, A., Oude Wolbers, M.P., van Veggel, F.C.J.M., Reinhoudt, D.N., Klink, S.I., Hebbink, G.A., Grave, L., Oude Wolbers, P., Peters, F.G.A., van Beelen, E.S.E., Chem Soc Perkin, J., Snel-link-Rue, È.B.H.M., Chem, J., Klink, I., Balzani, V., Barigelleti, F., Campagna, S., Belser, P., von Zelewsky, A., Vignola, R.: A novel iron-based catalyst for the biphasic oxidation of benzene to phenol with hydrogen peroxide. **39** (2000)

46. Baráth, E.: Hydrogen transfer reactions of carbonyls, alkynes, and alkenes with noble metals in the presence of alcohols/ethers and amines as hydrogen donors. Catalysts **8**(12), 671 (2018)

47. Rylander, P.N.: Hydrogenation Methods. Academic Press (1990)

48. Noyori, R.: Asymmetric catalysis: science and opportunities (Nobel Lecture 2001). Adv. Synth. Catal. **345**(1–2), 15–32 (2003)

49. Knowles, W.S.: Asymmetric hydrogenations (Nobel Lecture 2001). Adv. Synth. Catal. **345**(1–2), 3–13 (2003)

50. De Wildeman, S.M.A., Sonke, T., Schoemaker, H.E., May, O.: Biocatalytic reductions: from lab curiosity to "first choice." Acc. Chem. Res. **40**(12), 1260–1266 (2007)

51. Somorjai, G.A., McCrea, K.: Roadmap for catalysis science in the 21st century: a personal view of building the future on past and present accomplishments. Appl. Catal. A Gen. **222**(1–2), 3–18 (2001)

52. Somorjai, G.A., McCrea, K.R., Zhu, J.: Active sites in heterogeneous catalysis: development of molecular concepts and future challenges. Top. Catal. **18**(3), 157–166 (2002)

53. Newhall, W.F.: Derivatives of (+)-limonene. I. Esters of *trans*-p-menthane-1, 2-diol1. J. Org. Chem. **23**(9), 1274–1276 (1958)

54. Jackman, L.M., Webb, R.L., Yick, H.C.: Synthesis and chiroptical properties of some piperidin-2-ones. J. Org. Chem. **47**(10), 1824–1831 (1982)

55. Zartman, W.H., Adkins, H.: The variations in the behavior of phenylethenes and ethanes during catalytic hydrogenation. J. Am. Chem. Soc. **54**(4), 1668–1674 (1932)

56. Adkins, H., Billica, H.R.: The preparation of Raney nickel catalysts and their use under conditions comparable with those for platinum and palladium catalysts. J. Am. Chem. Soc. **70**(2), 695–698 (1948)

References 257

57. Motoyama, R., Nishimura, S., Imoto, E., Murakami, Y., Hari, K., Ogawa, J.: Nippon Kagaku Zasshi 1957, 78, 954. Chem. Abstr. **54**, 74585 (1960)
58. Oliver, R.G., Wells, P.B.: The hydrogenation of alkadienes: VIII. Deuterium tracer study of alkane formation in the palladium-catalyzed hydrogenation of propadiene and of 1, 2-butadiene and its implications concerning the breakdown of selectivity in ethyne hydrogenation. J. Catal. **47**(3), 364–370 (1977)
59. Borodziński, A., Bond, G.C.: Selective hydrogenation of ethyne in ethene-rich streams on palladium catalysts. Part 1. Effect of changes to the catalyst during reaction. Catal. Rev. **48**(02), 91–144 (2006)
60. Marvell, E.N., Thomas, L.I.: Catalytic semihydrogenation of the triple bond. Synthesis (Stuttg) **1973**(08), 457–468 (1973)
61. List, B., Lerner, R.A., Barbas, C.F.: Proline-catalyzed direct asymmetric aldol reactions. J. Am. Chem. Soc. **122**(10), 2395–2396 (2000)
62. Mukherjee, S., Yang, J.W., Hoffmann, S., List, B.: Asymmetric enamine catalysis. Chem. Rev. **107**(12), 5471–5569 (2007)
63. Zou, Y.-Q., Hörmann, F.M., Bach, T.: Iminium and enamine catalysis in enantioselective photochemical reactions. Chem. Soc. Rev. **47**(2), 278–290 (2018)
64. Hao, L., Zhao, Y., Yu, B., Yang, Z., Zhang, H., Han, B., Gao, X., Liu, Z.: Imidazolium-based ionic liquids catalyzed formylation of amines using carbon dioxide and phenylsilane at room temperature. ACS Catal. **5**(9), 4989–4993 (2015)
65. Boyle, P.H., Keating, M.T.: Asymmetric hydrogenation of a carbon-nitrogen double bond in folic acid. J. Chem. Soc. Chem. Commun. **10**, 375–376 (1974)
66. Levi, A., Modena, G., Scorrano, G.: Asymmetric reduction of carbon-nitrogen, carbon-oxygen, and carbon-carbon double bonds by homogeneous catalytic hydrogenation. J. Chem. Soc. Chem. Commun. **1**, 6–7 (1975)
67. Oppolzer, W., Wills, M., Starkemann, C., Bernardinelli, G.: Chiral toluene-2, α-sultam auxiliaries: preparation and structure of enantiomerically pure (2R)-and (S)-ethyl-2, 1′-sultam. Tetrahedron Lett. **31**(29), 4117–4120 (1990)
68. Burk, M.J., Feaster, J.E., Nugent, W.A., Harlow, R.L.: Preparation and use of C2-symmetric bis (phospholanes): production of. Alpha.-amino acid derivatives via highly enantioselective hydrogenation reactions. J. Am. Chem. Soc. **115**(22), 10125–10138 (1993)
69. Hosseini, M., Grau, J.S., Sørensen, K.K., Søtofte, I., Tanner, D., Murray, A., Tønder, J.E.: Short and efficient diastereoselective synthesis of pyrrolidinone-containing dipeptide analogues. Org. Biomol. Chem. **5**(14), 2207–2210 (2007)
70. Roe, A., Montgomery, J.A.: Kinetics of the catalytic hydrogenation of certain Schiff bases1. J. Am. Chem. Soc. **75**(4), 910–912 (1953)
71. Maitlis, P.M., Haynes, A., Sunley, G.J., Howard, M.J.: Methanol carbonylation revisited: thirty years on. J. Chem. Soc. Dalt. Trans. **11**, 2187–2196 (1996)
72. Beller, M., Indolese, A.F.: Advances in the carbonylation of aryl halides using palladium catalysts. Chim. Int. J. Chem. **55**(9), 684–687 (2001)
73. Roessler, F.: Catalysis in the industrial production of pharmaceuticals and fine chemicals. Chim. Int. J. Chem. **50**(3), 106–109 (1996)
74. Gördes, D., Neumann, H., von Wangelin, A.J., Fischer, C., Drauz, K., Krimmer, H., Beller, M.: Synthesis of *N*-Acetyl-α-aminobutyric Acid *via* Amidocarbonylation: a case study. Adv. Synth. Catal. **345**(4), 510–516 (2003)
75. Beller, M., Eckert, M., Moradi, W.A., Neumann, H.: Palladium-catalyzed synthesis of substituted hydantoins—a new carbonylation reaction for the synthesis of amino acid derivatives. Angew. Chem. Int. Ed. **38**(10), 1454–1457 (1999)
76. De Meijere, A., Meyer, F.E.: Fine feathers make fine birds: the heck reaction in modern garb. Angew. Chem. Int. Ed. Engl. **33**(23/24), 2379–2411 (1995)
77. Beletskaya, I.P., Cheprakov, A.V.: The heck reaction as a sharpening stone of palladium catalysis. Chem. Rev. **100**(8), 3009–3066 (2000)
78. Oestreich, M.: The Mizoroki-Heck Reaction. Wiley (2009)

79. Cabri, W., Candiani, I.: Recent developments and new perspectives in the Heck reaction. Acc. Chem. Res. **28**(1), 2–7 (1995)
80. Jagtap, S.: Heck reaction—state of the art. Catalysts **7**(9), 267 (2017)
81. De Vries, J.G.: The Heck reaction in the production of fine chemicals. Can. J. Chem. **79**(5–6), 1086–1092 (2001)
82. Blaser, H., Indolese, A., Naud, F., Nettekoven, U., Schnyder, A.: Industrial R&D on catalytic C–C and C–N coupling reactions: a personal account on goals approaches and results. Adv. Synth. Catal. **346**(13/15), 1583–1598 (2004)
83. Zapf, A., Beller, M.: Fine chemical synthesis with homogeneous palladium catalysts: examples status and trends. Top. Catal. **19**(1), 101–109 (2002)
84. Larsen, R.D., King, A.O., Chen, C.Y., Corley, E.G., Foster, B.S., Roberts, F.E., Yang, C., Lieberman, D.R., Reamer, R.A., Tschaen, D.M.: Efficient synthesis of losartan, a nonpeptide angiotensin II receptor antagonist. J. Org. Chem. **59**(21), 6391–6394 (1994)
85. Chinchilla, R., Nájera, C.: The Sonogashira reaction: a booming methodology in synthetic organic chemistry. Chem. Rev. **107**(3), 874–922 (2007)
86. Giese, M.W., Moser, W.H.: Construction of the benzindenoazepine skeleton via cyclopentan-nulation of fischer aminocarbene complexes: total synthesis of bulgaramine. J. Org. Chem. **70**(16), 6222–6229 (2005)
87. Sumi, S., Matsumoto, K., Tokuyama, H., Fukuyama, T.: Stereocontrolled total synthesis of (−)-aspidophytine. Tetrahedron **59**(43), 8571–8587 (2003)
88. Sumi, S., Matsumoto, K., Tokuyama, H., Fukuyama, T.: Enantioselective total synthesis of aspidophytine. Org. Lett. **5**(11), 1891–1893 (2003)
89. Haas, D., Hammann, J.M., Greiner, R., Knochel, P.: Recent developments in Negishi cross-coupling reactions. ACS Catal. **6**(3), 1540–1552 (2016)
90. Phapale, V.B., Cárdenas, D.J.: Nickel-catalysed Negishi cross-coupling reactions: scope and mechanisms. Chem. Soc. Rev. **38**(6), 1598–1607 (2009)
91. Manley, P.W., Acemoglu, M., Marterer, W., Pachinger, W.: Large-scale Negishi coupling as applied to the synthesis of PDE472, an inhibitor of phosphodiesterase type 4D. Org. Process Res. Dev. **7**(3), 436–445 (2003)
92. Nolasco, L., Perez Gonzalez, M., Caggiano, L., Jackson, R.F.W.: Application of Negishi cross-coupling to the synthesis of the cyclic tripeptides OF4949-III and K-13. J. Org. Chem. **74**(21), 8280–8289 (2009)
93. Kolb, H.C., Finn, M.G., Sharpless, K.B.: Click chemistry: diverse chemical function from a few good reactions. Angew. Chem. Int. Ed. **40**(11), 2004–2021 (2001)
94. Meldal, M., Tornøe, C.W.: Cu-catalyzed azide—alkyne cycloaddition. Chem. Rev. **108**(8), 2952–3015 (2008)
95. Agrahari, A.K., Bose, P., Jaiswal, M.K., Rajkhowa, S., Singh, A.S., Hotha, S., Mishra, N., Tiwari, V.K.: Cu (I)-catalyzed click chemistry in glycoscience and their diverse applications. Chem. Rev. **121**(13), 7638–7956 (2021)
96. Tiwari, V.K., Mishra, B.B., Mishra, K.B., Mishra, N., Singh, A.S., Chen, X.: Cu-catalyzed click reaction in carbohydrate chemistry. Chem. Rev. **116**(5), 3086–3240 (2016)
97. Zhang, L., Chen, X., Xue, P., Sun, H.H.Y., Williams, I.D., Sharpless, K.B., Fokin, V.V., Jia, G.: Ruthenium-catalyzed cycloaddition of alkynes and organic azides. J. Am. Chem. Soc. **127**(46), 15998–15999 (2005)
98. Sareen, N., Singh, A.S., Tiwari, V.K., Kant, R., Bhattacharya, S.: A dinuclear copper (I) thiodiacetate complex as an efficient and reusable 'click'catalyst for the synthesis of glycoconjugates. Dalt. Trans. **46**(37), 12705–12710 (2017)
99. Sheldon, R.A., Downing, R.S.: Heterogeneous catalytic transformations for environmentally friendly production. Appl. Catal. A Gen. **189**(2), 163–183 (1999)
100. Corma, A., García, H.: Lewis acids: from conventional homogeneous to green homogeneous and heterogeneous catalysis. Chem. Rev. **103**(11), 4307–4366 (2003)
101. Hoelderich, W.F.: Environmentally benign manufacturing of fine and intermediate chemicals. Catal. Today **62**(1), 115–130 (2000)

References

102. Mitsutani, A.: Future possibilities of recently commercialized acid/base-catalyzed chemical processes. Catal. Today **73**(1), 57–63 (2002)
103. Ballantine, J.A., Davies, M., Patel, I., Purnell, J.H., Rayanakorn, M., Williams, K.J., Thomas, J.M.: Organic reactions catalysed by sheet silicates: ether formation by the intermolecular dehydration of alcohols and by addition of alcohols to alkenes. J. Mol. Catal. **26**(1), 37–56 (1984)
104. Adams, J.M., McCabe, R.W.: Clay minerals as catalysts. Dev. Clay Sci. **1**, 541–581 (2006)
105. Bhorodwaj, S.K., Dutta, D.K.: Activated clay supported heteropoly acid catalysts for esterification of acetic acid with butanol. Appl. Clay Sci. **53**(2), 347–352 (2011)
106. Reddy, C.R., Iyengar, P., Nagendrappa, G., Prakash, B.S.J.: Esterification of succinic anhydride to di-(p-cresyl) succinate over Mn^+-montmorillonite clay catalysts. J. Mol. Catal. A Chem. **229**(1–2), 31–37 (2005)
107. McCabe, R.W., Adams, J.M., Martin, K.: Clay-and zeolite-catalysed cyclic anhydride formation. J. Chem. Res. Synopses **11**, 356–357 (1985)
108. Villemin, D., Labiad, B., Loupy, A.: Clay catalysis: a convenient and rapid formation of anhydride from carboxylic acid and isopropenyl acetate under microwave irradiation. Synth. Commun. **23**(4), 419–424 (1993)
109. Laszlo, P.: Catalysis of organic reactions by inorganic solids. Pure Appl. Chem. **62**(10), 2027–2030 (1990)
110. Ponde, D.E., Deshpande, V.H., Bulbule, V.J., Sudalai, A.: Selective catalytic transesterification, transthiolesterification, and protection of carbonyl compounds over natural kaolinitic clay. J. Org. Chem. **63**(4), 1058–1063 (1998)
111. Ma, Y., Tong, W., Zhou, H., Suib, S.L.: A review of zeolite-like porous materials. Microporous Mesoporous Mater. **37**(1–2), 243–252 (2000)
112. Rangnekar, N., Mittal, N., Elyassi, B., Caro, J., Tsapatsis, M.: Zeolite membranes—a review and comparison with MOFs. Chem. Soc. Rev. **44**(20), 7128–7154 (2015)
113. Kim, W., Kim, J.-C., Kim, J., Seo, Y., Ryoo, R.: External Surface catalytic sites of surfactant-tailored nanomorphic zeolites for benzene isopropylation to cumene. ACS Catal. **3**(2), 192–195 (2013)
114. Ventura, M., Domine, M.E., Chávez-Sifontes, M.: Catalytic processes for lignin valorization into fuels and chemicals (aromatics). Curr. Catal. **8**(1), 20–40 (2019)
115. Gunnewegh, E.A., Hoefnagel, A.J., van Bekkum, H.: Zeolite catalysed synthesis of coumarin derivatives. J. Mol. Catal. A Chem. **100**(1–3), 87–92 (1995)
116. Corma, A., JoséCliment, M., García, H., Primo, J.: Design of synthetic zeolites as catalysts in organic reactions: acylation of anisole by acyl chlorides or carboxylic acids over acid zeolites. Appl. Catal. **49**(1), 109–123 (1989)
117. Chiche, B., Finiels, A., Gauthier, C., Geneste, P., Graille, J., Pioch, D.: Friedel-crafts acylation of toluene and p-xylene with carboxylic acids catalyzed by zeolites. J. Org. Chem. **51**(11), 2128–2130 (1986)
118. Kim, J.-C., Cho, K., Lee, S., Ryoo, R.: Mesopore wall-catalyzed Friedel-crafts acylation of bulky aromatic compounds in MFI zeolite nanosponge. Catal. Today **243**, 103–108 (2015)
119. Choudary, B.M., Sateesh, M., Lakshmi Kantam, M., Koteswara Rao, K., Ram Prasad, K.V., Raghavan, K.V., Sarma, J.A.R.P.: Selective nitration of aromatic compounds by solid acid catalysts. Chem. Commun. **1**, 25–26 (2000)
120. Ratnasamy, P., Singh, A.P., Sharma, S.: Halogenation over zeolite catalysts. Appl. Catal. A Gen. **135**(1), 25–55 (1996)
121. Bertea, L., Kouwenhoven, H.W., Prins, R.: Vapour-phase nitration of benzene over modified mordenite catalysts. Appl. Catal. A Gen. **129**(2), 229–250 (1995)
122. Losch, P., Kolb, J.F., Astafan, A., Daou, T.J., Pinard, L., Pale, P., Louis, B.: Eco-compatible zeolite-catalysed continuous halogenation of aromatics. Green Chem. **18**(17), 4714–4724 (2016)
123. Singh, M.P., Baghel, G.S., Titinchmi, S.J.J., Abbo, H.S.: Zeolites: smart materials for novel, efficient, and versatile catalysis. Adv. Catal. Mater. 385–410 (2015)

124. Hayashi, D., Narisawa, T., Masui, Y., Onaka, M.: H-type zeolite-catalyzed 1, 4-addition of benzene derivatives to labile acrolein. Bull. Chem. Soc. Jpn. **89**(4), 460–471 (2016)
125. Yu, W., Bian, F., Gao, Y., Yang, L., Liu, Z.-L.: Y-Zeolite-catalyzed cyclizations of terpenols. Adv. Synth. Catal. **348**(1–2), 59–62 (2006)
126. Kim, H.-J., Seo, G., Kim, J.-N., Choi, K.-H.: HY Zeolite catalyzed one-pot synthesis of 2, 3-dihydro-2, 2-dimethylbenzofurans from aryl methallyl ethers. Bull. Korean Chem. Soc. **25**(11), 1726–1728 (2004)
127. Sun, N., Huang, P., Wang, Y., Mo, W., Hu, B., Shen, Z., Hu, X.: Zeolite-catalyzed synthesis of 2,3-unsubstituted benzo[b]furans via the intramolecular cyclization of 2-aryloxyacetaldehyde acetals. Tetrahedron **71**(29), 4835–4841 (2015)
128. Hegedüs, A., Hell, Z.: Zeolite-catalyzed simple synthesis of isochromans via the oxa-pictet–spengler reaction. Org. Biomol. Chem. **4**(7), 1220–1222 (2006)
129. Cwik, A., Hell, Z., Hegedüs, A., Finta, Z., Horváth, Z.: A simple synthesis of 2-substituted oxazolines and oxazines. Tetrahedron Lett. **43**(22), 3985–3987 (2002)
130. Hegedues, A., Vigh, I., Hell, Z.: Zeolite-catalyzed simple synthesis of different heterocyclic rings, part 2. Heteroat. Chem. An Int. J. Main Gr. Elem. **15**(6), 428–431 (2004)
131. Hegedüs, A., Hell, Z., Vigh, I.: Convenient one-pot heterogeneous catalytic method for the preparation of 3, 4-dihydropyrimidin-2 (1H)-ones. Synth. Commun. **36**(1), 129–136 (2006)
132. Hegedüs, A., Hell, Z., Potor, A.: A simple environmentally-friendly method for the selective synthesis of 1, 5-benzodiazepine derivatives using zeolite catalyst. Catal. Lett. **105**(3–4), 229–232 (2005)
133. Jones, S.B., Simmons, B., Mastracchio, A., MacMillan, D.W.C.: Collective synthesis of natural products by means of organocascade catalysis. Nature **475**(7355), 183–188 (2011)

Chapter 7
Organocatalysis: A Versatile Tool for Asymmetric Green Organic Syntheses

1 Introduction

An organocatalyst can be defined as small organic molecules, which can catalyze a wide range of important asymmetric organic transformations without use of transition metals and enzymes. The concept of organocatalysis emerged to develop such catalysts which could be environmentally benign, reduce the limitations associated with metal and enzyme-catalyzed reactions and of course, equally productive in terms of yield and stereoselectivity. Transition metal catalysts have wider range of applicability in chemical transformations but are toxic in nature. Enzymes, on the other hand, are environmentally benign but have limited substrate scope. Addressing these shortcomings, organocatalysts are relatively nontoxic, stable to air and water, easily handled, eco-friendly, and readily separable from crude reaction mixture which makes them a fantastic tool for organic synthesis in pharmaceutical industries and academia. Furthermore, they are inexpensive in comparison to transition metal catalysts and have wide substrate scope too. For instance, organocatalysts have broad applicability in multicomponent and multistep domino or tandem reactions, which can afford small and complex molecules like natural products with high stereoselectivity.

The term organocatalysis was coined in 1990s but the use metal-free catalysis in organic syntheses has a history long back over a century. The earliest documented example of organocatalysis was accidently appeared in 1860, when Justus von Liebig noticed that the conversion of dicyan **1** to oxamide **2** in the presence of aqueous solution of acetaldehyde (Scheme 1a) [1]. The enzymes and their functionalities have been used for a long time in asymmetric organic syntheses. The first asymmetric reaction was noticed by Pasteur when he performed decarboxylative kinetic resolution of racemic solution of ammonium tartrate by microorganism *Penicillium glauca*. It was found that due to the enzymatic activity of microorganism one of the enantiomers, i.e., the *d*-form destroyed rapidly compared to the other form [2]. The first asymmetric C–C bond formation is attributed to George Breding. He was motivated to investigate the chemical origin of enzymes and also performed asymmetric

© The Author(s), under exclusive license to Springer Nature Singapore Pte Ltd. 2022
V. K. Tiwari et al., *Green Chemistry*,
https://doi.org/10.1007/978-981-19-2734-8_7

Scheme 1 **a** Von Liebig's oxamide synthesis. **b** Catalysis through quinine and quinidine

synthesis of α-hydroxybenzeneacetonitrile **4** from benzoic acid **3** in presence of pseudoenantiomeric quinine **5** or quinidine **6** like alkaloids as catalysts (Scheme 1b) [3].

The reinvestigation of Breding's work on asymmetric cyanohydrin synthesis by Prelog during 1950s undoubtedly sow the seed of asymmetric synthesis and led the way for new discoveries in the field of catalysis and more efficient reactions [4]. The progress of enantioselectivity in such type of reactions reached a significant level in late 1950s when Pracejus reported the synthesis of (–)-α-phenyl methyl-propionate **9** with 74% *ee* from methyl phenyl ketene **7** using *O*-acetylquinine **8** as catalyst (Scheme 2a) [5]. This impressive result inspired other researchers to explore such reactions using the cinchona catalyst system. Bergson and Långström, in 1973, reported first ever Michael addition reaction by converting β-keto ester to acrolein using 2-(hydroxymethyl)-quinuclidine as organocatalyst [6]. Amino acids (including short oligopeptides) and other nitrogen containing alkaloids particularly strychnine, brucine, etc. were investigated as organic catalysts in several asymmetric version of organic reactions [1]. In early 1970s, the discovery of L-proline **11**-mediated asymmetric Robinson annulation considered as a key event in the history of organocatalytic reactions. The intramolecular aldol reaction also called as Hajos–Parrish–Eder–Sauer–Wiechert reaction led to formation of some key intermediates for the synthesis of natural products of chemotherapeutic values (Scheme 2b) [7–9].

The milestone of organocatalytic reactions were led between 1970 to 1980s, when several reactions proceeded via ion-pair mechanism were investigated thoroughly. Reinvestigation of the Hajos–Parrish–Eder–Sauer–Wiechert reaction by List and Barbas opened a new arena for many established organic named reactions such as intermolecular cross aldol reaction, Mannich reaction, Michael reaction, Diels–Alder type reaction, and other related reactions [10–12]. Since then, by the efforts of different research groups, the field of organocatalysis has been enriched with different types of catalysts and the reactions catalyzed by them. Now, we have various derivatives of proline, cinchona alkaloids, quinine, and DMAP as well as some totally new catalysts which catalyze a wide range of important reactions.

2 Classification of Organocatalysis

Scheme 2 **a** Enantioselective ester synthesis from phenyl methyl ketene. **b** L-Proline-mediated Robinson annulations

2 Classification of Organocatalysis

The classification of organocatalysts was done by Benjamin List which was based on the action of catalyst on to the substrate and the reactive intermediate formed. List categorized the organocatalysts, except few, into four broad classes, namely, Lewis bases, Lewis acids, Brønsted bases and Brønsted acids. According to simplified catalytic cycles depicted in Fig. 1, the Lewis base catalysts (**B:**) undergo nucleophilic addition reaction to substrates (**S**) and form adduct (**B$^+$-S$^-$**) which further releases the product (**P**) and the catalyst again get ready for another turnover. In the same way, nucleophilic substrates (**S:**) get activated by Lewis acid catalysts (**A**). Brønsted base and acid catalytic cycles initiated via partial deprotonation or protonation, respectively.

The four classes of organocatalysts can be briefly described as follows.

2.1 Lewis Base Catalysis

Lewis base catalysts are in general those molecules which contains atoms possessing-free lone pair of electrons such as nitrogen, oxygen, sulfur, carbon, and phosphorus

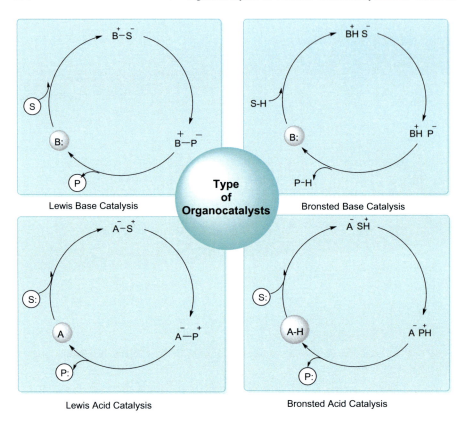

Fig. 1 Four main types of organocatalytic cycles [13]

which reacts and convert the substrate either into activated nucleophiles or electrophiles. These activated substrates are the intermediates which may be iminium ions, enamines, acyl ammonium ions. N-heterocyclic carbenes and S-ylides are another class of Lewis base catalysts (Fig. 2).

2.2 Lewis Acid Catalysis

An elegant example of this class of catalysis includes the epoxidation of olefins. Epoxidation of stilbenes [14], terminal olefins [15], and α,β-unsaturated esters [16] were enantioselectively done by using carbohydrate-based ketone catalyst derived from D-fructose (Scheme 3a). Another example of this class of catalysis is the epoxidation of *trans*-stilbene by *N*-acetylglucosamine-based ketone organocatalyst (Scheme 3b) [17].

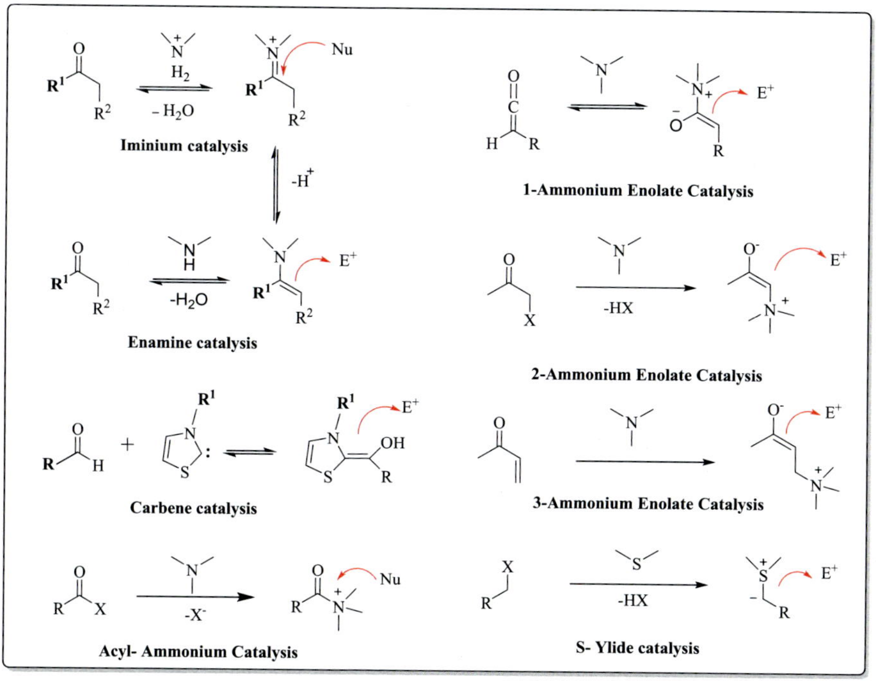

2 Classification of Organocatalysis

Fig. 2 Some examples of Lewis base organocatalysis [13]

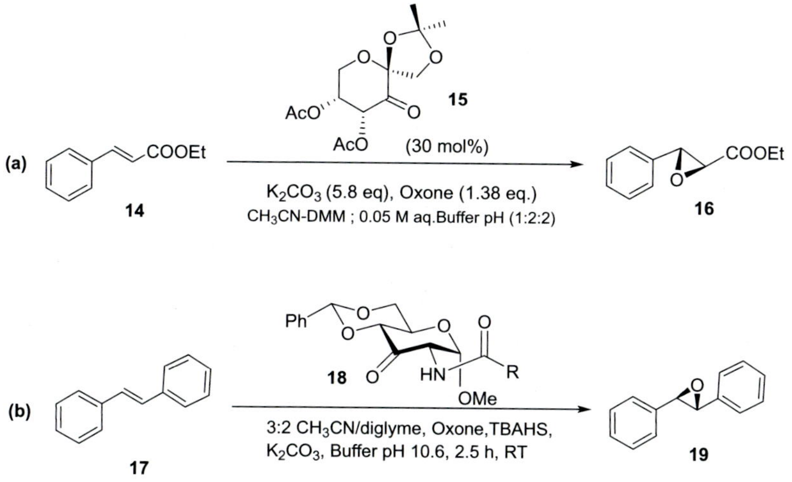

Scheme 3 Selected examples of Lewis acid organocatalysis

2.3 Brønsted Base Catalysis

Some elegant examples of Brønsted base catalysis are hydrocyanation reactions including Strecker reaction and cyanohydrin synthesis. The use of chiral bicyclic guanidine **21** as a catalyst was reported by Corey and Grogan for Strecker reaction which lead to the formation of (R)-**22** with excellent enantiomeric excess (Scheme 4a) [18]. Inoue and coworkers showed the utility of cyclopeptides as catalysts in the reaction of HCN with various aldehydes with high stereoselectivity [19]. Similarly, Lipton and coworkers reported the use of cyclopeptide catalyst **23** for the addition of various N-benzhydryl imines **20** to hydrogen cyanide to afford α-aminonitriles (S)-**22** (Scheme 4b) [20]. The plausible mechanism acting in these reactions was supposed to be the interaction of hydrogen cyanide with nitrogen containing bases via hydrogen bonding interactions to produce cyanide ions that can then add to carbonyl compound or with imines coordinated with peptides or with guanidine hydrogen. Another pertinent example of Brønsted base catalysis is the Michael reaction of a prochiral glycine derivative, tert-butyl diphenyliminoacetate **24** performed under solvent-free conditions by the help of modified guanidine catalyst **26** (Scheme 4c) [21].

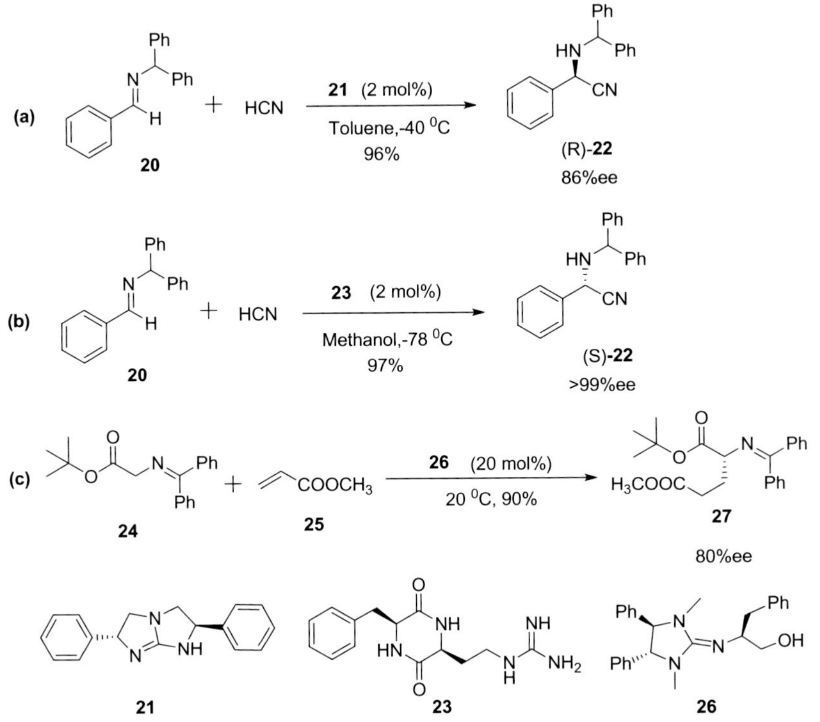

Scheme 4 Some selected examples of Brønsted base catalysis

2 Classification of Organocatalysis

2.4 Brønsted Acid Catalysis

Mechanistically, the Brønsted acid catalysis is very similar to the enzyme catalysis where hydrogen bonding to the transition state is very common. For example, the bifunctional organocatalyst, the Takemoto catalyst **29**, developed by using chiral thiourea derivatives with neighboring tertiary amino groups is found to activate the nitro compounds for the enantioselective Michael and aza-Henry reactions (Scheme 5a) [22, 23]. In case of such reactions, the nucleophile is being activated by the tertiary amino group and with the help of hydrogen bonding activation the thiourea moiety interacts with the nitro group. This can be well explained by the example of the reaction of nitro olefins **32** with malonates **31** to form the corresponding Michael adducts in good yields and in 93% *ee* in the presence of the chiral thiourea-derived catalyst **33** [24]. Similarly, another group of scientists have reported several other imine-based reactions such as Strecker [25], Pictet–Spengler [26], hydrophosphonylation [27], and Mannich reactions [28], using the chiral urea and thiourea-derived catalyst. The mode of activation of imine substrate through these catalysts is supposed to be via the hydrogen bonds that are formed in a bridging fashion from the urea hydrogens to the imine nitrogen [24]. Some examples of this type of organocatalysis are given in Scheme 5.

The succeeding section covers the details of various organocatalysts and their applications in chemical transformations.

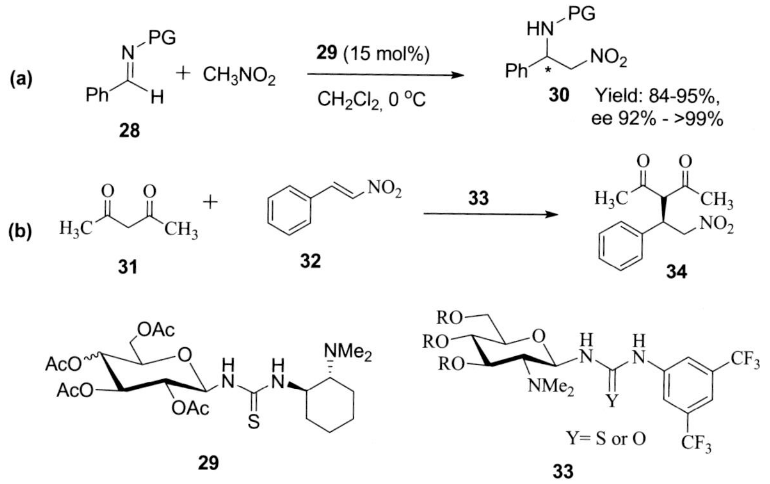

Scheme 5 Some selected examples of Brønsted acid catalysis

3 Iminium Catalysis

In 1864, H. Schiff reported the synthesis of imine which was resulted from the condensation of aldehydes or ketones with primary amines [29]. These imines are basic in nature and exists as iminium ions in acidic medium. The primary amine-derived imines are popularly known as Schiff bases. With secondary amines, aldehydes or ketones form iminium cations which cannot undergo deprotonation and exist in the form of salts of strong acids (Scheme 6). Both primary and secondary amines are used in iminium catalysis, but secondary amine dominated the field. Primary amines always need external acids as cocatalyst for activation; secondary amine may or may not require external acid for the reaction.

Iminium ion activates aldehydes and ketones toward nucleophilic attack. In fact, the iminium ions are stronger electrophiles then corresponding aldehydes and ketones. Interestingly, the iminium ion formation leads to various type of electrophile–nucleophile interactions which includes, cycloaddition, nucleophilic addition, retro-Aldol type reactions, and attack by bases (deprotonation results in enamine formation). Examples of different modes of iminium activation are depicted in Fig. 3 [30]. The last and important step in iminium catalysis is the removal of amine catalyst in hydrolysis or elimination step; otherwise, it will not be considered as amine-catalyzed reaction.

The earliest known iminium-catalyzed reaction was supposed to be Knoevenagel condensation reaction mediated by primary and secondary amines [31–33]. The mechanism which supports the role of iminium ion in Knoevenagel condensation was emerged slowly. Blanchard et al. reported the role of cationic species derived from secondary amine in Knoevenagel condensation [34]. Further, Crowell and Peck after a couple of decades came with kinetic studies which supported the role of imine/iminium ions as activated species in Knoevenagel type condensation reactions [35]. In 1907, Pollak postulated imines/iminium ions as active intermediates in decarboxylation of β-ketocarboxylic acids. Widmark and Jeppsson, in 1922, reported the aniline-catalyzed decarboxylation of β-ketocarboxylic acids [36]. In 1934, Pedersen suggested the involvement of iminium ion as active intermediate in

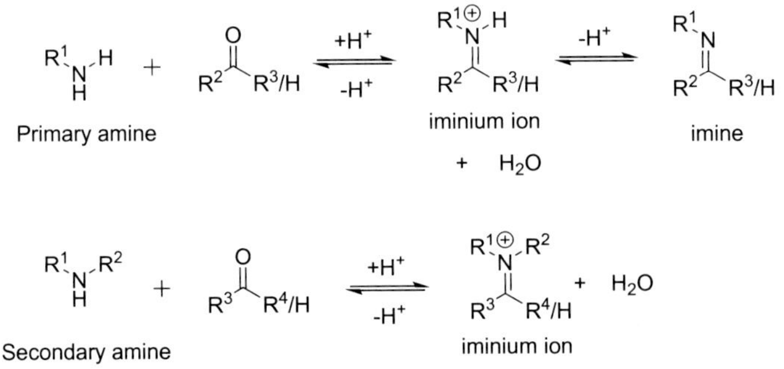

Scheme 6 Formation of imine and iminium ions

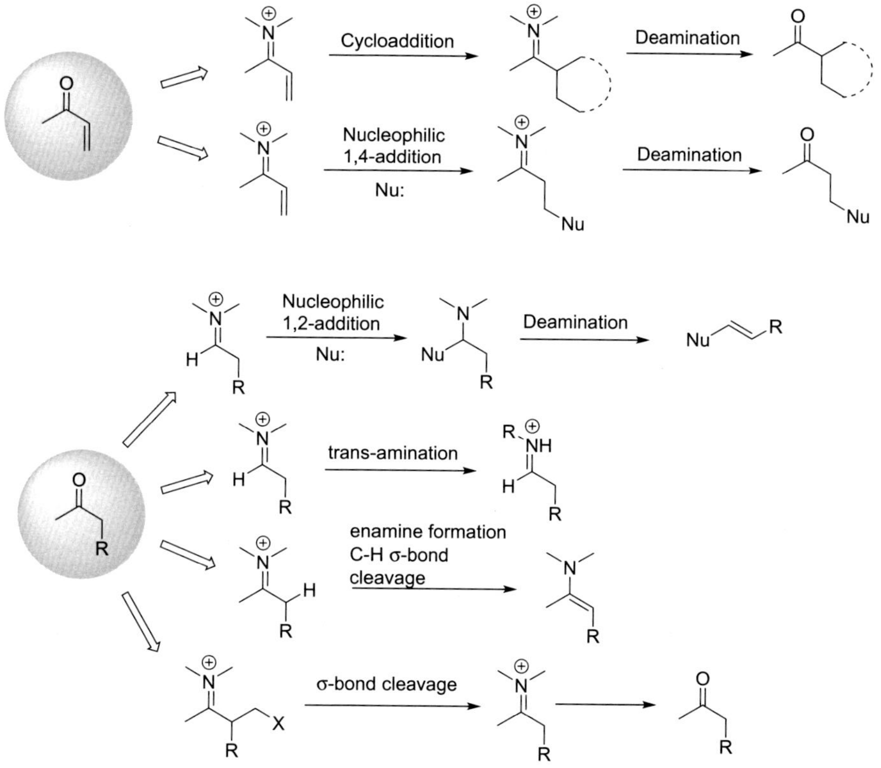

3 Iminium Catalysis

Fig. 3 Activation modes of iminium catalysts

decarboxylation of β-keto acids. In 1937, Langenbeck et al. reported for the first time the iminium-catalyzed conjugate addition reaction. Another landmark in the history of iminium-catalyzed reaction was led by Cordes and Jencks in 1962 who discovered the iminium-catalyzed transimination reaction [37]. The deprotonation of iminium intermediate and the formation of enamine attracted the attention of researchers in 1960s and 1970s [38–41]. Iminium-catalyzed cycloaddition reaction was another landmark discovery in this field which was reported by Baum and Viehe in 1976 [42]. They reported iminium-catalyzed Diels–Alder reaction. In 2000, MacMillan and coworkers reported more generalized form of Diels–Alder reaction via enantioselective iminium catalysis [43]. Till date, a number of amines are studied which readily used as iminium catalysts and can accelerate several fundamentally important organic transformations with high stereocontrol. Some of them are depicted in Fig. 4.

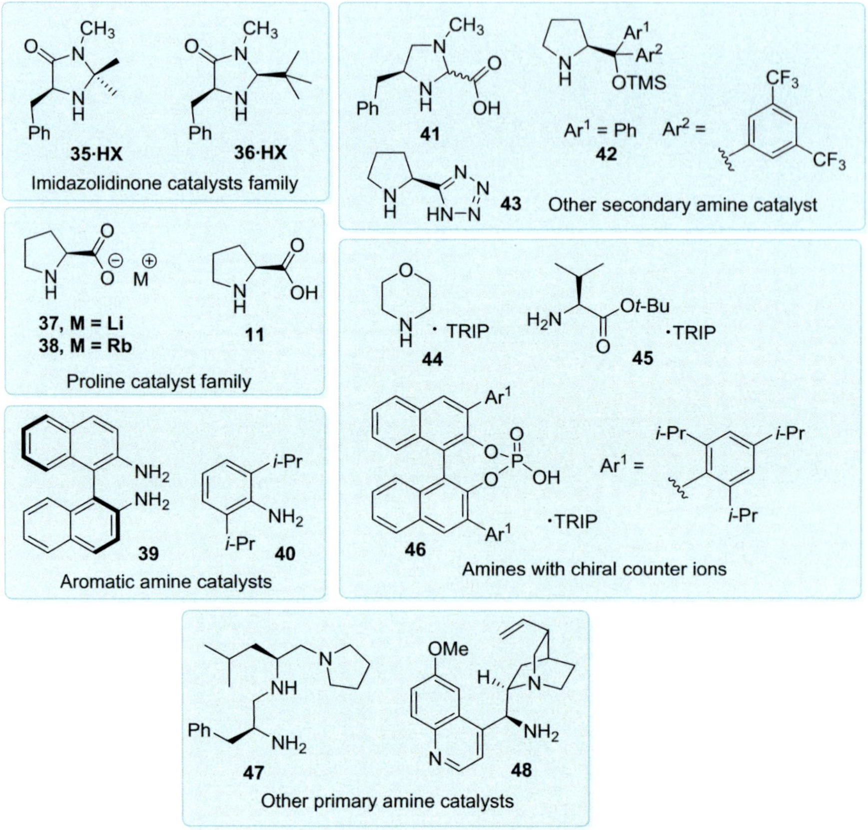

Fig. 4 Different classes of amines used as catalysts

3.1 Application of Iminium Catalysts in Organic Synthesis

The major reactions which are catalyzed by primary or secondary amine-derived imine/iminium salts are listed below.

1. Cycloaddition reactions—Diels–Alder reaction
2. Conjugate addition
3. Domino process

3.1.1 Cycloaddition Reactions

The first organocatalyzed cycloaddition reaction was reported by MacMillan and coworkers in 2000 which was the turning point in the field of iminium-catalyzed Diels–Alder reaction [43]. MacMillan shared the Nobel Prize with B. List in year 2021 for the contribution in the field of organocatalyst [44]. MacMillan and coworkers

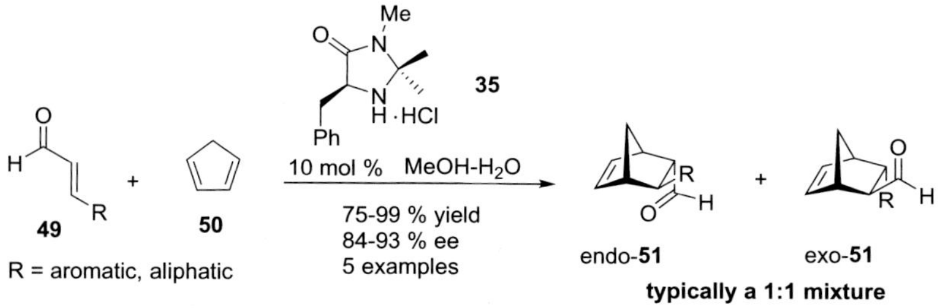

3 Iminium Catalysis

Scheme 7 Iminium-catalyzed Diels–Alder reaction

reacted cinnamaldehyde **49** and cyclopentadiene **50** in the presence of chiral imidazolidinone catalyst which afforded *exo*-**51** and *endo*-**51** cyclic products with good enantioselectivity (Scheme 7).

They proposed the role of iminium ion catalyst in this reaction. Mechanistic studies by MM3 force field simulation showed that the Z-iminium ion formed by reaction of amine catalyst and aldehyde, due to unfavorable interaction between iminium α-hydrogen atom and methyl group of the catalyst, turn to E-iminium ion which is sterically more favorable. The diene approaches iminium ion from more favorable *Si* face (Scheme 8a). It was proposed that the position of benzyl group of catalyst is favored by π-stacking between electron rich benzene ring and the electron-deficient iminium ion. Houk and coworkers on the basis of density functional theory (DFT) ab initio calculations (B3LYP/6-31G(d)) denied MacMillan's theory. In their study, they showed that the position of benzyl group is orthogonal to the iminium ion and the interaction of benzyl ring of the catalyst is C–H π interaction with one of the geminal methyl group (Scheme 8b). However, Houk's calculation supported the MacMillan's conclusion [45, 46].

Cozzi and coworkers and subsequently different research groups exploited the covalent nature of iminium catalyst and attached them on different polymeric and solid supports. Cozzi and coworkers prepared the imidazolidinone catalyst derivative **58** using (*S*)-tyrosine as the starting material which consequently linked to polyethylene glycol (PEG) (Scheme 9) [47]. These solid/polymer supported catalysts are readily recovered from the reaction mixture. The polymer supported imidazolidinone catalyst was used in cycloaddition of acrolein **55** and cyclohexadiene **56** which leads to *endo*-**57** in good yield with high enantioselectivity. Several other research groups derivatized the imidazolidinone catalyst **35** and allowed them to link with solid support. For example, Pihko and coworkers derivatized imidazolidinone **35** from nitrogen atom at third position of the ring and linked it to two different polymers, Jandajel polymer and silica gel which afforded **59** and **60** (Scheme 9). The silica gel supported catalyst **60** was found to be better catalyst in terms of yield and diastereoselectivity [48].

Furthermore, recyclable fluorous catalyst **61** was reported by Zhang and coworkers (Scheme 10) [49]. After the completion of reaction, both the final product

Scheme 8 a Mechanism of iminium-catalyzed Diels–Alder reaction, b theories of interaction of benzyl group to stabilize the *E*-iminium intermediate

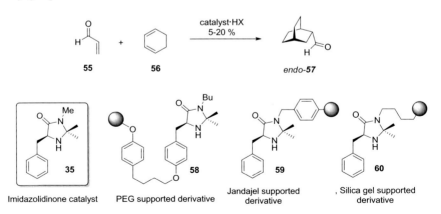

Scheme 9 Polymer supported imidazolidinone catalysts

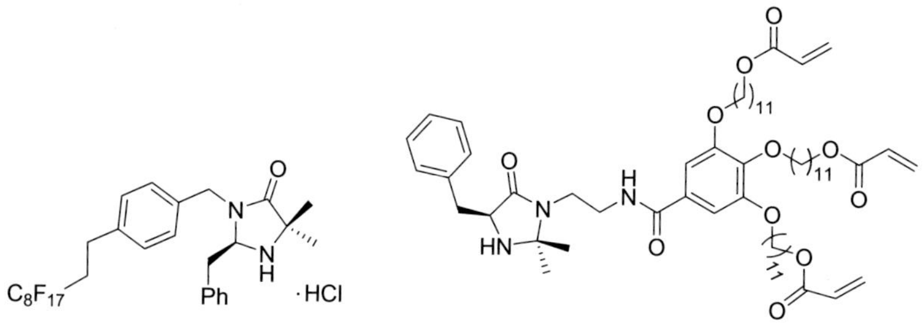

3 Iminium Catalysis

273

Scheme 10 Structure of fluorous and nanostructured organocatalyst

and catalyst were recovered by fluorous solid phase separation technique. The catalyst was reported to recover in 86% yield and with 97% purity. A nanostructured solid organocatalyst **62** was introduced by Gin and coworkers for the Diels–Alder reaction (Scheme 10) [50, 51]. Ying and coworkers reported solid mesocellular foam (MCF) supported imidazolidinone catalyst and used it in Diels–Alder reaction between cinnamaldehyde and cyclopentadiene [52].

In 2002, Northrup and Macmillan introduced an asymmetric Diels–Alder reaction for α,β unsaturated ketones [53]. When catalyst **35** was tested, the yield remained low and no enantioselectivity was found. They modified the existing catalyst **35** by introducing (5-methyl)furyl at the second position replacing two geminal methyl groups. The new catalyst **63** catalyzed the reaction between different α,β-unsaturated ketones **64** and cyclopentadiene **50** which resulted in good yields and high selectivity (Scheme 11).

The earliest intramolecular organocatalytic Diels–Alder reaction was reported by Selkälä and Koskinen in 2005, using triene-aldehyde as the starting material

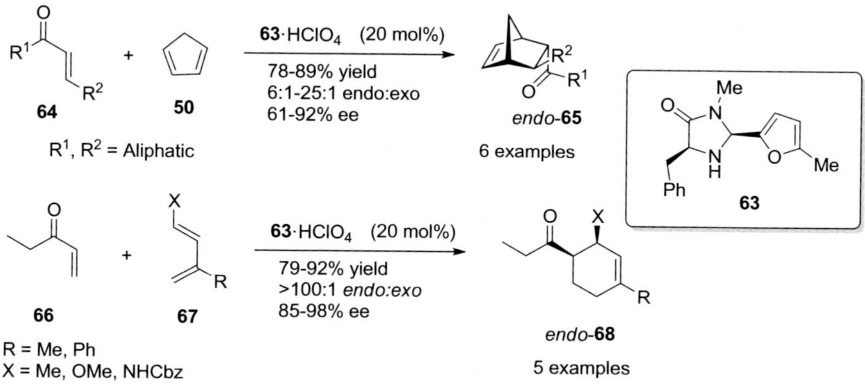

Scheme 11 Diels–Alder reaction of enones and dienes

274   7 Organocatalysis: A Versatile Tool for Asymmetric Green …

in the presence of various imidazolidinone catalysts **35**, **36**, **71**, **72** which afforded bicyclo[4.3.0]nonane **70** in high yield (Scheme 12) [54]. Catalyst **35** and **71** gave the desired product in satisfactory yield but enantioselectivity was poor. The best result was shown by catalyst **36** with remarkable yield of 99% and with high selectivity (72% *ee*).

MacMillan and coworkers, in the same year, reported the organocatalytic intramolecular Diels–Alder reaction using same trienes **73** as starting material to afford corresponding bicyclic products **74** (Scheme 13). The enantioselectivity of the product obtained by MacMillan was greater than Selkälä and Koskinen despite the same starting material was used. The reason sought for such difference in selectivity was the use of different acid cocatalysts. Selkälä and Koskinen used HCl whereas MacMillan and coworkers used TFA and $HClO_4$ as cocatalyst [55].

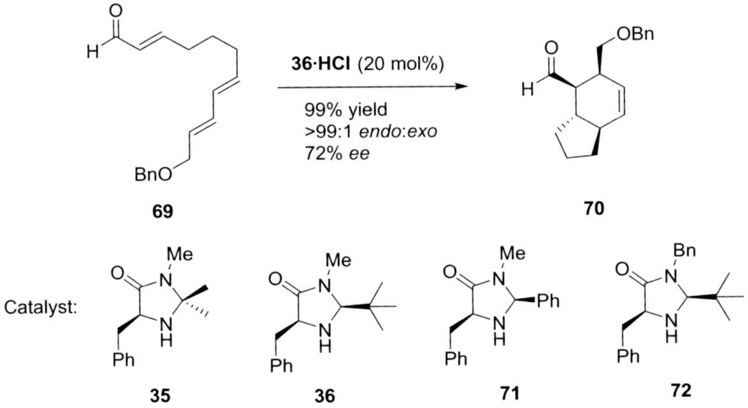

Scheme 12 First asymmetric intramolecular Diels–Alder reaction reported by Selkälä and Koskinen

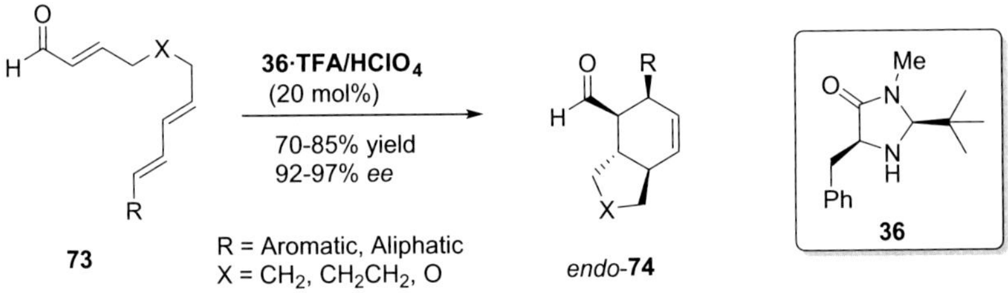

Scheme 13 Asymmetric organocatalytic Diels–Alder reaction reported by MacMillan

3 Iminium Catalysis

3.1.2 Conjugate Addition Reactions

The first asymmetric iminium-catalyzed conjugate addition reaction was reported by Yamaguchi and coworkers in 1991 [56]. They found that the reaction between dimethyl malonate **76** and hexenal **75** is readily promoted by the use of pyrrolidine and other amino acids like proline. The most active catalyst was lithium salt of proline **37** (Scheme 14). The use of lithium salt of proline afforded good yield but enantioselectivity was poor. Later, the catalyst was further optimized and Rb salt of proline **38** was found be better substitute than Li salt of proline **37** [57]. The Rb salt of proline accelerated the selective addition of diisopropyl malonate **78** to both cyclic and acyclic β-substituted enones which afforded the desired product **81** and **82** in good yields and better enantioselectivities depending upon structure of enones (Scheme 15).

Another chiral proline-derived catalyst, the proline ammonium hydroxides **83**, was explored by Kawara and Taguchi [58]. The catalyst **83** catalyzed the reaction between cycloenones **79**, and 4-phenylbutanone **84** and dibenzyl malonates **85** to afford Michael adducts **86** and **87** in moderate yield and enantioselectivities (Scheme 16). Catalyst **83a** gave better results. Acyclic enones gave poor results with both the catalysts.

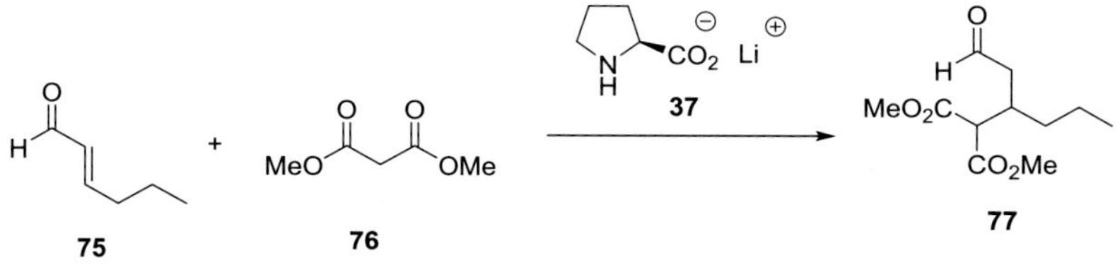

Scheme 14 First asymmetric iminium-catalyzed conjugate addition

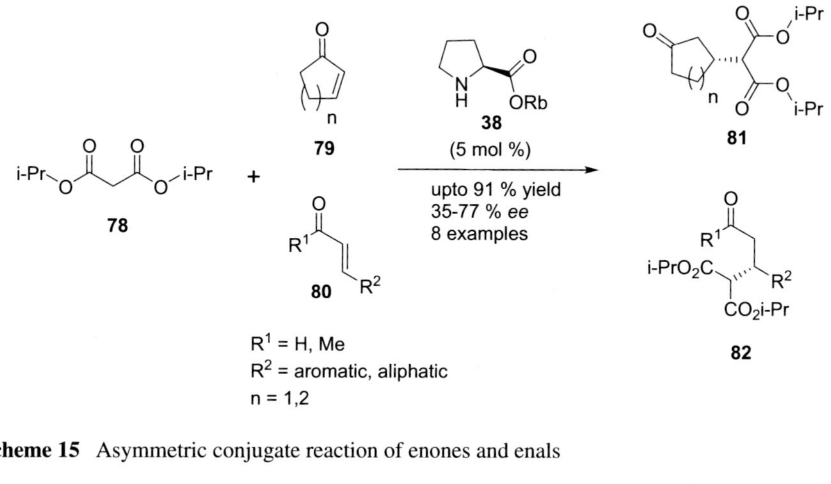

Scheme 15 Asymmetric conjugate reaction of enones and enals

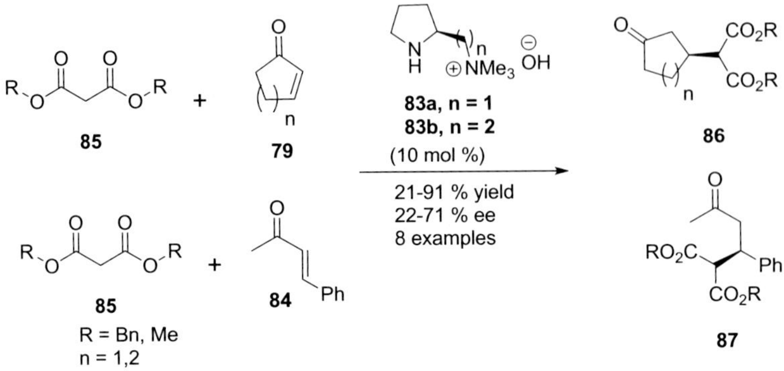

Scheme 16 Asymmetric conjugate addition of enones to malonate

The screening of various L-proline metal and tetraalkylammonium salts as catalyst by Yamaguchi group and Kawara and Taguchi revealed some interesting trends [58, 59]. While rubidium prolinate provided highest asymmetric induction, Lithium prolinate afforded fast reaction rates. Furthermore, it was found that the size of catalyst's counter ions played crucial role in facial selectivity of the reactions. Increase in the size of counter ions, i.e., from Li to Rb lead to reversal of absolute configuration of the product from "*S*" to "*R*." Similarly, the increase in the size of alkyl chain of ammonium salts of proline, reversed the absolute configuration from "*R*" to "*S*."

It should be noted that the absolute configuration of proline catalyst used by both, Yamaguchi group and Kawara and Taguchi, was same, but the facial selectivity was opposite. It was due to the face control of approaching malonate nucleophile (Scheme 17). In the case of proline ammonium hydroxide catalyst, the incoming nucleophile favors the vicinity of ammonium ion due to electrostatic interaction

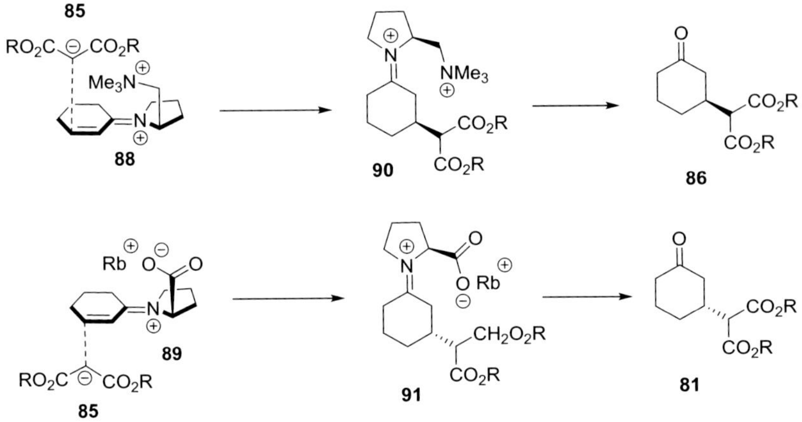

Scheme 17 Facial selectivity of approaching malonate nucleophile

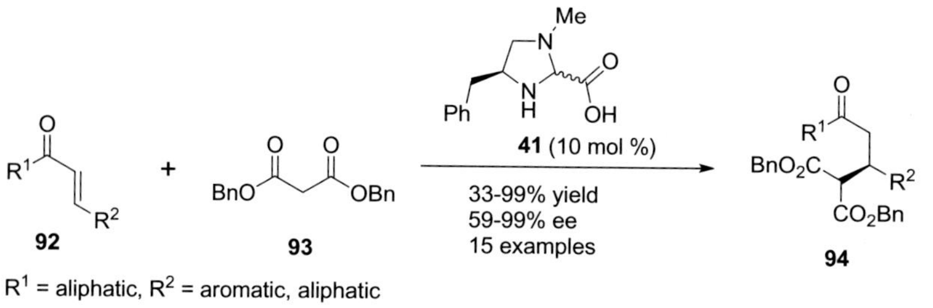

3 Iminium Catalysis

Scheme 18 Asymmetric addition of malonates to enones

and thus approaches iminium double bond through *si* face. In contrast, in case of rubidium prolinate, the malonate nucleophile approaches from *re* face to avoid the steric hindrance of side chain.

Jørgensen and coworkers reported a phenylalanine-derived catalyst **41** which promoted the addition of malonates **93** to enones **92** [60]. The size of ester functionality of malonates governs the yield and enantioselectivities. While the bulkier malonates reacted at lower pace and afforded poor yield, moderate sized malonates like dibenzyl malonate afforded excellent yield and high enantioselectivity (Scheme 18).

Ley and coworkers reported the application of tetrazole derivative of proline, 5-Pyrrolidin-2-yltetrazole, **43** in asymmetric conjugate addition of wide variety of cyclic and acyclicenones **95** to methyl and ethyl malonates **85** with high enantioselectivities [61]. The tetrazole catalyst **43** had already been used in conjugate addition of nitroalkanes to cyclic and acyclic enones [62]. The catalyst displayed an improved efficiency and selectivity which afforded Michael adducts from a variety of aromatic, heteroaromatic enones, and malonates (Scheme 19).

The asymmetric conjugated addition of enals to malonates with the aid of proline-based catalyst **42** was reported by Jørgenson and colleagues in 2006. His group successfully transformed α,β-unsaturated aldehyde to β-substituted Michael adduct in good yield and enantioselectivity. This methodology was applied for total syntheses of (−)-paroxetine **100** and (+)-femexotine **101** [63]. Addition of cinnamaldehyde or *p*-fluorocinnamaldehyde to dibenzyl malonate afforded required

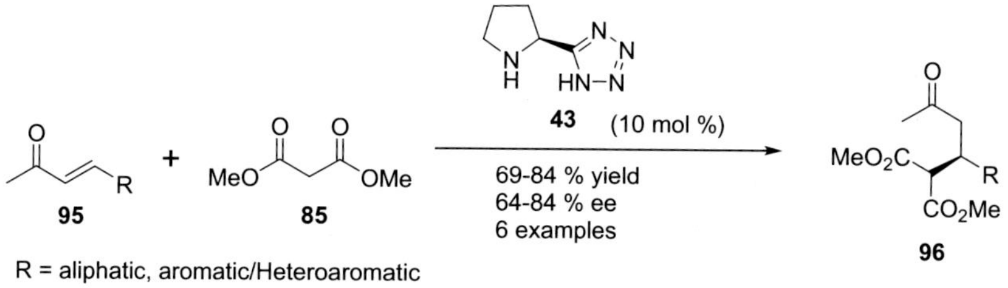

Scheme 19 Tetrazole analog of proline-catalyzed addition reaction

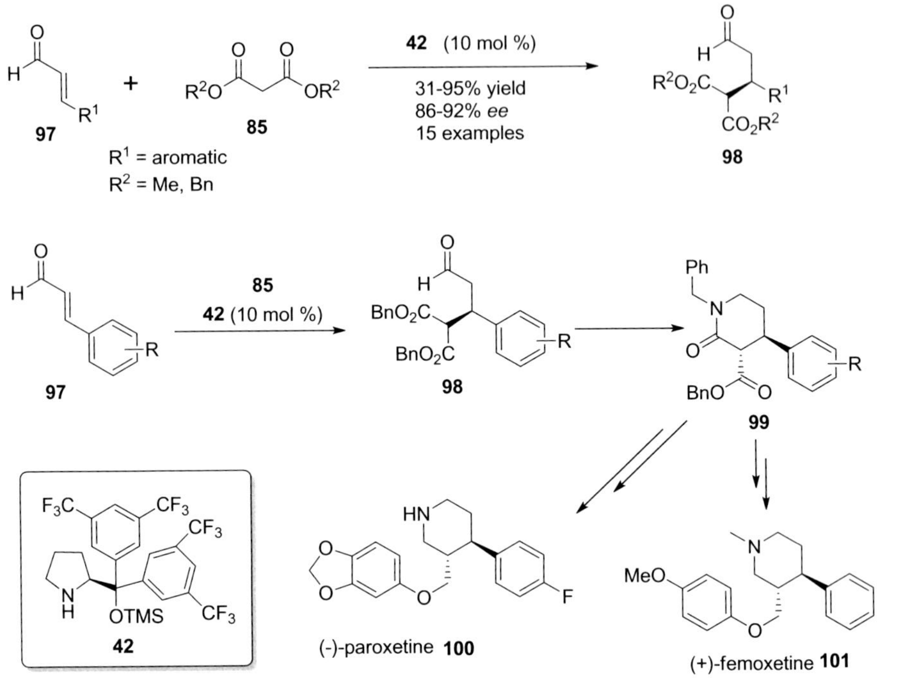

Scheme 20 Asymmetric conjugated addition of α,β-unsaturated aldehydes to malonates

stereocenter in building block in good yield. These intermediate products further underwent through reductive amination-cyclization sequence to afford corresponding chiral lactam. This whole process of total synthesis was new, short, and simple methodology for the synthesis of chiral lactams and lactones (Scheme 20).

Jørgensen and coworkers reported one step organocatalytic enantioselective procedure for the synthesis of one of the most widely used anticoagulant, warfarin **106**, using different acyclic β-substituted enones **104** and 4-hydroxycoumarin **105** (Scheme 21) [64]. The authors investigated four catalysts including, (S)-proline **11**, imidazolidine catalysts **41** and **102**, the condensation product of 1,2-diphenylethanediamine and glyoxylic acid, **103**. However, the catalyst **103** was found superior to other tested amine catalysts. Uniformly good yield and enantioselectivity was obtained.

Jørgensen investigated the catalytic activity and stereochemical outcomes of the reactions using PM3 calculations. The authors proposed that aminal, a bicyclic intermediate **105** rather than iminium ion intermediate **104,** is responsible for the enantioselectivity of the reaction (Scheme 22a) [64]. Later, Chin and coworkers suggested the breakdown of catalyst and formation of diamine **106** which leads to the formation of nucleophile-acid complex **107** is responsible for the selectivity (Scheme 22b) [65]. Their suggestion was further supported by DFT calculations.

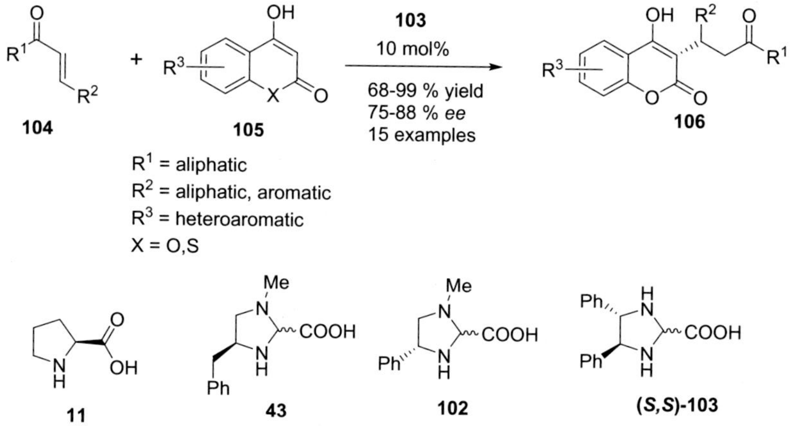

3 Iminium Catalysis

Scheme 21 One step asymmetric synthesis of warfarin using imidazolidine as organocatalyst

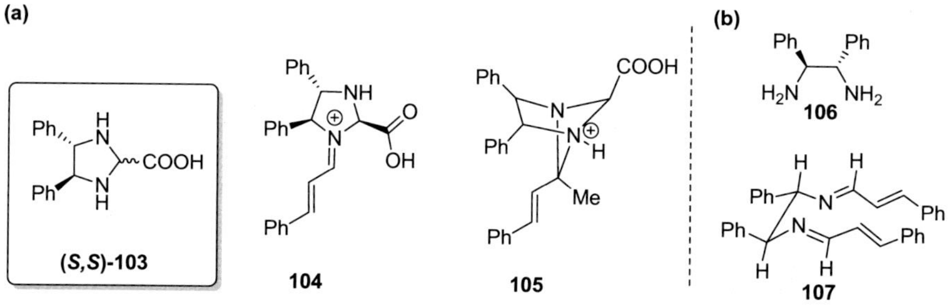

Scheme 22 Reaction intermediates involved in imidazolidine-catalyzed reaction proposed by **a** Jørgensen et al. and **b** Chin et al.

Chen and colleagues reported Michael addition product **110** by the addition of 4-hydroxycoumarin **108** to β-substituted enones **109** with high yield and enantios-electivity employing the organocatalyst **48** [64]. The catalyst gave excellent results when TFA was used as cocatalyst. In addition to 4-hydroxycoumarin (X=O), 4-hydroxythiocoumarin (X=S) and methylhydroxycarbostyril (X=NMe) were used as reaction substrates (Scheme 23).

3.1.3 Domino Reactions

A domino process is defined as the sequence of consecutive reactions in which the subsequent reaction depends upon the functionality generated by the bond formation or cleavage in the previous reaction. A pertinent example of domino

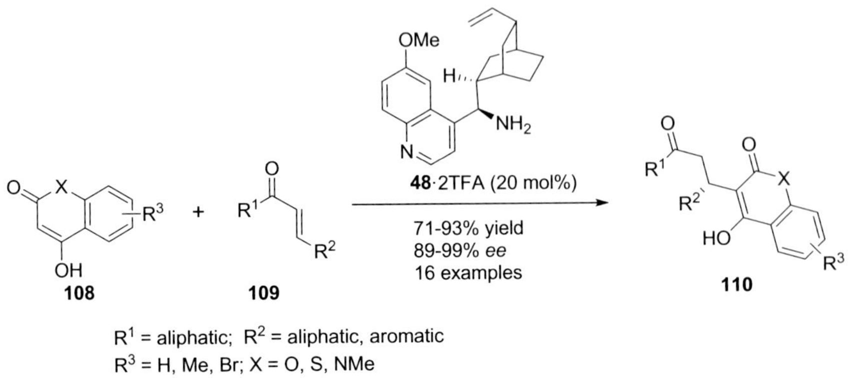

R^1 = aliphatic; R^2 = aliphatic, aromatic
R^3 = H, Me, Br; X = O, S, NMe

Scheme 23 Asymmetric addition of 4-hydroxycoumarin and enones

reaction in organocatalysis is the involvement of active iminium-enamine intermediate. Depending upon the mechanism and sequence of involvement of intermediates, domino reactions could be categorized into two broad classes–inter and intramolecular domino process.

Intramolecular domino reaction further divided into distinct classes based on the mechanism involved. For example, in enamine-iminium type I domino process, the final cyclization step take place via iminium intermediate. An alternative route followed is the enamine—Diels–Alder pathway. In iminium-enamine type II domino process, the final cyclization step follows the typical *exo*-enol aldol type mechanism. In type III, double iminium/enamine, both steps are iminium mediated in contrast to type I where the final step is iminium activated (Scheme 24). However, several intramolecular reactions do not clearly fall in any of the above categories.

Cyclopropanation Reaction

Iminium ion-catalyzed cyclopropanation reaction was reported by Kunz and MacMillan [66]. The iminium ion formed by the reaction of α,β unsaturated aldehydes **111** and the amine catalyst **112** reacts with nucleophilic ylides **113** to afford enamine. The enamine, subsequently, attack to close the cyclopropane ring with the expulsion of dimethylsulfide in the reaction (Scheme 25). The reaction afforded various cyclopropanes **114** with good yield and high stereoselectivity. It was further concluded that the stereocontrol of the reaction was exhibited by virtue of electrostatic interaction between the polarities present on the molecule. Similarly, another pertinent example of iminium-catalyzed cyclopropanation reaction was reported by Ley and coworkers. The authors used proline tetrazole catalyst **117** for the cyclopropanation of cyclohexenone **115** with bromonitromethane **116** to produce **118** with moderate enantioselectivity (Scheme 26) [67].

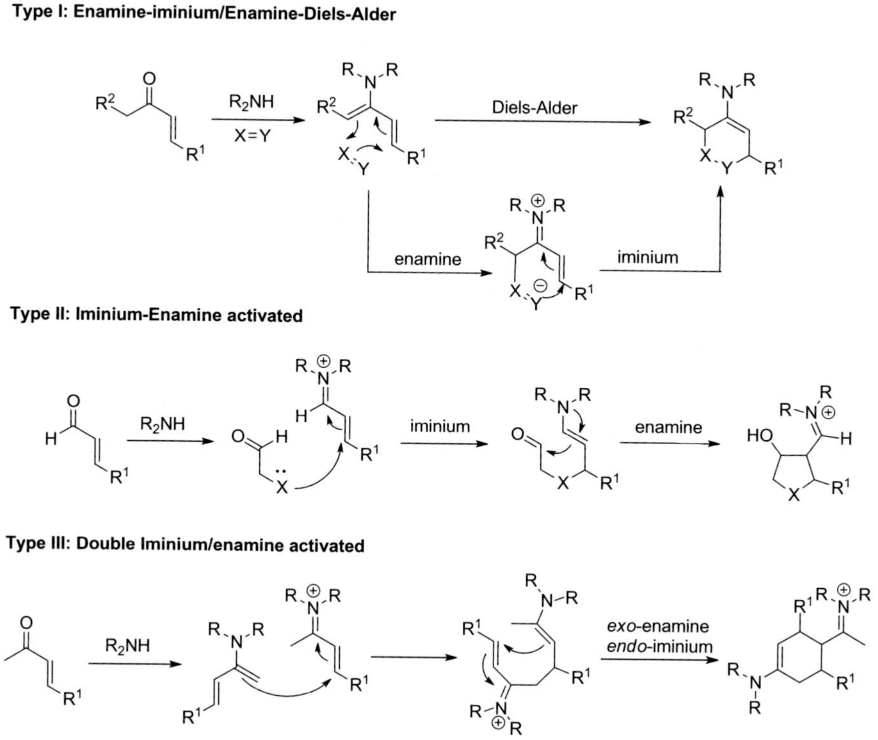
Type I: Enamine-iminium/Enamine-Diels-Alder
R2NH
X=Y
Diels-Alder
enamine
iminium
Type II: Iminium-Enamine activated
R2NH
iminium
enamine
Type III: Double Iminium/enamine activated
R2NH
exo-enamine
endo-iminium

Scheme 24 Various modes of iminium-catalyzed intramolecular Domino process

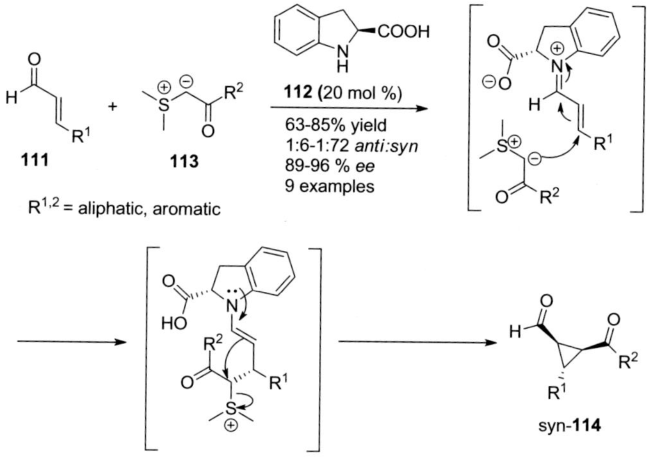
COOH
112 (20 mol %)
63-85% yield
1:6-1:72 anti:syn
89-96 % ee
9 examples
111
113
R1,2 = aliphatic, aromatic
syn-114

Scheme 25 Iminium-catalyzed domino cyclopropanation of β-substituted aldehydes

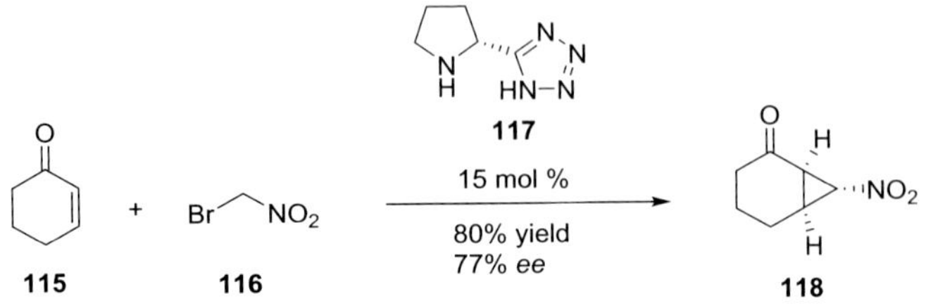

Scheme 26 Iminium-catalyzed cyclopropanation using bromonitromethane

Epoxidation Reaction

The earliest iminium-catalyzed epoxidation reaction was reported by Jørgensen et al. [68]. The epoxidation process was performed with hydrogen peroxide and α,β-unsaturated aldehydes **119** in the presence of proline-derived catalyst **42** with moderate yield and high stereoselectivities (Scheme 27a). Hydrogen peroxide may be replaced with organic peroxides like *tert*-Butyl hydroperoxide in the presence of different catalytic systems **40** and **124** without loss of enantioselectivity (Scheme 27b, c). Macmillan and coworkers [69] proposed imidazolidinone-catalyzed epoxidation reaction in which in situ formed nucleophilic iodosobenzene reacts to the iminium ion formed by reaction of β-substituted enals and imidazolidinone catalyst with excellent yield and high enantioselectivity. They optimized slow release of iodosobenzene from iminoiodinane source NsNIPh for high degree of enantiomeric control and efficiency (Scheme 28a). The plausible mechanism of this reaction is depicted in Scheme 28b.

Aziridination Reaction

Córdova and coworkers reported aziridine ring formation with α,β-unsaturated aldehydes **131** and acylated hydroxy carbamates in the presence of prolinol derivative catalyst (Scheme 29) [70]. The reaction proceed via iminium ion formation with enals and the prolinol catalyst followed by conjugate addition of hydroxycarbamate to iminium ion which afford enamine. The aziridine ring get closed analogous to the epoxidation reaction mechanism (see, Scheme 28b) with the elimination of acetate group.

Five and Six-Membered Carbocycle Formation

List and coworkers reported an efficient in situ iminium conjugate reduction followed by enamine-catalyzed intramolecular Michael cyclization [71]. They showed imidazolidinone **36**-catalyzed cyclization of enone-enals **136** in which the first step is iminium conjugate reduction by Hantzsch ester **137**, participating as hydride ion

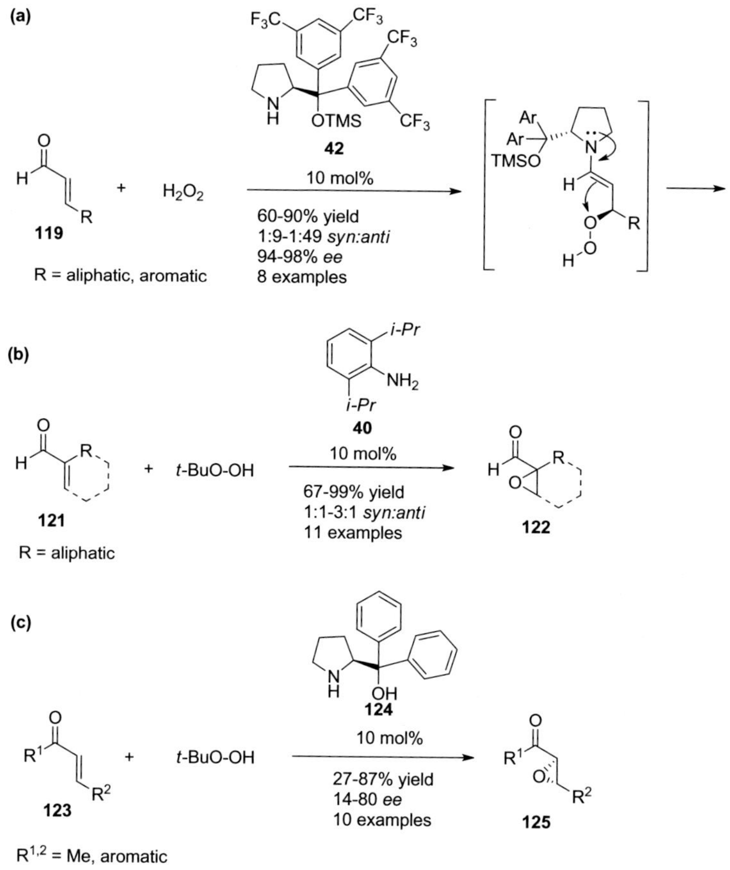

3 Iminium Catalysis

(a)

(b)

(c)

Scheme 27 Asymmetric epoxidation reactions

donor. This resulted in enamine intermediate **138** formation. The reductively generated enamine attack ketone moiety to afford bicyclic product **139** with high yield and enantioselectivity (Scheme 30). The reaction is equally effective in six-membered ring formation and aliphatic substrates.

Bui and Barbas reported a single step proline-catalyzed Robinson annulation of enones **140** and cyclic diketones **141** (Scheme 31) [72]. Robinson annulation is at wo-step process, i.e., base-mediated conjugate addition is followed by cyclization reaction. Similar approach was employed by Swaminathan et al. for the synthesis of spiroenediones from methyl vinyl ketone and cyclic ketoaldehydes [73].

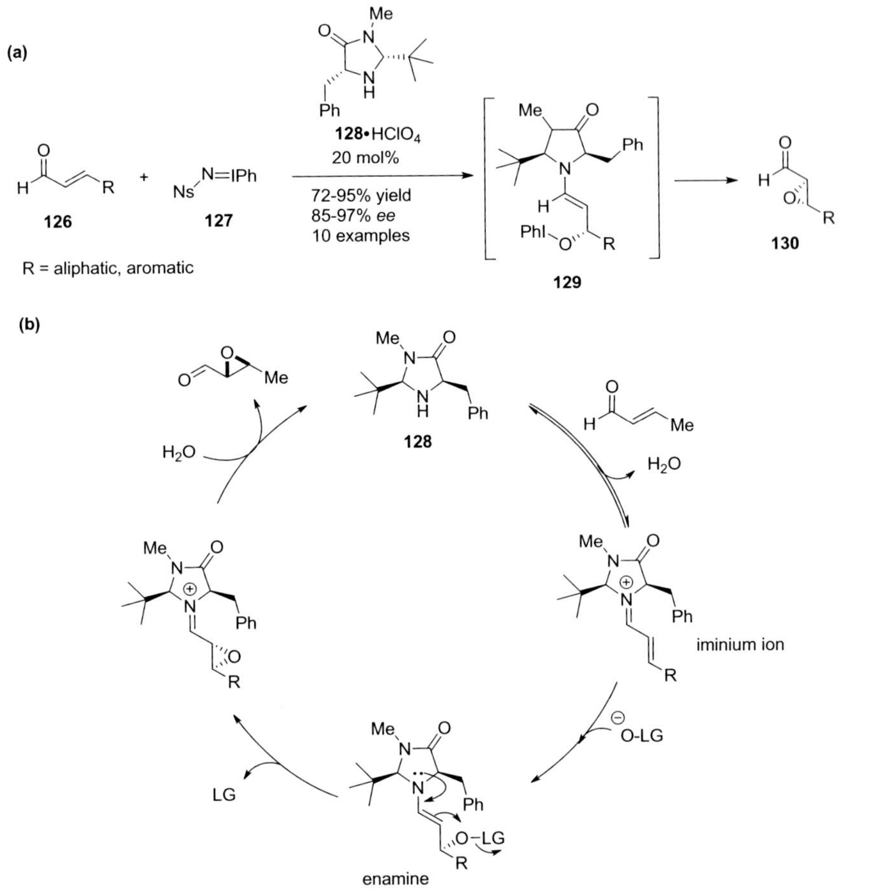

Scheme 28 **a** Iminium-catalyzed domino epoxidation reaction. **b** Plausible mechanism supporting the involvement of iminium and enamine intermediates

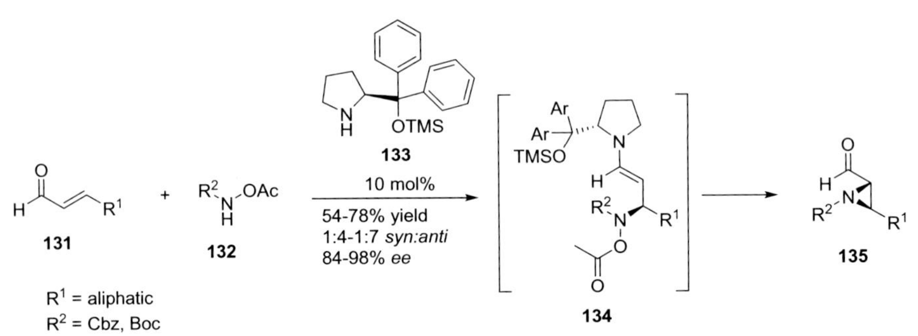

Scheme 29 Asymmetric aziridination ring formation

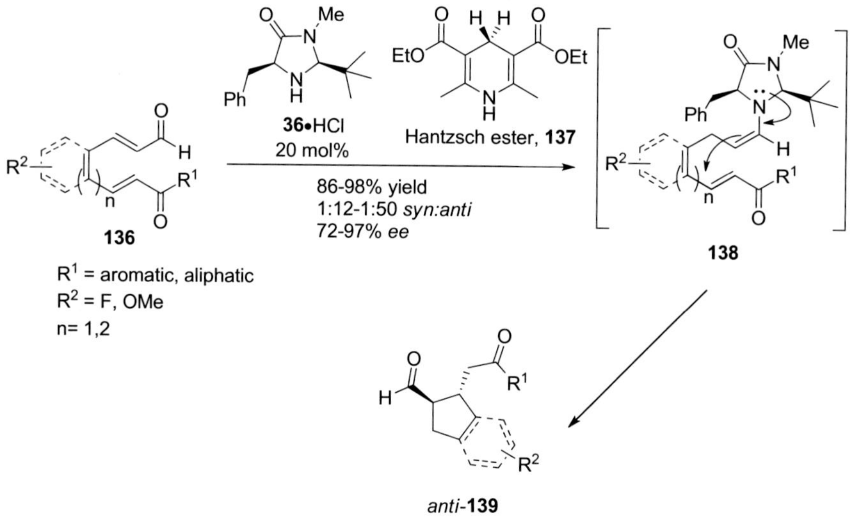

4 Enamine Catalysis

285

Scheme 30 Iminium-enamine activated five-membered ring carbocycle formation

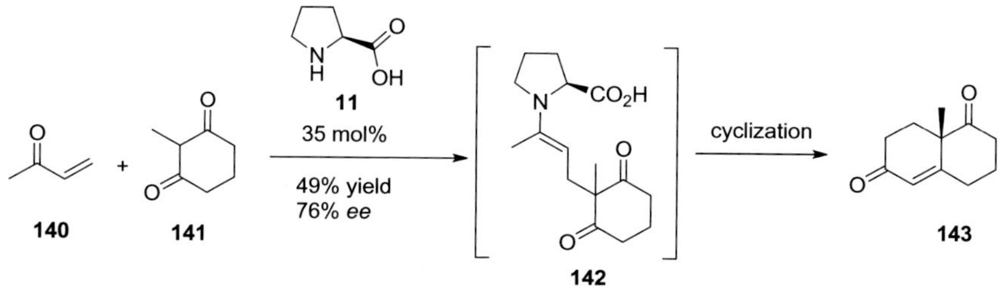

Scheme 31 Proline-catalyzed asymmetric Robinson annulation reaction

4 Enamine Catalysis

In enamine catalysis, the carbonyl group first reacts with primary or secondary amines followed by dehydration which leads to the generation of enamine, an intermediate, with highest occupied molecular orbital (HOMO) greater in energy than the HOMO of corresponding carbonyl compound. Therefore, it reacts with lowest unoccupied (LUMO) or singly occupied molecular orbital (SOMO) of the electrophile more readily [74, 75]. In simple words, it is the typical reaction of enamine with an electrophile or electrophilic radical. The overall mechanism of enamine catalysis is depicted in Scheme 32.

The enamine catalysis has showed a great development in last two decades in the field of asymmetric synthesis due to its almost ideal atom economic and step economic approach. The key step in enamine catalysis is the reversible generation of enamine resulting from the reaction between carbonyl compound and catalytic

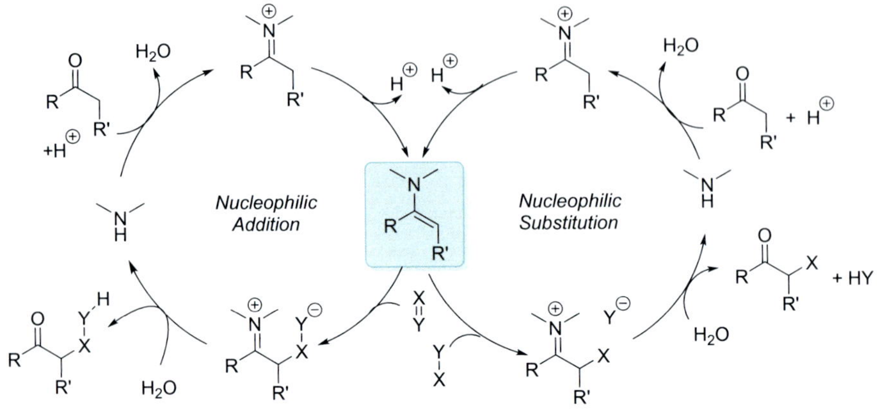

Scheme 32 Catalytic cycle of enamine-catalyzed reactions

amount of amine. Addition to the LUMO lowering effect, the C–H acidity increases with the formation of initial iminium ion which leads to the formation of enamine intermediate. Basically, there are two types of enamine catalytic cycles depending upon the nature of electrophile used. While the double bond containing electrophiles including, aldehydes, imines, Michael acceptors, etc. insert into the C–H bond situated at α-position of carbonyl compound via nucleophilic addition reaction, the single bond containing electrophiles like alkyl halides react in a nucleophilic substitution fashion which in turn leads to stoichiometric byproducts (Scheme 32).

Although the origin of enamine-catalyzed reactions can be witnessed by the work of Stork and others [76], where the enamine was used stoichiometrically as nucleophile in organic syntheses, catalytic synthesis by aldolases [77] and proline-catalyzed aldol condensation by Hajos–Parrish–Eder–Sauer–Wiechert reaction [9], the explosive growth of asymmetric catalysis by enamine was explored in the beginning of this century by List and coworkers [12]. Since then, an enormous growth in the field of organocatalysis, particularly, asymmetric amine-catalyzed reactions expanded at a great pace. Furthermore, new catalysts have been developed, several important organic transformations have been modified, new reactions are being discovered and applied, and the mechanism behind the asymmetric approach is becoming clearer day by day.

4.1 Asymmetric Aldol Reactions

4.1.1 Intramolecular Aldolization

Aldol reaction is well known and most commonly applied for C–C bond forming reaction [78]. The reaction is robust and versatile for construction of chiral building

4 Enamine Catalysis

Scheme 33 Types of intramolecular aldolization

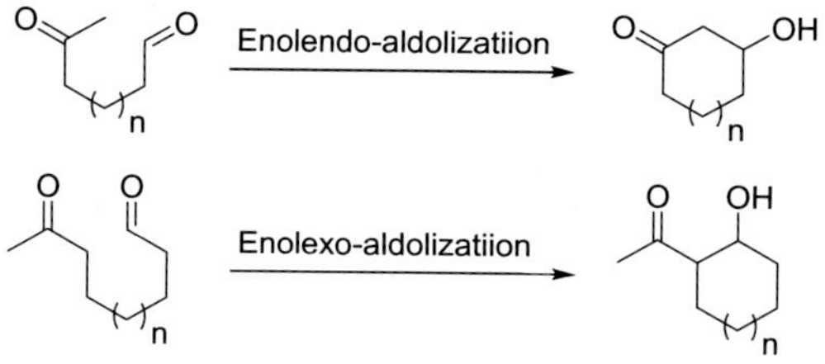

blocks which could be used in synthesis of complex molecules like natural products and various drug products. Aldol reaction can be catalyzed by both Lewis and Brønsted acids and bases [79].

Asymmetric aldol reaction can be categorized as intra and intermolecular aldolization. Further, the intramolecular aldol reaction is categorized as enolendo and enolexo aldolizations (Scheme 33).

The earliest example of aminocatalytic asymmetric aldolization was Hajos–Parrish–Eder–Sauer–Wiechert cyclization reaction discovered in early 1970s. The reaction is organocatalytic and results in highly enantioselective products, but its mechanism was not well understood. This reaction is an excellent example of 6-enolendo-aldolization resulting in the cyclization of triketones **10** and **144** to furnish aldols **12** and **145** in good yield and with high enantioselectivity using proline **11** as an excellent catalyst (Scheme 34a, b) [7]. Further, the addition products in the presence of acid dehydrated to produce condensed product **13** and **146**. This reaction was further independently modified by Eder, Sauer, and Wiechert using an acid as co-catalyst (Scheme 34c, d). This reaction was used in synthesis of several other substrates, preferably steroid [80–82] and other natural products [83].

Several different amino catalysts were employed for such enolendo aldolization, proline derivatives were found to be most preferred catalyst. In several studies, primary amino acid catalyst, phenylalanine **147** was found to be more advantageous than its counterpart, i.e., proline catalyst **11**, especially in case of non-methyl ketones used as substrates. For instance, Danishefsky and coworkers [84] reported only 27% *ee* when proline was used as catalyst in cyclization reaction of triketones **148** to afford **149**, whereas 86% *ee* was found in case of phenylalanine (Scheme 35a). Similar results were obtained by Agami et al. in cyclization of ketones (Scheme 35b) [85].

Amine-catalyzed non-asymmetric enolexo reactions are very common but proline-catalyzed asymmetric enolexo aldolization reaction was recently reported by List and coworkers [86]. They found that a variety of achiral dialdehydes **152** and 7-oxoheptanals in the presence of catalytic amount of *S*-proline **11** afford antialdols **153** in moderate to good yield with high stereoselectivity (Scheme 36). Interestingly, the amine-catalyzed non-asymmetric enolexo aldolization give condensation product, proline-catalyzed aldolization give addition product. Differently substituted dialdehydes and ketoaldehydes can be used as substrates. An elegant example of proline-catalyzed 6-enolexo-aldolization reaction is the desymmetrization of *meso*-dialdehyde **154** in order to obtain (+)-cocaine **156** (Scheme 37) [87].

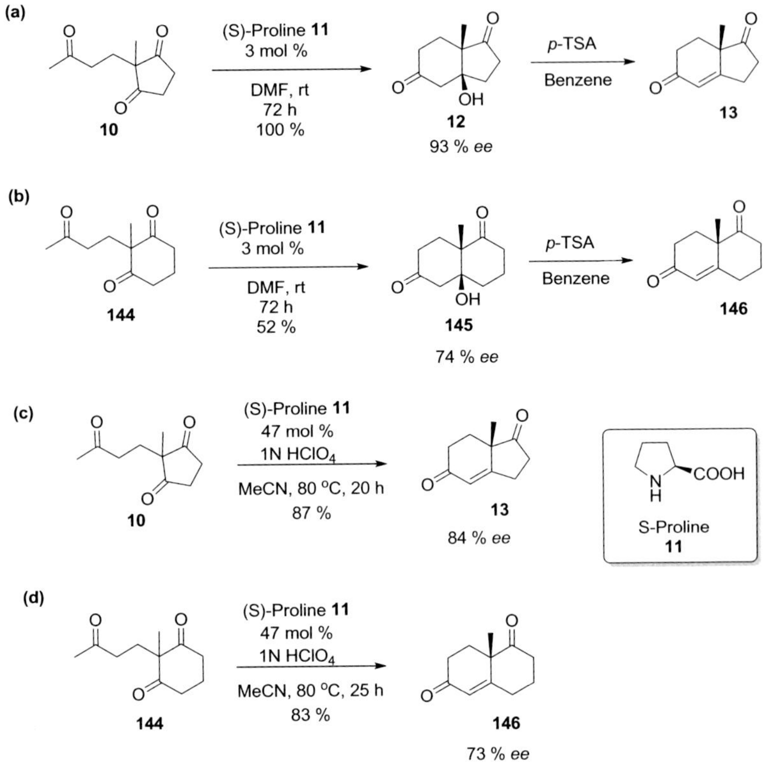

Scheme 34 Hajos–Parrish–Eder–Sauer–Wiechert cyclization reaction

Enders and coworkers [88] reported proline-catalyzed asymmetric intramolecular *cis* 5-enolexo aldolization of dicarbonyls **157** that results in the formation of 2,3-dihydrobenzofuranols **158**, a core moiety present in several antimicrobials (Scheme 38). The reaction was conducted in the presence of 30 mol% (*S*)-proline in 0.1 M DMF which leads to *cis*-selective intramolecular aldolization with considerable diastereoselectivity and high enantioselectivity.

4.1.2 Intermolecular Aldol Reactions

The first amine-catalyzed asymmetric intermolecular Aldol reaction was discovered by List and coworkers in 2000 [11, 12]. Excess of acetone **159** reacts with aromatic or α-branched aldehydes **160** in the presence of catalytic amount (usually 20–30 mol%) of (*S*)-proline **11** in DMSO to afford aldol products **161** with high yield and enantioselectivities (Scheme 39). Several primary and secondary, cyclic

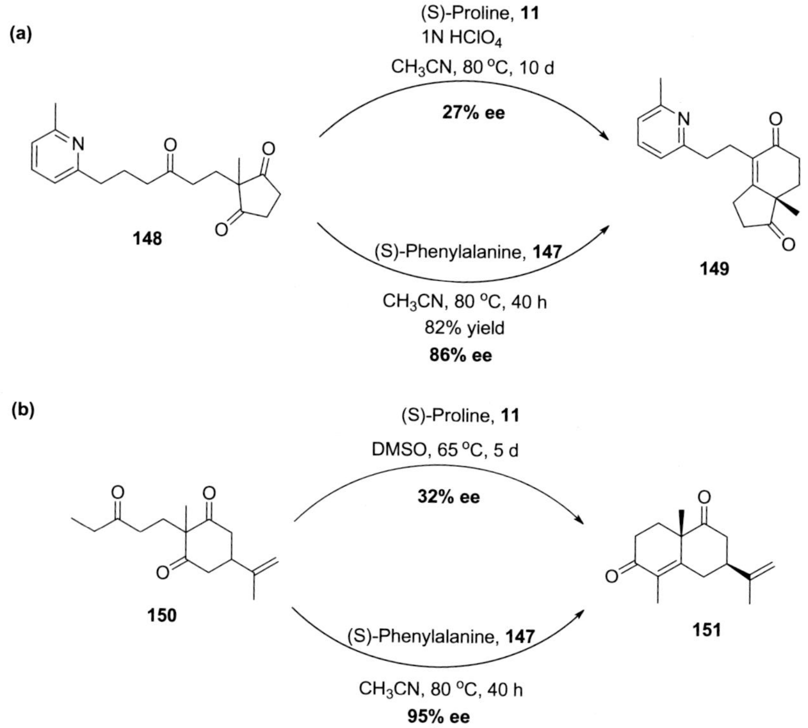

4 Enamine Catalysis

(a)

(S)-Proline, 11
1N HClO₄
CH₃CN, 80 °C, 10 d

27% ee

148

(S)-Phenylalanine, 147

CH₃CN, 80 °C, 40 h
82% yield
86% ee

149

(b)

(S)-Proline, 11
DMSO, 65 °C, 5 d

32% ee

150

(S)-Phenylalanine, 147

CH₃CN, 80 °C, 40 h
95% ee

151

Scheme 35 Comparative study of (*S*)-proline and (*S*)-phenylalanine-catalyzed cyclization of triketones

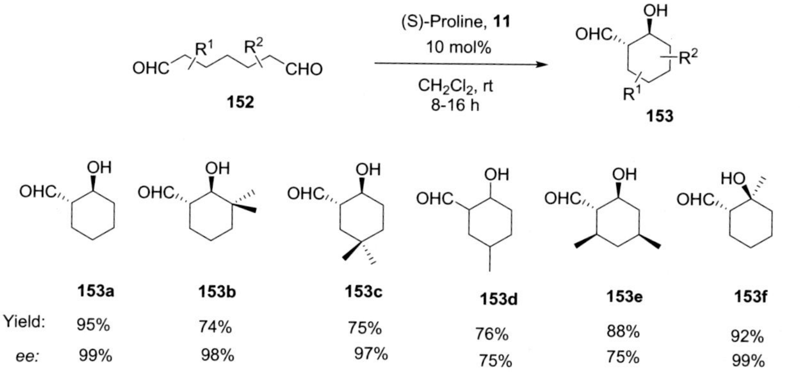

152

(S)-Proline, 11
10 mol%

CH₂Cl₂, rt
8-16 h

153

	153a	153b	153c	153d	153e	153f
Yield:	95%	74%	75%	76%	88%	92%
ee:	99%	98%	97%	75%	75%	99%

Scheme 36 Proline-catalyzed 6-enolexo-aldolization reaction

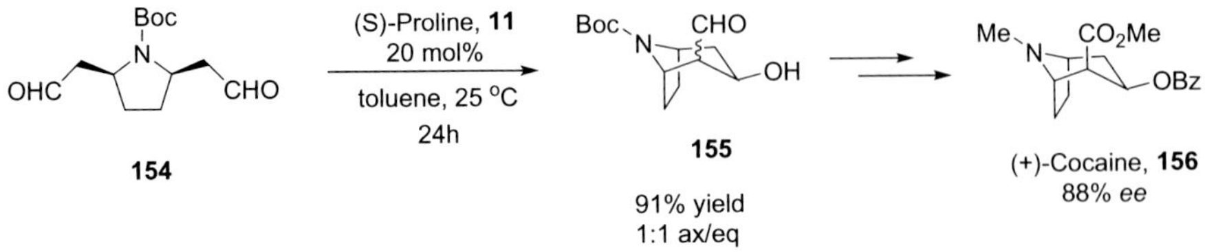

Scheme 37 Proline-catalyzed desymmetrization of dialdehyde in total synthesis of (+)-Cocaine

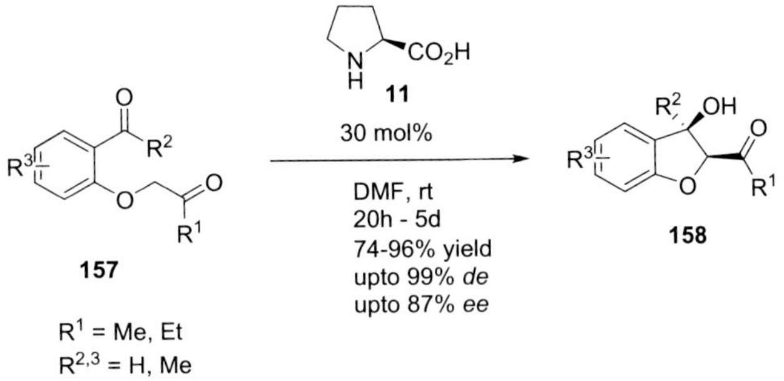

Scheme 38 Proline-catalyzed asymmetric 5-enolexo aldolization of dicarbonyls

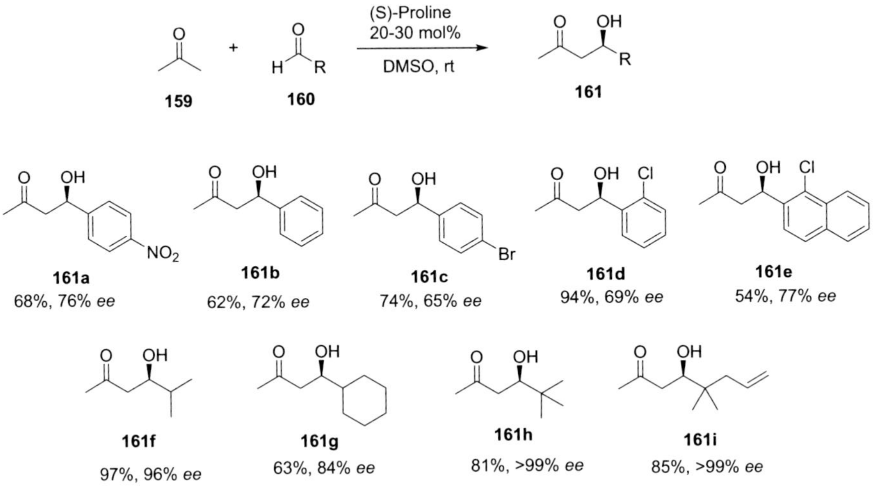

157

R^1 = Me, Et
R^{2,3} = H, Me

161a
68%, 76% ee

161b
62%, 72% ee

161c
74%, 65% ee

161d
94%, 69% ee

161e
54%, 77% ee

161f
97%, 96% ee

161g
63%, 84% ee

161h
81%, >99% ee

161i
85%, >99% ee

Scheme 39 Proline-catalyzed asymmetric intermolecular Aldol reaction

and acyclic amino acid derivatives were tested but the best result was displayed by (S)-proline. Simple primary amino acids showed poor catalytic activity, although phenylalanine, a primary amino acid displayed good results when used as catalyst in Hajos–Parrish–Eder–Sauer–Wiechert reactions (Scheme 35). This showed that secondary amines are required for intermolecular aldol reactions. Acyclic secondary amino acids, for instance, N-methylvaline, have no catalytic property. Different cyclic ring sizes were also screened for catalysis, for example, six-membered picolinic acid was found to be inactive. Some pioneer work on enamine chemistry by Stork suggested that pyrrolidine forms better enamine with carbonyl compounds than piperidine [89]. The enamine formed by pyrrolidine was found to be more nucleophilic compared to the enamine formed by piperidine. In context to the covalent bonding formation between proline and carbonyl compounds, it was found that N-methylproline remains inactive for intermolecular aldolization reactions. Furthermore, the importance of carboxylate ion can be realized by the fact that prolinamide displayed poor results both in terms of yield and stereoselectivity. Hence, proline was turned out be best catalyst for intermolecular aldol reactions.

Extensive research has been conducted by different research groups to insight into the mechanism of proline-catalyzed intermolecular aldol reactions [7, 10, 90–95]. Similar to class I aldolases, proline also operate via enamine mechanism. However, there are debate on several mechanistic aspects of this reaction and various models were proposed to validate the mechanism. Hajos and coworkers [8] proposed the mechanism in which one of the enantiotopic acceptor carbonyl groups is activated as carbinolamine (Scheme 40a). However, the stereochemistry of this model was questioned by Jung very soon after its release [96]. Agami and coworkers, based on nonlinear studies, proposed a side chain enamine mechanism which involves two proline molecules in C–C bond forming transition state. One of them forms enamine and other act as a proton transfer mediator (Scheme 40b) [95]. Swaminathan and colleagues proposed heterogenous aldolization mechanism on a crystalline surface (Scheme 40c) [97]. However, it is now widely accepted that proline-catalyzed reaction follows homogenous aldolization. List et al. challenged the widely accepted

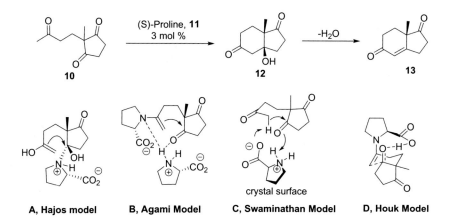

Scheme 40 Various proposed models for asymmetric proline-catalyzed Hajos–Parrish–Eder–Sauer–Wiechert Reaction

Agami's two proline mechanism with their homogenous-one proline-enamine intermediate mechanism with intermolecular variants in which the proton is transferred by proline's carboxylic acid functionality (Scheme 41) [12]. Similar results were obtained by Houk et al. in their DFT calculations with intramolecular variants (Scheme 40d). Unfortunately, very few experimental data are present to validate any of the proposed mechanisms. With handful of experimental and theoretical data, the mechanism of proline-catalyzed reaction was proposed to involve intermediates like carbinolamine, iminium, and enamine. The carboxylic acid functionality of proline act as general Brønsted cocatalyst. The mechanism is depicted in Scheme 41.

Reinvestigation of the Hajos–Parrish–Eder–Sauer–Wiechert reaction by List and Barbas inspired the organic chemists to think about the utility of organocatalysts in many established organic named reactions such as intermolecular cross aldol reaction, Mannich reaction, Michael reaction, Diels–Alder type reaction and other related reactions. Till date, by the continuous efforts of different research groups,

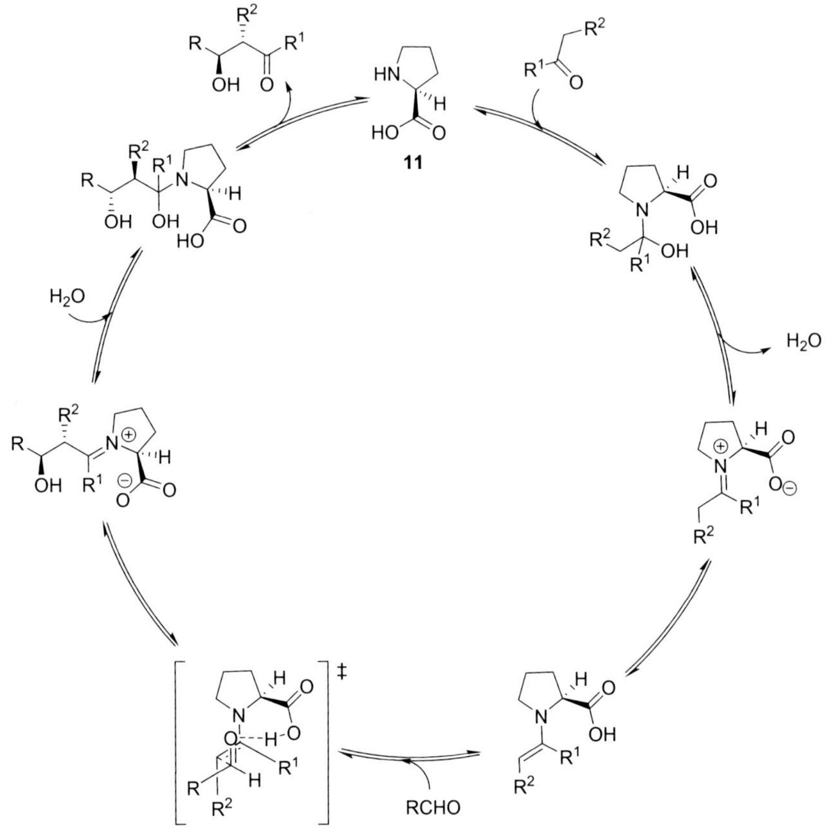

Scheme 41 Proposed mechanism for proline-catalyzed intermolecular aldolization by List et al. [10, 12, 75]

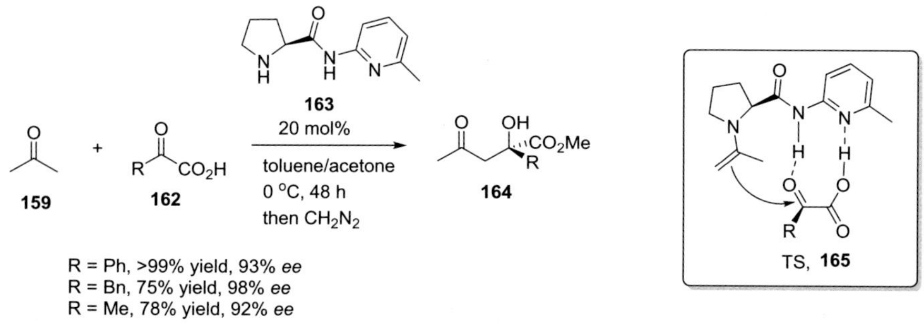

4 Enamine Catalysis

R = Ph, >99% yield, 93% ee
R = Bn, 75% yield, 98% ee
R = Me, 78% yield, 92% ee

Scheme 42 Asymmetric aldol reaction by novel prolinamide organocatalyst

the field of organocatalysis has been enriched with different types of catalysts, and the reactions catalyzed by them. For instance, Gong et al. [98] came with novel 2-aminopyridine-derived prolinamide catalyst **163** for asymmetric aldol reaction between ketone **159** and α-keto acid **162** (Scheme 42). The rationale behind the designing the catalyst was molecular recognition and the activation of α-keto acid via formation of hydrogen bond with the catalyst in transition state **165**. Acetone was the main ketone component and 20 mol% of the catalyst **163** afforded the α-hydroxy carboxylicacids **164** with good yield and excellent enantioselectivity.

Dondoni and coworkers [99] reported homoaldolization of ethyl pyruvate **166** in the presence of diamine TFA salt **167** as catalyst which afforded isotetronic acid analogs **168** after treatment with an acidic resin. Further, the enol moiety was protected by *tert*-butyldimethylsilyl chloride (TBDMS) to give high yield of compound **169** (Scheme 43).

Samanta and Zhao [100] reported proline-catalyzed enantioselective aldol reaction between α-keto phosphonates **170** and ketones **159** to afford α-hydroxy phosphonates **171** in good yield (Scheme 44). α-Keto phosphonates are another class of ketone acceptors and pose challenges as substrates due to susceptibility toward nucleophilic attack and leaving nature of phosphonate group. The products, hydroxy phosphonates and their corresponding acids are biologically active. The reaction has wide range of substrate scope including aryl, alkyl, and alkenyl α-keto phosphonates.

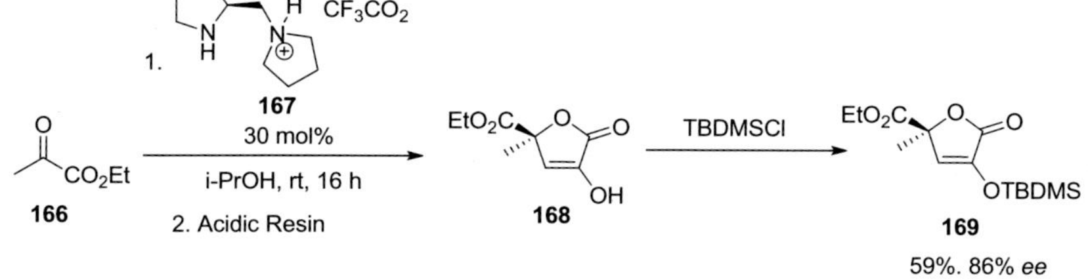

Scheme 43 Homoaldol reaction of ethyl pyruvate using diamine TFA salt

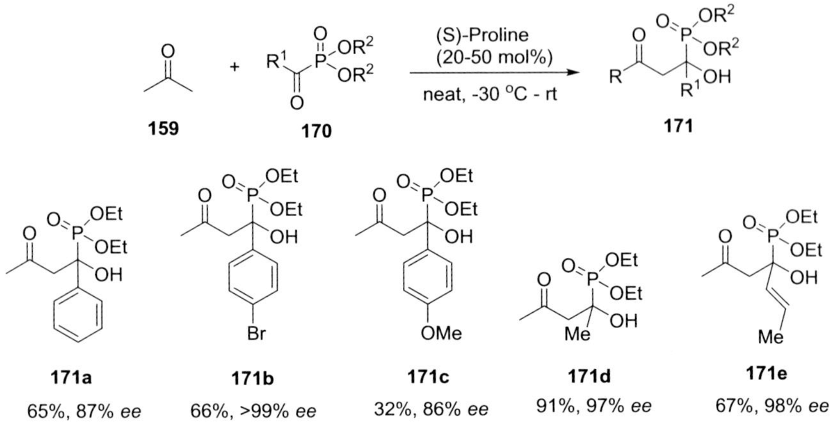

Scheme 44 Proline-catalyzed enantioselective aldolization of α-keto phosphonates and ketones

Another proline-catalyzed enantioselective aldolization reaction was reported by Kotsuki et al. [101] where cyclopentanone **172** and aliphatic aldehydes **173** were reacted to afford **174** which further underwent multistep synthesis to produce (−)-(5R,6S)-6-acetoxyhexadecanolide **175**, an oviposition pheromone which attracts female *Culex* mosquito (Scheme 45). In the reaction, (*S*)-proline was used in solvent-free conditions to afford syn adduct **174** as the major diastereomer and with 83% *ee*.

Another most elegant example of proline-catalyzed enantioselective synthesis of prostaglandin was reported by Aggrawal and coworkers in 2012 [102]. They synthesized miraculous molecule, prostaglandin $PGF_{2\alpha}$ **185**, in just seven steps. Prostaglandins, like $PGF_{2\alpha}$, are hormone like chemical messengers that regulates a wide range of physiological metabolism, including circulation of blood, inflammatory actions, menstruations cycle, digestion, reproduction, etc. [103]. Prostaglandin derivatives have a great use in pharmaceutical industries. For instance, latanoprost,

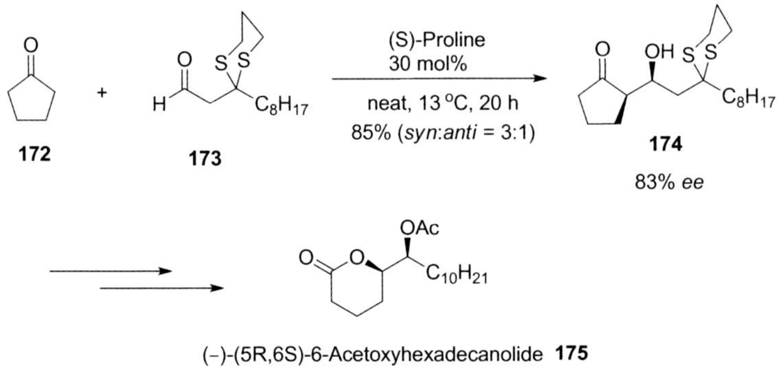

Scheme 45 Proline-catalyzed asymmetric synthesis of (−)-(5R,6S)-6-acetoxyhexadecanolide

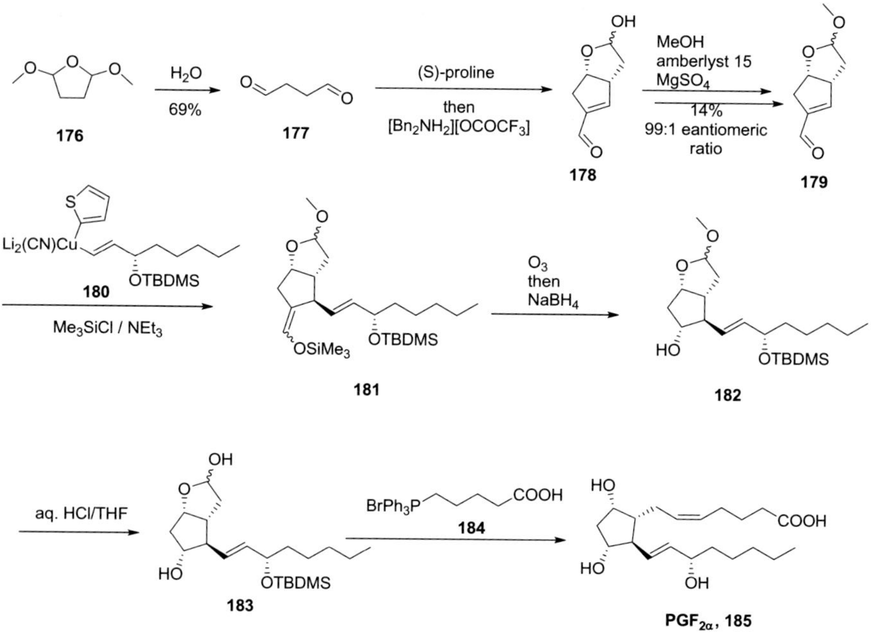

4 Enamine Catalysis

Scheme 46 Proline-catalyzed concise asymmetric synthesis of prostaglandin, $PGF_{2\alpha}$ [105]

a prostaglandin analog, which is used in glaucoma treatment have become million-dollar drugs [104]. Previous synthetic procedures in literatures are lengthy, chemical and energy consuming, and unwanted waste generation lead to material loss. Aggrawal and coworkers reduced the step without compromising with yield and stereochemistry. The key step in the synthesis of such a complex molecule is an aldol cascade reaction of succinaldehyde **177** using (S)-proline which afford a bicyclic enal **178** in one step with 98% *ee*. The total synthesis of prostaglandin $PGF_{2\alpha}$ is outlined in Scheme 46.

Beside several important asymmetric reactions are catalyzed by *S*-proline, various other small organic molecules including primary amino acids, substituted α-amino acids, thiazolidinium-4-carboxylates, substituted pyrrolidine, etc. were tested for direct asymmetric Aldol reactions. For instance, 20 mol% of 5,5-dimethyl thiazolidinium-4-carboxylate was used in direct asymmetric aldol reaction by Barbas and coworkers for aromatic aldehydes and ketones [105]. Similarly, 30 mol% of acyclic α-amino acid, specially(*S*)-Alanine **186** and (*S*)-valine were exploited as catalyst for direct asymmetric aldolization of butanone and cyclic ketones **187** with electron-deficient aldehydes **188** in DMSO (Scheme 47). Antialdol products **189** were obtained with high enantioselectivities and good diastereocontrol but yield remains moderate to good. Authors proposed the role of water and Houk-type transition state **190** for high stereoselectivity of the reaction [106, 107].

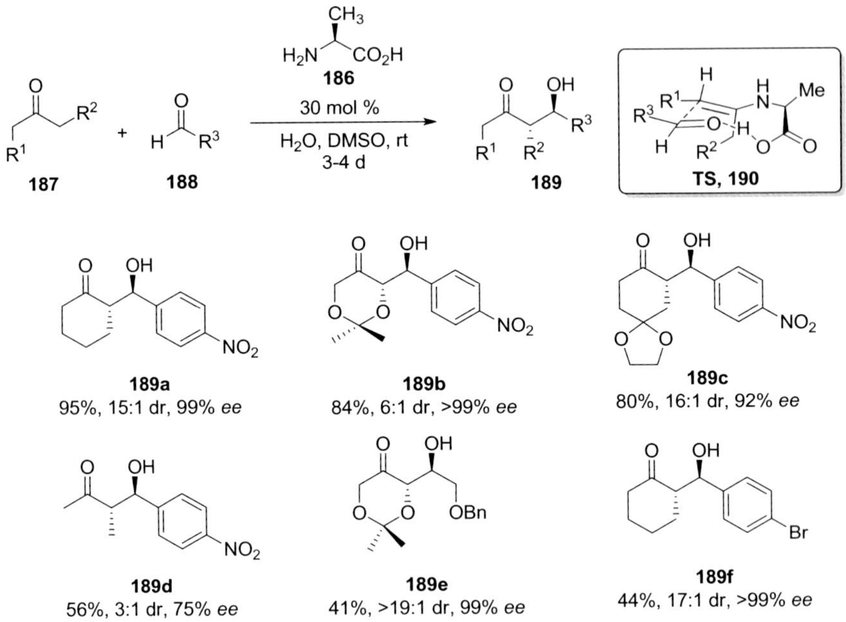

Scheme 47 Alanine-catalyzed direct asymmetric Aldolization of ketones and aldehydes

Gong, Wu and coworkers reported first example of amino alcohol amides as organocatalyst derived from (S)-proline for enantioselective direct aldolization of aldehydes with neat acetone [108, 109]. After screening of several proline-based amides, the hydroxyl amide **191** displayed the excellent enantioselectivity. 20 mol% of the catalyst afforded the aldol product **195** upto 93% *ee* in case of aromatic aldehydes and upto >99% *ee* when aliphatic aldehydes were used as substrate (Scheme 48). Interestingly, S-prolinamides were found inactive for direct aldol reactions, S-prolinamide with hydroxyl group have showed quite impressive results [110]. The catalyst **191** was used in ionic liquid [bmim][BF$_4$] and interestingly better results in terms of yield and enantioselectivities were obtained at 0 °C and 20 mol% of catalyst loading [111]. The improved catalytic activity of amide **191** was attributed to either stabilization of the iminium ion intermediate or increased nucleophilicity of enamine. Gong et al. continuously worked on the development of improved amide catalysts, reported that the amides possessing electron withdrawing group exhibit better results in terms of catalytic activity and enantioselectivity than their analogs with electron-donating groups [112]. Only 2 mol% of the catalyst **192** was found sufficient for catalyzing direct asymmetric aldol reaction with wide range of aromatic and aliphatic aldehydes and acetone and butanones.

Proline-derived tetrazole catalyst **196** was reported by Saito, Yamamoto, and coworkers [113, 114], which was found to be efficient catalyst for direct aldol reaction of ketones and reactive aldehydes. Here, the reaction was assisted by water and almost no products were obtained when the reaction was performed in anhydrous

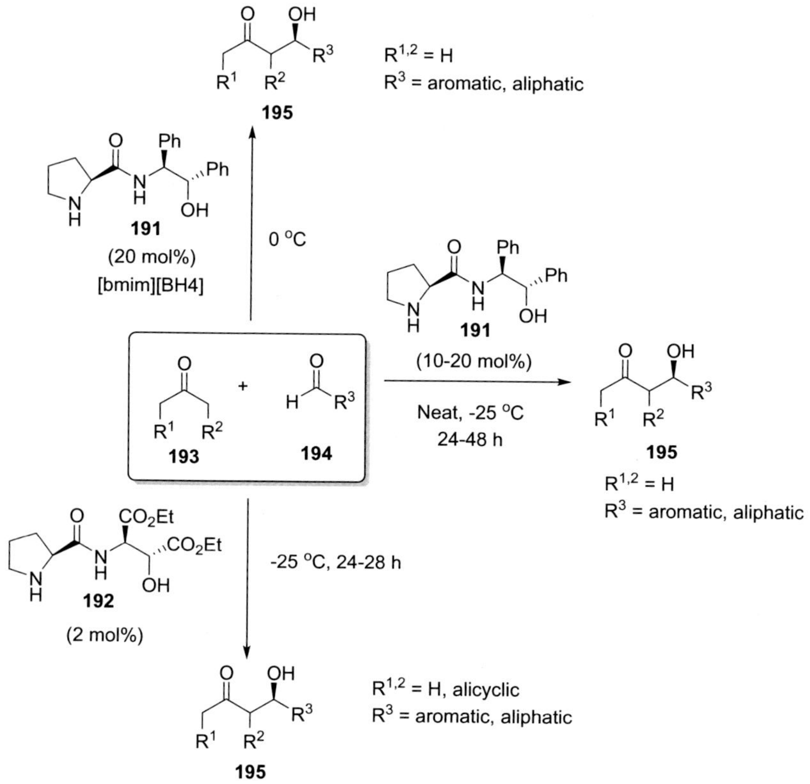

4 Enamine Catalysis

Scheme 48 Proline-based hydroxylamide-catalyzed asymmetric aldol reactions

condition or in catalytic amount. Interestingly, enantioselectivities of aldol products were increases as the amount of water is increased. The reactions have wide range of substrate scope with high enantioselectivities and diastereoselectivities (Scheme 49).

Singh and coworkers [115] developed Gong like small prolinamide molecule **200a** and **200b** which catalyzes intermolecular acetone aldolization reaction with best known enantioselectivity till date. Due to its promising results, catalyst is well known as Singh's catalyst in honor of its founder. The authors reported the direct aldol reaction between acetone and various aldehydes **201** with high enantioselectivity (upto >99%) especially in case of various aromatic aldehydes and cyclohexane carbaldehydes (Scheme 50). The high enantioselectivity is attributed to the presence of *gem*-diphenyl group at the β-carbon. Higher reactivity of the catalyst is rationalized by formation of double hydrogen bonds with substrate and catalyst. The transition state is very similar to the model proposed by Gond and Wu [109] (Fig. 5).

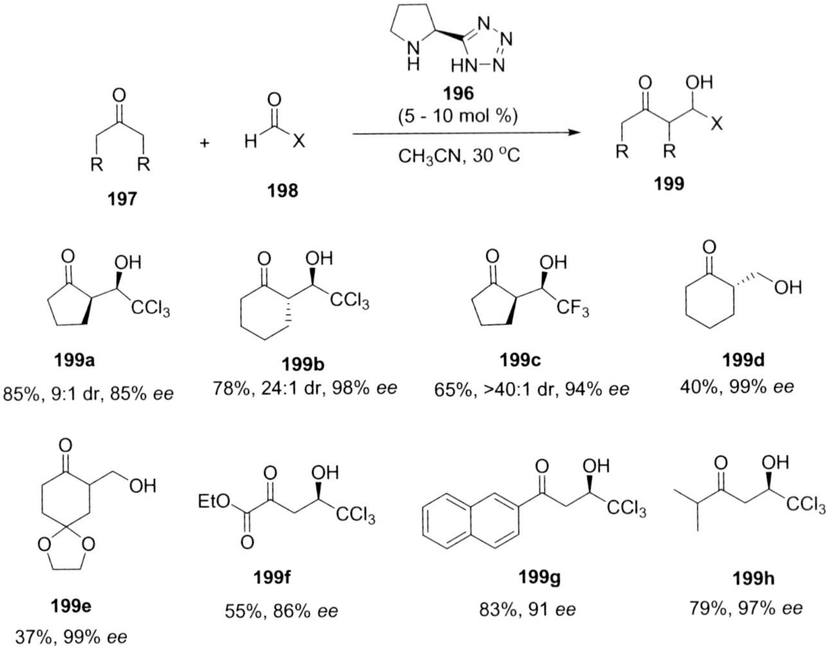

Scheme 49 Proline-derived tetrazole-catalyzed asymmetric aldol reaction

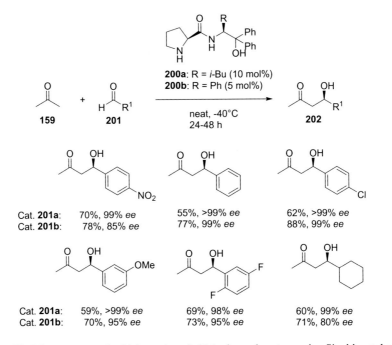

Scheme 50 Direct asymmetric aldol reaction of aldehydes and acetone using Singh's catalyst

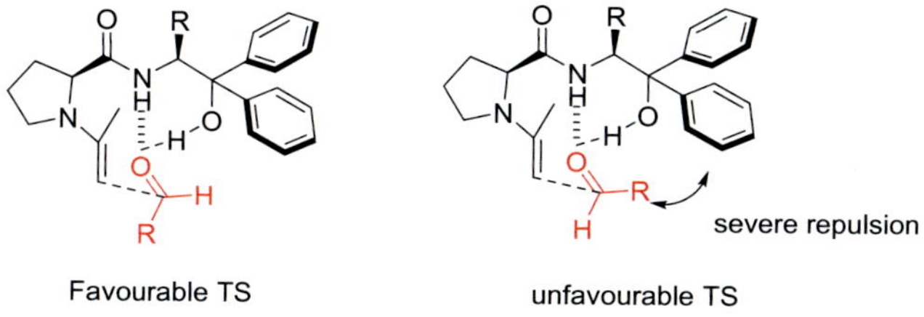

4 Enamine Catalysis

Favourable TS   unfavourable TS

Fig. 5 Proposed transition state using catalyst **200** for aldolization [115]

4.2 Asymmetric Michael Reaction

The conjugate addition of nucleophile at the β position of α,β-unsaturated carbonyl compounds is well known in organic synthesis due to its importance in the formation of C–C bond, referred as Michael reaction [116, 117]. The increasing demand of optically active compounds in pharmaceutical and agrochemical industries enhanced the importance of Michael reaction as the stereogenic center may be created in due course of the reaction. Thus, it is much required to make efforts to develop novel and efficient catalytic stereoselective methods [118]. Although there are many reports in which metal catalysts were employed for obtaining stereoselective products via Michael reaction, only recently, small organic molecules as catalyst have been employed for asymmetric synthesis [119, 120]. In this context, amino catalysts gained wide attention as they are able to catalytically activate the Michael donor via enamine or enolate formation (Fig. 6a, b). Additionally, Michael acceptors also get activated via iminium ion formation (Fig. 6c).

In general, earlier attempts in the asymmetric synthesis used active methylene group as Michael donors and α,β-unsaturated carbonyl compounds as Michael acceptor where chiral aminocatalysts activated them via iminium ion formation. Chiral tertiary amine form *in situ* enolates. Asymmetric Michael addition reactions via enamine formation recently gained considerable attention after the work by Stork [89]. Since then a number of aminocatalysts were developed for asymmetric and non-asymmetric Michael reactions [121, 122]. The overall proposed mechanism for enamine-catalyzed Michael reaction is outlined in Scheme 51.

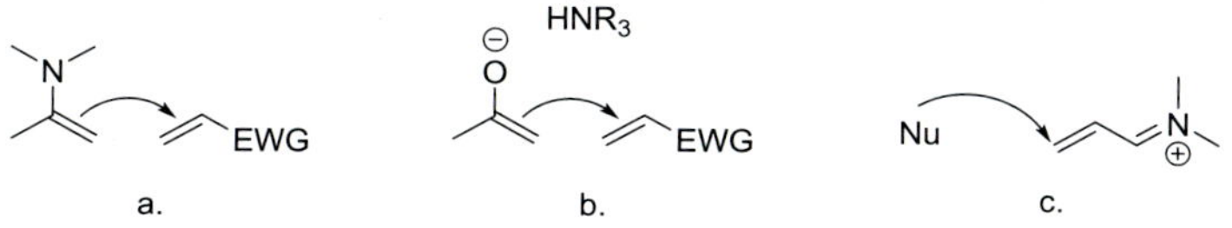

Fig. 6 **a** Enamine formation, **b** enolate formation, **c** iminium ion formation, in Michael reaction

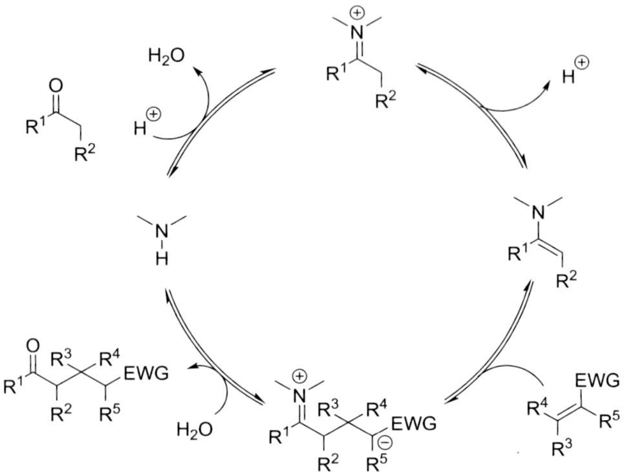

Scheme 51 Enamine-catalyzed Michael reaction [75]

4.2.1 Intramolecular Michael Reaction

In 2004, Fonseca and List reported first asymmetric amino catalytic intramolecular Michael reaction of various aldehydes [123]. The authors employed 10 mol% of commercially available Macmillan imidazolidinone **35** catalyst for cyclization of formyl enones **203** to afford *trans*-disubstituted cyclic five-membered ketoaldehydes **204** in excellent yields and high stereoselectivities (up to 49:1 *dr* and 97% *ee*). The reaction displayed good substrate scope as different aromatic and aliphatic enones afforded good yield and stereoselectivity (Scheme 52).

Scheme 52
Imidazolidinone-catalyzed
asymmetric intramolecular
Michael reaction

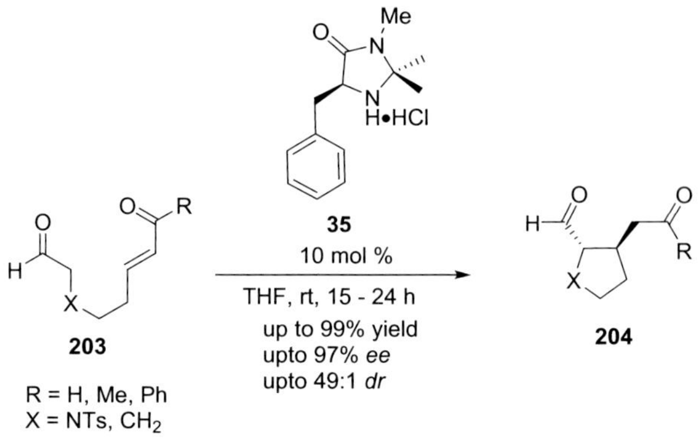

4.2.2 Intermolecular Michael Reaction

List et al. in 2001 [124], reported first enamine-catalyzed asymmetric intermolecular Michael reaction in which nitroolefins **205** were used as Michael acceptor and unactivated symmetric ketones **206** as donors. The reaction was performed in the presence of catalytic amount of (S)-proline **11** in DMSO which furnished desired γ-nitro ketones **207** with in general, high yields, good diastereoselectivities but low enantioselectivities (Scheme 53a). The rationale behind low enantioselectivity is the less efficient formation of hydrogen bonding between catalyst and Michael acceptor or different mechanism is being operative. The enantioselectivity of the reaction could be increased by using methanol as solvent (76% *ee*, Method B, Scheme 53a). The *syn* diastereoselectivity and absolute configuration of the product obtained was rationalized by Enders and coworkers. They proposed acyclic synclinal transition state (Scheme 53b) based on Seebach's model [125, 126].

Scheme 53 **a** (*S*)-Proline-catalyzed intermolecular asymmetric Michael reaction of ketones and nitroolefins. **b** Proposed transition state (TS)

5 Carbohydrate-Based Asymmetric Organocatalysis

Carbohydrates constitute a plethora of structurally and functionally diverse class of molecules, well known for their role in various fundamental cell functions including, energy storage, transportation, adhesion, modulation of protein function, signal transduction, bacterial and viral cell recognition, etc. [127–130]. Besides biological relevance, carbohydrates are equally important for pharmaceuticals [131–133] and chiral synthesis [134–136]. They occur naturally in the form of monomers, oligomers, and polymers possessing several chiral centers and hydroxy group that make them a fantastic substrate for structural variations. Their rigid conformation and known stereochemistry enhance their synthetic utility as potential building blocks, auxiliary reagents in nature's chiral pool, ligands, and catalysts in organic syntheses [137, 138].

Carbohydrates emerged as promising scaffold for the development of organocatalysts. It is a relatively cheaper and most abundant source of chiral nonracemic members of natural chiral pool. Chirality and functional diversity are the key features due to which sugar moiety might prove to be potent scaffold for metal-free catalysis. A wide range of stereospecific reactions reportedly accelerated by carbohydrate containing chiral molecules. Carbohydrates' role in asymmetric synthesis in the form of chiral auxiliaries, chiral reagents, and chiral ligands has brought the attention of researchers toward its utilization in the emerging field of organocatalysis [139]. Due to these reasons, carbohydrate moiety gained attention from different groups and emerged from relative obscurity to prominence as highly promising scaffold for the development of novel organocatalysts which afford products with desired stereochemistry [140–142].

5.1 Enantioselective Epoxidation

Shi and coworkers [14] developed fructose-based ketones **209** for epoxidation of *trans*-olefins **210** with high enantiomeric excess. Their newly designed catalyst **209** possesses some salient features like, (a) stereogenic center situated in vicinity to the reaction center for better stereochemical communication between the substrates and catalyst, (b) presence of a quaternary center and a fused ring adjacent to carbonyl group to avoid epimerization, (c) one face of catalyst is blocked thus avoiding the possibility of competitive approach. The fructose-based ketone **209** used as catalyst in presence of excess potassium peroxomonosulfate (oxone) as oxidant to execute epoxidation reaction in a wide variety of *trans*- and trisubstituted olefins (Scheme 54). To avoid the loss of enantiomeric excess due to decomposition, the ketone was used in excess and for shorter reaction time. This approach was employed to both functionalized and unsubstituted *trans*-olefins to obtain the product with high stereoselectivity (Fig. 7) [143, 144]. The plausible mechanism of asymmetric epoxidation reaction by the catalyst **209** is outlined in Scheme 55.

5 Carbohydrate-Based Asymmetric Organocatalysis

Scheme 54 Asymmetric epoxidation of *trans*-olefins with fructose-derived ketone

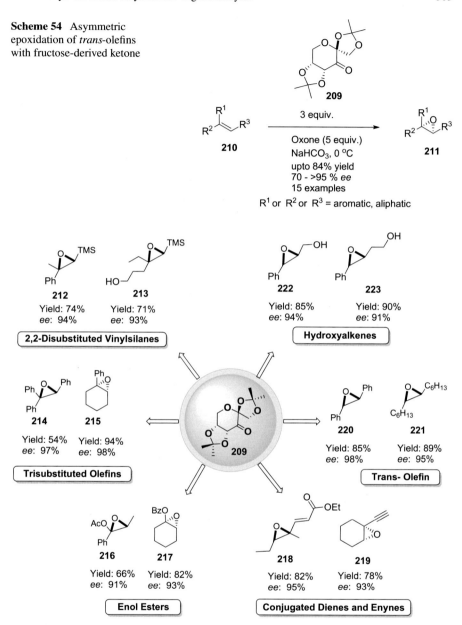

Fig. 7 Substrate scope for asymmetric epoxidation catalyzed by D-Fructose-derived ketone **209** [145]

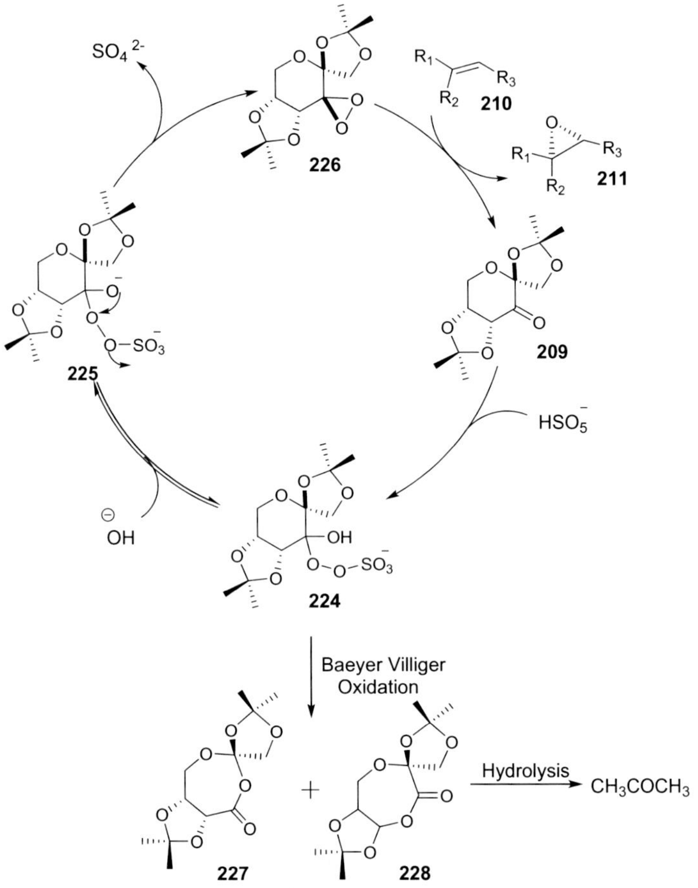

Scheme 55 Plausible mechanism of asymmetric epoxidation of olefins with ketone **209** [145]

The oxone reacts with catalyst **209** via nucleophilic addition to form adduct **224**. The base used in the reaction abstract proton from intermediate **224** and the elimination of sulfate ion afford dioxirane **226** which further reacts with *trans*-olefins **210** to yield the epoxy products **211** with high enantioselectivity. It is here to be noted that the intermediate **224** undergo decomposition to compound **227** and **228** via Bayer–Villiger oxidation process which ultimately decomposed to acetone (Scheme 55). To avoid such decomposition, the *pH* of the reaction should be maintained at 10.5–12 [146, 147]. Another benefit of maintaining this range of *pH* is that the nucleophilicity of the oxone also increases which reduces the reaction time.

Several reports advocate the use of catalyst **209** in enantioselective and chemoselective epoxidation of enynes. Interestingly, the epoxidation occurs chemoselectively at the olefinic bond with high enantioselectivity [148, 149].

5 Carbohydrate-Based Asymmetric Organocatalysis

5.2 Sugar-Based Prolinamide Catalysts

Proline, as already discussed in previous sections, is an organocatalyst of choice in wide range of asymmetric syntheses including aldol reaction, Mannich reaction, and Michael addition [11, 13, 75]. However, enantioselectivity reduces drastically when proline-catalyzed reactions are performed in aqueous medium. Studies showed that water interrupts the formation of hydrogen bonding between carboxylic group of proline and the substrate which in turn leads to the destabilization of transition state and hence reduces the enantioselectivity [150]. Tsutsui and colleagues [151] developed carbohydrate-based prolinamide to catalyze asymmetric aldol reaction in aqueous media. They successfully synthesized methyl-2-deoxy-2-(L-prolyl)amido-α-D-glucopyranoside (**229a**) and its isomer 2-deoxy-2-(D-prolyl)amido-α-D-glucopyranoside (**229b**) and studied its utility regarding yield and enantioselectivity in aldol reaction between acetone and 4-nitrobenzaldehydes (Scheme 56). Like other amino acid-based organocatalystsin terms of selectivitesfor aldol reaction, **229a** gave R-aldol, i.e., (4R)-4-hydroxy-4-(4-nitrophenyl)-butan-2-one whereas **229b** gave S-aldol product.

Later, Agarwal and Peddinti reported the reactions with sugar-based prolinamide catalysts for solvent-free aldol reaction [152]. Their idea was to block one face of aldehyde effectively by α-alkyloxy group present on anomeric position of the sugar scaffold and probable hydrogen bonding between carbonyl group of aldehyde and amide proton of prolinamide catalyst to induce asymmetry. Encouraged by these ideas, compounds **232–234** were screened (Fig. 8) [152, 153]. Catalyst screening results showed that 20 mol% of compound **234** gave the best results when 3-nitrobenzaldehyde **231**and cyclohexanone afforded the product with 96% yield, 91:9 (*anti:syn*) diastereomeric ratio and 93% *ee*. The reaction was further tested for substrate scope with a variety of substituted aromatic aldehydes (Scheme 57).

Nisco et al. reported a new sugar-based prolinamide **235** which was found to be very efficient for enantioselective aldol addition of cyclohexanone with a variety of substituted benzaldehydes (Scheme 56) [154].

5.3 Carbohydrate Based Pyrrolidine Catalysts

Sugar moiety attached to pyrrolidine ring was found to be an efficient catalysts for Michael addition reactions [155, 156]. Wang et al. designed and synthesized carbohydrate-based pyrrolidine catalysts **236** which catalyzed the Michael addition reaction between nitrostyrenes **237**and various ketones **238** with high stereoselectivity. The reaction between nitrostyrene and ketones leads to adduct **239** appreciable yield (up to 98%), high enantioselectivity, and excellent diastereoselectivity (upto >99:1 *dr*) under neat condition (Scheme 58) [156].

Similarly, Kumar and Balaji synthesized sugar-based pyrrolidine compound **240** and **241** which catalyzes asymmetric Michael addition of nitroolefins with various

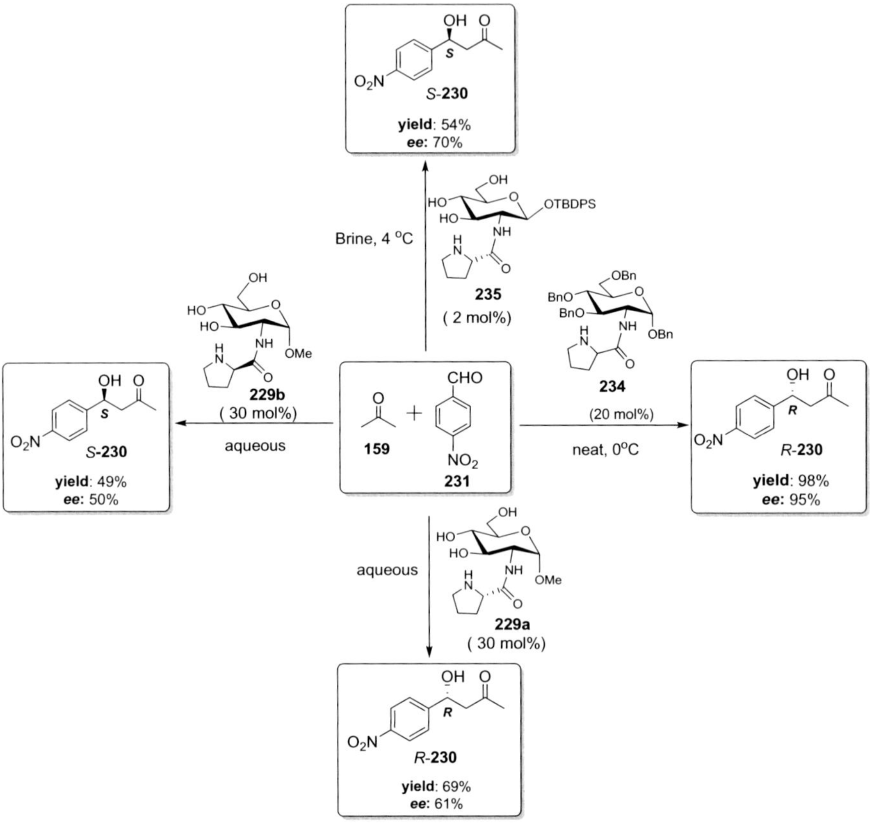

Scheme 56 Aldol reaction between *p*-nitro-benzaldehyde and acetone catalyzed by carbohydrate-based prolinamides

ketone substrates (Scheme 59) [155]. They performed the Michael addition reaction between various substituted β-nitrostyrene **242** and cyclohexanone which leads to the corresponding adduct with high yield (upto 96%), excellent diastereomeric ratio (*syn:anti*, 93:7 to 99:1) and moderate to good enantioselectivity (85–91%). Their studies advocate compound **241** as better catalyst.

6 Conclusion

The field of organocatalysis is dedicated to the development of green, robust and sustainable asymmetric synthetic protocols to produce chiral molecules as building block for construction of complex drugs and natural products. Earlier the field of asymmetric synthesis was dominated by metal-catalyzed reactions and enzymes.

6 Conclusion

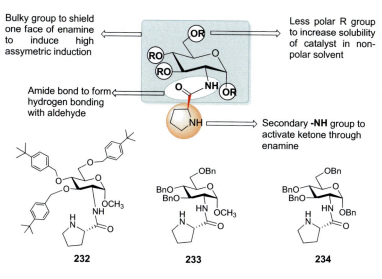

Fig. 8 Structural aspects of prolinamide catalysts **232–234** [145]

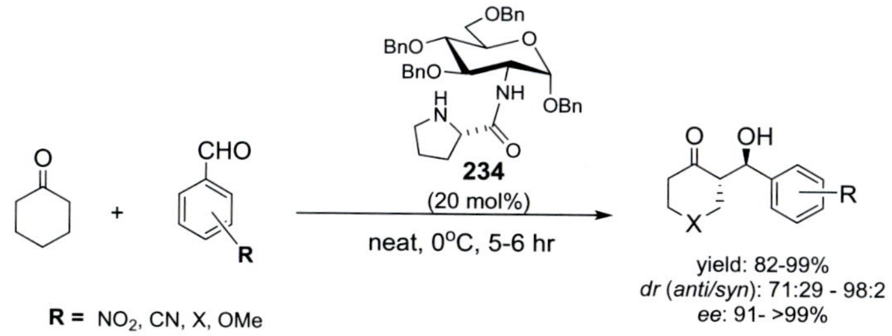

Scheme 57 Asymmetric aldol reaction using prolinamide **234** as a catalyst

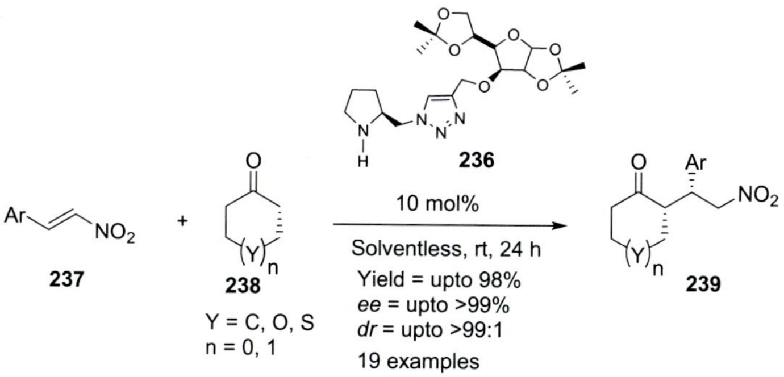

Scheme 58 Asymmetric Michael addition of nitrostyrene **237** and ketones **238** using sugar-derived pyrrolidine **236** as catalyst

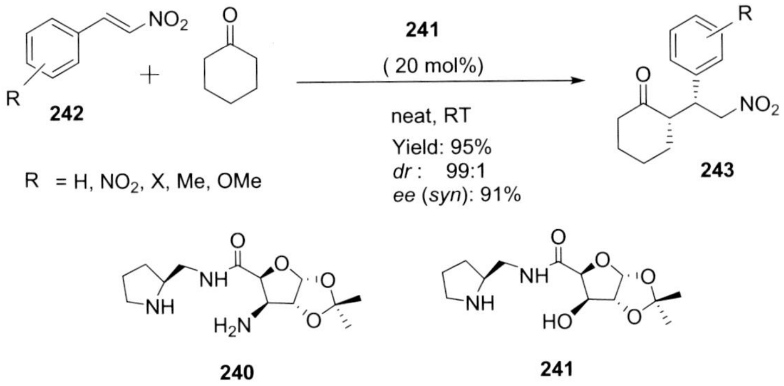

Scheme 59 Asymmetric Michael addition of nitroolefins with various ketone substrates

After the initial work of Benzamin List on proline-catalyzed aldolization reactions and David Macmillan's work on Diels–Alder reaction, the field of asymmetric organocatalysis got revolutionized due to very generic mode of activation. Several groups came with different modes of action of organocatalysts which provides in depth knowledge about the mechanism.

Asymmetric organocatalysis has now became the third vertical for catalysis in organic synthesis. With very attentive and huge expansion of research efforts in the field of organocatalysis across the globe, novel concepts of reactivity for the transformation of important organic reaction have been identified. Although a lot of research are going on and the field is growing day by day, development of new catalysts with novel mode action is always required. Very well said by Professor Benjamin List, one of the Nobel Laureates of 2021—*"the research in organocatalysis constitutes the tip of the iceberg of a novel catalytic principle, of which the entire scope still remains to be fully uncovered."*

References

1. Dalko, P.I.: Asymmetric organocatalysis: a new stream in organic synthesis. Enantioselective Organocatalysis React. Exp. Proced. (2007)
2. Jacobsen, E.N., Pfaltz, A., Yamamoto, H.: Comprehensive Asymmetric Catalysis: Supplement 1, vol. 1. Springer Science & Business Media (2003)
3. Kampen, D., Reisinger, C.M., List, B.: Chiral Brønsted acids for asymmetric organocatalysis. Asymmetric Organocatalysis 1–37 (2010)
4. Prelog, V., Wilhelm, M.: Untersuchungen Über Asymmetrische Synthesen VI. Der Reaktionsmechanismus Und Der Sterische Verlauf Der Asymmetrischen Cyanhydrin-Synthese. Helv. Chim. Acta **37**(6), 1634–1660 (1954)
5. Pracejus, H.: Organische Katalysatoren, LXI. Asymmetrische Synthesen Mit Ketenen, I. Alkaloid-Katalysierte Asymmetrische Synthesen von α-Phenyl-Propionsäureestern. Justus Liebigs Ann. Chem. **634**(1), 9–22 (1960)
6. Långström, B., Bergson, G., Hjeds, H., Songstad, J., Norbury, A.H., Swahn, C.-G.: Asymmetric induction in a Michael-Type reaction. Acta Chem. Scand. **27**, 3118–3119 (1973)

References

7. Hajos, Z.G., Parrish, D.R.: Synthesis and conversion of 2-Methyl-2-(3-Oxobutyl)-1,3-cyclopentanedione to the isomeric Racemic Ketols of the [3.2.1]bicyclooctane and of the perhydroindane series. J. Org. Chem. **39**(12), 1612–1615 (1974)
8. Hajos, Z.G., Parrish, D.R.: Asymmetric synthesis of bicyclic intermediates of natural product chemistry. J. Org. Chem. **39**(12), 1615–1621 (1974)
9. List, B., Turberg, M.: The Hajos–Parrish–Eder–Sauer–Wiechert reaction. Synfacts **15**(2), 0200 (2019)
10. List, B., Hoang, L., Martin, H.J.: New mechanistic studies on the Proline-Catalyzed Aldol reaction. Proc. Natl. Acad. Sci. U. S. A. **101**(16), 5839 (2004)
11. List, B.: Asymmetric aminocatalysis. Synlett **11**, 1675–1686 (2001)
12. List, B., Lerner, R.A., Barbas, C.F.: Proline-catalyzed direct asymmetric aldol reactions. J. Am. Chem. Soc. **122**(10), 2395–2396 (2000)
13. Seayad, J., List, B.: Asymmetric organocatalysis. Org. Biomol. Chem. **3**(5), 719–724 (2005)
14. Tu, Y., Wang, Z.-X., Shi, Y.: An efficient asymmetric epoxidation method for trans-olefins mediated by a fructose-derived ketone. J. Am. Chem. Soc. **118**(40), 9806–9807 (1996)
15. Tian, H., She, X., Xu, J., Shi, Y.: Enantioselective epoxidation of terminal olefins by Chiral Dioxirane. Org. Lett. **3**(12), 1929–1931 (2001)
16. Wu, X.-Y., She, X., Shi, Y.: Highly enantioselective epoxidation of α, β-unsaturated esters by Chiral Dioxirane. J. Am. Chem. Soc. **124**(30), 8792–8793 (2002)
17. Boutureira, O., McGouran, J.F., Stafford, R.L., Emmerson, D.P.G., Davis, B.G.: Accessible sugars as asymmetric olefin epoxidation organocatalysts: glucosaminide ketones in the synthesis of terminal epoxides. Org. Biomol. Chem. **7**(20), 4285–4288 (2009)
18. Corey, E.J., Grogan, M.J.: Enantioselective synthesis of α-amino nitriles from N-benzhydryl imines and HCN with a chiral bicyclic guanidine as catalyst. Org. Lett. **1**(1), 157–160 (1999)
19. Tanaka, K., Mori, A., Inoue, S.: The cyclic dipeptide cyclo [(S)-phenylalanyl-(S)-histidyl] as a catalyst for asymmetric addition of hydrogen cyanide to aldehydes. J. Org. Chem. **55**(1), 181–185 (1990)
20. Iyer, M.S., Gigstad, K.M., Namdev, N.D., Lipton, M.: Asymmetric catalysis of the Strecker amino acid synthesis by a cyclic dipeptide. Amino Acids **11**(3–4), 259–268 (1996)
21. Ishikawa, T., Araki, Y., Kumamoto, T., Seki, H., Fukuda, K., Isobe, T.: Modified guanidines as Chiral superbases: application to asymmetric michael reaction of glycine imine with acrylate or its related compounds. Chem. Commun. **3**, 245–246 (2001)
22. Okino, T., Nakamura, S., Furukawa, T., Takemoto, Y.: Enantioselective Aza-Henry reaction catalyzed by a bifunctional organocatalyst. Org. Lett. **6**(4), 625–627 (2004)
23. Okino, T., Hoashi, Y., Takemoto, Y.: Enantioselective Michael reaction of malonates to nitroolefins catalyzed by bifunctional organocatalysts. J. Am. Chem. Soc. **125**(42), 12672–12673 (2003)
24. Nugent, B.M., Yoder, R.A., Johnston, J.N.: Chiral proton catalysis: a catalytic enantioselective direct Aza-Henry reaction. J. Am. Chem. Soc. **126**(11), 3418–3419 (2004)
25. Sigman, M.S., Jacobsen, E.N.: Schiff base catalysts for the asymmetric Strecker reaction identified and optimized from parallel synthetic libraries. J. Am. Chem. Soc. **120**(19), 4901–4902 (1998)
26. Taylor, M.S., Jacobsen, E.N.: Highly enantioselective catalytic Acyl-Pictet−Spengler reactions. J. Am. Chem. Soc. **126**(34), 10558–10559 (2004)
27. Joly, G.D., Jacobsen, E.N.: Thiourea-catalyzed enantioselective hydrophosphonylation of imines: practical access to enantiomerically enriched α-amino phosphonic acids. J. Am. Chem. Soc. **126**(13), 4102–4103 (2004)
28. Wenzel, A.G., Jacobsen, E.N.: Asymmetric Catalytic Mannich Reactions catalyzed by urea derivatives: enantioselective synthesis of β-aryl-β-amino acids. J. Am. Chem. Soc. **124**(44), 12964–12965 (2002)
29. Schiff, H.: Sur Quelques Dérivés Phéniques Des Aldéhydes. Ann. Chim. **131**, 118 (1864)
30. Erkkilä, A., Majander, I., Pihko, P.M.: Iminium catalysis. Chem. Rev. **107**(12), 5416–5470 (2007)

31. Tietze, L.F., Beifuss, U.: In: Trost, B.M. (ed.) Comprehensive Organic Synthesis, vol. 2. Pergamon Press, Oxford, UK (1991)
32. Tietze, L.F., Beifuss, U., Trost, B.M.: Comprehensive organic. Synthesis **II**, 341–394 (1991)
33. Knoevenagel, E.: Ueber Eine Darstellungsweise Der Glutarsäure. Berichte der Dtsch. Chem. Gesellschaft **27**(2), 2345–2346 (1894)
34. Blanchard, K.C., Klein, D.L., MacDonald, J.: Positive Ion catalysis in the Knoevenagel Reaction. J. Am. Chem. Soc. **53**(7), 2809–2810 (1931)
35. Crowell, T.I., Peck, D.W.: Kinetic evidence for a Schiff base intermediate in the Knoevenagel condensation1. J. Am. Chem. Soc. **75**(5), 1075–1077 (1953)
36. Widmark, E., Jeppsson, C.A.: Ein Definierter Organischer Katalysator Mit Wasserstoffionoptimum 1. Skand. Arch. Physiol. **42**(1), 43–61 (1922)
37. Cordes, E.H., Jencks, W.P.: Nucleophilic catalysis of semicarbazone formation by Anilines. J. Am. Chem. Soc. **84**(5), 826–831 (1962)
38. Roberts, R.D., Ferran, H.E., Jr., Gula, M.J., Spencer, T.A.: Superiority of very weakly basic amines as catalysts for Alpha-Proton abstraction via iminium ion formation. J. Am. Chem. Soc. **102**(23), 7054–7058 (1980)
39. Williams, A., Bender, M.L.: Studies on the mechanism of oxime and ketimine formation1. J. Am. Chem. Soc. **88**(11), 2508–2513 (1966)
40. Fridovich, I., Westheimer, F.H.: On the mechanism of the enzymatic decarboxylation of acetoacetate II. J. Am. Chem. Soc. **84**(16), 3208–3209 (1962)
41. Hupe, D.J., Kendall, M.C.R., Spencer, T.A.: Amine catalysis of elimination from. Beta.-acetoxy ketone. Catalysis via iminium ion formation. J. Am. Chem. Soc. **94**(4), 1254–1263 (1972)
42. Baum, J.S., Viehe, H.G.: Synthesis and cycloaddition reactions of acetylenic iminium compounds. J. Org. Chem. **41**(2), 183–187 (1976)
43. Ahrendt, K.A., Borths, C.J., MacMillan, D.W.C.: New strategies for organic catalysis: the first highly enantioselective organocatalytic Diels−Alder reaction. J. Am. Chem. Soc. **122**(17), 4243–4244 (2000)
44. https://www.nobelprize.org/prizes/chemistry/2021/prize-announcement/
45. Gordillo, R., Houk, K.N.: Origins of stereoselectivity in Diels−Alder cycloadditions catalyzed by Chiral imidazolidinones. J. Am. Chem. Soc. **128**(11), 3543–3553 (2006)
46. Allemann, C., Gordillo, R., Clemente, F.R., Cheong, P.H.-Y., Houk, K.N.: Theory of asymmetric organocatalysis of aldol and related reactions: rationalizations and predictions. Acc. Chem. Res. **37**(8), 558–569 (2004)
47. Benaglia, M., Celentano, G., Cinquini, M., Puglisi, A., Cozzi, F.: Poly (Ethylene Glycol)-supported chiral imidazolidin-4-one: an efficient organic catalyst for the enantioselective Diels-Alder Cycloaddition. Adv. Synth. Catal. **344**(2), 149–152 (2002)
48. Selkälä, S.A., Tois, J., Pihko, P.M., Koskinen, A.M.P.: Asymmetric organocatalytic Diels-Alder reactions on solid support. Adv. Synth. Catal. **344**(9), 941–945 (2002)
49. Chu, Q., Zhang, W., Curran, D.P.: A recyclable fluorous organocatalyst for Diels-Alder reactions. Tetrahedron Lett. **47**(52), 9287–9290 (2006)
50. Gin, D.L., Lu, X., Nemade, P.R., Pecinovsky, C.S., Xu, Y., Zhou, M.: Recent advances in the design of polymerizable lyotropic liquid crystal assemblies for heterogeneous catalysis and selective separations. Adv. Funct. Mater. **16**(7), 865–878 (2006)
51. Pecinovsky, C.S., Nicodemus, G.D., Gin, D.L.: Nanostructured, solid-state organic, chiral Diels−Alder catalysts via acid-induced liquid crystal assembly. Chem. Mater. **17**(20), 4889–4891 (2005)
52. Zhang, Y., Zhao, L., Lee, S.S., Ying, J.Y.: Enantioselective catalysis over chiral imidazolidin-4-one immobilized on siliceous and polymer-coated mesocellular foams. Adv. Synth. Catal. **348**(15), 2027–2032 (2006)
53. Northrup, A.B., MacMillan, D.W.C.: The first general enantioselective catalytic Diels−Alder reaction with simple α, β-unsaturated ketones. J. Am. Chem. Soc. **124**(11), 2458–2460 (2002)
54. Selkälä, S.A., Koskinen, A.M.P.: Preparation of bicyclo [4. 3. 0] nonanes by an organocatalytic intramolecular Diels-Alder reaction. Eur. J. Org. Chem. **2005**(8), 1620–1624 (2005)

References

55. Wilson, R.M., Jen, W.S., MacMillan, D.W.C.: Enantioselective organocatalytic intramolecular Diels−Alder reactions. The asymmetric synthesis of solanapyrone D. J. Am. Chem. Soc. **127**(33), 11616–11617 (2005)
56. Yamaguchi, M., Yokota, N., Minami, T.: The Michael addition of dimethyl malonate to α, β-unsaturated aldehydes catalysed by Proline Lithium salt. J. Chem. Soc. Chem. Commun. **16**, 1088–1089 (1991)
57. Yamaguchi, M., Shiraishi, T., Hirama, M.: A catalytic enantioselective Michael addition of a simple malonate to prochiral α, β-unsaturated ketoses and aldehydes. Angew. Chemie Int. Ed. Eng. **32**(8), 1176–1178 (1993)
58. Kawara, A., Taguchi, T.: An enantioselective michael addition of soft nucleophiles to prochiral enone catalyzed by (2-pyrrolidyl) alkyl ammonium hydroxide. Tetrahedron Lett. **35**(47), 8805–8808 (1994)
59. Yamaguchi, M., Shiraishi, T., Hirama, M.: Asymmetric Michael addition of malonate anions to prochiral acceptors catalyzed by l-proline rubidium salt. J. Org. Chem. **61**(10), 3520–3530 (1996)
60. Halland, N., Aburel, P.S., Jørgensen, K.A.: Highly enantioselective organocatalytic conjugate addition of malonates to acyclic α, β-unsaturated enones. Angew. Chemie **115**(6), 685–689 (2003)
61. Knudsen, K.R., Mitchell, C.E.T., Ley, S.V.: Asymmetric organocatalytic conjugate addition of malonates to enones using a proline tetrazole catalyst. Chem. Commun. **1**, 66–68 (2006)
62. Mitchell, C.E.T., Brenner, S.E., Ley, S.V.: A versatile organocatalyst for the asymmetric conjugate addition of nitroalkanes to enones. Chem. Commun. **42**, 5346–5348 (2005)
63. Brandau, S., Landa, A., Franzén, J., Marigo, M., Jørgensen, K.A.: Organocatalytic conjugate addition of malonates to α, β-unsaturated aldehydes: asymmetric formal synthesis of (−)-paroxetine, chiral lactams, and lactones. Angew. Chemie Int. Ed. **45**(26), 4305–4309 (2006)
64. Halland, N., Hansen, T., Jørgensen, K.A.: Organocatalytic asymmetric Michael reaction of cyclic 1, 3-dicarbonyl compounds and α, β-unsaturated ketones—a highly atom-economic catalytic one-step formation of optically active warfarin anticoagulant. Angew. Chem. **115**(40), 5105–5107 (2003)
65. Kim, H., Yen, C., Preston, P., Chin, J.: Substrate-directed stereoselectivity in vicinal diamine-catalyzed synthesis of Warfarin. Org. Lett. **8**(23), 5239–5242 (2006)
66. Kunz, R.K., MacMillan, D.W.C.: Enantioselective organocatalytic cyclopropanations. The identification of a new class of iminium catalyst based upon directed electrostatic activation. J. Am. Chem. Soc. **127**(10), 3240–3241 (2005)
67. Hansen, H.M., Longbottom, D.A., Ley, S.V.: A new asymmetric organocatalytic nitrocyclopropanation reaction. Chem. Commun. **46**, 4838–4840 (2006)
68. Marigo, M., Franzen, J., Poulsen, T.B., Zhuang, W., Jørgensen, K.A.: Asymmetric organocatalytic epoxidation of α, β-unsaturated aldehydes with hydrogen peroxide. J. Am. Chem. Soc. **127**(19), 6964–6965 (2005)
69. Lee, S., MacMillan, D.W.C.: Enantioselective organocatalytic epoxidation using hypervalent iodine reagents. Tetrahedron **62**(49), 11413–11424 (2006)
70. Vesely, J., Ibrahem, I., Zhao, G., Rios, R., Córdova, A.: Organocatalytic enantioselective aziridination of α, β-unsaturated aldehydes. Angew. Chemie **119**(5), 792–795 (2007)
71. Yang, J.W., Hechavarria Fonseca, M.T., List, B.: Catalytic asymmetric reductive Michael cyclization. J. Am. Chem. Soc. **127**(43), 15036–15037 (2005)
72. Bui, T., Barbas, C.F., III.: A proline-catalyzed asymmetric Robinson annulation reaction. Tetrahedron Lett. **41**(36), 6951–6954 (2000)
73. Rajagopal, D., Narayanan, R., Swaminathan, S.: Asymmetric one-pot Robinson annulations. Tetrahedron Lett. **42**(29), 4887–4890 (2001)
74. Zou, Y.-Q., Hörmann, F.M., Bach, T.: Iminium and enamine catalysis in enantioselective photochemical reactions. Chem. Soc. Rev. **47**(2), 278–290 (2018)
75. Mukherjee, S., Yang, J.W., Hoffmann, S., List, B.: Asymmetric enamine catalysis. Chem. Rev. **107**(12), 5471–5569 (2007)
76. Rappoport, Z.: The Chemistry of Enamines. Wiley & Sons (1994)

77. Machajewski, T.D., Wong, C.: The catalytic asymmetric Aldol Reaction. Angew. Chemie Int. Ed. **39**(8), 1352–1375 (2000)
78. Trost, B.M., Schreiber, S.L.: Comprehensive Organic Synthesis Selectivity, Strategy & Efficiency in Modem Organic Chemistry Vol 1. Pergamon Press
79. Mahrwald, R.: Modern Aldol Reactions. Wiley-VCH Verlag GmbH & Co. KGaA: Weinheim, vol. 1 and 2 (2004)
80. Terashima, S., Sato, S., Koga, K.: Regioselective intramolecular asymmetric cyclization of symmetrical open chain triketone. Tetrahedron Lett. **20**(36), 3469–3472 (1979)
81. Takano, S., Kasahara, C., Ogasawara, K.: Enantioselective synthesis of the Gibbane framework. J. Chem. Soc. Chem. Commun. **13**, 635–637 (1981)
82. Blazejewski, J.: The Angular Trifluoromethyl Group. Part 2. Synthesis of (+) 2, 3, 7, 7a-Tetrahydro-7a-Trifluoromethyl-1H-Indene-1, 5-(6H)-Dione. J. Fluor. Chem. **46**(3), 515–519 (1990)
83. Danishefsky, S.J., Masters, J.J., Young, W.B., Link, J.T., Snyder, L.B., Magee, T.V., Jung, D.K., Isaacs, R.C.A., Bornmann, W.G., Alaimo, C.A.: Total synthesis of baccatin III and taxol. J. Am. Chem. Soc. **118**(12), 2843–2859 (1996)
84. Danishefsky, S., Cain, P.: Optically specific synthesis of estrone and 19-norsteroids from 2, 6-lutidine. J. Am. Chem. Soc. **98**(16), 4975–4983 (1976)
85. Agami, C., Meynier, F., Puchot, C., Guilhem, J., Pascard, C.: Stereochemistry-59: new insights into the mechanism of the proline-catalyzed asymmetric Robinson cyclization; structure of two intermediates asymmetric dehydration. Tetrahedron **40**(6), 1031–1038 (1984)
86. Pidathala, C., Hoang, L., Vignola, N., List, B.: Direct catalytic asymmetric enolexo aldolizations. Angew. Chemie Int. Ed. **42**(24), 2785–2788 (2003)
87. Mans, D.M., Pearson, W.H.: Total synthesis of (+)-cocaine via desymmetrization of a meso-dialdehyde. Org. Lett. **6**(19), 3305–3308 (2004)
88. Enders, D., Niemeier, O., Straver, L.: Asymmetric organocatalytic synthesis of Cis-substituted dihydrobenzofuranols via intramolecular aldol reactions. Synlett **20**, 3399–3402 (2006)
89. Stork, G., Brizzolara, A., Landesman, H., Szmuszkovicz, J., Terrell, R.: The enamine alkylation and acylation of carbonyl compounds. J. Am. Chem. Soc. **85**(2), 207–222 (1963)
90. List, B.: Enamine catalysis is a powerful strategy for the catalytic generation and use of carbanion equivalents. Acc. Chem. Res. **37**(8), 548–557 (2004)
91. Clemente, F.R., Houk, K.N.: Computational evidence for the enamine mechanism of intramolecular aldol reactions catalyzed by Proline. Angew. Chemie Int. Ed. **43**(43), 5766–5768 (2004)
92. Hoang, L., Bahmanyar, S., Houk, K.N., List, B.: Kinetic and stereochemical evidence for the involvement of only one proline molecule in the transition states of proline-catalyzed intra- and intermolecular aldol reactions. J. Am. Chem. Soc. **125**(1), 16–17 (2003)
93. Agami, C., Levisalles, J., Puchot, C.: A new diagnostic tool for elucidating the mechanism of enantioselective reactions. Application to the Hajos–Parrish reaction. J. Chem. Soc. Chem. Commun. **8**, 441–442 (1985)
94. Agami, C., Levisalles, J., Sevestre, H.: Extension of the Proline-catalysed asymmetric annelation to diketones. A new case of kinetic resolution. J. Chem. Soc. Chem. Commun. **7**, 418–420 (1984)
95. Puchot, C., Samuel, O., Dunach, E., Zhao, S., Agami, C., Kagan, H.B.: Nonlinear effects in asymmetric synthesis. Examples in asymmetric oxidations and aldolization reactions. J. Am. Chem. Soc. **108**(9), 2353–2357 (1986)
96. Jung, M.E.: A review of annulation. Tetrahedron **32**(1), 3–31 (1976)
97. Rajagopal, D., Moni, M.S., Subramanian, S., Swaminathan, S.: Proline mediated asymmetric Ketol cyclization: a template reaction. Tetrahedron: Asymmetry **10**(9), 1631–1634 (1999)
98. Tang, Z., Cun, L.-F., Cui, X., Mi, A.-Q., Jiang, Y.-Z., Gong, L.-Z.: Design of highly enantioselective organocatalysts based on molecular recognition. Org. Lett. **8**(7), 1263–1266 (2006)
99. Dambruoso, P., Massi, A., Dondoni, A.: Efficiency in isotetronic acid synthesis via a Diamine−Acid couple catalyzed ethyl pyruvate homoaldol reaction. Org. Lett. **7**(21), 4657–4660 (2005)

References

313

100. Samanta, S., Liu, J., Dodda, R., Zhao, C.-G.: C 2-symmetric bisprolinamide as a highly efficient catalyst for direct aldol reaction. Org. Lett. **7**(23), 5321–5323 (2005)
101. Ikishima, H., Sekiguchi, Y., Ichikawa, Y., Kotsuki, H.: Synthesis of (−)-(5R, 6S)-6-acetoxyhexadecanolide based on l-proline-catalyzed asymmetric aldol reactions. Tetrahedron **62**(2–3), 311–316 (2006)
102. Coulthard, G., Erb, W., Aggarwal, V.K.: Stereocontrolled organocatalytic synthesis of prostaglandin PGF 2α in seven steps. Nature **489**(7415), 278–281 (2012)
103. Funk, C.D.: Prostaglandins and leukotrienes: advances in eicosanoid biology. Science **294**(5548), 1871–1875 (2001)
104. Collins, P.W., Djuric, S.W.: Synthesis of therapeutically useful prostaglandin and prostacyclin analogs. Chem. Rev. **93**(4), 1533–1564 (1993)
105. Notz, W., Tanaka, F., Barbas, C.F.: Enamine-based organocatalysis with proline and diamines: the development of direct catalytic asymmetric aldol, Mannich, Michael, and Diels−Alder Reactions. Acc. Chem. Res. **37**(8), 580–591 (2004)
106. Córdova, A., Zou, W., Ibrahem, I., Reyes, E., Engqvist, M., Liao, W.-W.: Acyclic amino acid-catalyzed direct asymmetric aldol reactions: alanine, the simplest stereoselective organocatalyst. Chem. Commun. **28**, 3586–3588 (2005)
107. Markert, M., Mulzer, M., Schetter, B., Mahrwald, R.: Amine-catalyzed direct aldol addition. J. Am. Chem. Soc. **129**(23), 7258–7259 (2007)
108. Tang, Z., Jiang, F., Yu, L.-T., Cui, X., Gong, L.-Z., Mi, A.-Q., Jiang, Y.-Z., Wu, Y.-D.: Novel small organic molecules for a highly enantioselective direct aldol reaction. J. Am. Chem. Soc. **125**(18), 5262–5263 (2003)
109. Tang, Z., Jiang, F., Cui, X., Gong, L.-Z., Mi, A.-Q., Jiang, Y.-Z., Wu, Y.-D.: Enantioselective direct aldol reactions catalyzed by L-prolinamide derivatives. Proc. Natl. Acad. Sci. **101**(16), 5755–5760 (2004)
110. Sakthivel, K., Notz, W., Bui, T., Barbas, C.F.: Amino acid catalyzed direct asymmetric aldol reactions: a bioorganic approach to catalytic asymmetric carbon−carbon bond-forming reactions. J. Am. Chem. Soc. **123**(22), 5260–5267 (2001)
111. Guo, H.-M., Cun, L.-F., Gong, L.-Z., Mi, A.-Q., Jiang, Y.-Z.: Asymmetric direct aldol reaction catalyzed by an L-prolinamide derivative: considerable improvement of the catalytic efficiency in the ionic liquid. Chem. Commun. **11**, 1450–1452 (2005)
112. Tang, Z., Yang, Z.-H., Chen, X.-H., Cun, L.-F., Mi, A.-Q., Jiang, Y.-Z., Gong, L.-Z.: A highly efficient organocatalyst for direct aldol reactions of ketones with aldedydes. J. Am. Chem. Soc. **127**(25), 9285–9289 (2005)
113. Saito, S., Yamamoto, H.: Design of acid−base catalysis for the asymmetric direct aldol reaction. Acc. Chem. Res. **37**(8), 570–579 (2004)
114. Torii, H., Nakadai, M., Ishihara, K., Saito, S., Yamamoto, H.: Asymmetric direct aldol reaction assisted by water and a proline-derived tetrazole catalyst. Angew. Chemie Int. Ed. **43**(15), 1983–1986 (2004)
115. Raj, M., Vishnumaya Ginotra, S.K., Singh, V.K.: Highly enantioselective direct aldol reaction catalyzed by organic molecules. Org. Lett. **8**(18), 4097–4099 (2006)
116. Rossiter, B.E., Swingle, N.M.: Asymmetric conjugate addition. Chem. Rev. **92**(5), 771–806 (1992)
117. Perlmutter, P.: Conjugate Addition Reactions in Organic Synthesis. Elsevier (2013)
118. Tsogoeva, S.B.: Recent advances in asymmetric organocatalytic 1, 4-conjugate additions. Eur. Org. Chem. **11**, 1701–1716 (2007)
119. Hermann, K., Wynberg, H.: Asymmetric induction in the Michael reaction. J. Org. Chem. **44**(13), 2238–2244 (1979)
120. Wynberg, H., Helder, R.: Asymmetric induction in the alkaloid-catalysed Michael reaction. Tetrahedron Lett. **16**(46), 4057–4060 (1975)
121. Yamada, S., Hiroi, K., Achiwa, K.: Asymmetric synthesis with amino acid I asymmetric induction in the alkylation of keto-enamine. Tetrahedron Lett. **10**(48), 4233–4236 (1969)
122. Seebach, D., Missbach, M., Calderari, G., Eberle, M.: [3+ 3]-carbocyclizations of nitroallylic esters and enamines with stereoselective formation of up to six new stereogenic centers. J. Am. Chem. Soc. **112**(21), 7625–7638 (1990)

123. Hechavarria Fonseca, M.T., List, B.: Catalytic asymmetric intramolecular Michael reaction of aldehydes. Angew. Chemie **116**(30), 4048–4050 (2004)
124. List, B., Pojarliev, P., Martin, H.J.: Efficient proline-catalyzed Michael additions of unmodified ketones to nitro olefins. Org. Lett. **3**(16), 2423–2425 (2001)
125. Seebach, D., Goliński, J.: Synthesis of open-chain 2, 3-disubstituted 4-nitroketones by diastereoselective Michael-addition of (E)-enamines to (E)-nitroolefins. A topological rule for C, C-bond forming processes between prochiral centres. Preliminary communication. Helv. Chim. Acta **64**(5), 1413–1423 (1981)
126. Blarer, S.J., Seebach, D.: Asymmetrische Michael-Additionen. Stereoselektive Alkylierungen Des (R)-und (S)-Enamins Aus Cyclohexanon Und 2-(Methoxymethyl) Pyrrolidin Durch A-(Methoxycarbonyl) Zimtsäure-methylester. Chem. Ber. **116**(6), 2250–2260 (1983)
127. Jelinek, R., Kolusheva, S.: Carbohydrate biosensors. Chem. Rev. **104**(12), 5987–6016 (2004)
128. Varki, A.: Biological roles of oligosaccharides: all of the theories are correct. Glycobiology **3**(2), 97–130 (1993)
129. Tiwari, V.K., Kumar, A., Schmidt, R.R.: Disaccharide-containing macrocycles by click chemistry and intramolecular glycosylation. Eur. J. Org. Chem. **15**, 2945–2956 (2012)
130. Bertozzi, C.R., Kiessling, L.L.: Chemical glycobiology. Science **291**(5512), 2357–2364 (2001)
131. Galan, M.C., Benito-Alifonso, D., Watt, G.M.: Carbohydrate chemistry in drug discovery. Org. Biomol. Chem. **9**(10), 3598–3610 (2011)
132. Marradi, M., Huang, X., Molinaro, A., Silipo, A., Penades, S., Roy, R., Schweizer, F., Vincent, S., Mulard, L., Rauter, A.P.: Carbohydrates in Drug Design and Discovery. Royal Society of Chemistry (2015)
133. Tiwari, V.K.: Carbohydrates in Drug Discovery and Development: Synthesis and Application. Elsevier (2020)
134. Laschat, S., Kunz, H.: Carbohydrates as chiral templates: diastereoselective synthesis of N-glycosyl-N-homoallylamines and beta.-amino acids from imines. J. Org. Chem. **56**(20), 5883–5889 (1991)
135. Kunz, H., Rück, K.: Carbohydrates as chiral auxiliaries in stereoselective synthesis. New synthetic methods (90). Angew. Chemie Int. Ed. Eng. **32**(3), 336–358 (1993)
136. Nicolaou, K.C., Mitchell, H.J.: Adventures in carbohydrate chemistry: new synthetic technologies, chemical synthesis, molecular design, and chemical biology. Angew. Chemie Int. Ed. **4**(9), 1576–1624 (2001)
137. Blaser, H.U.: The chiral pool as a source of enantioselective catalysts and auxiliaries. Chem. Rev. **92**(5), 935–952 (1992)
138. Hultin, P.G., Earle, M.A., Sudharshan, M.: Synthetic studies with carbohydrate-derived chiral auxiliaries. Tetrahedron **53**(44), 14823–14870 (1997)
139. Dalko, P.I., Moisan, L.: Enantioselective organocatalysis. Angew. Chemie Int. Ed. **40**(20), 3726–3748 (2001)
140. Faisca Phillips, A.M.: Applications of carbohydrate-based organocatalysts in enantioselective synthesis. Eur. J. Org. Chem. **33**, 7291–7303 (2014)
141. Minuth, T., Boysen, M.M.K.: Bis (oxazolines) based on glycopyranosides-steric, configurational and conformational influences on stereoselectivity. Beils. J. Org. Chem. **6**(1), 23 (2010)
142. Dalko, P.I., Moisan, L.: In the golden age of organocatalysis. Angew. Chemie Int. Ed. **43**(39), 5138–5175 (2004)
143. Shi, Y.: Organocatalytic asymmetric epoxidation of olefins by Chiral Ketones. Acc. Chem. Res. **37**(8), 488–496 (2004)
144. Wong, O.A., Shi, Y.: Organocatalytic oxidation. Asymmetric epoxidation of olefins catalyzed by Chiral Ketones and iminium salts. Chem. Rev. **108**(9), 3958–3987 (2008)
145. Mishra, A., Mishra, N.K., Tiwari, V.: Carbohydrate-based organocatalysts: recent developments and future perspectives. Curr. Org. Synth. **13**(2), 176–219 (2016)
146. Wang, Z.-X., Tu, Y., Frohn, M., Zhang, J.-R., Shi, Y.: An efficient catalytic asymmetric epoxidation method. J. Am. Chem. Soc. **119**(46), 11224–11235 (1997)

References

147. Wang, Z.-X., Shi, Y.: A PH study on the chiral ketone catalyzed asymmetric epoxidation of hydroxyalkenes. J. Org. Chem. **63**(9), 3099–3104 (1998)
148. Wang, Z.-X., Cao, G.-A., Shi, Y.: Chiral ketone catalyzed highly chemo-and enantioselective epoxidation of conjugated enynes. J. Org. Chem. **64**(20), 7646–7650 (1999)
149. Cao, G.-A., Wang, Z.-X., Tu, Y., Shi, Y.: Chemo-and enantioselective epoxidation of enynes. Tetrahedron Lett. **39**(25), 4425–4428 (1998)
150. Lindström, U.M.: Stereoselective organic reactions in water. Chem. Rev. **102**(8), 2751–2772 (2002)
151. Tsutsui, A., Takeda, H., Kimura, M., Fujimoto, T., Machinami, T.: Novel enantiocontrol system with aminoacyl derivatives of glucoside as enamine-based organocatalysts for aldol reaction in aqueous media. Tetrahedron Lett. **48**(30), 5213–5217 (2007)
152. Agarwal, J., Peddinti, R.K.: Highly efficient and solvent-free direct aldol reaction catalyzed by glucosamine-derived prolinamide. Tetrahedron: Asymmetry **21**(15), 1906–1909 (2010)
153. Agarwal, J., Peddinti, R.K.: Synthesis and characterization of monosaccharide derivatives and application of sugar-based prolinamides in asymmetric synthesis. Eur. J. Org. Chem. **2012**(32), 6390–6406 (2012)
154. De Nisco, M., Pedatella, S., Bektaş, S., Nucci, A., Caputo, R.: D-glucosamine in a chimeric prolinamide organocatalyst for direct asymmetric aldol addition. Carbohydr. Res. **356**, 273–277 (2012)
155. Kumar, T.P., Balaji, S.V.: Sugar amide-pyrrolidine catalyst for the asymmetric Michael addition of ketones to nitroolefins. Tetrahedron: Asymmetry **25**(5), 473–477 (2014)
156. Wang, L., Liu, J., Miao, T., Zhou, W., Li, P., Ren, K., Zhang, X.: Sugar-based pyrrolidine as a highly enantioselective organocatalyst for asymmetric Michael addition of ketones to nitrostyrenes. Adv. Synth. Catal. **352**, 2571–2578 (2010)

Chapter 8
Enzymes in Organic Synthesis

1 Introduction

In the context of "green chemistry" as sustainable science, "CATALYSIS" is one of the exceptionally significant disciplines that facilitate an efficient and cost-effective protocol [1]. This greatly applied with a high innovative potential for a wide range of different agrochemicals, fine chemicals, materials, and moreover as life-saving drug candidates useful for humankind. In organic synthesis, "reactivity" is only one issue, while "selectivity" is another but even more important criterion to precede any reaction in laboratory and industry [2, 3]. A number of such convenient catalytic methods were explored since last 2–3 decades that remarkably not only increased the reaction rates but also bring the desired selectivities, e.g., regioselectivity (i.e., selectivity toward positional isomers), chemoselectivity (i.e., selectivity toward functional groups), as well the stereoselectivity (i.e., selectivity toward stereoisomers, may be enantioselectivity as well distereoselectivity) too and this way these methods displayed a very strong impact in academics and pharmaceutical industry [4, 5]. For an example, Cu(I)-catalyzed azide-alkyne cycloaddition (popularly known as "Click Chemistry") [6] to furnish high-to-excellent yield of respective 1,2,3-triazoles with excellent regioselectivity is widely explored catalytic method during the last twenty years in various emerging field of science particularly in carbohydrate chemistry [7, 8]. Although the list is too long, other notable catalytic methods, including Heck coupling, Negishi coupling, Suzuki coupling, Kumada coupling, Sonogashira coupling, Glaser coupling, Grubbs–Schrock reaction, and other important catalytic protocols, have been widely explored in industry for various purposes [9–14].

In addition, chiral induction through asymmetric synthesis is further growing area of investigation in recent years. Thus, in another fruitful green aspect, the organocatalytic methodologies applied for an easy access of enantio- and/or diastereo-enriched molecules are receiving special attention particularly for the development of new chemical entities and drug candidates that do not tolerate metal contamination [15]. A number of such organocatalysis strategies utilizing cost-effective air- and moisture-stable organocatalysts have been greatly explored for the asymmetric synthesis

---

© The Author(s), under exclusive license to Springer Nature Singapore Pte Ltd. 2022
V. K. Tiwari et al., *Green Chemistry*,
https://doi.org/10.1007/978-981-19-2734-8_8

317

of intermediates and biologically relevant molecules in optically pure form, total synthesis of bioactive natural products, and also commercially available drugs such as oseltamivir, paroxetine, baclofen, warfarin, maraviroc. Research in this direction is being recently gained popularity in the form of "Nobel Prize in Chemistry for the year 2021" awarded to Prof. Benjamin List from Max Planck Institute Germany and Prof. David Macmillan from Princeton University USA for their notable contribution in the domain of "**asymmetric organocatalysis**" [16, 17]. This section is just described in detail in Chap. 6.

In addition to the enhanced reactivities of particular reaction, the specificity resulted in desire selectivities are even more desirable target. At this end, another very important catalysis protocol that can be considered is the "*biocatalysis*" [18–21], which is basically an enzyme-induced synthetic protocol. Despite its ease and great practical utilities, the biocatalysis route in earlier days received even little attention in comparison with homogeneous and heterogeneous catalyses. Time has been changed, and nowadays enzymes have displayed wide applications in organic synthesis. In the late 1980s, just after the invention of recombinant DNA technology, the biocatalysis route changed way of catalysis significantly in terms of green concept. Recombinant DNA technology furthermore covered the approach for enzyme engineering, a required tool to tackle the limitations for synthetic utilities of enzymes that occur in nature. Even though finding suitable enzymes is rather challenging, once characterized they are beyond doubt easy to obtain and many times are even inexpensive. Although over hundreds of industrial enzymatic processes being implemented since then [3], despite the insight about incompetent enzymes still continue. Experiencing through progress in revolutionary technologies, e.g., genome sequencing and synthesis, bioinformatics, etc., that provided rapid access to a vast biodiversity at low cost, and moreover through their growing fruitful synthetic applications, we all do hope that this perception may be changed in the present scenario. Therefore, enzyme-mediated synthetic tool is rapidly expanding, and a number of such powerful enzyme engineering strategies help to find out promising enzymes that can provide ample opportunities in identifying a valuable biocatalyst to address their specific role in chemoenzymatic synthesis.

This chapter presents a brief overview of few selected enzyme-mediated synthetic protocols with highlighting their performance toward upgrading the selectivity and efficiency in some selected enzyme systems including lyases (aldolase), sialyltransferase, oxido-reductases, hydrolases, and isomerases well-utilized in the synthesis of some selected molecules of biological relevance. We hope that the exemplified presentation using designed enzymes may provide an ample opportunity in organic synthesis both in academic and industry.

2 Applications of Enzymes in Synthesis

D-amino acid oxidases (DAAOs) are very popular as enantiospecific enzymes for organic synthesis in laboratory as well as industry. The flavin adenine dinucleotide

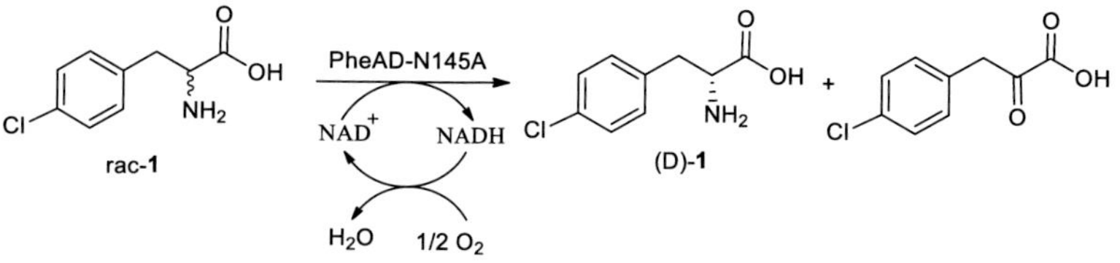

2 Applications of Enzymes in Synthesis

Scheme 1 Resolution of rac-4-chlorophenylalanine using N145A variant of L-Phenylalanine

(FAD)-containing biocatalyst has displayed wide substrate scope on numerous amino acids [22]. Engel et al. have portrayed the resolution of racemic non-natural amino acids using wild-type enzyme L-phenylalanine dehydrogenases. In this structured way, the modified variant N145A along with NAD⁺ (cofactor) was implemented for the resolution of rac-4-chlorophenylalanine **1**, which resulted the pure D-isomer of amino acid **1** (Scheme 1) [23].

Lipases from *Pseudomonas aeruginosa* are frequently used in organic synthesis for the cleavage as well as ester formation. Pseudomonas aeruginosa lipase (PAL) was the first enzyme to which application of directed evolution to enhance the enantioselectivity was implemented [24]. Jaeger et al. have described a new technique, i.e., mutation via directed evolution method to develop more enantioselective biocatalysts. As a result, the selectivity factor for (S)-configured acid was enhanced substantially from E = 1.1 wild type to E = 51 for the best variant. The enantioselective kinetic resolution of racemic *p*-nitrophenyl2-methyl decanoate **2** in (S)-configured acid **3** and (R)-configured ester **2** was best demonstrated by the engineered lipases [25]. Moreover, the enhanced enantioselectivity was well-supported by the molecular dynamics simulations [26] (Scheme 2).

In another example, lipase of B. cepacia KWI-56 was utilized to hydrolyze racemic ethyl 3-phenylbutanoate (**4**) enantioselectivily in the (R)-configured acid **5** (selectivity factor Er = 38). Whereas wild-type lipase was yielded (S)-configured acid **5** (selectivity factor Es = 33) (Scheme 3) [27]. Moreover, semi-rational and semi-random combinatorial designs of a mutant are enthralling tool to modify the enantioselectivity of biocatalysts.

In order to increase the enantioselectivity of hyperthermophilc esterase (from archaeon *Aeropyrum pernix* K1 (APE1547), a directed evolution method is implemented to prepare a number of variants from native enzyme. The authors have identified a potential mutation (TBC26) after one time of epPCR. The enantioselectivity

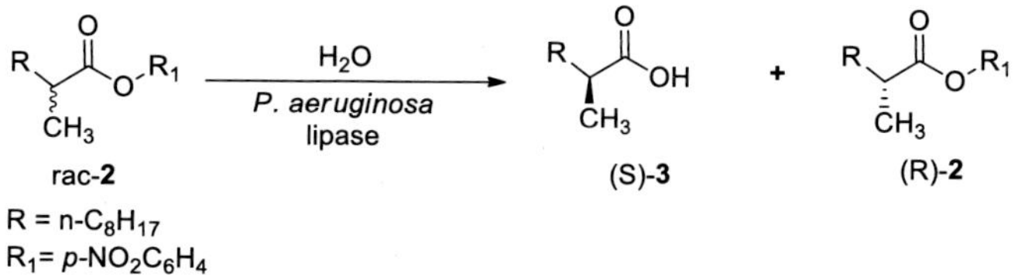

Scheme 2 Kinetic resolution of *rac*-2-methyldecanoic acid *p*-nitrophenyl ester **2** using lipases

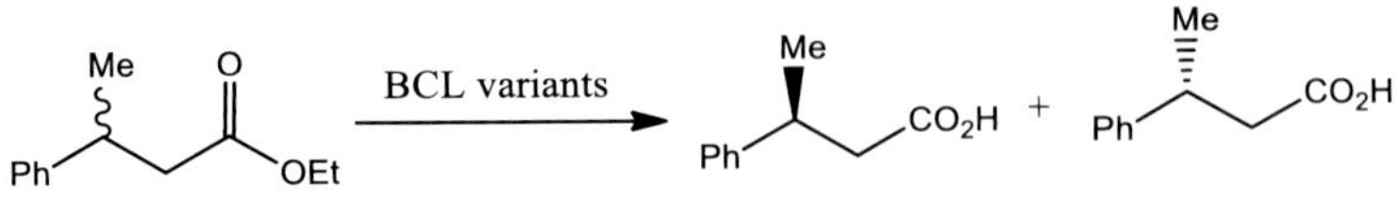

Scheme 3 Hydrolysis of *rac*-ethyl 3-phenylbutanoate **4** using BCL variant of lipase enzyme

Scheme 4 Kinetic resolution of 3-Octyl acetate and rac-3-Phenylbutanoate via enzymatic route

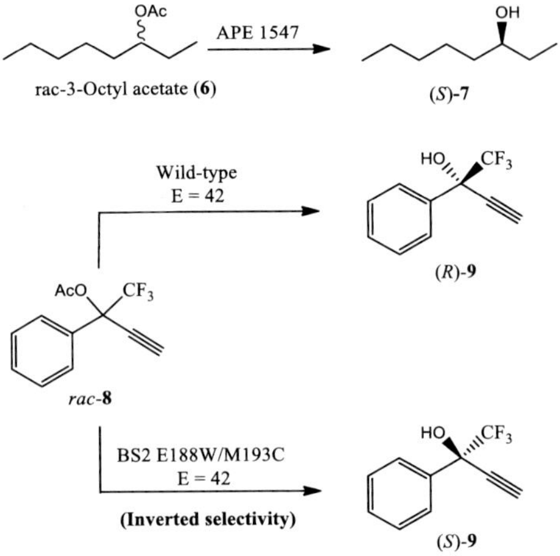

of TBC26 was enhanced ~ 2.6 times than the wild-type biocatalyst. The resulted biocatalyst was implemented for the kinetic resolution of rac-3-octyl acetate **6** to (S)-configured 3-hydroxyoctane **7** selectively. This outcome has the clear indication for the close relation of enantioselectivity with enzyme (Scheme 4) [28].

Chiral molecules are the persuasive starting scaffolds for the generation of natural products, pharmaceuticals, and enantiopure alcohols. In a structured way, Bornscheuer and coworkers have highlighted the kinetic resolution of tertiary alcohols **8** using suitable hydrolases [29]. After the comprehensive investigation of structure function relation of 92 microbial serine hydrolases, the authors have identified motif GGG(A)X a concurrence sequence in kinetic resolution of *tert* alcohols [30]. The same group has successfully innovated the highly enantioselective double mutant E188W/M193C of BS2 esterase, which facilitates the s-configured scaffold **9** opposite to that of wild type (Scheme 4) [31]. These findings were further defended by the docking study, and investigation of the substrate range, which facilitated the increased understanding of enantioselectivity of BS2 in the direction of acetylated tertiary alcohols.

Chiral epoxides and enantiopure vicinal diols are the most important intermediates in asymmetric synthesis. These scaffolds have been an important precursor for the production of costly anticancer agents [32]. Therefore, a highly enantioselective and regioselective mutant of epoxide hydrolase from *A. niger* (ANEH) was utilized to

2 Applications of Enzymes in Synthesis

explore the hydrolytic kinetic resolution of racemic glycidyl phenyl ether **10**. The wild-type enzyme has only $E = 4.6$ for (S)-enantiomer **11**, while its modified variant evolved in several steps depicted enhanced enantioselectivity ($E = 115$) (Scheme 5) [33, 34]. Furthermore, the reason behind the enhanced enantioselectivity was proven by kinetic study, MD calculations, molecular modeling, and X-ray structures.

In another investigation, Reetz and coworkers introduced nine mutations into epoxide hydrolase from *Aspergillus niger* (ANEH) for kinetic resolution of trans-epoxide **12** and cis-diols **13**, selectively (Scheme 6) [35].

In another investigation, Bornscheuer et al. have reported the generation of enantiopure vicinal diols **15** from racemic epoxide **14** in high yield using epoxide hydrolases (EHs) from *Aspergillus niger* (Scheme 7) [36].

Jansen and coworkers have employed a sequential kinetic resolution approach catalyzed by halohydrin dehydrogenase to achieve two treasure enantiopure architectures. Resolution of racemic substrate methyl 4-chloro-3-hydroxybutanoate methyl

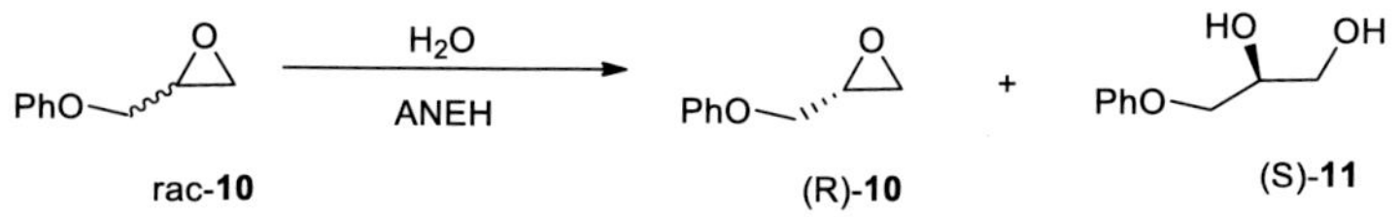

Scheme 5 Hydrolytic kinetic resolution of *rac*-glycidyl phenyl ether **10** using hydrolase from A. niger

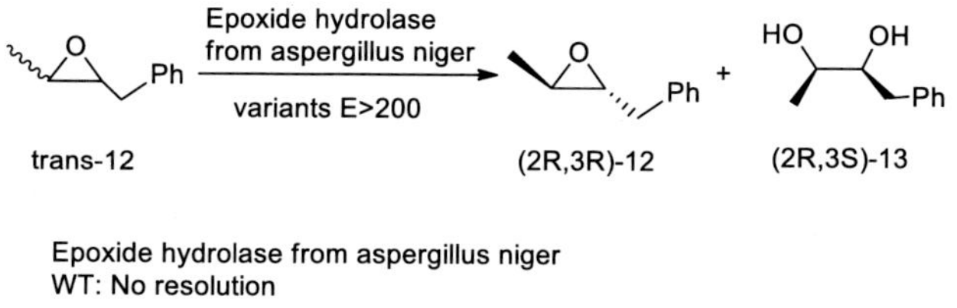

Scheme 6 Kinetic resolution of trans-epoxide with EH variants from aspergillus niger

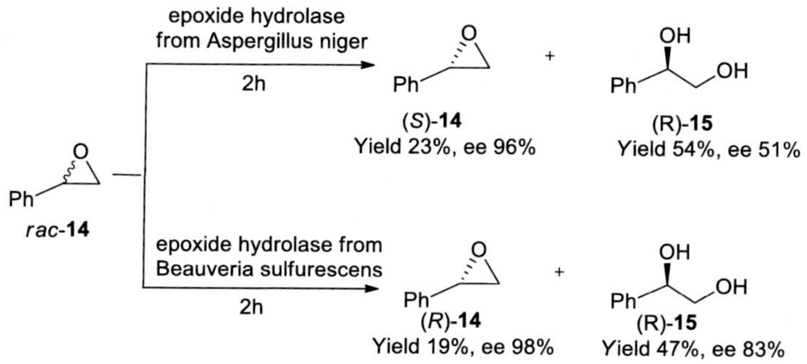

Scheme 7 Enantioconvergent kinetic resolution of an epoxide using two enantiocomplementary hydrolases

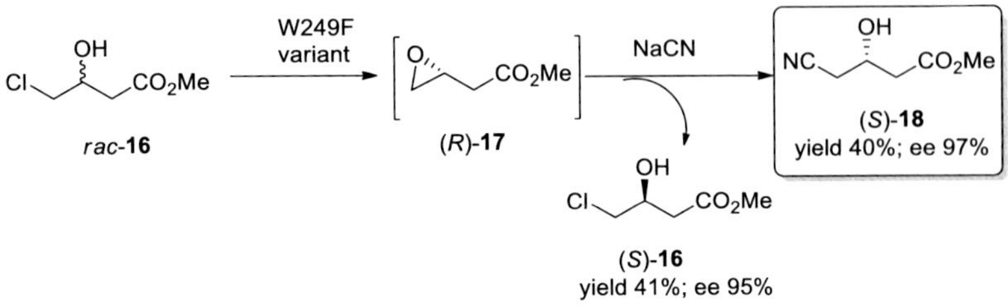

Scheme 8 Kinetic resolution of chlorohydrins using W249F variant

ester (R,S)-**16** was carried out using the variant W249F to afford enantiomeric pure compounds methyl 4-cyano-3-hydroxybutanoate methyl ester ((S)-**18**) with 97% ee (40% yield) and (S)-**16** with 95% ee (41% yield). The outcomes have displayed the versatility of chosen biocatalyst with other catalyst in kinetic resolution of *vic*-halohydrins, dynamic kinetic resolution, tandem ring closure, enantioselective epoxide ring-opening, etc. (Scheme 8) [37].

Iterative saturation mutagenesis (ISM) has been applied to widen the substrate scope of old yellow enzyme homologue YqiM, by directing the enantioselectivity in the bioreduction of library of cyclopentenone **19** and cyclohexenone **20** derivatives. Several mutations were engineered and subjected to bioreduction which yielded enhanced stereoselectivity as compared to wild-type YqiM. In this regard, the authors obtained both the enantiomers **19** and **20** with substantial enantiomeric excess (ee) by using the appropriate engineered YqiM biocatalyst (Scheme 9) [38].

Simply the conversion of ketones (**21** and **23**) to ester **22** or lactones **24** functionality through the oxygen atom insertion is well known as Baeyer–Villiger reaction [39, 40]. Most of the times, reactions are carried out in acids, bases, or metal complexes, except this asymmetric catalysis is also possible in certain cases [41]. Nowadays, these synthetically important scaffolds are constructed by both the approaches, i.e., chemical as well as enzymatic. Flavin-dependent enzymes like cyclohexanone monooxygenases (CHMOs) react with dioxygen (air) to generate an

Scheme 9 YqjM
Muteins-mediated reduction
of cyclic enones

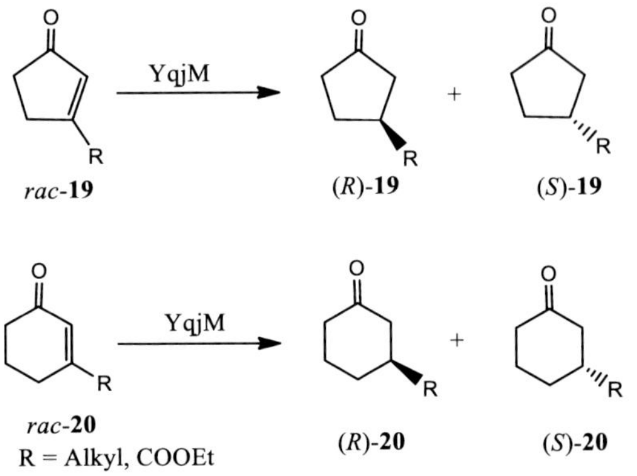

2 Applications of Enzymes in Synthesis 323

Scheme 10
Mono-oxygenase-mediated
Baeyer–Villiger oxidation of
a ketone to ester or cyclic
ketone to lactone

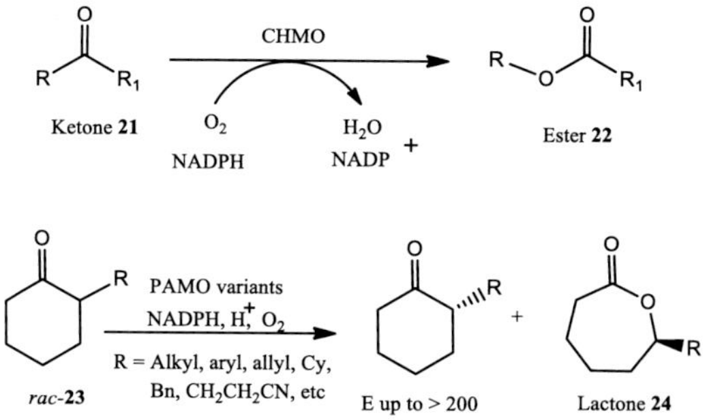

intermediate hydroperoxide, and by transferring the one oxygen atom to substrate resulted in the corresponding esters **22** or lactones **24** (Scheme 10) [42, 43].

Kayser and coworkers have applied the same directed evolution approach to emerge enantioselective cyclohexanone monooxygenases as catalysts in Baeyer–Villiger reactions, herein, the enantioselectivity ranges between 90 and 99%. [44, 45] However, the authors have applied the CHMO from *Acinetobacter sp.* NCIMB 9871, to invert the selectivity in the oxidation of 4-hydroxycyclohexanone **25** (Scheme 11) [46]. Without comprehending the three-dimensional enzymatic structure the procedure reached its successful outcome, which is not amendable with rational design approach [47]. However, till now a very limited protein sequence has been discovered as well as very less mutant screening (less than 20,000). Therefore, for sound theoretical analysis the crystallographic data of the wild-type CHMO and of the enantioselective mutants is a valuable piece in the puzzle.

The production of methyl (S)-4-bromo-3-hydroxybutyrate was efficiently achieved by asymmetric reduction of methyl 4-bromo-3-oxobutyrate with a mutant β-keto ester reductase (KER-L54Q) from Penicillium citrinum and a cofactor-regeneration enzyme such as glucose dehydrogenase (GDH) from a transformant cell system. The enzymatic mutant KER-L54Q performed high enantioselectivity, productivity, and thermal stability in various reactions than the

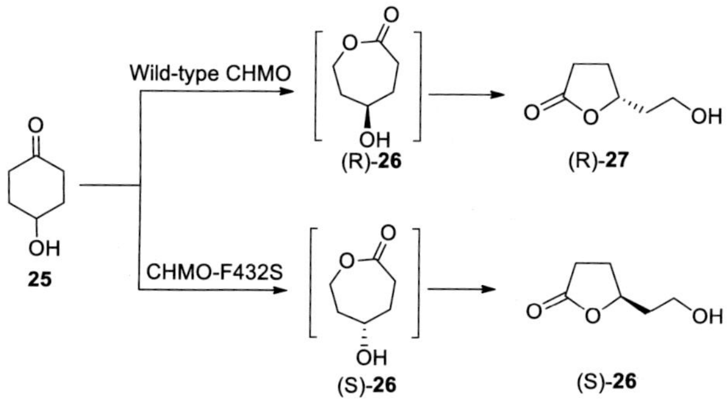

Scheme 11 Oxidation of 4-hydroxycyclohexanone using CHMO variant enzyme

wild-type enzyme [48]. The authors performed the optimization for biocatalytic synthesis of carbapenem derivatives **31** starting from the synthesis of methyl (2S,3R)-2-(benzamidomethyl)-3-hydroxybutyrate **29** which is an important intermediate reaction intermediate. Further, (3R,4R)-4-Acetoxy-3-[(R)-(tert-butyldimethylsilyloxy)ethyl]-2-azetidinone **30** obtained from **29** plays key role in synthesis of carbapenem derivatives **31** (Scheme 11). The performance of biocatalysts is dependent on the specific activity of the enzyme for the substrate. The (R)-specific enzymatic mutant Lactobacillus kefir produced **29** with excellent selectivity (de > 99%) and productivity and $NADP^+$ produced can be recycled to NADPH by using 2-propanol in a reaction as shown in (Scheme 12) [49].

Interestingly, a directed evolution approach was adopted to achieve more potent CHMO inhibitory efficiency against enantioselective sulfoxidation of prochiral thioethers in *Acinetobacter sp*. The hard work resulted in variants with > 95% ee for substrate **27** (Scheme 13) [50]. Surprisingly, the mutagenesis of toluene omonooxygenase (*Burkholderia cepacia* G4) and the analogous position I100 in toluene

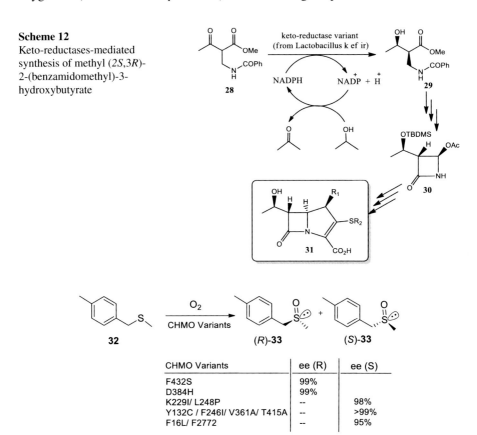

Scheme 12 Keto-reductases-mediated synthesis of methyl (2*S*,3*R*)-2-(benzamidomethyl)-3-hydroxybutyrate

32 (*R*)-**33** (*S*)-**33**

CHMO Variants	ee (R)	ee (S)
F432S	99%	
D384H	99%	
K229I/ L248P	--	98%
Y132C / F246I/ V361A/ T415A	--	>99%
F16L/ F2772	--	95%

Scheme 13 Stereoselective sulfoxidation of thioethers using various CHMO variants (from Acinetobacter sp.)

2 Applications of Enzymes in Synthesis

Scheme 14 Enzymatic
oxidation to
3-*tert*-Butylcatechol

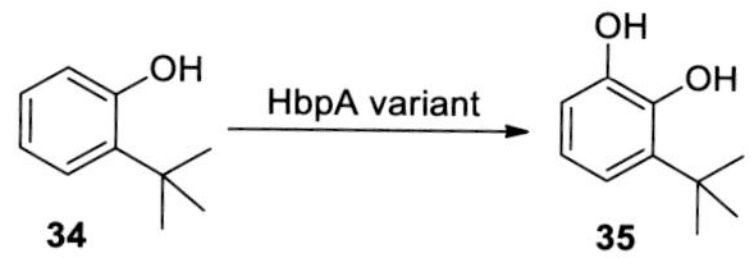

4-monooxygenase (*Pseudomonas mendocina* KR1) have exhibited positive influence over the rate and enantioselectivity of methyl p-tolyl sulfide **32** oxidation to respective sulfoxide process **33** [51].

2-Hydroxybiphenyl 3-monooxygenase (HbpA) is more often involved in biosynthesis of cholesterol and originated from Pseudomonas azelaica HBP1 (non-heme monooxygenase) [52]. By implementing the HbpA enzyme, Witholt et al. reported the oxidation of 2-*tert*-butylphenol **34** to 3-*tert*-butylcatechol **35** (Scheme 14) [53].

Since last few years, biocatalysts have attracted the greater attention among which cytochrome P450s a Fe-heme-dependent enzymes have displayed the wide roles in organic synthesis [54]. Cytochrome P450 BM-3 collected from *Bacillus megaterium* was engineered to increase its regioselectivity for the various substrates [45]. Therefore, Arnold and coworkers regioselectively hydroxylated the linear alkane regioselectively using the protein engineered P450 enzyme from Bacillus megaterium. The beginning of this evolution study started with the selective acceptance of alkanes of C3-C8 chain length **36** as a substrate by the P450 BM-3 variant to afford its hydroxylated alkane **37** (Scheme 15) [55, 56].

Raadt et al. reported the regioselective hydroxylation of compound **38** by using the engineered cytochrome P450 variant 1-12G and variant 139–3 to afford the respective enantioselective products **39** and **40** (Scheme 16) [57].

In another report, Schmid et al. have reported that the wild-type P450 monooxygenase which depicted the very low hydroxylation activity for β-ionone **40**. Therefore, the authors did site-directed mutagenesis in P450 BM-3 to improve its activity and

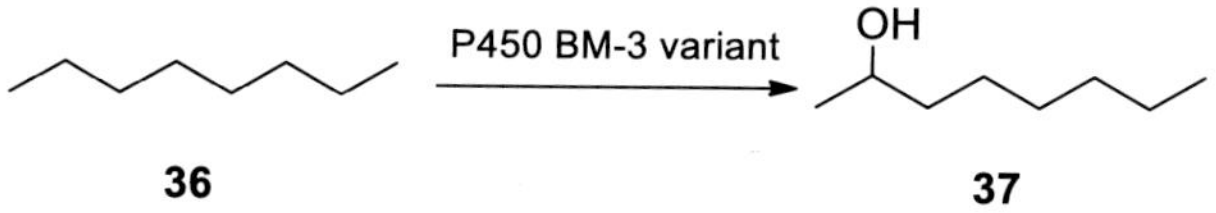

Scheme 15 Hydroxylation of alkane led to enantiomers (40% ee) using P450 BM-3 variants

Scheme 16 Asymmetric
hydroxylation of
2-cyclopentyl-benzooxazole
with P450 BM-3 Variants

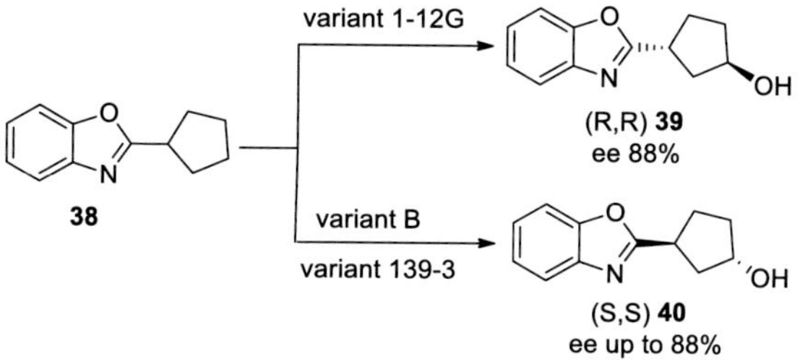

Scheme 17 Asymmetric hydroxylation of β-Ionone **40** using P450 BM-3 Variants

40

(S)-41
ee 39%

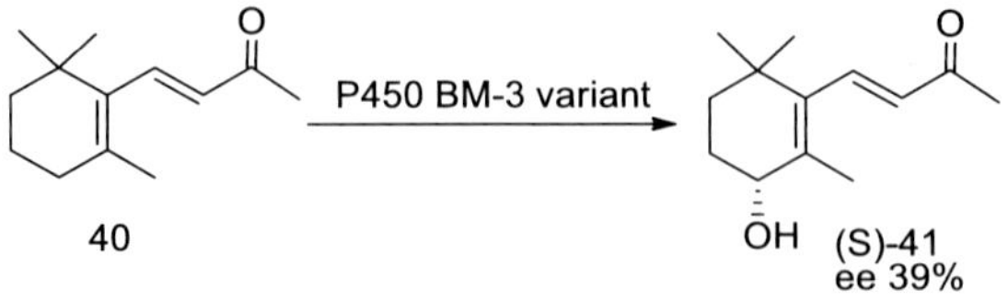

the resulted variant R47L/Y51F/F87V displayed 300 times enhance in activity along with high enantioselectivity for the S-enantiomer **41** (~39%) (Scheme 17) [58].

2-Aryl-2-hydroxyacetic acid derivatives are promising pharmaceutical constructing units for variety of drugs like cephalosporin, penicillins, and antiobesity agents [59]. Owing to increased demand of these compounds, numerous novel techniques were adopted, among which biocatalysts are quite interesting [60]. Therefore, Arnold group have employed a modified variant of biocatalyst P450- BM-3 for the enantioselective hydroxylation at the second position of aryl-esters **42** to afford the respective esters **43** with enhanced ee up to 95% selectivity (Scheme 18). [61].

Further, Wong et al. have reported the oxidation of naphthalene by utilizing the variant P450cam Y96F to afford the α- and β-naphthol selectively. The oxidation rate was further enhanced by 140 folds by the same group only replacing the Y96 of the heme monooxygenase cytochrome P450cam from P. putida with hydrophobic residue (Scheme 19) [62].

Kubo et al. have incorporated the various mutations to the modified P450 enzyme which resulted, two variants P450 BM-3 that transformed alkene **47** to their respective (R)- and (S)-epoxide **48** (~ 83% ee) with increased catalytic efficiency (~1370) and remarkable selectivity for epoxidation (Scheme 20) [63]. This report clearly revealed that the BM-3active site can easily architecture for enantio- and regioselective epoxidation of a variety of substrates.

Scheme 18 Asymmetric hydroxylation of phenyl acetic acid esters using P450 BM-3 Variants

42
R = alkyl

(S)-43
ee 56-95%

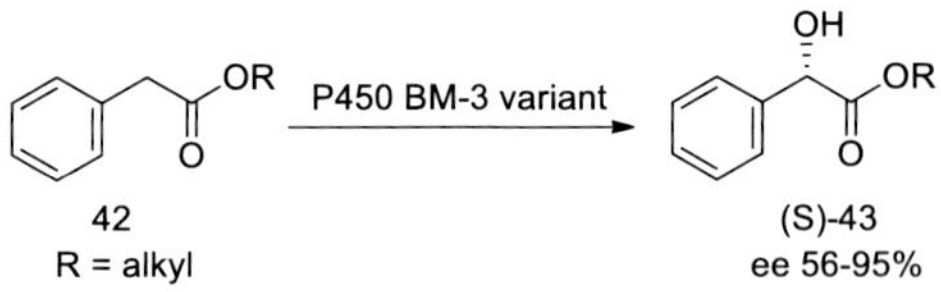

Scheme 19 Oxidation of naphthalene to naphthols using P450cam Variant

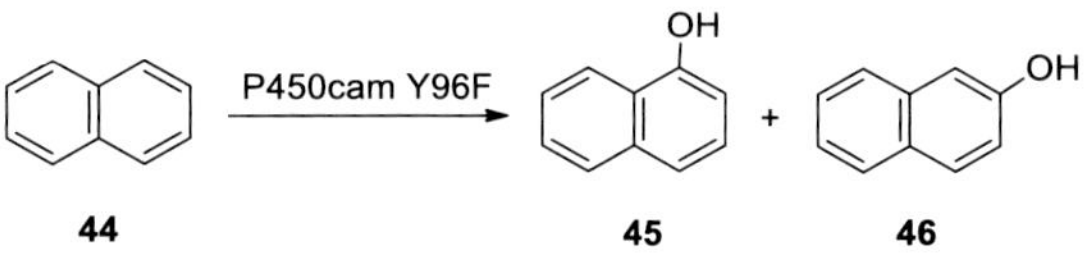

44

45

46

Scheme 20 Enantioselective epoxidation of terminal alkenes using engineered P450 BM-3 enzyme variants

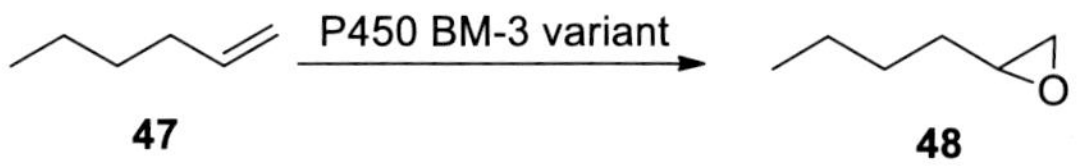

47

48

2 Applications of Enzymes in Synthesis

Enzyme variants "CYP107Z" possessing enhanced regiospecificity were achieved by an amalgamation of random mutagenesis, recombination of numerous natural and synthetic CYP107Z gene fragments along with protein-structure-directed mutagenesis. After the obtained CYP enzyme was utilized for the biocatalytic alteration of avermectin **49** to more potent 4'-oxo-avermectin **50** with regiospecificity (Scheme 21). The specificity at the particular regions was only possible because of its flexibility of the gene reassembly evolution method [64, 65].

In another study, Wong and coworkers have utilized the wild-type P450$_{cam}$ to generate the mixtures of monooxygenated terpenes by oxidation of monoterpenes (+)-R-pinene **51** and (S)-limonene **53**, which are excellent fragrance and flavoring chemicals. Further, the variant Y96F/V247L of monooxygenase cytochrome P450$_{cam}$ has displayed the enhanced activity for the oxidation of these two monoterpenes **51/53**. As a result, verbenol **52** and isopiperitenol **54** (as major products) were obtained with high regio- and nearly total stereoselectivity (Scheme 22) [66].

Scheme 21 Oxidation of Avermectin using CYP107Z variant enzyme

Scheme 22 Synthesis of Verbenol (**52**) and Isopiperitenol (**54**)

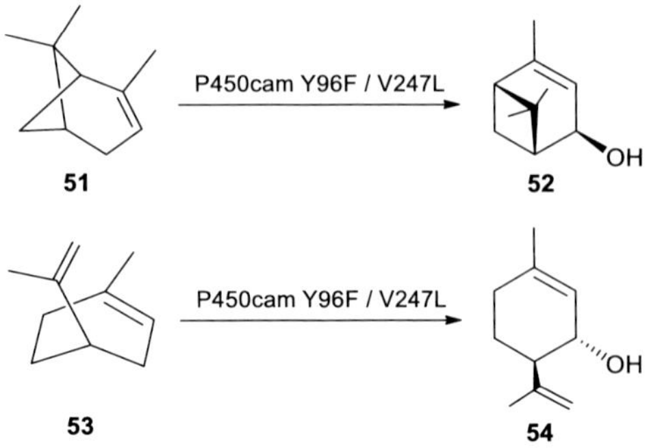

Scheme 23 Asymmetric hydroxylation of aliphatic asparagines and aspartic acid using AsnO variant

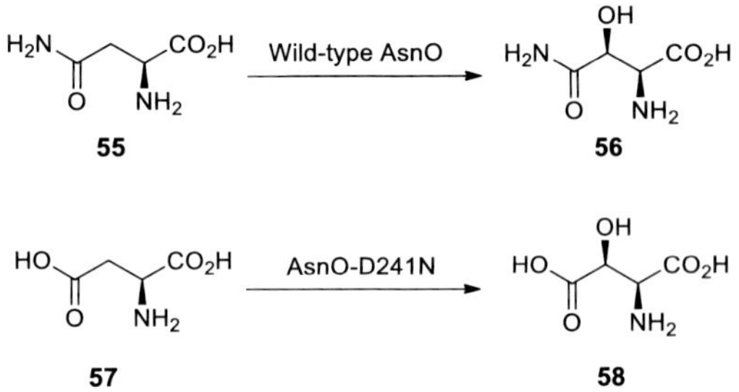

Marahiel et al. have highlighted a highly specific asparagines oxygenase (AsnO) enzyme that catalyzes the direct hydroxylation of aliphatic β-position of the aliphatic asparagines **55** to its L-threo-hydroxyasparagine **56**. Unlike to wild-type enzyme, variant D241N competently transform L-aspartic acid **57** to enatiomerically pure L-threo-hydroxyaspartic acid **58** of high biological importance (Scheme 23) [67].

Dioxygenases a NaD(P)H-dependent enzymes have showed the potential utility for the biosynthesis of secondary metabolites like flavonoids and alkaloids [68]. Wood and coworkers have employed the site-directed mutagenesis to increase such enzyme's activity and developed various kinds of potential variants. Wild-type naphthalene dioxygenase (NDO) obtained from *Ralstonia sp.* has no identifiable activity on 2-amino-2,6-dinitro-toluene **59**, while its variant F350T combined with G407S has afforded 2-amino-4,6-dinitrobenzyl alcohol **60** three times faster than the F350T variant itself. In addition, variant L225 had resulted 4-amino-2-nitrocresol **63** and 4-amino-2-nitrobenzylalcohol **64** from 4-amino-2-nitrotoluene 12 times faster than its wild-type NDO (Scheme 24) [69].

Hydantoinase is the designed biocatalyst for the C-N bond formation. The synthesis of chiral R-amino acids along with D or L configuration can easily be prepared via hydantoinase process. These enzymes have displayed the enthralled utility for the kinetic resolution with 100% conversion and enantioselectivity. Arnold et al. have implemented the engineered hydantoinase from Arthrobacter sp. DSM9771 to achieve the L-methionine **67** from D-5-(2-methylthioethyl) hydantoin (D-MTEH) **65** (Scheme 25) [70]. In all, a very interesting D-selective hydantoinase

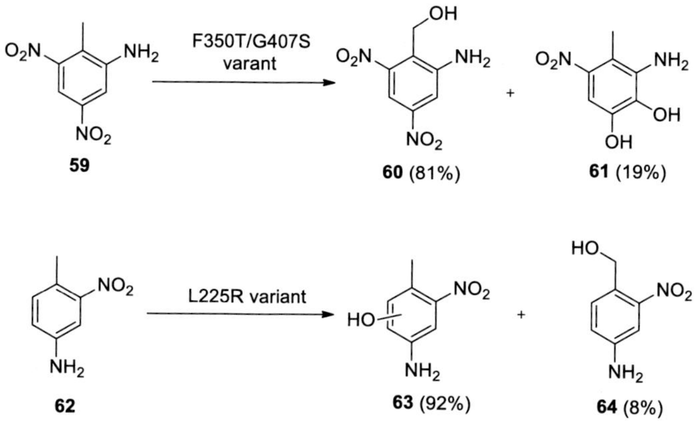

2 Applications of Enzymes in Synthesis

329

Scheme 24 NDO's variant-mediated oxidation of 2-aminodinitro toluene and 4-amino-2-nitro toluene

Scheme 25 Biosynthesis of L-Methionine using Hydantoinase procedure

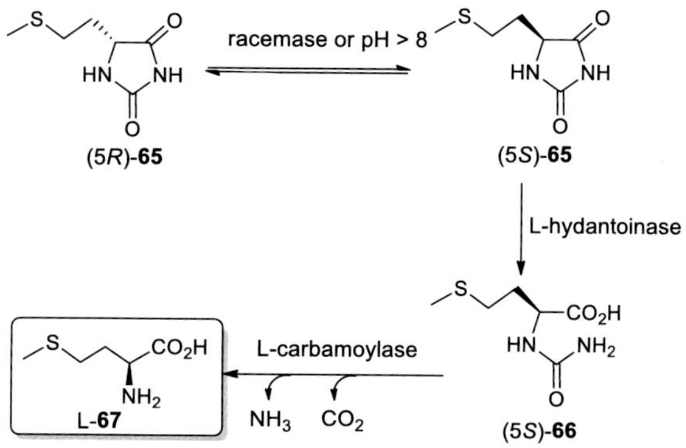

was evolved through the directed evolution method for the efficient generation of chiral scaffolds.

One of the enthralling synthetic alterations is the conversion of the corresponding nitrile functionality to the carboxylic acid by using the nitrilases (biocatalyst). However, the naturally occurring nitrilases are associated with many disadvantages like less activity, narrow enantioselectivity, and low substrate range. The foremost enantioselective nitrilase discovered during the desymmetrization process of prochiral 3-hydroxyglutardinitrile (**68**) to (R)-4-cyano-3-hydroxybutyric acid (**69**) (Scheme 26), a chiral intermediate of Lipitor; the well-known cholesterol-lowering drug [71].

Acylases are the remarkable biocatalysts for the biotechnological consideration, due to their pertinent utility in the pharma industry for the construction of semisynthetic antibiotics. These enzymes catalyze the cleavage of the amide bond between the appended chain of carboxylic acid and the β-lactam nucleus in cephalosporins and penicillins, departing the cyclic β-lactam amide bond intact [72, 73]. This was

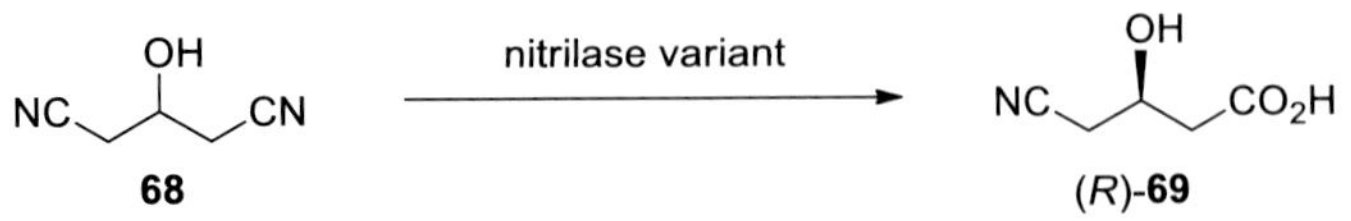

Scheme 26 Synthesis of R-configured 4-cyano-2-hydroxybutanoic acid **69** using nitrilase variant

a huge success in the scalable production of 7-aminocephalosporanic acid (7-ACA; **71**), 6-aminopenicillanic acid (6-APA), and 7-aminodeacetoxycephalosporanic acid (7-ADCA). In this regard, glutaryl-7-aminocephalosporanic acid acylase was engineered to accept the antibiotic cephalosporin C (CephC; **70**) as a natural substrate [74]. Further, engineered variant E423Y/E442Q/D445N of γ-glutamyltranspeptidase obtained from B. subtilis had revealed the 900 times faster hydrolysis of 7-(glutarylamino)cephalosporanic acid (**72**) to afford 7-ACA (**71**) (Scheme 27) [75].

Chiral amines are pharmaceutically very important scaffolds, and therefore, many techniques have been developed to construct these molecules, among which reductive amination is the oldest but the powerful and broadly used synthetic transformation to achieve various kinds of amines [76]. In this advancement, enzyme-catalyzed reductive amination is the most promising technique for the development of optically pure amines. The earlier reported amine dehydrogenases displayed the limited catalytic capacity for the bulky ketones [77, 78]. Therefore, Zheng and coworkers highlighted the substrate scope of a devised amine dehydrogenase GkAmDH obtained from *Geobacillus kaustophilus* for the reductive amination of library of ketones **73**. The authors successfully developed a series of structurally bulky amines **74** with high enantioselectivity (up to >99%), high yield >99%, and TON (up to 18,900) (Scheme 28). The above result revealed the enthralled applicability of the engineered biocatalysts [79].

The carbon–nitrogen bond formation is one of the focused themes in the advance organic chemistry. Thus, Zhen and coworkers have underlined the dual function biocatalyst P411 enzyme **L7_FL** to catalyze carbine N–H insertion to form the α-amino lactones **77** with appreciable enantioselectivity (98%) and yield (>99%)

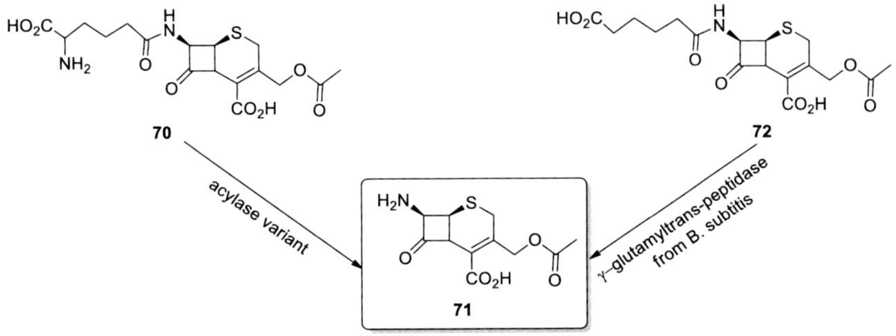

Scheme 27 Enzymatic cleavage of CephC (**70**) to 7-ACA (**71**)

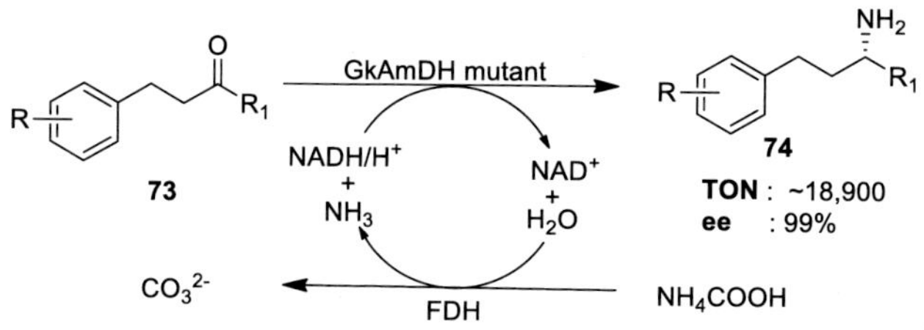

2 Applications of Enzymes in Synthesis 331

Scheme 28 Asymmetric reductive amination of ketones using engineered dehydrogenase GkAmDH from *Geobacillus kaustophilus*

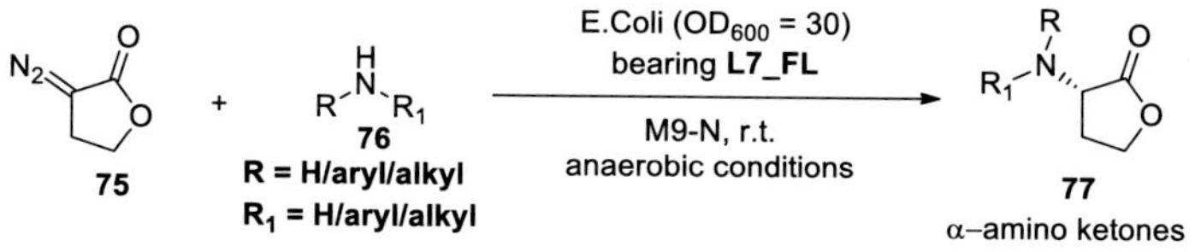

Scheme 29 Synthesis of α-amino lactones by means of enantioselective carbene N–H insertion

(Scheme 29). Moreover, a highly efficient system was implemented to prepare bioactive chiral amines for synthesis as well as drug discovery [80].

Novozyme®-435 lipase-mediated selective acetylation and deacetylation reaction has been widely explored for the chemo-enzymatic convergent synthesis of locked nucleic acid and C-4′-spirooxetanoribonucleo-sides. The antisense oligonucleotides (ASOs) have demonstrated tremendous potential in modulating the expression of target gene/gene product [81]. The ASOs are single-stranded oligonucleotides, ~12–22 nucleotides long, whose sequences are fully complementary to the target RNAs. The chemical modifications are required to improve the "drug-like" properties of ASOs such as binding affinity toward the target RNA and metabolic stability against nucleases [81, 82].

Locked nucleic acid (LNA) is a bicyclic modification in which the ribose sugar has a methylene bridge between 2'-*O* and 4'-*C* of ribose [1]. Incorporation of LNA modification pre-organizes the ASOs leading to energetically favorable duplexes and enhances their binding affinity toward the RNA/DNA targets. Further, the LNA modification provides high nuclease stability to ASOs. However, the synthesis of LNA monomers is challenging, low yielding and requires extensive protection and deprotection of the sugar functionalities.

Recently, Sharma et al. [83] devised a chemo-enzymatic approach to selectively manipulate the diastereotopic hydroxymethyl functions in 3-*O*-benzyl-4-*C*-hydroxymethyl-1,2-*O*-isopropylidene-α-D-ribofuranose **78** (Scheme 30). The greener methodology was successfully utilized for the convergent synthesis of LNA monomers. Further, the lipase Novozyme®-435 can be used for ten cycles of selective acylation reaction without any loss of selectivity and efficiency.

B	Overall Yields
uracil-1-yl	72.4%
thymin-1-yl	63.9%
adenin-9-yl	54.1%
cytosin-1-yl	55.9%

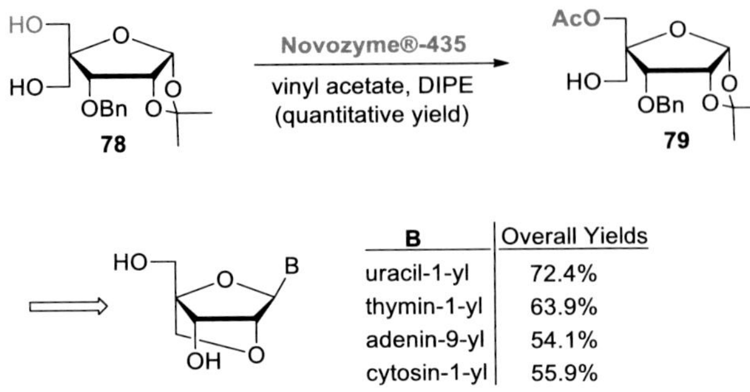

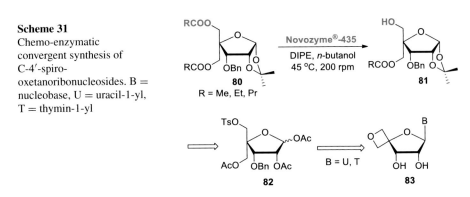

Scheme 30 Chemo-enzymatic convergent synthesis of LNA monomers. B = nucleobase

Scheme 31 Chemo-enzymatic convergent synthesis of C-4′-spiro-oxetanoribonucleosides. B = nucleobase, U = uracil-1-yl, T = thymin-1-yl

Interestingly, the same lipase, Novozyme®-435 when was tested on the per-acetylated sugar derivative, 5-*O*-acyl-4-*C*-acyloxymethyl-3-*O*-benzyl-1,2-*O*-isopropylidene-α-D-ribofuranose **80**, diastereoselective deacylation was achieved in appreciable yield [84]. The greener selective biocatalytic deacylation methodology was used to synthesize novel C-4′-spiro-oxetanoribonucleosides **83** (Scheme 31).

Molnupiravir is a nucleoside-approved drug whose chemical structure mimics naturally occurring cytidine, and used for the treatment of mild-to-moderate COVID-19 in adults by the U.K.'s Medicines and Healthcare Products Regulatory Agency. Considering the importance of this drug in this pandemic situation, McIntosh and coworkers of "Merck group" have highlighted the importance of mutagenesis in constructing a scalable, biocatalytic approach to develop molnupiravir, an orally bioavailable antiviral agent only in four steps depicted in Scheme 32 [85].

Several high-yielding synthetic methods for the construction of organic azides were reported since long back; unfortunately, their practical utility in industry and academia was limited. This was essentially due to its hazardous and low molecular weight; however, sugar azides are very stable in variety of reaction ambient. Since last two decades, the demands of azide functionalized sugars have been increased after the discovery of "Click Chemistry" in 2002 [7, 8]. Considering the importance of azides particularly in glycoscience, an enzymatic mode of selectivity for the introduction

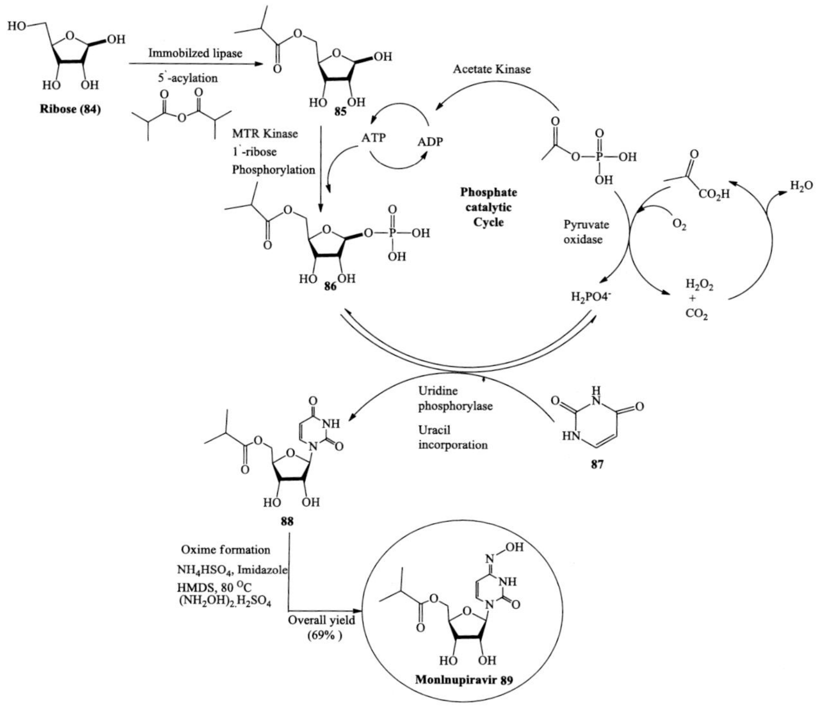

2 Applications of Enzymes in Synthesis

Scheme 32 Complete biocatalytic route for the synthesis of molnupiravir (developed by the Merck group)

of azido group to sugar unit was adopted by Kotik and coworkers [86]. For this, the authors utilized wild-type rutinosidases for the construction of rutinosyl β-azide (**91a**) from compound (**90**). However, rutinosyl α-azide (**91b**) was achieved by means of E 319A mutant (Scheme 33). [86].

Enzymatic routes significantly offer the preparative-scale synthesis of glycosyl azides with great ease by simply switching the mode of catalytic mutation of preserved glycosides although this protocol has not been employed for scale-up synthesis. Further in a recent literature, Gorantla et al. demonstrated the scale-up synthesis of 1-azido-β-D-glucose (sugar azide) from p-nitrophenyl-β-D-glucopyranoside (pNPGlc) by the action of thermoanaerobacterium xylanolyticus GH116 β-glucosidase in aqueous media [87]. Thus, 1-azido-β-D-glucose extensively synthesized by treatment of NaN_3 with pNPGlc in aqueous MES buffer (pH 5.5) at 55 °C and catalytic acid/base mutant of TxGH116D593A in a very good yield through transglycosylation. Finally, CuAAC protocol has been significantly furnished the desired triazolyl glycoconjugates by utilizing this developed glycosyl azide with diverse alkynes in a good yield [87].

The regioselective and stereospecificity enduring oligosaccharides have focal functionality in various biological systems involving, cell recognition, immune

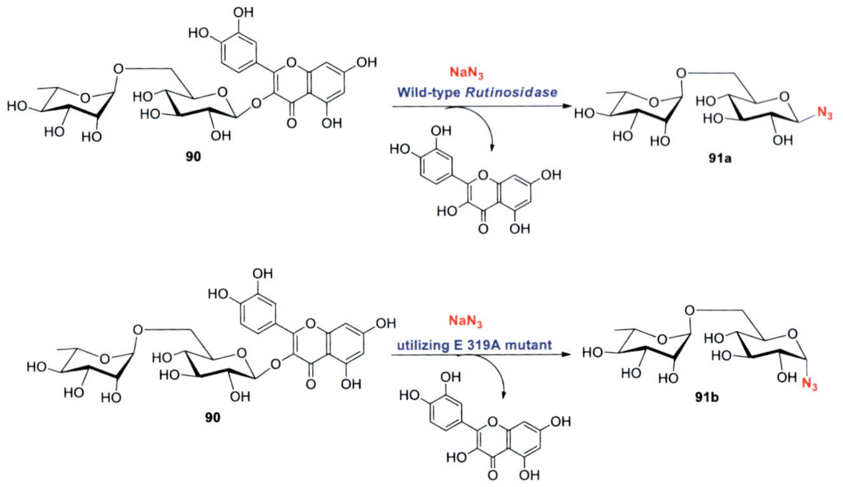

Scheme 33 Selective synthesis of α- and β-Rutinosylated Flavonoid from Rutin (**90**) by means of rutinosidase

response, etc. [88]. Thus, there is an increased demand for the development of carbohydrate-based molecules of chemotherapeutic potential. In recent years, therefore, the search of reliable and highly selective glycosylation tool is gaining the urge of research and development [89]. In glycosylation technology, the enzymatic glycosylation is an innovative dependable tool to build the desirable and targeted regio- and/or stereocontrolled glycosylating building blocks depending on the recognition of enzymes [20, 90–93].

In recent years, an efficient enzymatic chain-elongation process has been introduced to easily reach out for biologically relevant non-natural glycans like 6-deoxygenated α(1 → 4)-oligoglucans. Thereafter, 6-deoxygenated α(1 → 4)-oligoglucans were synthesized using 6-deoxy-α-D-glucose 1-phosphate **92** as a non-native substrate from maltotriose primer via dependable thermostable glucan phosphorylase (from *Aquifex aeolicus* VF5)-mediated chain-elongation process. A tetrasaccharide with 6-deoxy-α-D-glucose residue at non-reducing end was the major glycan product at the first step. The obtained glycan on further execution of enzymatic chain elongation (consecutive glycosylations) in the presence of a desired primer, like 6-deoxy-α-D-glucose 1-phosphate, α-D-glucose 1-phosphate and glucan phosphorylase (GP, EC 2.4.1.1) ended up achieving the varying degrees of polymerized oligosaccharides, e.g., 6-deoxygenated α(1 → 4)-oligoglucans **98** (both homo-type and hetero-type) (Scheme 34). [94].

In order to improve the desired selectivity (regio-, chemo-, and stereoselectivity) in products under environmentally benign condition, a number of enzymatic reactions are well-explored in various ionic liquids as green media or catalyst or both. Also, oxido-reductases and hydrolases preserve their activities in the suspensions of suitable ILs [95]. RTILs are mainly due to non-toxic, biocompatibility, environmental benign polar media receiving great attention in chemistry, biology, biomaterial, and

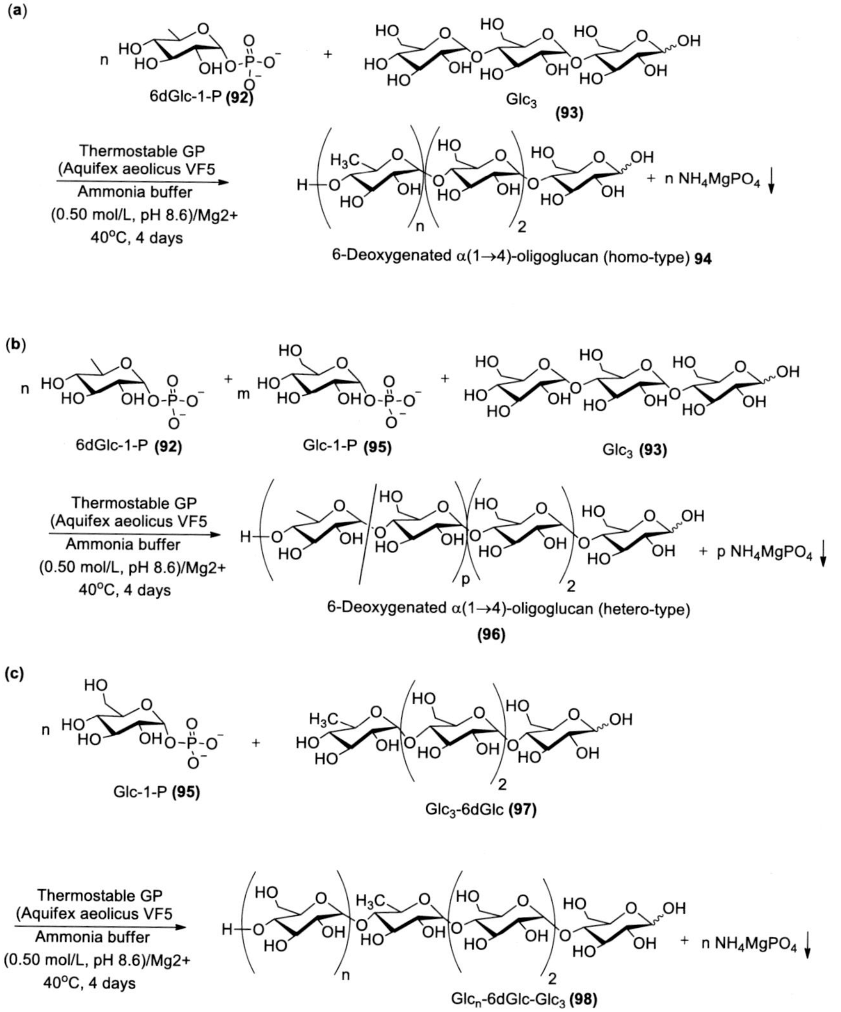

2 Applications of Enzymes in Synthesis

Scheme 34 Thermostable GP-mediated enzymatic chain elongation (consecutive glycosylation) from Glc$_3$ primer using **a** 6dGlc-1-P and **b** 6dGlc-1-P and Glc-1-P, and **c** from 6dGlc-Glc$_3$ using Glc-1-P

biomedical sciences [96, 97]. They are known to be compatibility with functional group in carbohydrate analogues, and thus their role glycoscience particularly for dissolution, modulation, glycosylation, functionalization, and modification of carbohydrates is ever-increasing [98–100]. Lipase-mediated enantio- and regioselective acylation of β-D-Glucose in dialkyl imidazolium IL [101], peracetylation of carbohydrates in IL [102], acylation of monoprotected glycosides in [bmim][PF$_6$] [103], and CAL-B-mediated acylation of monosaccharide in [bmim][BF$_4$] [104] are the notable examples. In addition, role of IL in a β-Galactosidase-induced enzymatic transgalactosylation reaction [105] and enzymatic glycosylation of resveratrol is well-known [106]. In addition to enzyme catalysis in ILs, the dissolution of unmodified cellulose in [bmIm][Cl] under *MW* condition [107] and IL-mediated unfolding and inactivation of cellulase from *T. reesei* [108] and glucose oxidase–peroxidase-mediated sulfoxidation are also investigated [109]. Impact of RTILs is already discussed in Chap. 4, and thus, this part is not covered in this topic.

Another very important aspect of enzymes and their role in organic synthesis may include the **Chemo-enzymatic synthesis of diverse Sialoconjugates,** biologically important scaffolds. One of the most important reactions in organic chemistry is certainly the C–C bond-forming methodology. Stereoselective version of C–C bond-forming methodology is widely explored. In the meantime, the enzyme-induced asymmetric variant of C–C bond-forming protocol is greatly explored in organic chemistry [110]. For a very first fundamental C–C bond formation in simple substrate include the BFD variants L476Q and M365L/L461S-induced chiral synthesis of (*S*)-2-hydroxy-1-(2-substituted)phenyl propan-1-one **101** from aldehydes **99** [111]. Further in continuation of BFD-mediated enantioselective C–C bond-forming reaction, a number of important such reactions have been explored in laboratory and industry by using Aldolase-mediated protocol. Next to the transition-metal catalysis as well organocatalysis, Aldolases are even more competent to widen the asymmetric synthesis tools and considered as the most motivating enzymes in organic synthesis particularly for the development of complex carbohydrate-containing biologically relevant molecules. Representative examples include the KDPG Aldolase-mediated C–C bond-forming methodology, where KDFG variants may accept non-phosphorylated D- and L-glyceraldehyde (Scheme 35) [112]. Interestingly, RhaD aldolase mutant can accept a non-phosphorylated donor [113, 114]. Furthermore, D-fructose-6-phosphate aldolase (FSA)-mediated synthesis of D-iminocyclitols is well-explored by using hydroxyacetone, and 1-hydroxy-2-butanone as suitable donor substrates [115]. A beautiful application includes the 2-deoxyribose-5-phosphate aldolase-mediated synthesis of statin. At the end, even more important use include the NAL Aldolase-mediated synthesis of sialic acid and their analogues starting from D-man/ManNAc and their derivatives reacting with pyruvic acid [116, 117]. Further, NAL variant may accept L-arabinose instead of *N*-acetylmannosamine [118]. Interestingly, the diastereoselectivity of an enzyme was changed by using Tagatose aldolase which can successfully transformed into a fructose aldolase [119]. Thus, Aldolase-mediated C–C bond-forming reaction is greatly exemplified in glycoscience including sialic acid chemistry [21, 120, 121].

In nature, all cells are covered with a multifaceted array of carbohydrates, namely sialic acids (commonly known as "Sia"), a negatively charged nine-carbon α-keto aldonic acids [122]. They are occupied at the terminal end and play a significant role on cell surfaces notably for the mediation and modulation of many important cellular interactions. The structural modification of sugar in nature generally resulted in their notable non-negligible biological significance [123]. For example, sulfated carbohydrates present in proteoglycans, glycolipids, and glycoprotein on the mammalian cell surface pivotal role in some specific molecular recognition events. In addition, some

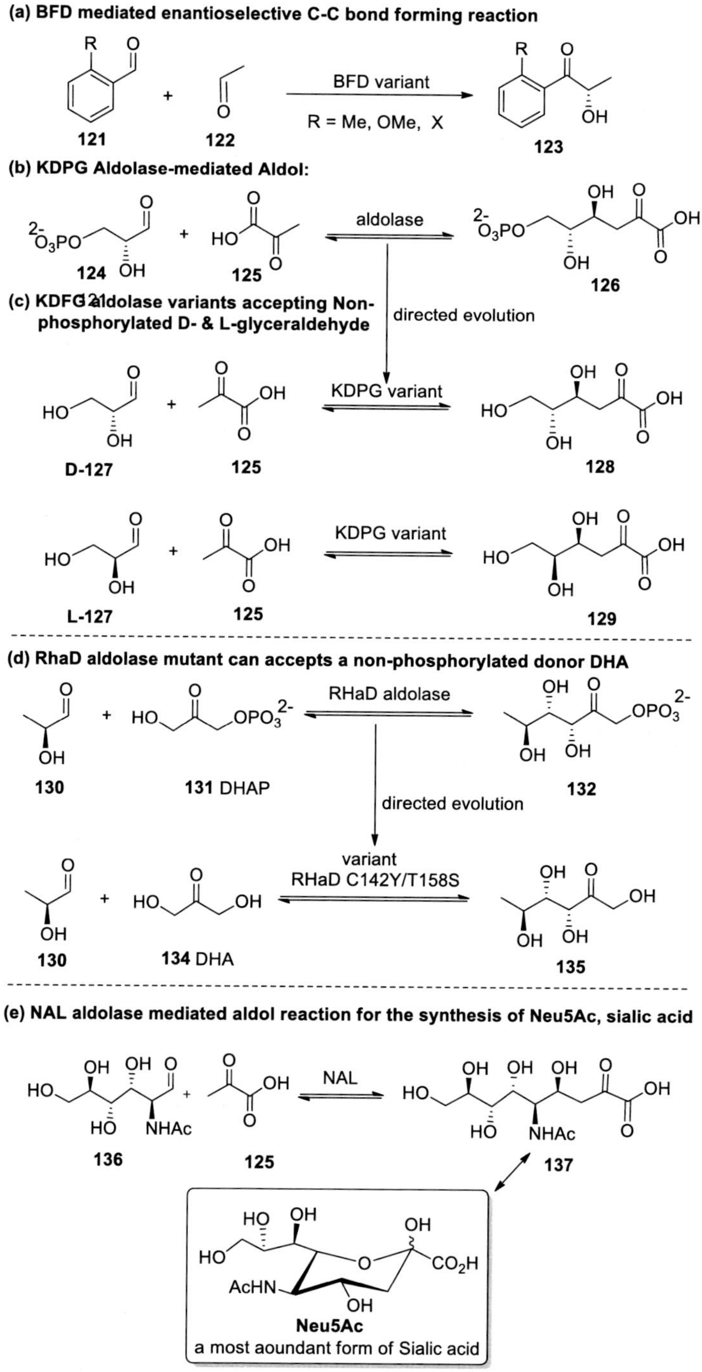

2 Applications of Enzymes in Synthesis

(a) BFD mediated enantioselective C-C bond forming reaction

(b) KDPG Aldolase-mediated Aldol:

(c) KDFG aldolase variants accepting Non-phosphorylated D- & L-glyceraldehyde | directed evolution

(d) RhaD aldolase mutant can accepts a non-phosphorylated donor DHA

(e) NAL aldolase mediated aldol reaction for the synthesis of Neu5Ac, sialic acid

Scheme 35 Enzyme-mediated some representative carbon–carbon bond-forming reactions simple modification at specific position of the sialic acids including acetylation, methylation, sulfation, phosphorylation, lactylation, and other common modification resulted in the diverse structurally modified analogues serving an imperative biochemical functions [124–127]. Despite devoted investigations in this emerging areas, only little is known about the SAR of such modified sialic acid analogues. In addition to other reason, the major one includes the technical difficulties in order to achieve homogeneous libraries of such structurally modified sialoconjugates.

Enzymatic synthesis is beyond doubt considered to be highly efficient and stereo- and regio specific. However, finding the right enzymes is not always easy and often considered as the key point for the successful enzymatic approach. Some important key points required to get the promising enzyme includes an easy expression of active and soluble enzyme in high yield (Obtainable) and also can tolerate substrate modification (Flexible) [128]. Due to the complex nature of sialic acid-containing glycans, their practical synthesis is inherently difficult. Their natural path of synthesis greatly encourages the investigators to see the sights of facile chemoenzymatic route needed for an easy and expeditious synthesis of sialic acids and their analogues (Scheme 36) [128].

Chen and coworkers explored a capillary electrophoresis assay to substrates specificity studies and direct characterization of the activities of NanAs in either direction like cleavage of Neu5Ac and synthesis of Neu5Ac. The investigation displayed that 5-O-methyl-ManNAc (**120**) can be employed efficiently as a suitable substrate by PmNanA for the high-yielding successful synthesis of 8-O-methyl Neu5Ac (**121**) (Scheme 37) [129]. In a similar way, ManNGc5Me (**122**) on PmNanA-catalyzed aldol addition reaction with sodium pyruvate (5 equiv.) in Tris–HCl buffer (100 mM, pH 7.5) at 37 °C for 24 h afforded high yield of the desired Neu5Gc8Me (**123**) (Scheme 37) [130]. The similar enzymatic synthesis by utilizing EcNanA was not

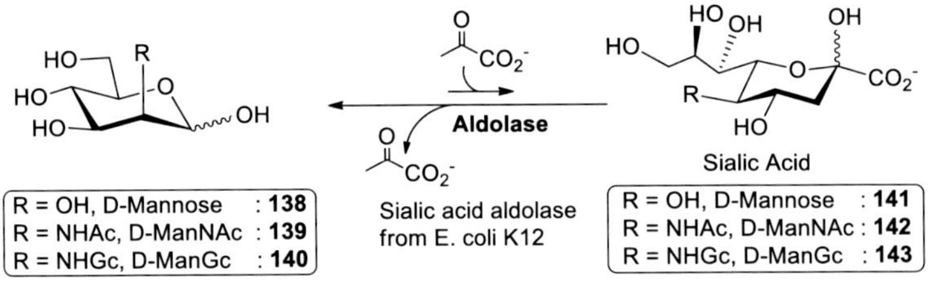

Scheme 36 Aldolase-mediated facile synthesis of sialic acid monosaccharides KDN (**117**), Neu5Ac (**118**) and Neu5Gc (**119**) from respective six-carbon monosaccharides (**114–116**)

Scheme 37 Aldolase-mediated high-yielding synthesis of 8-MeNeu5Ac (**121**) and 8-MeNeu5Gc (**123**)

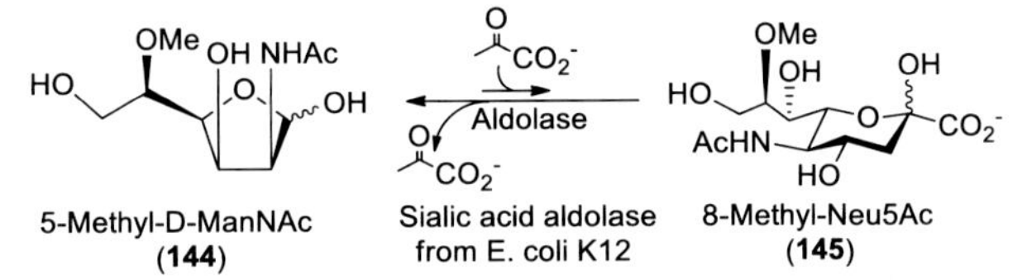

2 Applications of Enzymes in Synthesis

so efficient. Fascinatingly, the higher expression level and a broader substrate tolerance make PmNanA to be a superior biocatalyst over EcNanA for the chemoenzymatic synthesis of sialic acids. The developed sialic acid analogues could be purified through the combined anion exchange chromatography and gel filtration column.

Like the other bacterial lyases, it brings together into a homotetramer with each monomer folding into a classic $(\beta/\alpha)_8$ TIM barrel-shaped fold. Two wild-type structures, first in the absence of substrates and next trapped in a Schiff base intermediate between Lys164 and pyruvate, were determined. Three structures of the K164A variant, one in the absence of substrates and two binary complexes with Neu5Ac and Neu5Gc, were determined. Both sialic acids are known to bind to the active site of the open-chain form of ketone of the monosaccharide. The structures confirm that every hydroxyl group of the linear sugars put together hydrogen bonding interactions with an enzyme and the residues that determine the required specificity [131].

The C2- and/or C6-substituted Man, ManNAc, ManNGc analogues could be explored as suitable precursors for aldolase-mediated reactions which were successfully delivered in high-to-excellent yields of the respective sialic acid-containing molecular structures with C5 and/or C9-modification [132–134].

However, Neu5Gc-containing glycans have been found on human cancer cells, and in a range of human tissues, human cells are unable to synthesize Neu5Gc presumably due to its dietary incorporation. Over twenty different Neu5Gc analogues have been found in non-human vertebrates. To evaluate their biological roles, a number of reports available to focus on the synthesis of Neu5Gc and their derivatives [135]. Interestingly, a disaccharide, namely Galβ-1,2-ManGc (**124**) which has ManNGc at the reducing end on enzymatic reaction with pyruvic acid in the presence of EcNanA, gave desired Galβ1–5Neu5Gc (**125**) in 34% yield (Scheme 38) [136].

Similar EcNanA-catalyzed aldol reaction of six carbon sugars, e.g., ManNGc with 3-fluoropyruvate afforded 3F-Neu5GC analogues as mixture of 3F(axial)Neu5Gc and 3F (equatorial) Neu5Gc in ~ 1:1 ratio [137].

The resulted neuraminic acid analogues can be further explored to get an easy access of diverse and well-defined sialosides including α-2,3-linked sialosides (**131**) or α-2,6-linked sialosides (**132**) [136, 138–141]. This protocol is now emerged as a common method to evaluate the substrate or ligand specificity of sialic acid recognizing proteins. This can be easily estimated mainly by three ways. The first common method is accomplished by simple change in the structural core that can be easily

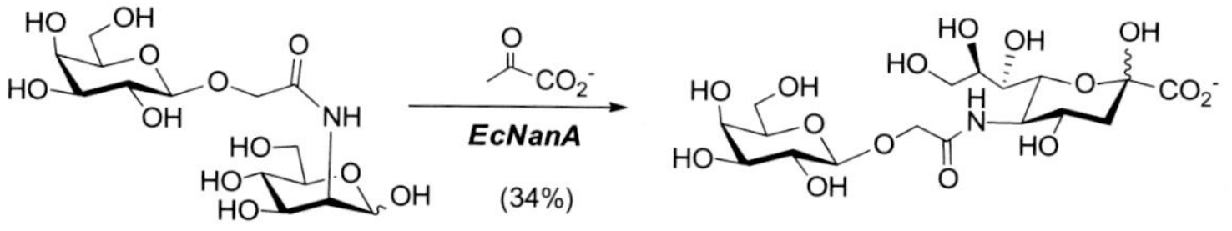

Galβ1-2ManNGc (**124**)   Galβ1-5Neu5Gc (**125**)

Scheme 38 EcNanA-mediated enzymatic synthesis of Galβ1–5Neu5Gc from Galβ-1,2-ManGc having ManNGc at the reducing end attained by using various donors, like Mannose or ManNAc or their derivatives (O-methyl, deoxy, azido, acetate, etc.) either at 2- or 4- or 5- or 6-position individually or as multiple substitution at these positions. The next even more valuable consideration could be the type of glycosidic linkage present (either α-2,3 or α-2,6 linkage of sialic acid with penultimate sugar) [138]. This one is widely explored as valuable system to evaluate the substrate/ ligand specificity of the sialic acid recognizing proteins. At the end, the third another feature is the structure of the penultimate sugar residue present (e.g., Gal or GalNAc or Lactose or their analogues). A libraries of homogenous α-2,3-linked sialosides (131) or α-2,6 linked sialosides (132) are outmost required for the successful investigation of the biological significance of complex sialic acid-containing glycans and their widespread applications in chemical biology.

Chemo-enzymatic synthesis is a powerful and reliable approach to get the complex sialic acid-containing molecules in high-to-excellent yields. Interestingly, Chen's chemo-enzymatic multi-enzyme method is widely explored for an easy access of biologically relevant sialic acid-containing glycans with required α-2,3 or α-2,6 linkages between sialic acids and penultimate sugars [136–141]. The method has several notable features including the regioselectivity, stereoselectivity (α-selectivity) in products, environmental benign condition (enzymes are non-toxic and biodegradable), one-pot condition (all the three enzymatic steps can be carried out in one-pot tandem fashion without the isolation of intermediates involved), and quite useful for the synthesis of diverse natural and unnatural "Sia" analogues with varied linkages. The OPME route begins with the sialic acid aldolases/N-acetylneuraminate lyases (NanAs)-mediated reversible aldol cleavage of Neu5Ac or their analogues to form pyruvate and ManNAc (Step 1). Then, a consequent activation of Sia donors (128) using a CMP-sialic acid synthetase (Step 2). At the end (in Step 3), transfer to suitable acceptors 130 (frequently, a Gal or Lac or their derivatives) to sia donors 129 by a respective sialyltransferase (ST), for example, tPm0188Ph (N-meningitidis-α-2,3ST) to afford high-to-excellent yield of desired α-2,3-linked sialosides (131) (Scheme 39a). Likewise, the transfer to acceptors 130 to sialic acid donors 129 by Photobacterium damsela-α-2,6- sialyltransferase afforded high-to-excellent yield of desired α-2,6-linked sialosides (132) (Scheme 39b).

Likewise, 5-azido modified hexoses on optimized OPME conditions afforded high yield of respective azidosialosides, such as 8-azido-α-2,3-linked sialic acids (using α-2,3-ST in third step) or 8-azido-α-2,6-linked sialyltrisaccharides (using α-2,6-ST in third step), an interesting azido analogues useful for the CuAAC click diversification. This OPME route now became an well-established tool in glycoscience and widely explored with different substitution patterns in both donor–acceptor systems for an easy access of diverse α-2,3-linked sialosides and α-2,6-linked sialosides of biological importance [130]. Interestingly, the OPME synthesis of α-2,3- and 2,6-linked sialosides work equally well for 8-deoxy KDN and Neu5Ac8Me [130]. Also, the optimized OPME approach is used to produce high yields of natural 8-O-methylated sialoconjugates (e.g., 8-O-Me-Neu5Ac and 8-O-Me-Neu5Gc) and Kdn8Me, 8-deoxy-Kdn and α-2,3- and α-2,6-linked sialyltrisaccharides containing Neu5Ac8Me and Kdn8Deoxy (136–139) starting from corresponding 5-O-modified

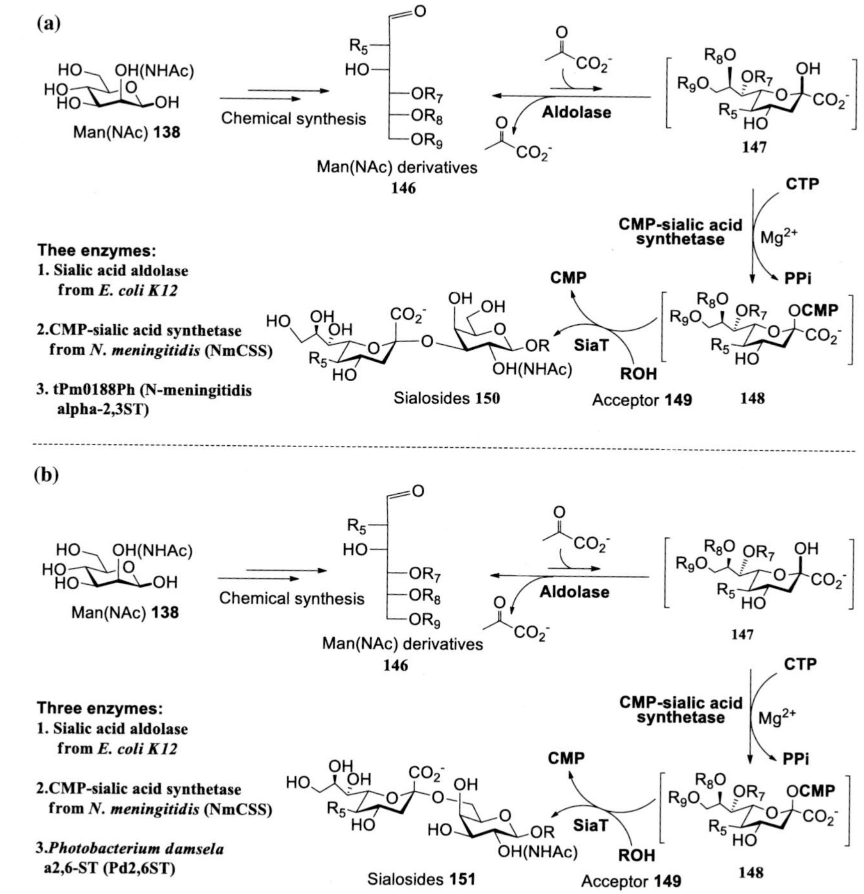

2 Applications of Enzymes in Synthesis

(a)

(b)

Scheme 39 Chen's regio- and stereoselective one-pot multi-enzymatic (OPME) synthesis of α-2,3-linked Sialosides (**131**) and α-2,6-linked Sialosides (**132**, α-selectivity in sialosides)

six-carbon monosaccharides (**134** and **135**) (Scheme 40) [130]. The strategy offers an expeditious tool to generate high yields of biologically relevant C8-modified sialic acid glycans.

In another very important investigation, Chen and coworkers reported an expeditious, straightforward, short and high-yielding two-step multi-enzyme approach for an easy access of GD3 ganglioside oligosaccharides and other disialyl glycans containing a terminal Siaα-2-8Sia component (Scheme 41) [142]. This OPME protocol works well high-yielding synthesis of various natural, semi-synthetic and synthetic sialic acid and their 2,8-linked analogues which were found as valuable molecules to understand the biological importance of sialic acid variations on disialyl structures in nature. In the investigation, it was invented that the α-2,8-sialyltransferase activity of a recombinant multifunctional CstIIΔ32I53S has notable

Scheme 40 OPME route for the synthesis of α-2,3- and 2,6-linked Neu5Ac8Me and 8-deoxy-Kdn sialosides

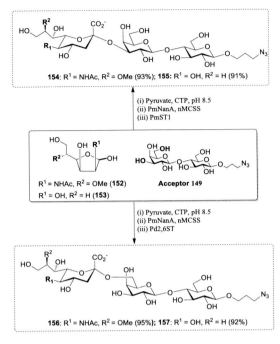

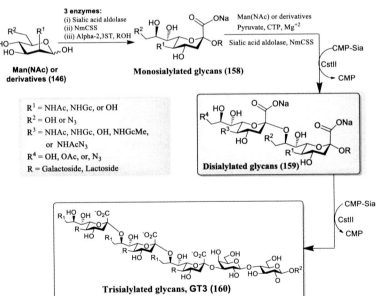

Scheme 41 Chemo-enzymatic synthesis of GD3 and GT3 Oligosaccharides (**142** and **143**) of potential application in the study of carbohydrate–protein interaction

2 Applications of Enzymes in Synthesis

promiscuous donor substrate specificity and also tolerate a number of suitable substitutions at C-5 or C-9 position of Sia in the CMP-sialic acid. On the other hand, the α-2,8-sialyltransferase activity of CstIIΔ32I53S is comparatively restricted with acceptor substrate specificity in glycosylation. While α-2,3- and also α-2,6-linked monosialic acids were the potential acceptors for CstIIΔ32I53S, the terminal sia residues in the adequate monosialyl oligosaccharide acceptors are inadequate to KDN, Neu5Ac, Neu5Gc, and some of their C-9-modified analogues.

The truncated CstII mutant CstIIΔ32I53S, cloned from a synthetic gene has displayed α-2,8-sialidase activity that catalyzes the specific cleavage of the α-2,8-sialyl linkage of GD3-type oligosaccharides. It also showed α-2,8-trans-sialidase activity with the purpose to catalyze the transfer of a sialic acid from a GD3 oligosaccharide to a GM3 oligosaccharide [142]. Thus, α-2,3- or α-2,6-linked monosialylated oligosaccharides obtained using OPME approach were then explored as suitable acceptor for the α-2,8-sialyltransferase activity of a recombinant truncated multifunctional *Campylobacter jejuni* sialyltransferase CstII mutant to develop high-to-excellent yield of desired GD3-type disialyl oligosaccharides **142** (e.g., Neu5Gc-α-2,8Neu5Gc-α-2,3LacProN$_3$ and KDN-α-2,8-KDN-α-2,3-LacProN$_3$) (Scheme 41) [142]. The protocol is again extremely attractive and widely useful in chemical biology particularly for the study of carbohydrate–protein interaction. These analogues may be useful as a precious probe and help to understand the biological importance of variable sialic acid residues on disialyl oligosaccharide structures present in nature.

Despite recognition of a number of glycan-binding proteins by few particular sialic acid or their analogues, the complete information about such molecular recognition is still not fully understood. There is an always a requirement of a thorough investigation about the sialic acid recognition process. Toward this end, Chen and coworkers exemplified 16 modified sialic acids on different glycan backbones and reported a novel sialylated glycan microarray (SGM) tool. Glycans having β-linked galactose at the non-reducing terminal end and alkyl amine having fluorophore at the reducing end (**145**) were subjected to sialylated using standard OPME protocol and afforded total of 77 α-2,3- and α-2,6-linked sialyl glycans **147** (Scheme 42). Interestingly, the glycans terminating either with Neu5Ac or Neu5Gc residue were originate to interact with Influenza A virus and three human para-influenza viruses. The investigation confirms the wide utility of this sialylated glycan microarray which is nowadays an established and important tool to examine the biological significance of the modified sialic acids in protein-carbohydrate interactions [143].

Recently, the authors extended the chemistry and exemplified diazido and triazido-mannose analogous as effective chemoenzymatic synthons for OPME sialylation to give high-to-excellent yield of azido appended sialosides. Azido groups were then converted to respective *N*-acetyl groups, and the resulted *N*-acetyl sialosides have been well-utilized as stable probes for sialic acid-binding proteins including plant lectin MAL II [144].

In addition, proteins having receptor to bind Sia are predicted to be implicated in a number of important biological processes and, therefore, the study about carbohydrate–protein interaction and their supplementary role in vertebrates may affect

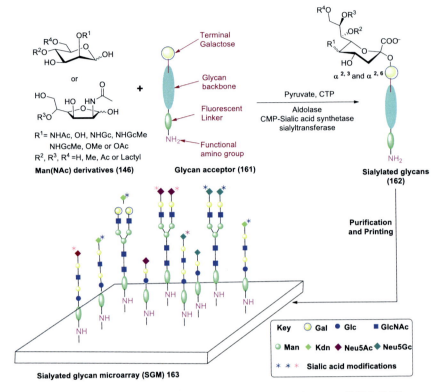

Scheme 42 The design and development of a sialylated glycan microarray (SGM) [143]

in the increased understanding of physiological and pathological variance and at the end for the evaluation of human evolution. Beyond doubt, sialic acids have a broad pharmacological significance in human physiology and are one of the widely explored scaffolds for over three decades.

3 Conclusion and Future Perspectives

Through the above-described examples, this is easy to conclude that the "enzyme-induced synthetic methods" have a great impact in "green chemistry" as the sustainable science. This "biocatalysis" tool is one of the exceptionally significant disciplines that facilitate an efficient, selective (regioselective, chemoselective, and/or stereoselective), and cost-effective high innovative tool in organic synthesis for an easy access of a wide range of fine chemicals, scaffolds, and biologically potent molecules. Just after the invention of recombinant DNA technology in the late 1980s, the biocatalysis route changed way of catalysis significantly in terms of green concept which nowadays has displayed wide utilities in synthesis.

Even though finding suitable enzymes is rather challenging, beyond doubt the enzyme-mediated synthetic tool is rapidly expanding and a number of such powerful enzyme engineering strategies provide ample opportunities in identifying a valuable biocatalyst to address their specific role in chemoenzymatic synthesis. We believe that this chapter dealing with enzyme-induced protocols with their performance toward upgrading the selectivity and efficiency in some selected enzyme systems including lyases (aldolase), sialyltransferase, oxido-reductases, hydrolases, isomerases, etc., certainly provide an ample opportunity in organic synthesis for an easy access of biologically relevant molecules.

References

1. Eigen, M., Hammes, G.G.: Elementary steps in enzyme reactions (as Studied by Relaxation Spectrometry). Adv. Enzymol. Rel. Areas Mol. Biol. **25**, 1–38 (1963)
2. Strohmeier, G.A., Pichler, H., May, O., Gruber-Khadjawi, M.: Application of designed enzymes in organic synthesis. Chem. Rev. **111**, 4141–4164 (2011)
3. Liese, A., Seelbach, K., Wandrey, C.: Industrial Biotransformations. Wiley-VCH, Weinheim, Germany (2000)
4. Walsh, P.J., Kozloski, M.C.: Fundamental of Asymmetric Catalysis. University Science Book, Sausalito, California (2008)
5. Jacobsen, E.N., Pfaltz, A., Yamamoto, H.: Comprehensive Asymmetric Catalysis. Springer Verlag, Newyork (1999)
6. Rostovtsev, V.V., Green, L.G., Fokin, V.V., Sharpless, K.B.: A stepwise huisgen cycloaddition process: Copper(I)-Catalyzed regioselective ligation of azides and terminal alkynes. Angew. Chem. Int. Ed. **41**, 2596–2599 (2002)
7. Tiwari, V.K., Mishra, B.B., Mishra, K.B., Mishra, N., Singh, A.S., Chen, X.: Cu(I)-Catalyzed click reaction in carbohydrate chemistry. Chem. Rev. **116**, 3086–3240 (2016)
8. Agrahari, A.K., Bose, P., Jaiswal, M.K., Rajkhova, S., Singh, A.S., Hotha, S., Mishra, N., Tiwari, V.K.: Cu(I)-Catalyzed click chemistry in glycoscience and their diverse applications. Chem. Rev. **12**, 7638–7956 (2021)
9. Chinchilla, R., Najera, C.: The Sonogashira reaction: a booming methodology in synthetic organic chemistry. Chem. Rev. **107**, 874–922 (2007)
10. Corbet, J.P., Mignani, G.: Selected patented cross-coupling reactions technologies. Chem. Rev. **106**, 2651–2710 (2006)
11. Hassan, J., Sevignon, M., Gozzi, C., Schulz, E., Lemaire, M.: Aryl−Aryl bond formation one century after the discovery of the ullmann reaction. Chem. Rev. **102**, 1359–1470 (2002)
12. Climent, M.J., Corma, A., Iborra, S.: Heterogeneous catalysts for the one-pot synthesis of chemicals and fine chemicals. Chem. Rev. **111**, 1072–1133 (2011)
13. Hong, L., Sun, W., Yang, D., Li, G., Wang, R.: Additive effects on asymmetric catalysis. Chem. Rev. **116**, 4006–4123 (2016)
14. Gawande, M.B., Goswami, A., Felpin, F.-X., Asefa, T., Huang, X., Silva, R., Zou, X., Zboril, R., Varma, R.S.: Cu and Cu-based nanoparticles: synthesis and applications in catalysis. Chem. Rev. **116**, 3722–3811 (2016)
15. Singh, G.S.: Greener approaches to selected asymmetric addition reactions relevant to drug development. Curr. Org. Chem. **25**, 1497–1522 (2021)
16. List, B., Lerner, R.A., Barbas, C.F.: Proline-catalyzed direct asymmetric aldol reactions. J. Am. Chem. Soc. **122**, 2395–2396 (2000)
17. Ahrendt, K., Borths, C., Macmillan, D.W.C.: New strategies for organic catalysis: the first highly enantioselective organocatalytic diels-alder reaction. J. Am. Chem. Soc. **122**, 4243–4244 (2000)

18. Brenna, E., Fuganti, C., Gatti, F.C., Serra, S.: Biocatalytic methods for the synthesis of enantioenriched odor active compounds. Chem. Rev. **111**, 4036–4072 (2011)
19. Schmaltz, R.M., Hanson, S.R., Wong, C.H.: Enzymes in the synthesis of glycoconjugates. Chem. Rev. **111**, 4259–4307 (2011)
20. Kadokawa, J.-I.: Precision polysaccharide synthesis catalyzed by enzymes. Chem. Rev. **111**, 4308 (2011)
21. Fessner, W-Dr., He, N., Yi, D., Unruh, P., Knorst, M.: Enzymatic generation of sialoconjugate diversity. In: Riva, S., Fessner, W.-D. (eds.) Cascade Biocatalysis: Integrating Stereoselective and Environmentally Friendly Reactions, Wiley-VCH (2014)
22. Pollegioni, L., Sacchi, S., Caldinelli, L., Boselli, A., Pilone, M., Piubelli, L., Molla, G.: Engineering the properties of D-Amino acid oxidases by a rational and a directed evolution approach. Curr. Protein Pept. Sci. **8**, 600–618 (2007)
23. Paradisi, F., Conway, P.A., Maguireb, A.R., Engel, P.C.: Engineered dehydrogenase biocatalysts for non-natural amino acids: efficient isolation of the d-enantiomer from racemic mixtures. Org. Biomol. Chem. **6**, 3611–3615 (2008)
24. Reetz, M.T., Gotor, V., Alfonso, I., García-Urdiales, E.: In: Asymmetric Organic Synthesis with Enzymes, pp. 21–63. Wiley-VCH, Weinheim, Germany (2008)
25. Reetz, M.T., Zonta, A., Schimossek, K., Liebeton, K., Jaeger, K.-E.: Creation of enantioselective biocatalysts for organic chemistry by In Vitro evolution. Angew. Chem. Int. Ed. Engl. **36**, 2830–2832 (1997)
26. Reetz, M.T., Puls, M., Carballeira, J.D., Vogel, A., Jaeger, K.-E., Eggert, T., Thiel, W., Bocola, M., Otte, N.: Learning from directed evolution: further lessons from theoretical investigations into cooperative mutations in lipase enantioselectivity. ChemBioChem **8**, 106–112 (2007)
27. Koga, Y., Kato, K., Nakano, H., Yamane, T.: Inverting Enantioselectivity of Burkholderia cepacia KWI-56 Lipase by combinatorial mutation and high-throughput screening using single-molecule PCR and In Vitro expression. J. Mol. Biol. **331**, 585–592 (2003)
28. Zhang, G., Gao, R., Zheng, L., Zhang, A., Wang, Y., Wang, Q., Feng, Y., Cao, S.: Study on the relationship between structure and enantioselectivity of a hyperthermophilic esterase from archaeon Aeropyrum pernix K1. J. Mol. Catal. B **38**, 148–153 (2006)
29. Henke, E., Pleiss, J., Bornscheuer, U.T.: Activity of lipases and esterases towards tertiary alcohols: insights into structure-function relationships. Angew. Chem. Int. Ed. **41**, 3211–3213 (2002)
30. Heinze, B., Kourist, R., Fransson, L., Hult, K., Bornscheuer, U.T.: Highly enantioselective kinetic resolution of two tertiary alcohols using mutants of an esterase from Bacillus subtilis. Protein Eng., Des. Sel. **20**, 125–131 (2007)
31. Bartsch, S., Kourist, R., Bornscheuer, U.T.: Complete inversion of enantioselectivity towards acetylated tertiary alcohols by a double mutant of a Bacillus Subtilis Esterase. Angew. Chem. Int. Ed . **47**, 1508–1511 (2008)
32. Archelas, A., Furstoss, R.: Synthetic applications of epoxide hydrolases. Curr. Opin. Chem. Biol. **5**, 112–119 (2001)
33. Cedrone, F., Niel, S., Roca, S., Bhatnagar, T., Ait-abdelkader, N., Torre, C., Krumm, H., Maichele, A., Reetz, M.T., Baratti, J.C.: Directed evolution of the epoxide hydrolase from aspergillus niger. Biocatal. Biotransform. **21**, 357–364 (2003)
34. Reetz, M.T., Torre, C., Eipper, A., Lohmer, R., Hermes, M., Brunner, B., Maichele, A., Bocola, M., Arand, M., Cronin, A., Genzel, Y., Archelas, A., Furstoss, R.: Enhancing the enantioselectivity of an epoxide hydrolase by directed evolution. Org. Lett. **6**, 177–180 (2004)
35. Reetz, M.T., Bocola, M., Wang, L.-W., Sanchis, J., Cronin, A., Arand, M., Zou, J., Archelas, A., Bottalla, A.-L., Naworyta, A., Mowbray, S.L.: Directed evolution of an enantioselective epoxide hydrolase: uncovering the source of enantioselectivity at each evolutionary stage. J. Am. Chem. Soc. **131**, 7334–7343 (2009)
36. Bornscheuer, U.T.: Trends and challenges in enzyme technology. Adv. Biochem. Eng. Biotechnol. **100**, 181–203 (2005)
37. Elenkov, M.M., Tang, L., Hauer, B., Janssen, D.B.: Sequential kinetic resolution catalyzed by halohydrin dehalogenase. Org. Lett. **8**, 4227–4229 (2006)

References 347

38. Bougioukou, D.J., Kille, S., Taglieber, A., Reetz, M.T.: Directed evolution of an enantiose-lective enoate-reductase: testing the utility of iterative saturation mutagenesis. Adv. Synth. Catal. **351**, 3287–3305 (2009)
39. Strukul, G.: Transition metal catalysis in the Baeyer-Villiger oxidation of ketones. Angew. Chem. Int. Ed. **37**, 1198–1209 (1998)
40. Bolm, C., Schlingloff, G., Weickhardt, K.: Optically active lactones from a Baeyer–Villiger-type metal-catalyzed oxidation with molecular oxygen. Angew. Chem. Int. Ed. Engl. **33**, 1848–1849 (1994)
41. Mihovilovic, M.D., Muller, B., Stanetty, P.: Monooxygenase-mediated Baeyer–Villiger Oxidations. Eur. J. Org. Chem. 3711–3730 (2002)
42. Taschner, M.J., Black, D.J.: The enzymatic Baeyer-Villiger oxidation: enantioselective synthesis of lactones from mesomeric cyclohexanones. J. Am. Chem. Soc. **110**, 6892–6893 (1988)
43. Alphand, V., Furstoss, R., Pedragosa-Moreau, S., Roberts, S.M., Willetts, A.J.: Comparison of microbiologically and enzymatically mediated Baeyer-Villiger oxidations: synthesis of optically active caprolactones. J. Chem. Soc. Perkin Trans. **1**, 1867–1872 (1996)
44. DeSantis, G., Wong, K., Farwell, B., Chatman, K., Zhu, Z., Tomlinson, G., Huang, H., Tan, X., Bibbs, L., Chen, P., Kretz, K., Burk, M.J.: Creation of a productive, highly enantioselective nitrilase through gene site saturation mutagenesis (GSSM). J. Am. Chem. Soc. **125**, 11476–11477 (2003)
45. Peters, M.W., Meinhold, P., Glieder, A., Arnold, F.H.: Regio-and enantioselective alkane hydroxylation with engineered cytochromes P450 BM-3. J. Am. Chem. Soc. **125**, 13442–13450 (2003)
46. Reetz, M.T., Brunner, B., Schneider, T., Schulz, F., Clouthier, C.M., Kayser, M.M.: Directed evolution as a method to create enantioselective cyclohexanone monooxygenases for catalysis in baeyer–villiger reactions. Angew. Chem., Int. Ed. **43**, 4075–4078 (2004)
47. Drauz, K., Waldmann, H.: In: Enzyme Catalysis in Organic Synthesis: A Comprehensive Handbook, vol. I-III, 2nd ed., VCH, Weinheim (2002)
48. (a) Asako, H., Shimizu, M., Makino, Y., Itoh, N.: Biocatalytic reduction system for the produc-tion of chiral methyl (R)/(S)-4-bromo-3-hydroxybutyrate. Tetrahedron Lett. **51**, 2664–2666 (2010). (b) Asako, H., Shimizu, M., Itoh, N.: Engineering of NADPH-dependent aldo-keto reductase from Penicillium citrinum by directed evolution to improve thermostability and enantioselectivity. Appl. Microbiol. Biotechnol. **80**, 805–812 (2008)
49. Campapiano, O.: WO-2009046153-A1—ketoreductase polypeptides for the production of azetidinone. Chem. Abstr. **150**, 416448 (2009)
50. Reetz, M.T., Daligault, F., Brunner, B., Hinrichs, H., Deege, A.: Directed evolution of cyclohexanone monooxygenases: enantioselective biocatalysts for the oxidation of prochiral thioethers. Angew. Chem., Int. Ed. **43**, 4078–4081 (2004)
51. Feingersch, R., Shainsky, J., Wood, T.K., Fishman, A.: Protein engineering of toluene monooxygenases for synthesis of chiral sulfoxides. Appl. Environ. Microbiol. **74**, 1555–1566 (2008)
52. Meyer, A., W€ursten, M., Schmid, A., Kohler, H.-P.E., Witholt, B.: Hydroxylation of indole by laboratory-evolved 2-Hydroxybiphenyl 3-Monooxygenase. J. Biol. Chem. **277**, 34161–34167 (2002)
53. Meyer, A., Schmid, A., Held, M., Westphal, A.H., R€othlisberger, M., Kohler, H.-P.E., Van Berkel, W.J.H., Witholt, B.: Changing the substrate reactivity of 2-Hydroxybiphenyl 3-Monooxygenase from Pseudomonas azelaica HBP1 by directed evolution. J. Biol. Chem. **277**, 5575–5582 (2002)
54. Gillam, E.M.J.: Engineering cytochrome P450 enzymes. Chem. Res. Toxicol. **21**, 220–231 (2008)
55. Farinas, E.T., Schwaneberg, U., Glieder, A., Arnold, F.H.: Directed evolution of a cytochrome P450 monooxygenase for alkane oxidation. Adv. Synth. Catal. **343**, 601–606 (2001)
56. Glieder, A., Farinas, E.T., Arnold, F.H.: Laboratory evolution of a soluble, self-sufficient, highly active alkane hydroxylase. Nat. Biotechnol. **20**, 1135–1139 (2002)

348   8   Enzymes in Organic Synthesis

57. M€unzer, D.F., Meinhold, P., Peters, M.W., Feichtenhofer, S., Griengl, H., Arnold, F.H., Glieder, A., de Raadt, A.: Stereoselective hydroxylation of an achiral cyclopentanecarboxylic acid derivative using engineered P450s BM-3. Chem. Commun. **20**, 2597–2599 (2005)
58. Urlacher, V.B., Makhsumkhanov, A., Schmid, R.D.: Biotransformation of β-ionone by engineered cytochrome P450 BM-3. Appl. Microbiol. Biotechnol. **70**, 53–59 (2006)
59. Furlemmeier, A., Quitt, P., Vogler, K., Lanz, P.: 6-Acyl derivatives of amminopencillanic acid. U.S. Patent 3, 957, 758 (1976)
60. Campbell, R.F., Fitzpatrick, K., Inghardt, T., Karlsson, O., Nilsson, K., Reilly, J.E., Yet L.: Enzymatic resolution of substituted mandelic acids. Tetrahedron Lett. **44**, 5477–5481 (2003)
61. Landwehr, M., Hochrein, L., Otey, C.R., Kasrayan, A., Backvall, J.-E., Arnold, F.H.: Enantioselective α-Hydroxylation of 2-Arylacetic Acid derivatives and buspirone catalyzed by engineered cytochrome P450 BM-3. J. Am. Chem. Soc. **128**, 6058–6059 (2006)
62. England, P.A., Harford-Cross, C.F., Stevenson, J.-A., Rouch, D.A., Wong, L.-L.: The oxidation of naphthalene and pyrene by cytochrome P450cam. FEBS Lett. **424**, 271–274 (1998)
63. Kubo, T., Peters, M.W., Meinhold, P., Arnold, F.H.: Enantioselective epoxidation of terminal alkenes to (R)- and (S)-epoxides by engineered cytochromes P450 BM-3. Chem. Eur. J. **12**, 1216–1220 (2006)
64. Molnar, I., Jungmann, V., Stege, J.T., Trefzer, A., Pachlatko, J.P.: Biocatalytic conversion of avermectin into 4''-oxo-avermectin: discovery, characterization, heterologous expression and specificity improvement of the cytochrome P450 enzyme. Biochem. Soc. Trans. **34**, 1236–1240 (2006)
65. Trefzer, A., Jungmann, V., Molnar, I., Botejue, A., Buckel, D., Frey, G., Hill, D.S., J€org, M., Ligon, J.M., Mason, D., Moore, D., Pachlatko, J.P., Richardson, T.H., Spangenberg, P., Wall, M.A., Zirkle, R., Stege, J.T.: Biocatalytic conversion of avermectin to 4''-Oxo-Avermectin: improvement of cytochrome P450 monooxygenase specificity by directed evolution. Appl. Environ. Microbiol. **73**, 4317 (2007)
66. Bell, S.G., Sowden, R.J., Wong, L.-L.: Engineering the haem monooxygenase cytochrome P450cam for monoterpene oxidation. Chem. Commun. **7**, 635–636 (2001)
67. Strieker, M., Essen, L.O., Walsh, C.T., Marahiel, M.A.: Non-heme hydroxylase engineering for simple enzymatic synthesis of L-threo-hydroxyaspartic acid. ChemBioChem **9**, 374–376 (2008)
68. Urlacher, V.B., Schmid, R.D.: Recent advances in oxygenase-catalyzed biotransformations. Curr. Opin. Chem. Biol. **10**, 156–161 (2006)
69. Keenan, B.G., Leungsakul, T., Smets, B.F., Mori, M., Henderson, D.E., Wood, T.K.: Protein engineering of the archetypal nitroarene dioxygenase of Ralstonia sp. Strain U2 for activity on Aminonitrotoluenes and Dinitrotoluenes through alpha-subunit residues Leucine 225, Phenylalanine 350, and Glycine 407. J. Bacteriol. **187**, 3302–3310 (2005)
70. May, O., Nguyen, P.T., Arnold, F.H.: Inverting enantioselectivity by directed evolution of hydantoinase for improved production of L-methionine. Nat. Biotechnol. **18**, 317–320 (2000)
71. DeSantis, G., Wong, K., Farwell, B., Chatman, K., Zhu, Z., Tomlinson, G., Huang, H., Tan, X., Bibbs, L., Chen, P., Kretz, K., Burk, M.J.: Creation of a productive, highly enantioselective nitrilase through gene site saturation mutagenesis (GSSM). J. Am. Chem. Soc. **125**, 11476–11477 (2003)
72. Arroyo, M., de la Mata, I., Hormigo, D., Castillon, M.P., Acebal, C.: Production and characterization of microbial β-lactam acylases. In: Mellado, E., Barredo, J.L. (eds.) Microorganisms for Industrial Enzymes and Biocontrol. Research Signpost, Trivandrum, pp 129–151. (2005)
73. Lopez-Gallego, F., Betancor, L., Sio, C.F., Reis, C.R., Jimenez, P.N., Guisan, J.M., Quax, W.J., Fernandez-Lafuente, R.: Adv. Synth. Catal. **350**, 343–348 (2008)
74. (a) Pollegioni, L., Lorenzi, S., Rosini, E., Marcone, G.L., Molla, G., Verga, R., Cabri, W., Pilone, M.S.: Evolution of an acylase active on cephalosporin C. Protein Sci. **14**, 3064–3076 (2005). (b) Sylvestre, J., Chautard, H, Cedrone, F., Delcourt, M.: Directed evolution of biocatalysts. Org. Process Res. Dev. **10**, 562–571 (2006)
75. Suzuki, H., Yamada, C., Kijima, K., Ishihara, S., Wada, K., Fukuyama, K., Kumagai, H.: Enhancement of glutaryl-7-aminocephalosporanic acid acylase activity of γ-glutamyltranspeptidase of Bacillus subtilis. Biotechnol. J. **5**, 829–837 (2010)

References

76. Tripathi, R.P., Verma, S.S., Pandey, J., Tiwari, V.K.: Recent development on catalytic reductive amination and applications. Curr. Org. Chem. **12**, 1093–1115 (2008)
77. Lalonde, J.: Highly engineered biocatalysts for efficient small molecule pharmaceutical synthesis. Curr. Opin. Biotechnol. **42**, 152–158 (2016)
78. Ghislieri, D., Turner, N.J.: Biocatalytic approaches to the synthesis of enantiomerically pure chiral amines. Top. Catal. **57**, 284–300 (2014)
79. Wang, D.-H., Chen, Q., Yin, S.-N., Ding, X.-W., Zheng, Y.-C., Zhang, Z., Zhang, Y.-H., Chen, F.-F., Xu, J.-H., Zheng, G.-W.: Asymmetric reductive amination of structurally diverse ketones with ammonia using a spectrum-extended amine dehydrogenase. ACS Catal. **11**, 14274–14283 (2021)
80. Liu, Z., Calvo-Tusell, C., Zhou, A.Z., Chen, K., Garcia-Borras, M., Arnold, F.H.: Dual-function enzyme catalysis for enantioselective carbon–nitrogen bond formation. Nature Commun. **13**, 1166–1172 (2021)
81. Sharma, V.K., Watts, J.K.: Oligonucleotide therapeutics: chemistry, delivery and clinical progress. Future Med. Chem. **7**, 2221–2242 (2015)
82. Sharma, V.K., Rungta, P., Prasad, A.K.: Nucleic acid therapeutics: basic concepts and recent developments. RSC Adv. **4**, 16618–16631 (2014)
83. Sharma, V.K., Kumar, M., Olsen, C.E., Prasad, A.K.: Chemoenzymatic convergent synthesis of 2′-O,4′-C-Methyleneribonucleosides. J. Org. Chem. **79**, 6336–6341 (2014)
84. Sharma, V.K., Kumar, M., Sharma, D., Olsen, C.E., Prasad, A.K.: Chemoenzymatic convergent synthesis of 2′-O,4′-C-Methyleneribonucleosides. J. Org. Chem. **79**, 8516–8521 (2014)
85. Isita, J., Meier, J.L.: Enzymatic catalysts to combat COVID-19. ACS Cent. Sci. **7**, 1963–1965 (2021)
86. Kotik, M., Brodsky, K., Halada, P., Javurkova, H., Pelantova, H., Konvalinkova, D., Bojarova, P., Kren, N.: Access to both anomers of rutinosyl azide using wild-type rutinosidase and its catalytic nucleophile mutant. Catalysis Commun. **149**, 106193 (2021)
87. Gorantla, J.N., Pengthaisong, S., Choknud, S., Kaewpuang, T., Manyum, T., Promarakb, V., Cairns, J.R.K.: Gram scale production of 1-Azido-b-D-Glucose *via* enzyme catalysis for the synthesis of 1,2,3-Triazole Glucosides. RSC Adv. **9**, 6211–6220 (2019)
88. Tiwari, V.K.: In: Carbohydrates in Drug Discovery and Development. Elsevier (2020)
89. Nielsen, M.M., Pedersen, C.M.: Catalytic glycosylations in oligosaccharide synthesis. Chem. Rev. **118**, 8285–8358 (2018)
90. Shoda, S., Uyama, H., Kadokawa, J., Kimura, S., Kobayashi, S.: Enzymes as green catalysts for precision macromolecular synthesis. Chem. Rev. **116**, 2307–2413 (2016)
91. Shoda, S.: Development of chemical and chemo-enzymatic glycosylations. Proc. Jpn. Acad., Ser. B. **93**, 125–145 (2017)
92. Loos, K., Kadokawa, J., Kobayashi, S., Uyama, H.: In: Enzymatic Polymerization towards Green Polymer Chemistry, pp. 47–87. Springer, Heidelberg (2019)
93. O'Neill, E.C., Field, R.A.: Enzymatic synthesis using glycoside phosphorylases. Carbohydr. Res. **403**, 23–37 (2015)
94. Kadokawa, J.-I., Lee, L.H., Yamamoto, K.: Thermostable α-Glucan Phosphorylase-catalyzed Enzymatic chain-elongation to produce 6-Deoxygenated α(1→4)-Oligoglucans. Curr. Org. Chem. **25**, 1345–1352 (2021)
95. Park, S., Kazlauskas, R.J.: Biocatalysis in ionic liquids: advantageous beyond green technology. Curr. Opin. Biotech. **14**, 432–437 (2003)
96. Qiao, Y., Ma, W., Theyssen, N., Chen, C., Hou, Z.: Temperature-responsive ionic liquids: fundamental behaviors and catalytic applications. Chem. Rev. **117**, 6881–6928 (2017)
97. Egorova, K.S., Gordeev, E.G., Ananikov, V.P.: Biological activity of ionic liquids and their application in pharmaceutics and medicine. Chem. Rev. **117**, 7132–7189 (2017)
98. Rajkhowa, S., Kale, R.R., Sarma, J., Kumar, A., Mohapatra, P.P., Tiwari, V.K.: Room temperature ionic liquids in glycoscience: opportunities and challenges. Curr. Org. Chem. **25**, 2542–2578 (2021)

99. Murugesan, S., Linhardt, R.J.: Ionic liquids in carbohydrate chemistry—current trends and future directions. Curr. Org. Syn. **2**, 437–451 (2005)
100. Farraan, A., Cai, C., Sandoval, M., Xu, Y., Liu, J., Hernaaiz, M.J., Linhardt, R.J.: Green solvents in carbohydrate chemistry: from raw materials to fine chemicals. Chem. Rev. **115**, 6811–6853 (2015)
101. Park, S., Kazlauskas, R.J.: Improved preparation and use of room-temperature ionic liquids in lipase-catalyzed enantio- and regioselective acylations. J. Org. Chem. **66**, 8395–8401 (2001)
102. Murugesan, S., Karst, N., Islam, T., Wiencek, J.M., Linhardt, R.J.: Dialkyl imidazolium benzoates- room temperature ionic liquids useful in the peracetylation and perbenzoylation of simple and sulfated saccharides. Synlett **9**, 1283–1286 (2003)
103. Kim, M.J., Choi, M.Y., Lee, J.K., Ahn, Y.: Enzymatic selective acylation of glycosides in ionic liquids: significantly enhanced reactivity and regioselectivity. J. Mol. Catal. B: Enzymatic **26**, 115–118 (2003)
104. Ganske, F., Bornscheuer, U.T.: Lipase-catalyzed glucose fatty acid ester synthesis in ionic liquids. Org. Lett. **7**, 3097–3098 (2005)
105. Kaftzik, N., Wasserscheid, P., Kragl, U.: Use of ionic liquids to increase the yield and enzyme stability in the β-galactosidase catalysed synthesis of N-Acetyllactosamine. Org. Proc. Res. Devlop. **6**, 553–557 (2002)
106. Winter, K.D., Verlindena, K., Krenb, V., Weignerovab, L., Soetaerta, W., Tom, D.: Ionic liquids as cosolvents for glycosylation by sucrose phosphorylase: balancing acceptor solubility and enzyme stability. Green Chem. **15**, 1949–1955 (2013)
107. Holbrey, J.D., Turner, M.B., Reichert, W.M., Rogers, R.D.: New ionic liquids containing an appended hydroxyl functionality from the atom-efficient, one-pot reaction of 1-methylimidazole and acid with propylene oxide. Green Chem. **5**, 731–736 (2003)
108. Turner, M.B., Spear, S.K., Huddleston, J.G., Holdbrey, J.D., Rogers, R.D.: Ionic liquid salt-induced inactivation and unfolding of cellulase from *Trichoderma reesei*. Green Chem. **5**, 443–447 (2003)
109. Okrasa, K., Guibe-Jampel, E., Therisod, M.: Tandem peroxidase-glucose oxidase catalysed enantioselective sulfoxidation of thioanisoles. J. Chem Soc. Perkin Trans-1 **7**, 1077–1079 (2000)
110. Brovetto, M., Gamenara, D., Méndez, P.S., Seoane, G.A.: C−C bond-forming lyases in organic synthesis. Chem. Rev. **111**, 4346–4403 (2011)
111. Kataoka, M., Yamamoto, K., Kawabata, H., Wada, M., Kita, K., Yanase, H., Shimizu, S.: Stereoselective reduction of ethyl 4-chloro-3-oxobutanoate by Escherichia coli transformant cells coexpressing the aldehyde reductase and glucose dehydrogenase genes. Appl. Microbiol. Biotechnol. **51**, 486–490 (1999)
112. Fong, S., Machajewski, T.D., Mak, C.C., Wong, C.-H.: Directed evolution of D-2-keto-3-deoxy-6-phosphogluconate aldolase to new variants for the efficient synthesis of D-and L-sugars. Chem. Biol. **7**, 873–883 (2000)
113. Royer, S.F., Haslett, L., Crennell, S.J., Hough, D.W., Danson, M.J., Bull, S.D.: Structurally informed site-directed mutagenesis of a stereochemically promiscuous aldolase to afford stereochemically complementary biocatalysts. J. Am. Chem. Soc. **132**, 11753–11758 (2010)
114. Gonzalez-Garcia, E., Helaine, V., Klein, G., Schuermann, M., Sprenger, G.A., Fessner, W.-D., Reymond, J.-L.: Fluorogenic stereochemical probes for transaldolases. Chem. Eur. J. **9**, 893–899 (2003)
115. Sugiyama, M., Greenberg, W., Wong, C.-H.: Recent advances in aldolase-catalyzed synthesis of unnatural sugars and iminocyclitols. J. Syn. Org. Chem., Japan. **66**, 605–615 (2008)
116. Wada, M., Hsu, C.-C., Franke, D., Mitchell, M., Heine, A., Wilson, I., Wong, C.-H.: Directed evolution of N-acetylneuraminic acid aldolase to catalyze enantiomeric aldol reactions. Bioorg. Med. Chem. **11**, 2091–2098 (2003)
117. Woodhall, T., Williams, G.J., Berry, A., Nelson, A.: Creation of a tailored aldolase for the parallel synthesis of sialic acid mimetics. Angew. Chem. Int. Ed. **44**, 2109–2112 (2005)
118. Hsu, C.-C., Hong, Z., Wada, M., Franke, D., Wong, C.-H.: Directed evolution of d-sialic acid aldolase to l-3-deoxy-manno-2-octulosonic acid (l-KDO) aldolase. Proc. Natl. Acad. Sci. U.S.A. **102**, 9122–9126 (2005)

References   351

119. Williams, G.J., Domann, S., Nelson, A., Berry, A.: Modifying the stereochemistry of an enzyme-catalyzed reaction by directed evolution. Proc. Natl. Acad. Sci. U.S.A. **100**, 3143–3148 (2003)

120. Yu, H., Chen, X.: One-pot multienzyme (OPME) systems for chemoenzymatic synthesis of carbohydrates. Org. Biomol. Chem. **14**, 2809–2818 (2016)

121. Knorst, M., Fessner, W.-D.: CMP-sialate synthetase from neisseria meningitidis—overexpression and application to the synthesis of oligosaccharides containing modified sialic acids. Adv. Syn. Cat. **343**, 698–710 (2001)

122. Blix, G., Gottschalk, A., Klenk, E.: Proposed nomenclature in the field of neuraminic and sialic acids. Nature **179**, 1088 (1957)

123. Varki, A.: Biological roles of oligosaccharides: all of the theories are correct. Glycobiology **3**, 97–130 (1993)

124. Dwek, R.A.: Glycobiology: toward understanding the function of sugars. Chem. Rev. **96**, 683–720 (1996)

125. Angata, T., Varki, A.: Chemical diversity in the sialic acids and related α-Keto acids: an evolutionary perspective. Chem. Rev. **102**, 439–469 (2002)

126. Honke, K., Taniguchi, N.: Sulfotransferases and sulfated oligosaccharides. Med. Res. Rev. **22**, 637–654 (2002)

127. Bose, P., Agrahari, A.K., Singh, A.S., Tiwari, V.K., Jaiswal, M.K.: Sialic acids in drug discovery and development. In: Tiwari, V.K. (ed) Carbohydrates in Drug Discovery and development, pp. 213–266. Elsevier, The Netherlands (2020)

128. Chen, X., Varki, A.: Advances in the biology and chemistry of sialic acids. ACS Chem. Biol. **5**, 163–176 (2010)

129. Li, Y., Yu, H., Cao, H., Lau, K., Muthana, S., Tiwari, V.K., Son, B., Chen, X.: Pasteurella multocida sialic acid aldolase: a promising biocatalyst. Appl. Microb. Biotech. **79**, 963–970 (2008)

130. Yu, H., Cao, H., Tiwari, V.K., Li, Y., Chen, X.: Chemoenzymatic synthesis of C8-modified sialic acids and related α2–3- and α2–6-linked sialosides. Bio. Org. Med. Chem. Lett. **21**, 5037–5040 (2011)

131. Huynh, N., Aye, A., Li, Y., Yu, H., Cao, H., Tiwari, V.K., Shin, D.-W., Chen, X., Fisher, A.J.: Structural basis for substrate specificity and mechanism of N-Acetyl-d-neuraminic acid lyase from Pasteurella multocida. Biochemistry **52**, 8570–8579 (2013)

132. Yu, H., Yu, H., Karpel, R., Chen, X.: Chemoenzymatic synthesis of CMP-sialic acid derivatives by a one-pot two-enzyme system: comparison of substrate flexibility of three microbial CMP-sialic acid synthetases. Bioorg Med Chem. **12**, 6427–6435 (2004)

133. Yu, H., Chokhawala, H., Karpel, R., Yu, H., Wu, B., Zhang, J., Zhang, Y., Jia, Q., Chen, X.: A multifunctional Pasteurella multocida sialyltransferase: a powerful tool for the synthesis of sialoside libraries. J. Am. Chem. Soc. **127**, 17618–17619 (2005)

134. Ding, L., Yu, H., Lau, K., Li, Y., Muthana, S., Wang, J., Chen, X.: Efficient chemoenzymatic synthesis of sialyl Tn-antigens and derivatives. Chem Commun. **47**, 8691–8693 (2011)

135. Kooner, A.S., Yu, H., Chen, X.: Synthesis of *N*-Glycolylneuraminic acid (Neu5Gc) and its glycosides. Front. Immunol. **2019**, 10 (2004). https://doi.org/10.3389/fimmu.2019.02004

136. Huang, S., Yu, H., Chen, X.: Disaccharides as sialic acid aldolase substrates: synthesis of disaccharides containing a sialic acid at the reducing end. Angew. Chem. Int. Ed. **46**, 2249–2253 (2007)

137. Chokhawala, H.A., Cao, H., Yu, H., Chen, X.: Enzymatic synthesis of fluorinated mechanistic probes for sialidases and sialyltransferases. J. Am. Chem. Soc. **129**, 10630–10631 (2007)

138. Yu, H., Chokhawala, H.A., Huang, S., Chen, X.: One-pot three-enzyme chemoenzymatic approach to the synthesis of sialosides containing natural and non-natural functionalities. Nature protocol **1**, 2485–2492 (2006)

139. Yu, H., Huang, S., Chokhawala, H., Sun, M., Zheng, H., Chen, X.: Highly efficient chemoenzymatic synthesis of naturally occurring and non-natural α-2,6-linked sialosides: a P. damsela α-2,6-Sialyltransferase with Extremely Flexible Donor–Substrate Specificity. Angew. Chem. Int. Ed. **45**, 3938–3944 (2006)

140. Cheng, J., Yu, H., Lau, K., Huang, S., Chokhawala, H.A., Li, Y., Tiwari, V.K., Chen, X.: Multifunctionality of Campylobacter jejunisialyltransferaseCstII: characterization of GD3/GT3 oligosaccharide synthase, GD3 oligosaccharide sialidase, and trans-sialidase activities. Glycobiology **18**, 686–697 (2008)
141. Xiao, A., Li, Y., Li, X., Santra, A., Yu, H., Li, W., Chen, X.: Sialidase-catalyzed one-pot multienzyme (OPME) synthesis of sialidase transition-state analogue inhibitors. ACS Catal. **8**, 43–47 (2018)
142. Yu, H., Cheng, J., Ding, L., Khedri, Z., Chen, Y., Lau, K., Tiwari, V.K.: Chemoenzymatic synthesis of GD3 oligosaccharides and other disialyl glycans containing natural and non-natural sialic acids. Chen, X. J. Am. Chem. Soc. **131**, 18467–18477 (2009)
143. Song, X., Yu, H., Chen, X., Lasanajak, Y., Tappert, M.M., Air, G.M., Tiwari, V.K., Cao, H., Chokhawala, H.A., Zheng, H., Cummings, R.D., Smith, D.F.: A sialylated glycan microarray reveals novel interactions of modified sialic acids with proteins and viruses. J. Biol. Chem. **286**, 31610–31622 (2011)
144. Kooner, A.S., Diaz, S., Yu, H., Santra, A., Varki, A., Chen, X.: Chemoenzymatic synthesis of sialosides containing 7-*N*- or 7,9-Di-*N*-acetyl sialic acid as stable *O*-Acetyl analogues for probing sialic acid-binding proteins. J. Org. Chem. **86**, 14381–14397 (2021)

Chapter 9
Application of Green Chemistry: Examples of Real-World Cases

1 Introduction

Green chemistry deals with the processes or products that reduce or replace the hazardous substances with greener alternatives. The major interest arose in the field of green chemistry when the USA passed Pollution Prevention Act in 1990 which established the green chemistry as a legitimate scientific field and subsequently became a formal focus of the US Environmental Protection Agency, EPA in 1991 [1, 2]. Before this act, several laws were passed for the command and control of pollution, but this act for the first time changed the regulatory policy from pollution control to pollution prevention as major strategy for combating environmental issues [2]. This act encouraged academics and industries to develop novel green technology replacing the old hazardous processes. In this venture, the place of chemistry is very crucial. This act inspired the chemists to devise greener reaction conditions (e.g., use of green solvents like water, ionic liquids in place of organic solvents), develop greener synthetic methodologies (i.e., encourage the use of biomass rather than using petrochemical feedstocks), develop catalytic processes rather than using stoichiometric reagents, and design new low toxic compounds having comparable desired properties as in existing compounds.

It is worth to say that chemistry has made the human life easier. The life expectancy of human being has been enhanced in last hundred years and this happened due to improved healthcare facilities, development of fast diagnostic techniques, and efficient pharmaceuticals. Chemistry has made the water and food safe to consume. It also made our television bigger and smarter, more efficient computers, faster and less polluting automobiles, several power alternatives like batteries and inverters. In a nutshell, one can say that chemistry has a continuous impact on our day-to-day life. Although the list is very long, these are some positive aspects of chemistry.

While chemistry has indisputably made the human life easier, it also caused several adverse environmental effects. For example, commodities made up of polyethylenes, like plastic bottles, computer hardwares, plastic bags, wire insulation, and toys have

---

© The Author(s), under exclusive license to Springer Nature Singapore Pte Ltd. 2022   353
V. K. Tiwari et al., *Green Chemistry*,
https://doi.org/10.1007/978-981-19-2734-8_9

become an integral part of our life. The murky side of the materials made of polyethylene is that they are not biodegradable, carcinogenic and take several decades to vanish naturally. They are responsible for pollution on land and smaller and larger aquatic bodies. Similarly, the use of DDT as pesticide had saved several lives by controlling the pests and insect which were responsible for deadly diseases. Later it was found that DDT got bioaccumulated and had several adverse effects in birds, for example, uncontrolled use of DDT resulted in extinction of bald eagle population and is a suspected carcinogen [3]. Another infamous example is the Bhopal gas leak tragedy, India. It was the largest industrial disaster on record, where life of millions of people got affected [4, 5]. The Cuyahoga River in Ohio is so polluted that is caught fire [6]. Pepcon disaster, in 1988, where a massive fire and explosions at a chemical plant killed and injured several people [7]. Falk Corporation Explosion, where a gas leak triggered a large explosion at a gear manufacturing facility in Milwaukee, Wisconsin [8]. These are few examples out of several disasters that the entire world has witnessed in human history.

Basically, a chemical can pose risk in terms of hazard and exposure. Earlier, several legislations were introduced to control the chemical exposure but unfortunately it was unable to furnish proper results. Green chemistry deals with less hazardous chemicals which do not pose any risk regardless of longer exposure. The green chemistry represents a fundamental shift from pollution control policies toward pollution prevention paradigm. A set of twelve principles (see Chap. 1) were developed by Anastas and Warner as a blueprint for the implementation of green chemistry in academia and chemical industries [9, 10].

In the early 1990s, European Community's Chemistry Council published some worthy works regarding green chemistry in *Chemistry for a Clean World.*" The first symposium regarding pollution prevention was held in 1994 at Chicago, sponsored by Environmental Chemistry Division, American Chemical Society (ACS). In 1995, US president Bill Clinton supported US EPA to establish an annual awards program *Presidential Green Chemistry Challenge awards* highlighting the innovative scientific findings from individuals, academic institutions, and industries [11, 12]. Since 1996, these awards are presented for innovative chemical technologies which incorporate the principles of green synthesis in chemical design, manufacturing, and use. EPA, honors this award in six fields including Academia, Small business, Greener Synthetic Pathways, Greener Reaction Conditions, Designing Greener Chemicals, and Specific Environmental Benefit (added in 2015). In 1997, Dr. Joe Breen and Dr. Dennis Hjeresen cofounded Green Chemistry Institute (GCI), which is an independent and non-profitable institute dedicated to advancing green chemistry. Since then, GCI started *Green Chemistry & Engineering Conference* that convene every year [13]. In 2001, the GCI merged with ACS, which is now the world's largest professional scientific chemical society and membership organization. ACS-GCI, in 2005, established the first roundtable for pharmaceutical industries to accelerate the involvement of green technology in chemical business [12].

To implement the principles of green chemistry at the grass root level, it is very important to educate the society and make them aware about the long-term consequences of pollution, various environmental issues and how to combat them

2 Selected Examples of Real-World Applications of Green Chemistry 355

using green chemistry. This is only possible by including the green chemistry in the curriculum in universities, colleges, and school levels. Both EPA and ACS realized the importance of educating the green chemistry in classroom and laboratories. Several campaigns have been launched together by EPA and ACS to develop green chemistry educational materials and organization of conferences and workshops [14]. This chapter discusses some real-world cases and groundbreaking innovations that followed the principle of green chemistry.

2 Selected Examples of Real-World Applications of Green Chemistry

2.1 Greener Synthetic Pathway for the Synthesis of Ibuprofen

Ibuprofen **9**, a well-known nonsteroidal anti-inflammatory drug (NSAID) used for pain relief and fever marketed under brand names Advil and Motrin [15, 16]. Andrew Dunlop, in 1960s developed this drug and tested on cures for hangover. Ibuprofen was first prescribed in 1974 in USA and soon became over-the-counter available drug. Due to its safety and efficacy, ibuprofen became one of the most leading pharmaceuticals for pain relief in a very short span of time [17]. The composition and synthetic process of ibuprofen was patented by the Boots Pure Drug Company in 1966 [18]. The initial synthetic procedure involves several complicated steps and elimination of considerable amount of chemical wastes (Scheme 1).

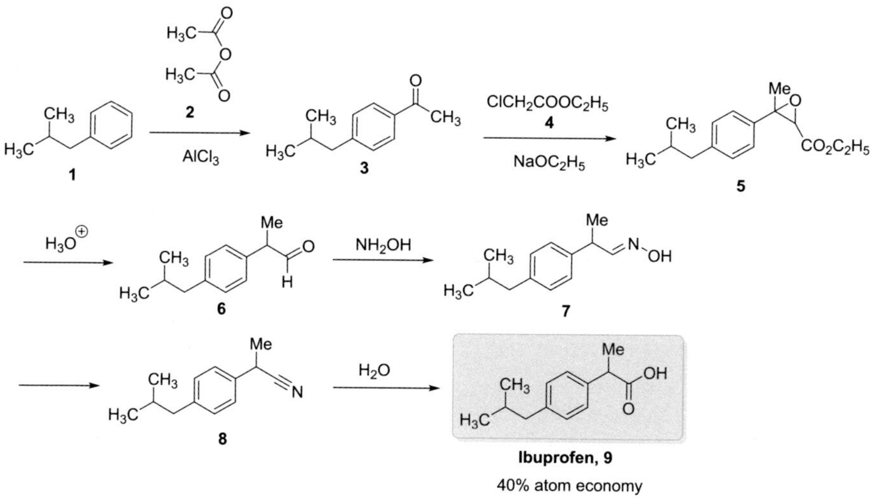

Scheme 1 Synthesis of Ibuprofen by boots pure drug company

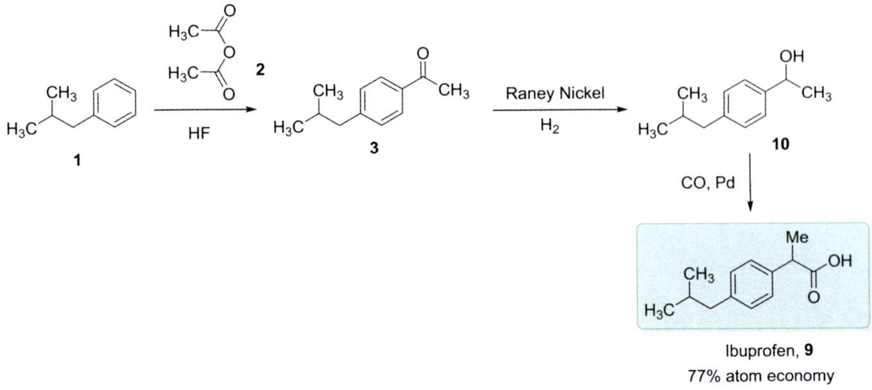

Scheme 2 Synthesis of Ibuprofen by BHC

In 1992, BHC Company unveiled an efficient method that was environmentally benign and a pertinent example of atom economy (about 77% atom utilization, Scheme 2). They replaced the previous technology having six stoichiometric steps to three catalytic steps and also recovered and recycled the waste by-product, virtually eliminated the large volume of aqueous salt wastes. The synthetic procedure is now owned by BASF corporation, one of the partner companies of BHC and produces 7.7 million pounds of ibuprofen per year representing approximately 20–25% of global production of ibuprofen **9**.

2.2 Application of Surfactants for Liquid Carbon Dioxide

The manufacturing and service industries use various organic solvents including volatile organic compounds (VOCs) and many halogenated compounds such as chlorofluorocarbons (CFCs), perchloroethylenes (PERCs), hydrochlorofluorocarbons (HCFCs) as cleansing agents, medical device fabrication, processing aids, and dispersants. These volatile solvents and halogenated solvents are very hazardous to the environment [19]. For instance, VOCs are responsible for photochemical smog formation and cause serious air pollution and breathing problems in human beings [20]. Similarly, halogenated organic compounds such as CFCs and HCFCs are responsible for ozone depletion and PERCs are well-known groundwater pollutant and cancer-causing agents [21].

An elegant alternative which can replace CFCs, HCFCs, and VOCs is using CO_2 as solvent [22–25]. The use of CO_2 has several benefits, such as it can be cheaply available as by-product of ammonia manufacturing industries and natural gas wells. Unlike organic solvents, CO_2 is non-inflammable and non-toxic. Furthermore, it is chemically less reactive and does not have any contribution in smog formation.

2 Selected Examples of Real-World Applications of Green Chemistry

However, CO_2 in gaseous form is a major contributor to global warming as its property to revert back the infrared radiation toward earth. The problem of global warming could be lessened by using carbon dioxide as a solvent in the form of liquid or supercritical CO_2. An additional advantage of using carbon dioxide as solvent is that after used as solvent or cleaning agents, it is very easy to recycle gaseous CO_2, leaving the impurities back by evaporation. The heat of vaporization of CO_2 is very low (15.3 kJ/mol at 215.7 K) compared to common solvents like several organic solvents and water. Thus, utilizing less amount of energy/heat and less consumption of fossil fuels for vaporization. The impurity-free gaseous CO_2 can be collected and transformed back to liquid or supercritical fluid for further use.

How the gaseous form of CO_2 can be changed to liquid or supercritical fluid? The transformation of any gaseous species to liquid state requires the temperature lower than critical temperature (T_c) and exerting critical pressure (P_c). The critical temperature of a substance is defined as the temperature above which it cannot be liquified with exerting any amount of pressure. The critical pressure is defined as the pressure above which a substance in gaseous phase transformed to liquid state. The T_c and P_c for CO_2 are 31 °C and 72.8 atm, respectively. However, if a substance allowed to stay at a temperature above critical temperature, T_c, and critical pressure, P_c, a supercritical fluid will be formed. A supercritical fluid is a state where the molecules pressed so close by exerted pressure that they behave almost like a liquid. But the molecules remain in excited state due to temperature above T_c, so that the intermolecular attraction cannot hold them like a liquid. Therefore, the supercritical fluids have density closer to a liquid state, but the viscosity is comparable to the gaseous state.

Smaller hydrocarbons (HCs), halogenated HCs, aldehydes, ketones, and esters are readily dissolved in liquid CO_2[26]. However, many industrial solvents cannot find liquid or supercritical CO_2 as their replacement due to low solubility of some common industrial materials like polymers, waxes, greases, and heavy oils. This issue could be lessened by using surfactant in liquid or supercritical CO_2. A surfactant works on the principle "Like dissolves like." A surfactant has two ends, one polar and other nonpolar. One end possesses similar polarity to the substance that has to be emulsified, and other end has polarity similar to the solvent. Therefore, surfactant molecules assemble themselves in a spherical shape called micelles. In a nutshell, one can define the function of a surfactant as to dissolve nonpolar solute in a polar solvent and vice-versa.

To enhance the utility of liquid and supercritical fluid CO_2 as a solvent or cleaning agents in various industries, Prof. Joseph M. DeSimone of University of North Carolina at Chapel Hill and North Carolina State University developed a high molar mass fluoropolymer which is soluble in liquid and supercritical fluid CO_2 [22]. The solubility of fluoropolymer is due to the existence of weak van der Waals interaction between CO_2 molecules, and similar van der Waals force of attraction presents between the fluorocarbon tails of the polymer as shown in Fig. 1.

Prof. DeSimone created a block and graft copolymer of which some parts are soluble and some parts are insoluble in CO_2. The surfactant contains the main chain with CO_2-phobic properties and side chains (fluorocarbon chains) with CO_2-philic

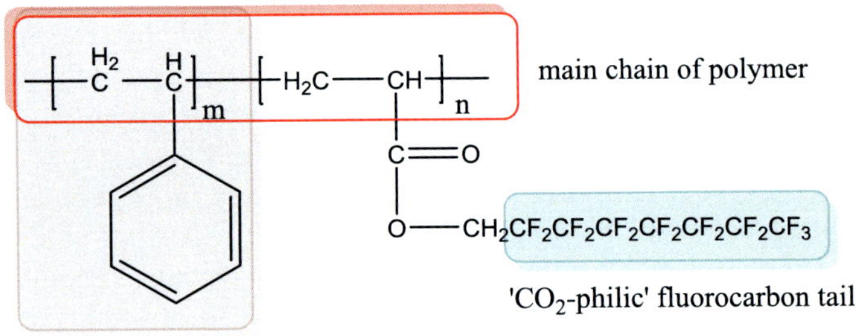

358   9 Application of Green Chemistry: Examples of Real-World Cases

Fig. 1 Copolymer used as surfactant in liquid CO_2

properties (Fig. 1). Depending upon the need, the CO_2-phobic segments could be made lipophilic or hydrophilic. Prof. DeSimone recently developed a copolymer which comprises polystyrene blocks which are insoluble in liquid CO_2 and fluorocarbon chains, poly(1,1dihydroperfluorooctylacrylate) which are soluble in CO_2 (CO_2-philic, Fig. 1) [22]. When this copolymer is placed in liquid or supercritical CO_2, it forms micelle structure and increases the solubility of some industrial wastes like grease, heavy oils, and waxes. The CO_2-surfactant technology is currently used on commercial level by Micell technology for manufacturing dry cleaning machines which uses liquid CO_2 and surfactant to dry clean the clothes replacing hazardous PERCs. For developing this innovative technology, Prof. DeSimone was honored with Academic Award in Presidential Green Chemistry Challenge program in 1997 [11, 27].

2.3 Development of Environmentally Benign Marine Antifoulant

Fouling mainly refers to the unwanted growth of marine plants and animals on submerged surface of ships and boats. The fouling causes hydrodynamic drag which costs approximately $3 billion per year to shipping industry due to increased fuel consumption [11]. In fact, according to a study, mere 1 mm layer of slime can increase the drag on ship by 80%. Furthermore, it decreases the average speed and increases the maintenance cost of the shipping industry. Additionally, the extra fuel consumption causes air pollution, global warming, acid rain and increases unnecessary burden on non-renewable resources.

To overcome the unnecessary build-ups of marine plants and animals, organotin compound, such as tributyltin oxide (TBTO, **11**, Fig. 2), was used as standard antifoulant across the world [28, 29]. TBTO is used as marine paints and coatings. While being effective on fouling, TBTO leaches out in surrounding and causes severe environmental problems. Organotin compounds persist in environment for longer

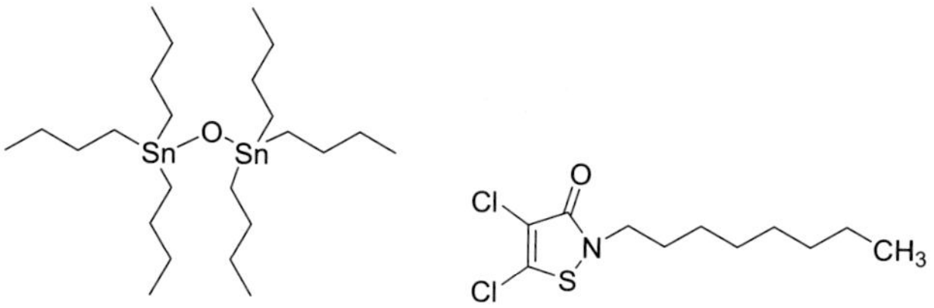

2 Selected Examples of Real-World Applications of Green Chemistry

Fig. 2 Structure of tributyltin oxide TBTO, **11** and sea nine (DCOI), **12**

time causing toxicity including bioaccumulation, acute toxicity to non-target marine organisms, reduces reproductive viability, increases shell thickness in shellfish, etc [30]. Toxicological studies of organotin compounds on marine organisms showed that due to bioaccumulation, the concentration of TBTO, in some cases, increased as high as 10,000 times greater than the concentration of TBTO in surrounding water. TBTO is also chronically toxic to several marine animals [28, 31]. Chronic toxicity refers to long-term exposure to any substance, although in small concentration, leads to long-term repercussions including cancer, mutation, reproductive dysfunctions, etc. For example, long-term exposure of organotin compounds leads to thickening of shell in shellfish and imposex which leads to infertility in sea snails [32, 33]. It also affects the immune system of fishes, dolphins, etc., which may cause severe disease to outbreak.

In 1988, in view of these harmful effects, Organotin Antifoulant Paint Control Act was passed to ban the use of TBTO antifoulant on ships and boats in USA [34, 35].

Rohm and Haas, an US-based company developed 4,5-dichloro-2-*n*-octyl-4-isothiazolin-3-one (DCOI, **12**, Fig. 2) as an alternative of TBTO [36]. DCOI is environmentally benign and popularly known as Sea Nine™ antifoulant [37]. Extensive research on the impact of DCOI on environment showed that it is non-persistent and degrade quickly [38]. The half-life of TBTO is 5–6 months in seawater and 6–9 months in sediments, while the half-life of Sea Nine is less than 24 h in seawater and 1 h in sediments. After leaching from the coatings of ship hull, DCOI bind tightly to the soil particle in the sediment. As Sea Nine settled to the sediments, less bioavailable for non-target marine organisms. At the sediment, Sea Nine is degraded irreversibly by microorganisms. All the component compounds generated after the biodegradation are found to be non-toxic. Therefore, the maximum allowable environmental concentration (MAEC) is 0.63 parts per billion (ppb) and 0.002 ppb for Sea Nine and TBTO, respectively. Additionally, TBTO has widespread chronic toxicity, DCOI antifoulant showed no chronic toxicity at all [39]. The biodegradation pathway of Sea Nine is outlined in Scheme 3[40].

In 1995, Rohm and Haas commercialized the DCOI-containing antifoulant under the brand name Sea Nine 211. Every year hundreds of heavy marine vehicles have been coated with paints containing Sea Nine 211 antifoulant. In view of increasing demand of Sea Nine antifoulant, Rohm and Haas opened new facility in Bayport, TX, in 1996.

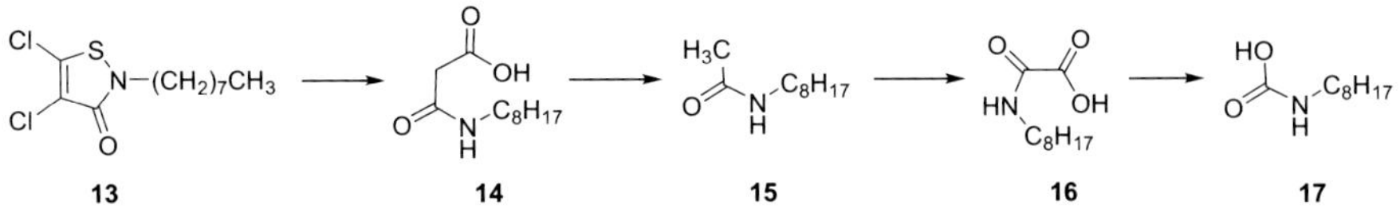

360   9 Application of Green Chemistry: Examples of Real-World Cases

Scheme 3 Biodegradation of sea nine 211 antifoulant

2.4 Use of Genetically Engineered Microbes as Environmentally Benign Catalyst

Adipic acid, an aliphatic dicarboxylic acid, is used for the production of a wide variety of industrially important materials including plasticizers, lubricants, polyurethane, etc. [41]. Therefore, large quantities-approx. 1.9 billion kg of adipic acid is produced every year globally [42]. Nylon-6,6 is manufactured by step growth polymerization using adipic acid and hexamethylenediamine as prime starting materials [43, 44]. Similarly, catechol is used as intermediate in various flavoring, fragrance, agrochemical and pharmaceutical industries, and annual production is about 20,000 tons globally [43].

The industrial production of adipic acid involves the hydrogenation of benzene **18** over Ni or Pd catalyst to form cyclohexane **19** which further air-oxidized to yield cyclohexanone **20** and cyclohexanol **21** in the presence of catalyst. Cyclohexanol **21** and cyclohexanone **20** oxidized catalytically in the presence of nitric acid to afford adipic acid **22** (Scheme 4a) [41]. Similarly, catechol **26** is produced using benzene as starting material. Benzene undergoes Friedel–Crafts alkylation with propene **23** in acidic condition to form cumene **24** which further oxidized to form phenol **25** and acetone is produced as by-product. Phenol, in the presence of hydrogen peroxide, oxidized to afford catechol **26** and hydroquinone **27** (Scheme 4b).

The existing industrial procedures for the syntheses of adipic acid **22** and catechol **26** have several drawbacks in context of health and environment. Benzene **18**

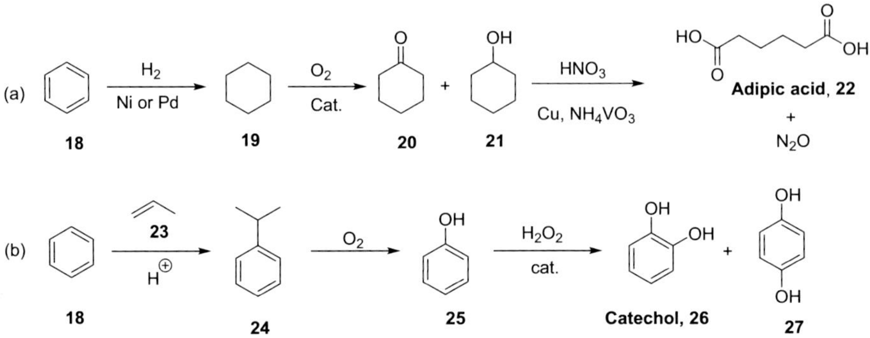

Scheme 4 Industrial synthesis of adipic acid and catechol using benzene as starting material

2 Selected Examples of Real-World Applications of Green Chemistry 361

is precursor of both adipic acid **22** and catechol **26**. Benzene is volatile organic compound, vaporizes readily at room temperature, and therefore poses environmental and health hazard. Benzene can cause chronic toxicity which lead to leukemia and cancer [45]. Additionally, benzene is obtained as a by-product of petroleum manufactures. Its production depends upon the non-renewable resources. Furthermore, the last step in the synthesis of adipic acid employs nitric acid for the oxidation of cyclohexanone and cyclohexanol. The by-product obtained in this step, nitrous oxide (N_2O) is the chemical of concern because it enters into stratosphere and destroys the ozone layer. Due to large-scale industrial synthesis of adipic acid, it contributes nearly 10% of nitrous oxide in the atmosphere [11]. N_2O is also responsible for enhancing global warming as it is active greenhouse gas.

Another compound possessing immense utility in chemical industries is butylated hydroxytoluene **33** (BHT). It prevents organic unsaturated compounds from autoxidation [46]. It has wide application in cosmetic, food, and industrial solvents to inhibit oxidation and free radical formation. BHT is manufactured in industries using toluene as the starting material which in the presence of chlorine afford *p*-chlorotoluene **29**. In basic condition, *p*-chlorotoluene **29** is transformed to *p*-cresol **30** which further alkylated under acidic condition to afford BHT **33** [46]. An alternative pathway involves the reaction of propylene and toluene to form cumene **32** followed by oxidation to yield cresol **30** (Scheme 5). The starting material of BHT, toluene is also obtained from the petroleum products and harmful to human health. However, it is considered to be less carcinogenic than benzene. Continuous inhalation of toluene leads to liver, kidney, vision, and brain damage.

Dr. Karen M. Draths and Prof. John W. Frost from Michigan State University used genetically engineered *E. Coli* for the synthesis of adipic acid, catechol, and DHS (dehydroshikimic acid, a potential alternate of BHT) from glucose [42, 47]. Draths and Frost, employing genetic engineering technology, altered the metabolic pathway of glucose in *E. Coli*. A single, genetically engineered strain of *E. Coli* is capable transforming the glucose into *cis,cis*-muconic acid which further hydrogenated to afford adipic acid [48] unmodified *E. Coli* follows the natural biocatalytic pathway, transforming glucose to DHS followed by the formation of amino acids like

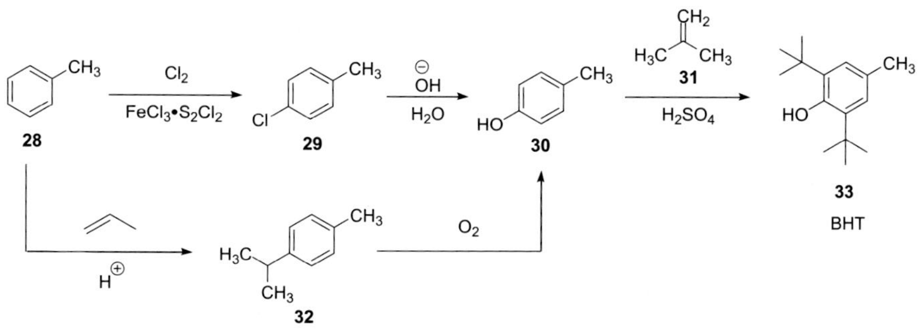

Scheme 5 Synthesis of BHT from toluene

L-phenylalanine, L-tyrosine, and L-tryptophan (Scheme 6, Path B). Intermediates including E4P, DAHP, DHQ, and shikimic acid are also formed [49–51]. Genetically engineered microbes follow the unnatural biocatalytic pathway to form protocatechic acid (PCA) from DHS followed by the formation of catechol and then catechol to

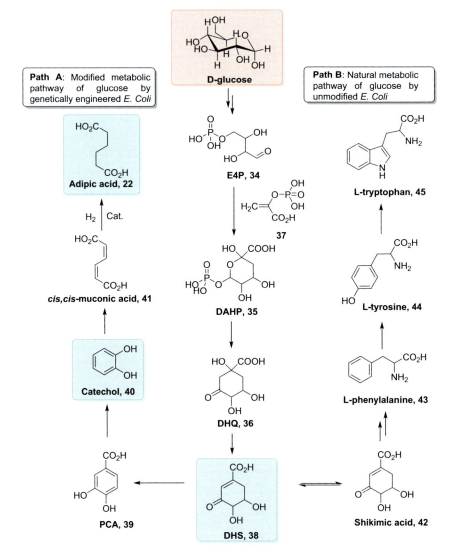

Scheme 6 Metabolic pathways of D-glucose: *Path A*: by genetically altered *E. Coli* forming adipic acid and catechol; *Path B*: by unmodified *E. Coli*. (E4P: Erythrose 4-phosphate; DAHP: 3-Deoxy-D-arabino-heptulosonic acid-7-phosphate; DHQ: 3-Dehydroquinic acid; DHS: 3-Dehydroshikimic acid; PCA: Protocatechic acid

2 Selected Examples of Real-World Applications of Green Chemistry 363

cis,cis-muconic acid (Scheme 6, Path A). *Cis,cis*-muconic acid with the aid of catalyst, hydrogenated to form adipic acid. Draths and Frost developed another strain of *E. Coli* which is capable of quenching the synthesis at either DHS or catechol [50].

The biosynthetic process of adipic acid, DHS, and catechol involves environmentally benign glucose as starting material instead of benzene which is potentially carcinogenic and toluene which is toxic on longer exposure. Additionally, glucose can be obtained from starch and cellulose feedstocks, derived from agricultural wastes, waste papers, etc. while benzene and toluene are derived from petroleum products and therefore exploit the non-renewable resources.

2.5 *Polylactic Acids as Green Alternate of Plastics*

Polyethylene **47** and polypropylene **49** derived commodities are lightweighted, durable, water-resistant, and economical. However, they pose several environmental problems due to their longer half-life, high resistant to water and microbial biodegradation [52]. The raw materials used for the preparation of polyethylene and polypropylene are ethane and propane, derived from natural gas plants, a non-renewable resource. Ethane and propane undergo cracking process in which they are treated with high temperature to afford, ethylene **46** and propylene **48** monomers. Ethylene and propylene in the presence of metal catalysts converted to polymeric chain-like structure (Scheme 7).

The annual production of plastic is about 299 million metric tons in 2013 which exceeded to 368 metric tons in 2019. The destination of about 60% of used plastics ends up in a landfill, aquatic bodies and other natural environments [53]. This creates excessive environmental hazard and life threat to terrestrial and aquatic animals and has long-lasting adverse effect on human too.

NatureWorks LLC has developed a new polymeric material polylactide (PLA) originated from annually renewable resources which compete head-to-head with conventional plastic packaging material and fiber on performance and economical ground. PLA is a biodegradable linear aliphatic thermoplastic polyester derived from annually renewable and abundant sources like corn [54, 55]. Initially, PLA found

Scheme 7 Synthesis of polyethylene and polypropylene

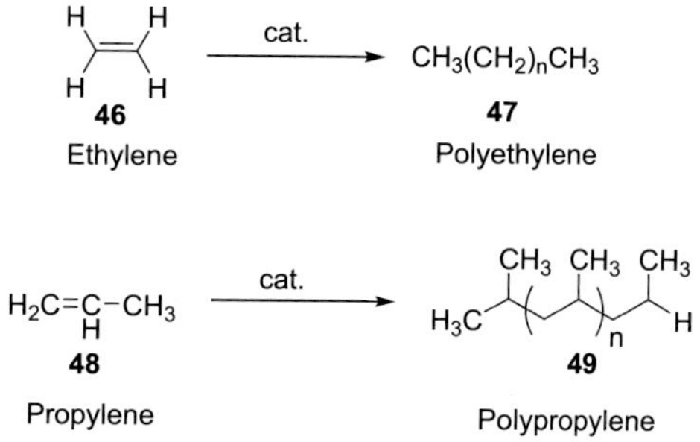

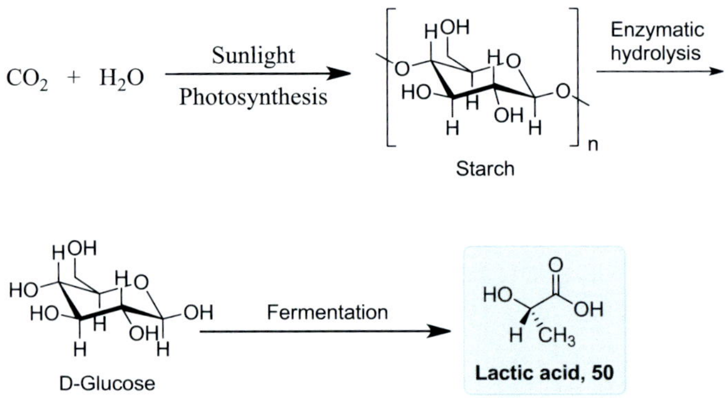

Scheme 8 Synthesis of lactic acid from starch

limited applications in biomedicals such as sutures and drug delivery system due to low availability and high manufacturing costs [56, 57]. In recent years, NatureWorks LLC initiated large-scale economical production of PLA and extended its application to packaging and apparel production [58, 59]. Currently, fiber and textile producing industries depend upon poly(ethyleneterephthalate) (PET) which accounts for 40% textiles consumption globally in the form of food and liquid containers, fibers for clothing, thermoforming for manufacturing, etc. PET is primarily obtained from non-renewable fossil fuels and disposed in landfills and aquatic bodies as they are non-biodegradable and not even readily recyclable.

The building block of PLA is lactic acid **50** which is readily obtained from renewable sources like corn crops and potatoes. Plants are the major source glucose in the form of starch and cellulose which are formed by photosynthesis in the presence sunlight, carbon dioxide, and water. The starch readily extracted from plants undergoes enzymatic hydrolysis to form glucose which further fermented to afford the simplest chiral hydroxyacid, the lactic acid-monomer of PLA (Scheme 8). The most abundant and cheapest source of sugar is dextrose obtained from corn crops.

PLA can be polymerized either by (a) condensation of lactic acid or (2) by ring-opening polymerization of lactides (a cyclic dimer of lactic acid) (Scheme 9). Former route involves the removal of water generated during condensation of lactic acid and solvents under high vacuum and temperature. This process affords low-to-medium molecular weight polymers only due to impurities and difficulty in evaporation. Additionally, in this route large reactor is required for evaporation and recovery of solvents.

Ring-opening polymerization is better route for obtaining high molecular weight polymers and adopted commercially as the advancing fermentation process reduced the production cost of lactic acid **50** drastically. An additional benefit in the production of lactic acid via fermentation over chemical process is that the fermentation process affords a specific isomer (i.e., L-lactic acid, 99.5%) in majority while chemical synthesis furnishes racemic mixture of lactic acid. In this process, water is

2 Selected Examples of Real-World Applications of Green Chemistry 365

Scheme 9 Polymerization routes for PLA

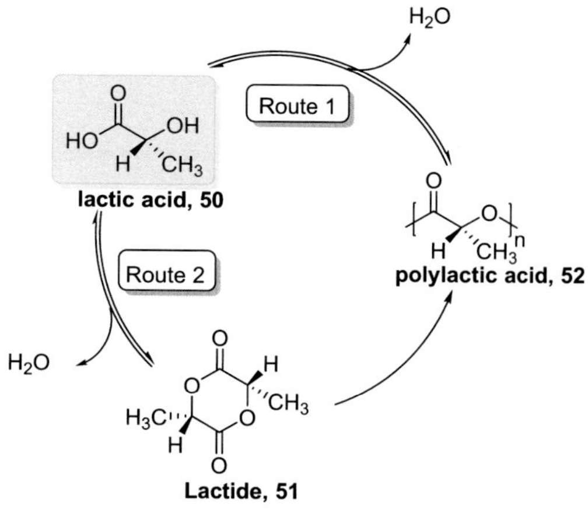

removed under mild condition without using any solvent to obtain cyclic intermediate, lactides **51** which further purified via vacuum distillation. The lactide obtained may exist in three forms, L,L-lactides (also known as L-lactide), D,D-lactide (or D-lactide), and D,L-lactide (or mesolactide). While L and D forms of lactide are optically active, mesolactide is not active and possesses different properties than the remaining two [60].

Cargill Dow LLC, a leading biopolymer manufacturing company, is based on lactide intermediate process developed a low-cost continuous production of polymers based on lactic acid. This synthetic route is a pertinent example of both greener and economical approaches for the synthesis of lactide and PLA **52** in melts rather than in solution which leads to the production of compostable commodity polymer obtained from annually renewable resources. The process initiates with synthesis of low molecular weight prepolymer PLA via condensation of lactic acid. The prepolymer is then treated with tin catalyst to form mixture of lactide stereoisomers through intramolecular cyclization process (Scheme 10). The molten mixture of lactides is purified by vacuum distillation and undergo tin-catalyzed ring-opening polymerization to form high molecular weight polymers. The catalyst used in small amount enhances the reaction rate, selectivity and reduces energy consumption. The unused monomers were removed from the final product by vacuum and recycled again. The best part of this synthesis is that the organic solvent is not used at any stage.

The NatureWorks™ PLA process is highly environment-friendly and possesses outstanding polymer production capacity incorporating all the principles of green chemistry. In addition to the production of lactic acid from annually renewable resources, it consumes less fossil fuel compared to petroleum-based plastic production. PLA is readily biodegraded to lactic acid for recycling back into polymerization process. Due to the continuous improvement in the physical properties of PLA, it firmly competes with petroleum-based plastic commodities and used for the

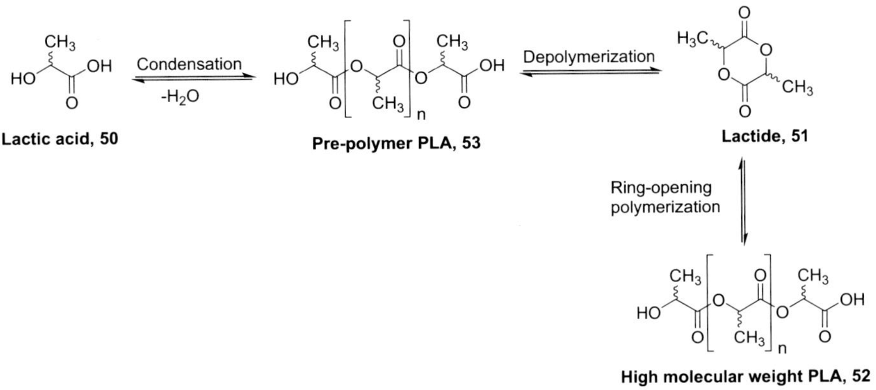

Scheme 10 High molecular weight PLA production using lactide and pre-polymer as intermediates

production of quality apparels, carpets, furniture, packing containers for liquids and foods.

2.6 Rightfit™ Pigments: A Green Replacement of Toxic Organic and Inorganic Pigments

Pigments are the colored substance which impart color to different materials such as food, materials of aesthetic values, garments, and decorative items. The descriptions of colors are found from very beginning of human evolution. In that period, minerals were used as colorant. Pigments that used to have in prehistoric periods include ochre (natural clay pigment composed of ferric oxide), charcoal (lightweighted carbon black residues), and lapis (deep blue metamorphic rock).

Pigments are generally insoluble or partially soluble in water. They absorb a particular set of wavelengths and transmit or scatter light with other set of wavelengths in visible region (complimentary colors). Based on the composition, pigments may be classified into inorganic and organic pigments.

Inorganic pigments are derived from the compounds having metallic origin that includes metal complexes, salts, and minerals such as metal oxides, sulfates, and chromates. Inorganic pigments contain one or more transition metals with incompletely filled d orbitals and different oxidation states. The electronic transitions within the metal (d-d transition), transition between the metal and ligands (through charge transfer process, MLCT or LMCT) or metal to metal charge transfer (intervalence charge transfer) are responsible for colors in transition metal complexes. Additionally, crystal defects are also responsible for imparting color in some metals. The major metallic pigments include cadmium pigments, chromium pigments, iron oxides, titanium dioxides, etc. Inorganic pigments are cost-effective, excellent fade-resistant but possess poor tonality and high toxicity.

2 Selected Examples of Real-World Applications of Green Chemistry

On the other hand, organic pigments are primarily derived from plants. The color in organic compounds is due to light absorbed by delocalized electrons in conjugated systems. Furthermore, the chromophores enhance the color in organic molecules. Several organic compounds act as pigments are azo pigments, alizarin, phthalocyanines, quinacridones, diketopyrrolopyrrole, dioxazine, etc. Organic pigments have good tonality, transparent and have bright color but expensive and fade easily over the time.

Inorganic pigments are frequently used in industrial applications such as making paints, coloring garments, fabrics, foods, and cosmetics. Earlier, industries applied heavy metal pigments for various purposes. Red, orange, and yellow colors are basically obtained from heavy metals like lead, hexavalent chromium, and cadmium. These metals are hazardous to human health and environment. They easily get bioaccumulated into the food chain and cause several ill effects. After the regulatory laws enacted by EPA on the use of heavy metal-based pigments, color formulators replaced the heavy metal pigments with high-performance organic pigments. Although these organic pigments well-replaced their inorganic counterparts but have several drawbacks: (a) the synthesis of organic pigments are expensive, (b) organic solvents like VOCs are used in large volumes, (c) some pigments syntheses require the use of polyphosphoric compounds which results in phosphates as by-product, and (d) some pigments are based on polychlorinated compounds such as dichlorobenzidine or polychlorinated phenyls.

Engelhard company has developed a variety of environmentally benign azo pigments popularly known as Rightfit™ pigments which contain Ca, Sr, and Ba replacing highly toxic heavy metal-based conventional pigments which contain Pb, Cr (VI), and Cd. These pigments have no adverse effect on environment and eliminate the risk to human health from exposure to heavy metals. Also, Rightfit™ pigments have low solubility in hydrophobic substances as they have very low octanol/water partition coefficients which reduce the risk of bioaccumulation. They have been approved for indirect food contact applications by both US Food and Drug Administration (FDA) and Canadian Health Protection Board (HPB). Additionally, Rightfit™ pigments are synthesized in aqueous medium, eliminating the risk of exposure to polychlorinated compounds and hazardous organic solvents (Scheme 11).

Rightfit™ pigments meet the necessary performance qualities at very low cost than the high-performance organic pigments. Being environmentally benign, Rightfit™ pigments possess several other qualities such as good dispersibility, enhanced

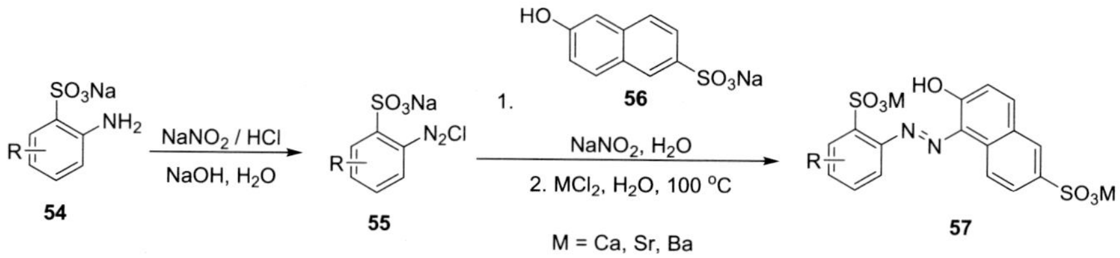

Scheme 11 Synthesis of Rightfit™ pigments in aqueous medium

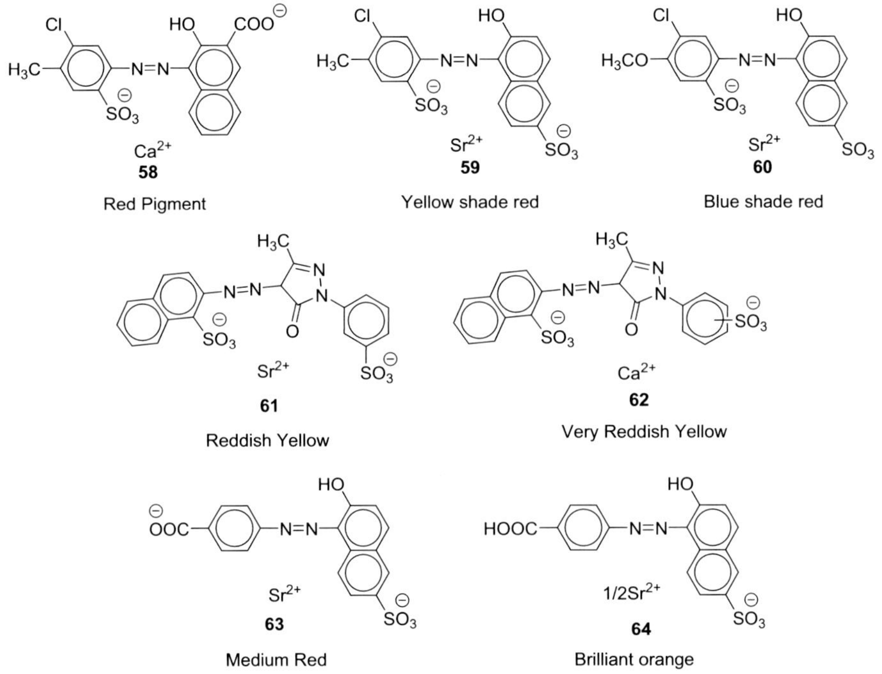

368 9 Application of Green Chemistry: Examples of Real-World Cases

58 Red Pigment

59 Yellow shade red

60 Blue shade red

61 Reddish Yellow

62 Very Reddish Yellow

63 Medium Red

64 Brilliant orange

Fig. 3 Structural similarity of different Rightfit™ azo pigments

dimensional stability, better heat resistance, and improved color strength. Due to high color strength, a small amount of color is sufficient to get same color value. Additionally, being structurally closer, they are compatible enough to achieve different intermediate color shades by mixing two or more pigments in different proportions (Fig. 3). These benefits helped the color formulators to switch over from inorganic and organic pigments to Rightfit™ pigments for coloration of various products like packaging of food, beverages, petroleum products, and other household goods.

2.7 *Healthier Fats and Oils by Enzymatic Interesterification*

Natural oils and fats are very common in our daily diets. It is mainly composed (~95%) of triacylglycerols (TAGs). The molecular structure of TAG contains a glycerol molecule esterified with three fatty acids. These fatty acids may be fully saturated (SFA) **65**, monounsaturated (MUFA) **66** or polyunsaturated (PUFA) **67** depending upon the degree of unsaturation present in the long hydrocarbon chain (Fig. 4). Each fatty acid occupies one of the three positions present on glycerol, referred by stereospecific numbering system (*sn*). The outer positions are referred as *sn−1* and *sn−3* and the central position is referred as *sn−2* (Fig. 5) [61] Several physical and

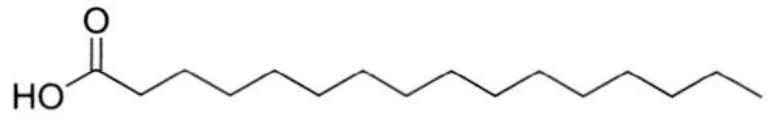

2 Selected Examples of Real-World Applications of Green Chemistry

65 Palmitic Acid (saturated fatty acid, SFA)

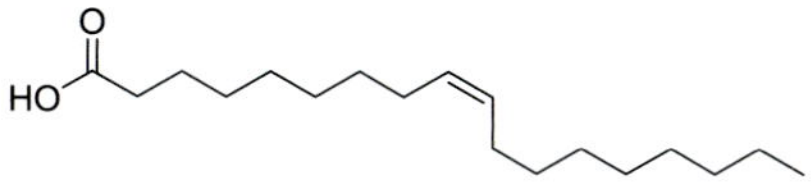

66 Oleric acid (Monounsaturated fatty acid, MUFA)

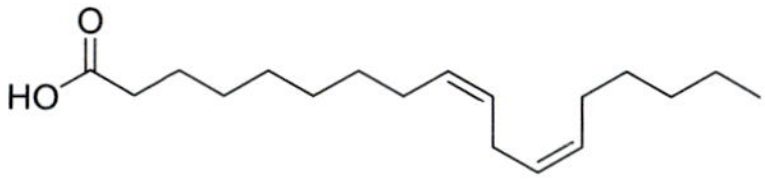

67 Linoleic acid (Polyunsaturated fatty acid, PUFA)

Fig. 4 Examples of SFA, MUFA, and PUFA

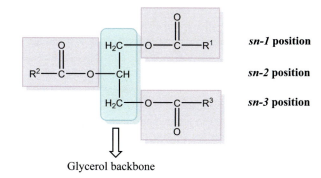

Fig. 5 Generic molecular structure of PGA with stereospecific numbering system (*sn*)

biochemical properties of fats depend upon the positioning, chain length, degree of unsaturation, and configuration of component fatty acids. The foods containing high concentration of SFA results in increased level of serum low-density lipoprotein cholesterol (LDL-C) which may raise the risk of the development of atherosclerosis and cardiovascular disease (CVD) [62]. Replacing SFA with unsaturated fatty acid (UFA), especially with cis-configured PUFA reduces the serum LDL-C level. Additionally, the replacement of SFA with UFA has other benefits including lower risk of endothelial function, inflammation, and platelet activity [63].

UFAs are less stable than saturated fatty acids. UFAs are prone to oxidation called rancidity, which leads to unusual odor and taste. To avoid rancidity and to enhance the stability and applicability, natural oils need some structural modifications. The oils rich in unsaturated fatty acids are partially hydrogenated to furnish semi-solid fats which have improved oxidative stability. In view of ill effects of SFA, guidelines to limit their use in diet were first included in 1961. Consequently, the use of partially hydrogenated vegetable oils was increased compared to the animal-derived fats in between 1960 and 1980s to due low cost, availability and some unique features like better spreadability compared to butter [61]. During partial hydrogenation, some of the *cis*-configured fatty acids transformed to *trans*-fatty acids. Industrially manufactured *trans*-fatty acids have several negative health impacts, especially related to heart and vascular diseases. In view of the seriousness and negative health issues related with consumption of *trans*-fatty acids led the US food industries to explore alternatives of partial hydrogenation, there are several fat modification methods evolved to avoid partial hydrogenation which include interesterification, blending, fractionation, and full hydrogenation. Out of the available methods, interesterification emerged as a revolutionary technology which reduces the trans-fatty acid contents from the food without compromising with the qualities of partially hydrogenated vegetable oils. Additionally, interesterified fats find applications in clinical nutrition including, infant formula milk fat and specialized fats for patients suffering from fat malabsorption disorders. Furthermore, structured lipids used in enteral and parenteral are also prepared by interesterification methodology [64]. This strategy is now adopted by most of the US food and ingredient industries.

Interesterification is the process in which a TAG possessing saturated fatty acid exchange one or two SFA, in either random or specific manner, with TAG having unsaturated fatty acids to furnish TAG without any *trans*-fatty acid and SFA. By interesterification, some physical properties of fats get altered such as melting point and crystalline structure which provide suitable functionality to the fat. Interesterification may be achieved chemically or enzymatically.

Chemical interesterification process involves sodium ethoxide as catalyst which randomly hydrolyses and redistributes the fatty acids on the glycerol backbone in a TAG mixture. On the other hand, enzymatic interesterification uses lipases enzyme derived from microbes (e.g., *Candida rugosa*) for the redistribution of fatty acids on glycerol backbone in random or in a specific fashion depending upon the lipase specificity. While chemical interesterification is an older method commercialized in 1940s, enzymatic interesterification was first developed in 1980s and became predominant interesterification method in USA and Canada [65]. Enzymatic interesterification method uses expensive lipase enzymes compared to sodium methoxide catalyst used in chemical interesterification; however, several other benefits are associated with enzymatic method including low equipment cost, high selectivity, lower operating temperature, reduced neutral oil loss and preserved oxidative stability [66].

Archer Daniels Midland Company (ADM) and Novozymes developed cost-effective immobilized enzyme for interesterification of natural oils and commercialized it in 2002. Since then, about 15 million pounds of interesterified oil have been produced by ADM. Additionally, their enzymatic interesterification method poses several environmental benefits such as the removal of potentially toxic chemicals and elimination of several by-products and waste streams. For example, margarines and shortenings use about 10 billion pounds of hydrogenated soybean oil. Using their methodology, ADM/Novozymes saves 400 million pounds of soybean oil and eliminates 116 million pounds of soap, 20 million pounds of sodium methoxide catalyst, 50 million pounds of bleaching clays, and 60 million gallons of water annually. The enzymatic method significantly eliminated the *trans*-fatty acids from the dietary fats and reduced the cardiovascular disorders [11, 67].

2.8 Green Approach Toward the Synthesis of Sertraline Hydrochloride (Zoloft)

Sertraline **71**, an antidepressant, marketed under the brand name Zoloft® by Pfizer was synthesized by employing green chemistry principles, not only minimized the waste generation but also reduced the cost and energy consumption [68]. Earlier, tetralone derivative **68** was used for the synthesis of sertraline **71** employing three different steps (Scheme 12a) that involved the formation of intermediates (**69** and **70**) which were isolated and purified at each step and then further used in next step [69]. Overall, the reaction requires isolation at each step which involves different solvents such as toluene, hexane, THF, and ethanol. Additionally, the generation of TiO_2 as waste material is another drawback associated with this method. Therefore, Pfizer modified the synthetic route and developed a greener synthetic route for the formation of sertraline where the synthesis of sertraline **71** was achieved in one-pot operation using ethanol as single solvent (Scheme 12b). Interestingly, no intermediate was isolated and total elimination of waste material like TiO_2 [70].

The improved route established by Pfizer increased the product yield, energy efficiency and saved the time. Furthermore, this optimized synthetic route reduces toxic waste generation either in the form of used solvents or by-products. Considering the immense importance of this method, it was awarded US EPA Presidential Green Chemistry Challenge Award in the year 2002 [11].

3 Conclusion

Green chemistry is the chemistry for "pollution prevention." Green chemistry is a step toward the sustainable delivery of services and goods to growing population without compromising environmental quality. According to an estimate by United Nations, the world population will cross 10.7 billion by 2050. This growing population will demand chemical goods and services about twofold that what we have today. This will result in the establishment of new chemical industries around the world which may lead to pollution of water and air, depletion of ozone layer, loss of biological species in forest and aquatic bodies, introduction of persistent organic chemicals in ecosystem, and climate change.

With the passage of Pollution Prevention Act in 1990, several initiatives were taken by US government and non-government organizations which attracted the attention of scientist across the globe. The office of US Environmental Protection Agency (EPA) launched the research program "*Alternative Synthetic Pathways for Pollution Prevention*" to encourage the researchers to think about the growth and innovations toward sustainable chemical research. In 1995, Presidential Green Chemistry Challenge program was launched to promote the environmental and economic benefits by developing and using novel green chemistry. This was the only presidential level award launched specifically in the field of chemistry. Throughout the

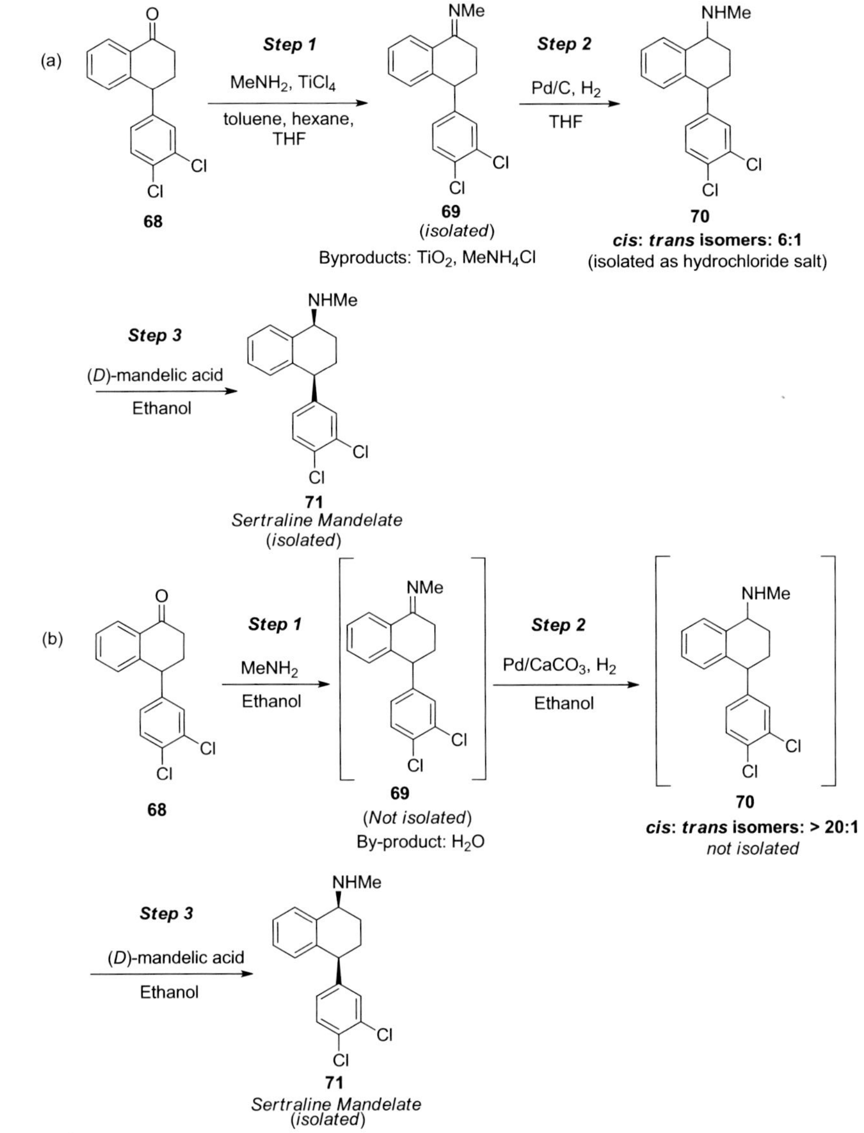

Scheme 12 Synthesis of sertraline; **a** Multistep synthesis involving isolation of intermediate after each step, **b** One-pot synthesis without isolation of intermediates

25 years of this program, EPA announced 128 winners and got over 1800 nominations. Since its inception, the winning technologies have put forth huge progress and direct implementation of green technology in industries. Furthermore, ACS in collaboration of GCI convene scientific programs and symposia to aware and share the progress achieved in the field of green chemistry. Additionally, they advocate for

the implementation of teaching and learning materials of green chemistry for undergraduate and postgraduate students so that a progressive thinking about sustainable development could be developed at very beginning of their adulthood.

The Presidential Green Chemistry Challenge Award not only put forth the live action, but also brings valuable and environmental sustainability of green chemistry. Industries are now embracing green technologies for their manufactures; it makes a good public relation. People should get aware of positive consequences of green chemistry and appreciate industrial products which are produced by sustainable approach, which may further encourage industries to embrace green technologies. Additionally, governments of developing countries provide incentives for start-ups which follow the set of green principles for manufacturing their products. A collaborative effort by government, industry, and academia is needed to promote the thrust area of chemistry implementing greener approach to achieve a sustainable society.

References

1. Johnson, S.M.: From reaction to proaction: the 1990 pollution prevention act. Colum. J. Envtl. L. **17**, 153 (1992)
2. Burnett, M.L.: The pollution prevention act of 1990: a policy whose time has come or symbolic legislation? Environ. Manage. **22**(2), 213–224 (1998)
3. Grier, J.W.: Ban of DDT and subsequent recovery of reproduction in bald eagles. Science, **218**(4578), 1232–1235 (1982)
4. Sriramachari, S.: The Bhopal gas tragedy: an environmental disaster. Curr. Sci. **86**(7), 905–920 (2004)
5. Gupta, J.P.: The Bhopal gas tragedy: could it have happened in a developed country? J. Loss Prev. Process Ind. **15**(1), 1–4 (2002)
6. Opheim, T.: Fire on the cuyahoga. EPA J. **19**(2), 44–45 (1993)
7. Mniszewski, K.R.: The pepcon plant fire/explosion: a rare opportunity in fire/explosion investigation. J. Fire Prot. Eng. **6**(2), 63–78 (1994)
8. Cheeda, V.K., Kumar, A., Ramamurthi, K.: Influence of height of confined space on explosion and fire safety. Fire Saf. J. **76**, 31–38 (2015)
9. Anastas, P.T., Warner, J.C.: Green chemistry. Frontiers (Boulder). **640**, 1998 (1998)
10. Anastas, P., Eghbali, N.: Green chemistry: principles and practice. Chem. Soc. Rev. **39**(1), 301–312 (2010)
11. https://www.epa.gov/sites/default/files/2016-10/documents/award_recipients_1996_2016. pdf.
12. https://www.acs.org/content/acs/en/greenchemistry/what-is-green-chemistry/history-of-green-chemistry.html.
13. Horváth, I.T., Anastas, P.T.: Innovations and green chemistry. Chem. Rev. **107**(6), 2169–2173 (2007)
14. Hjeresen, D.L., Boese, J.M., Schutt, D.L.: Green chemistry and education. J. Chem. Educ. **77**(12), 1543 (2000)
15. Davies, N.M.: Clinical pharmacokinetics of Ibuprofen. Clin. Pharmacokinet. **34**(2), 101–154 (1998)
16. Brain, P., Leyva, R., Doyle, G., Kellstein, D.: Onset of analgesia and efficacy of ibuprofen sodium in postsurgical dental pain: a randomized, placebo-controlled study versus standard Ibuprofen. Clin. J. Pain **31**(5), 444 (2015)
17. Halford, G.M., Lordkipanidzé, M., Watson, S.P.: 50th anniversary of the discovery of Ibuprofen: an interview with Dr Stewart Adams. Platelets **23**(6), 415–422 (2012)

374 9 Application of Green Chemistry: Examples of Real-World Cases

18. Stuart, N.J., Sanders, A.S.: Compositions and method for treating symptoms of inflammation, pain and fever. Google Patents (1966)
19. Bowen, H.J.M.: In: Environmental Chemistry of the Elements. Academic Press (1979)
20. Vallero, D.: In: Fundamentals of Air Pollution. Academic Press (2014)
21. Manahan, S.: In: Environmental Chemistry. CRC Press (2017)
22. DeSimone, J.M., Guan, Z., Elsbernd, C.S.: Synthesis of fluoropolymers in supercritical carbon dioxide. Science **257**(5072), 945–947 (1992)
23. Du, L., Kelly, J.Y., Roberts, G.W., DeSimone, J.M.: Fluoropolymer synthesis in supercritical carbon dioxide. J. Supercrit. Fluids **47**(3), 447–457 (2009)
24. Guan, Z., Combes, J.R., Elsbernd, C.S., DeSimone, J.M.: Synthesis of Fluoropolymers in Supercritical Carbon Dioxide. American Chemical Society, Washington, DC (United States) (1993)
25. Cooper, A.I., DeSimone, J.M.: Polymer synthesis and characterization in liquid/supercritical carbon dioxide. Curr. Opin. solid state Mater. Sci. **1**(6), 761–768 (1996)
26. Hyatt, J.A.: Liquid and supercritical carbon dioxide as organic solvents. J. Org. Chem. **49**(26), 5097–5101 (1984)
27. Varma, R.S.: In: The Presidential Green Chemistry Challenge Awards Program, Summary of 1997 Award Entries and Recipients; EPA744-S-97–001. US Environmental Protection Agency, Office of Pollution … (1997)
28. Clark, E.A., Sterritt, R.M., Lester, J.N.: The fate of tributyltin in the aquatic environment. Environ. Sci. Technol. **22**(6), 600–604 (1988)
29. Champ, M.A., Seligman, P.F.: An Introduction to Organotin Compounds and Their Use in Antifouling Coatings, pp. 1–25. Springer, In Organotin (1996)
30. Maguire, R.J.: Environmental aspects of tributyltin. Appl. Organomet. Chem. **1**(6), 475–498 (1987)
31. Thain, J.E., Waldock, M.J.: The impact of tributyl tin (TBT) antifouling paints on Molluscan Fisheries. Water Sci. Technol. **18**(4–5), 193–202 (1986)
32. Gibbs, P.E., Pascoe, P.L., Burt, G.R.: Sex change in the female Dog-Whelk, Nucella Lapillus, induced by Tributyltin from antifouling paints. J. Mar. Biol. Assoc. United Kingdom **68**(4), 715–731 (1988)
33. Bryan, G.W., Gibbs, P.E., Hummerstone, L.G., Burt, G.R.: The decline of the Gastropod Nucella Lapillus around South-West England: evidence for the effect of Tributyltin from Antifouling paints. J. Mar. Biol. Assoc. United Kingdom **66**(3), 611–640 (1986)
34. Champ, M.A., Wade, T.L.: Regulatory Policies and Strategies for Organotin Compounds, pp. 55–94. Springer, In Organotin (1996)
35. Dafforn, K.A., Lewis, J.A., Johnston, E.L.: Antifouling strategies: history and regulation, ecological impacts and mitigation. Mar. Pollut. Bull. **62**(3), 453–465 (2011)
36. Onduka, T., Ojima, D., Ito, M., Ito, K., Mochida, K., Fujii, K.: Toxicity of the antifouling biocide Sea-Nine 211 to Marine Algae, Crustacea, and a Polychaete. Fish. Sci. **79**(6), 999–1006 (2013)
37. Jacobson, A.H., Willingham, G.L.: Sea-nine Antifoulant: an environmentally acceptable alternative to Organotin Antifoulants. Sci. Total Environ. **258**(1–2), 103–110 (2000)
38. Cima, F., Bragadin, M., Ballarin, L.: Toxic effects of new antifouling compounds on Tunicate Haemocytes: I. Sea-Nine 211™ and Chlorothalonil. Aquat. Toxicol. **86**(2), 299–312 (2008)
39. Alzieu, C.: Tributyltin: case study of a chronic contaminant in the coastal environment. Ocean Coast. Manag. **40**(1), 23–36 (1998)
40. Callow, M.E., Willingham, G.L.: Degradation of antifouling biocides. Biofouling **10**(1–3), 239–249 (1996)
41. Castellan, A., Bart, J.C.J., Cavallaro, S.: Industrial production and use of adipic acid. Catal. Today **9**(3), 237–254 (1991)
42. Draths, K.M., Frost, J.W.: Environmentally compatible synthesis of adipic acid from D-Glucose. J. Am. Chem. Soc. **116**(1), 399–400 (1994)
43. Chenier, P.J.: In: Survey of Industrial Chemistry. Springer Science & Business Media, (2012)
44. Anastas, P.T.: In: Benign by Design Chemistry (1994)

References

45. Loomis, D., Guyton, K.Z., Grosse, Y., El Ghissassi, F., Bouvard, V., Benbrahim-Tallaa, L., Guha, N., Vilahur, N., Mattock, H., Straif, K.: Carcinogenicity of benzene. Lancet Oncol. **18**(12), 1574–1575 (2017)
46. Babich, H.: Butylated hydroxytoluene (BHT): a review. Environ. Res. **29**(1), 1–29 (1982)
47. Draths, K.M., Frost, J.W.: Environmentally compatible synthesis of catechol from D-Glucose. J. Am. Chem. Soc. **117**(9), 2395–2400 (1995)
48. Niu, W., Draths, K.M., Frost, J.W.: Benzene-free synthesis of adipic acid. Biotechnol. Prog. **18**(2), 201–211 (2002)
49. Draths, K.M., Knop, D.R., Frost, J.W.: Shikimic acid and Quinic acid: Replacing isolation from plant sources with recombinant microbial biocatalysis. J. Am. Chem. Soc. **121**(7), 1603–1604 (1999)
50. Draths, K.M., Pompliano, D.L., Conley, D.L., Frost, J.W., Berry, A., Disbrow, G.L., Staversky, R.J., Lievense, J.C.: Biocatalytic synthesis of aromatics from D-Glucose: the role of transketolase. J. Am. Chem. Soc. **114**(10), 3956–3962 (1992)
51. Chandran, S.S., Yi, J., Draths, K.M., Daeniken, R. von, Weber, W., Frost, J.W.: Phosphoenolpyruvate availability and the biosynthesis of shikimic acid. Biotechnol. Prog.**19**(3), 808–814 (2003)
52. Getor, R.Y., Mishra, N., Ramudhin, A.: The role of technological innovation in plastic production within a circular economy framework. Resour. Conserv. Recycl. **163**, 105094 (2020)
53. Martin, C., Baalkhuyur, F., Valluzzi, L., Saderne, V., Cusack, M., Almahasheer, H., Krishnakumar, P.K., Rabaoui, L., Qurban, M.A., Arias-Ortiz, A.: Exponential increase of plastic burial in mangrove sediments as a major plastic sink. Sci. Adv. **6**(44), eaaz5593 (2020)
54. Tsuji, H., Ikada, Y.: Blends of Aliphatic Polyesters. II. hydrolysis of solution-cast blends from Poly (L-lactide) and Poly (E-caprolactone) in Phosphate-buffered solution. J. Appl. Polym. Sci. **67**(3), 405–415 (1998).
55. Gruber, P.R., Drumright, R.E., Henton, D.E.: Polylactic acid technology. Adv. Mater **12**(23), 1841–1846 (2000)
56. Vert, M., Schwarch, G., Coudane, J.: Present and future of PLA polymers. J. Macromol. Sci. Part A Pure Appl. Chem. **32**(4), 787–796 (1995)
57. Lipinsky, E.S., Sinclair, R.G.: Is lactic acid a commodity chemical. Chem. Eng. Prog. **82**(8), 26–32 (1986)
58. Vink, E.T.H., Glassner, D.A., Kolstad, J.J., Wooley, R.J., O'Connor, R.P.: The eco-profiles for current and near-future NatureWorks® Polylactide (PLA) production. Ind. Biotechnol. **3**(1), 58–81 (2007)
59. Vink, E.T.H., Rabago, K.R., Glassner, D.A., Gruber, P.R.: Applications of life cycle assessment to NatureWorks^{TM} Polylactide (PLA) production. Polym. Degrad. Stab. **80**(3), 403–419 (2003)
60. Vink, E.T.H., Rábago, K.R., Glassner, D.A., Springs, B., O'Connor, R.P., Kolstad, J., Gruber, P.R.: The sustainability of NatureWorks^{TM} Polylactide Polymers and Ingeo^{TM} polylactide fibers: an update of the future. Macromol. Biosci. **4**(6), 551–564 (2004)
61. Berry, S.E., Bruce, J.H., Steenson, S., Stanner, S., Buttriss, J.L., Spiro, A., Gibson, P.S., Bowler, I., Dionisi, F., Farrell, L., Glass, A., Lovegrove, J.A., Nicholas, J., Peacock, E., Porter, S., Mensink, R.P., Hall, W. L.: Interesterified fats: what are they and why are they used? a briefing report from the roundtable on interesterified fats in foods. Nutr. Bull. **44**(4), 363–380 (2019). https://doi.org/10.1111/nbu.12397
62. Mensink, R.P.: Effects of saturated fatty acids on serum lipids and lipoproteins: a systematic review and regression analysis. World Heal. Organ. 1–63 (2016)
63. Stanner, S., Coe, S., Frayn, K.N.: In: Cardiovascular Disease: Diet, Nutrition and Emerging Risk Factors. Wiley (2018)
64. Karupaiah, T., Sundram, K.: Effects of stereospecific positioning of fatty acids in Triacylglycerol structures in native and randomized fats: a review of their nutritional implications. Nutr. Metab. (Lond) **4**(1), 1–17 (2007)
65. Dayton, C.L.G.: Enzymatic interesterification. Green Veg. Oil Process 205–224 (2014)
66. Rousseau, D., Marangoni, A.G.: 10: chemical interesterification of food lipids: theory and practice. Food lipids Chem. Nutr. Biotechnol. **267**, 267 (2008)

376   9 Application of Green Chemistry: Examples of Real-World Cases

67. Sigsgaard, T.: Implementation of green chemistry: real-world case studies. Green Chem. Beginners **205** (2021)
68. Manley, J.B., Anastas, P.T., Cue, B.W.: Frontiers in green chemistry: meeting the grand challenges for sustainability in R&D and manufacturing. J. Clean. Prod. 16(6), 743–750 (2008)
69. Quallich, G.J.: Development of the commercial process for Zoloft®/Sertraline. Chirality **17**(S1), S120–S126 (2005)
70. Vukics, K., Fodor, T., Fischer, J., Fellegvári, I., Lévai, S.: Improved industrial synthesis of Antidepressant Sertraline. Org. Process Res. Dev. **6**(1), 82–85 (2002)

Printed in the United States
by Baker & Taylor Publisher Services